Food Coatings and Preservation Technologies

Scrivener Publishing
100 Cummings Center, Suite 541J
Beverly, MA 01915-6106

Publishers at Scrivener
Martin Scrivener (martin@scrivenerpublishing.com)
Phillip Carmical (pcarmical@scrivenerpublishing.com)

Food Coatings and Preservation Technologies

Edited by

Mousumi Sen
Department of Chemistry, Amity University, Noida, Uttar Pradesh, India

Scrivener
Publishing

WILEY

This edition first published 2025 by John Wiley & Sons, Inc., 111 River Street, Hoboken, NJ 07030, USA and Scrivener Publishing LLC, 100 Cummings Center, Suite 541J, Beverly, MA 01915, USA

For more information about Scrivener publications please visit www.scrivenerpublishing.com.

Wiley Global Headquarters
111 River Street, Hoboken, NJ 07030, USA

For details of our global editorial offices, customer services, and more information about Wiley products visit us at www.wiley.com.

Library of Congress Cataloging-in-Publication Data

ISBN 978-1-394-23758-6

Front cover images supplied by Adobe Firefly
Cover design by Russell Richardson

Set in size of 11pt and Minion Pro by Manila Typesetting Company, Makati, Philippines

Printed in the USA

10 9 8 7 6 5 4 3 2 1

Contents

Preface

Edible coating is globally used to prevent the contamination of food products from harmful microorganisms and pathogens. They should be formulated in a manner that makes them easily degradable while consumed and cooked. Generally, coating is classified as Hydrocolloid Coating, Composite Coating, and Lipid-based Coatings. These categories can be further classified. These coatings enhance the storage duration by retarding the ripening rate of the product. Edible coating should not contain such chemical constituents that decrease the unique nutrients of the food substances below a desired level and control the moisture loss, providing a fresh and glossy appearance to the food product. Recent studies and research in this field have given rise to environmentally friendly products and cause less risk to individual health. Food preservation technologies face important challenges in extending the shelf life of perishable food products (e.g., meat, fish, milk, eggs, and many raw fruits and vegetables) that help to meet the daily nutrient requirement demand. In addition, food preservation has gone beyond only preservation; the current techniques focus on the fulfillment of two additional objectives, the suitability of the used processes and the generation of environmentally friendly products with the non-presence of any side effects on health. Moreover, they also look for additional nutritional properties. One of these preservation protocols deals with the use of edible coatings.

Food Coatings and Preservation Technologies aims to highlight the developments made toward designing new edible coatings (including antioxidants, preservatives, nanoparticles, or other additives to improve foods' mechanical integrity, handling, and quality, and to change surface gloss) that take advantage of organic synthesis to improve the quality and safety aspects of fresh produce and thus extend shelf life. This information will be helpful for processors to select the best coating material and its effective concentration for different fresh and minimal processed vegetables. This book also discusses the current state-of-the-art alternatives to conventional packaging, providing the main features necessary for these biodegradable packaging to meet the specific uses for the conservation and improvement of various food products. Herein, particular attention has been paid to the main components (e.g., biopolymers, additives, bioactive, and probiotic components), manufacturing methods (for edible films or coatings), and their application to specific products.

The book begins by focusing on the application of biopolymer-based edible coatings and films in food packaging and preservation. The first chapter addresses the environmental concerns associated with conventional plastic packaging and highlights biopolymers as sustainable alternatives. The study explores various biopolymers, their properties, and their incorporation of bioactive compounds. Different methods of film formation and their impact on food attributes are discussed, along with challenges in implementation.

This section underscores the essential role of edible preservation technology, often referred to as "green technology" in the contemporary era. Compared to conventional food preservation technologies, these methods not only prevent food deterioration but also preserve the wholesomeness of the food. Utilizing bio-edible coatings emerges as a more efficient approach for improving the quality and safety of fresh produce in comparison to existing methods. These coatings, comprising an ingestible layer that encases the food, control the transmission of oxygen, water vapor carbon dioxide, leading to a prolonged shelf life without compromising quality and safety standards.

The third chapter aims to provide a comprehensive understanding of the current state and critical developments in biodegradable polymer packaging systems for food applications.

As technology continues to advance, the future of active and intelligent antimicrobial coating systems is promising. Chapter four highlights the ongoing research focusing on optimizing coating effectiveness, expanding the range of antimicrobial agents used, and enhancing intelligent features for greater adaptability. The development of eco-friendly and sustainable materials for these coatings aligns with the growing consumer demand for environmentally conscious products. Furthermore, advancements in nanotechnology may lead to even more efficient antimicrobial coatings with increased surface area and targeted action.

Chapter five focuses on innovative coating technologies, presenting a promising avenue to enhance the delivery, stability, and efficacy of medical foods and nutraceuticals. By overcoming challenges related to stability, bioavailability, and patient acceptance, coatings contribute to better treatment outcomes and patient experiences.

The objective of the sixth chapter is to identify the utility of edible materials as a supplement to packaged foods, which has recently been the subject of various research. Therefore, the chapter's objective is to combine the most recent information available to provide a comprehensive understanding of the formulation method and the many approaches to enhancing the properties of the coatings applied to food products.

Studies on incorporating nanoparticles in edible films enhance the material strength and increase the shelf life. Food processors, environmental protection agencies, and consumers should promote using biodegradable packaging films to overcome the impacts of modern plastic packaging-based ecological issues. Chapter seven is aimed to cover how food safety and stability during processing, storage, transportation, and sales are significantly influenced by microbial contamination and biochemical deterioration.

Chapter eight tries to elucidate methods based on linear kinetic models by reviewing recent non-linear approaches. In conclusion, this significant area of accelerated shelf life testing is explored to suggest future directions. Moreover, it discusses what should be expected when developing novel practical and reliable tools that are applicable in the industrial process.

Chapter nine discuss in detail how the utilization of biodegradable coatings and films is poised to transform the realm of food packaging by offering sustainable substitutes for customary materials. The advancement, exploration, and partnership amongst academic institutions, businesses, and governing bodies will pave the way toward formulating efficacious, secure, and ecologically conscious coating remedies. This will facilitate the enhancement of food preservation standards while promoting quality assurance and sustainability in forthcoming years.

Chapter ten focuses on investigating the composition, properties, preparation methods, characterization, challenges associated with nanoemulsion-based edible coatings, potential scalability in the food sector, and beneficial health effects.

The eleventh chapter discusses encapsulation techniques and their significance in the food industry. Readers will gain insights into appropriate encapsulation materials for various food applications, enabling the formulation of efficient and targeted encapsulation strategies. The second section focuses on the mechanisms of encapsulation for preventing food deterioration. By addressing the underlying causes of food spoilage, including oxidation, moisture uptake, and enzymatic reactions, this section clarifies how encapsulation-based coatings act as protective barriers, thereby effectively limiting these deterioration mechanisms.

Chapter twelve provides a succinct overview of the overall concept of encapsulation and the mechanism, followed by the types of bioactive carriers included in the edible films along with their uses and limitations. It also focuses on technologies and methods of nutrient retention, their level of retention, and their ability to improve shelf life.

The natural polymers commonly found and utilized include polysaccharides like alginate, fats like glycerol proteins like zein. An important benefit of this packaging type is its ingestible nature along with the food item. The varied properties of the composite matrix, shaped by each component, result in differing functionalities of edible film materials, as discussed in chapter thirteen.

Chapter fourteen enables multilayer implementations and is widely used in commercial networks. Fluidized bed coating technique is great for coating baked goods since the food item is suspended in a fluidized bed of coating particles, ensuring full coverage while preventing the formation of clusters. Panning, on the other hand, uses the polishing motion of a pan to create an even and uniform surface on sweets, nuts, as well as certain processed fruits. The right deposition process should be chosen based on the coating materials' properties, the food product's intended effect, and economic considerations.

Chapter fifteen explains how to devise optimal formulations and compositions for individual polymers, employing diverse processing techniques for various bio-nanocomposites; how to examine interactions between food components and antimicrobial nanocarriers; and how to determine suitable co-nanoencapsulation systems for incorporating multiple bioactive agents within a single packaging system.

The sixteenth chapter discusses the futuristic aspects of prebiotics and probiotics and how the emerging technologies can produce prebiotics, as well as the delivery of probiotics into the human gut. It is necessary to determine the safety and physicochemical properties of nano prebiotics and probiotics to support tactile conclusions and policies of regulatory agencies. Further research on the antiviral potential of pre- and probiotic-based medications in the prevention and treatment of COVID-19 is warranted. Preclinical and clinical experiments ought to be conducted in the future to identify the benefits in the management of the COVID-19 pandemic.

Functional products are ideal food options because they strive to increase life quality by preventing nutrition-related disorders, as evidenced by their rising demand, the steady rise of life expectancy, and the overall desire for a better quality of life. The organoleptic qualities of the finished products and, consequently, consumer approval are frequently altered by new advancements. Chapter seventeen provides an overview of functional food innovation in terms of quality, safety regulations, problems, and prospects.

The last chapter gives a broader concept of Bacterial Cellulose (BC). BC, a non-biotoxic, chemically inert, biocompatible, biodegradable, and pure exopolysaccharide produced by bacteria, is one of the frontrunners in the search to replace plastics. BC has high tensile strength and mechanical robustness and can be combined with other biodegradable polymers and bioactive agents to form industrial-grade food packaging materials. This chapter focuses on the recent developments and current frontiers in BC production and highlights the industrial application of BC in the food packaging and functional food production sector.

I am very grateful for all the hard work and efforts put forth by the many contributors to this book. I thank all the authors for sharing their insightful research and information with us. I am very thankful to Aarushi Sen for her unending encouragement and support throughout the making of this book—her help is greatly appreciated. I am also most grateful to Martin Scrivener of Scrivener Publishing who helped to make this book possible. I thank him for his patience and consistent support throughout the journey. Finally, I express my sincere thanks to the Department of Chemistry, Amity University, Uttar Pradesh for all the help and support I have received.

Dr. Mousumi Sen

1

Edible Food Coatings are Biopolymers: Packaging and Preservation

Mousumi Sen* **and Hemendra Pratap Singh**

Department of Chemistry, Amity Institute of Applied Sciences, Amity University, Uttar Pradesh, India

Abstract

This chapter focuses on the application of biopolymer-based edible coatings and films in food packaging and preservation. It addresses the environmental concerns associated with conventional plastic packaging and highlights biopolymers as sustainable alternatives. The study explores various biopolymers, their properties, and their incorporation of bioactive compounds. Different methods of film formation and their impact on food attributes are discussed, along with challenges in implementation. Another innovative method that has shown to offer a safe and advantageous way to extend the shelf life of food products is edible coatings. This coating enhances the quality and extends the shelf life of minimally processed fruits, vegetables, and other food products. It has multiple applications in this regard. The use of these materials in food applications—particularly for highly perishable products like horticultural goods—depends on a number of particular qualities, such as cost, accessibility, usefulness, mechanical properties (tension and flexibility), optical properties (brightness and opacity), the effect of gas movement barriers, structural resilience to moisture and microorganisms, and sensory credibility. Many researchers have looked into the possibilities of using a variety of unique materials in the creation of edible coatings and films. The biopolymer's s is the most important factor in defining the final functional characteristics and attributes of biopolymer films. Overall, biopolymer-based edible coatings offer potential for sustainable food preservation, but further research is needed for optimization and cost-effective manufacturing.

Keywords**:** Food crisis, edible coating, food coating, nanocoating, biopolymers, packaging & preservation

**Corresponding author*: mousumi1976@gmail.com

Mousumi Sen (ed.) Food Coatings and Preservation Technologies, (1–26) © 2025 Scrivener Publishing LLC

1.1 Introduction

The consumptions of fresh and natural food increases because of increasing the population day by day in this way people also increases the production drastically. Although most of the Food items are perishable in nature, additionally, it results in a rapid decline in quality during both storage and transportation. Since we have the limited amount of land or soil, so people would have to reduce the postharvest deficit [1]. A study by Galanakis shed light on the alarming fact that approximately one-third of the world's food is squandered, with around 14% of this wastage occurring at various stages of processing, including harvest, farming, and slaughtering [1, 2]. Therefore, there is a significant need in the food sector to develop new, creative methods for extending the shelf life of fresh items, such as fruits and vegetables. The shelf life of foods is significantly influenced by the packaging industry [1–3]. When it comes to thickness, edible coatings are those that are less than 0.025 mm thick, whereas edible films and sheets are those that are thicker than 0.050 mm. The year 1992 marked a significant milestone in the field of edible coatings as it witnessed the introduction of the first officially recognized application of wax on the fruit surface. This breakthrough paved the way for the practical implementation of edible coatings [1–4]. An edible coating refers to a delicate layer of solution that is applied to the surface of food items, serving as a protective barrier against rapid deterioration. Furthermore, these coatings also play a vital role as primary packaging in the food industry; preserving the textural and sensory characteristics of the products they encompass [1–7].

The production of edible coatings involves the utilization of various types of biopolymers, including lipids, proteins, polysaccharides, and composite substances. This ensures that the coating material remains entirely edible. Notably, essential oils derived from plants offer valuable attributes such as antimicrobial and antioxidant properties, further enhancing the functionality of these coatings [1–10]. Meat and dairy products' quality and safety heavily depend on their packaging. The consumption of meat and dairy products is widespread across the world. Considering that they are loaded with substances that are high in vitamins, fatty acids, energy and proteins. These foods are sufficient for meeting the human diet's needs for fatty acids, protein, and other micronutrients [1–13]. The degradation of meat initiates upon the slaughtering of the animal, triggering a series of intricate chemical reactions. These reactions give rise to unfavourable outcomes such as discoloration, off-flavors, and a range of undesirable chemical and structural alterations in the meat [14].

In response to the environmental issues raised by conventional packaging materials, edible biopolymer coatings have become a viable option for food packaging. These coatings, which come from living things like plants, animals, and microbes, have several benefits. They provide an effective barrier against oxygen, moisture, and microbial contamination, thereby extending the shelf life of perishable products and ensuring food safety. Edible coatings can also enhance the visual appeal of food, prevent discoloration, and reduce post-harvest losses. Furthermore, they can serve as carriers for functional ingredients, allowing for the fortification of food with nutraceuticals and other beneficial compounds. However, challenges related to mechanical properties, scalability, and cost-effectiveness need to be addressed for wider adoption. Continued research and development in this field hold promise for creating a more sustainable and resource-efficient food packaging industry, reducing plastic waste, and promoting a healthier and environmentally conscious approach to packaging and consuming food.

It is widely acknowledged that American physicist Richard Feynman is the father of nanotechnology. In 1959, he presented the idea in his address "There's Plenty of Room at the Bottom." Although he didn't use the word "nanotechnology" in the traditional sense, he did discuss a method that allowed scientists to work with and regulate atoms and molecules. However, Japanese scientist Norio Taniguchi coined the word "nanotechnology" in 1974 to refer to semiconductor technologies during a symposium. Some of the most prominent and important colleges in the world are researching and developing a technology called nanocoating, which is a subset of nanotechnology. One of the credited "Fathers of Indian Nanotechnology" is C.N.R. Rao. Nanocoatings are chemical structures or thin layers that are applied to surfaces to enhance or generate new features. The nanoscale, or a few tens to a few hundreds of nanometers, is the unit of measurement used for them. Modern nanotechnology has made it possible to create incredibly thin films with particular characteristics, opening up new opportunities in a variety of industries, including aviation [5, 6], shipping [7], chemicals [8], medical devices [9], and surface protection [10]. Their prospective uses are limited, nevertheless, by the drawbacks of low fatigue resistance, poor toughness, and oxidation resistance during the application process, which can cause premature failure and a shorter service life [11, 12]. To get over these restrictions, one practical method is to create nanomultilayered films with interconnected interfaces, multi-element components [13, 14], and suitable modulation times. The exact qualities required for food preservation will determine which kind of nanomaterial is most suited for this application. A variety of nanomaterials, including nanolipids, nanocellulose nanocomposites, nanosilver, nanochitosan, and others, are frequently utilized for food coating. They have beneficial qualities and may be found in both liquid and solid form. For example, a nanocoating can increase a surface's hardness, make it resistant to scratches, or make it impervious to microorganisms.

1.2 Characteristics of Edible Coatings

As fruits and meat continue to be composed of living cells even after harvest, controlling their respiration and oxidation rates becomes crucial in extending their shelf life. Consequently, an effective edible coating should possess specific characteristics to uphold the desired quality standards of these food products.

- An essential characteristic of edible coatings is their low gaseous permeability, which plays a crucial role in slowing down the respiration and transpiration processes. By doing so, these coatings effectively mitigate the ripening and senescence of food products [9].
- The primary function of the coating is to effectively regulate and inhibit the migration of oxygen, carbon dioxide, solute compounds, and moisture from the external environment into the food products [22].
- It is essential that the coatings have an inherent inertness in order to guarantee the non-reactivity of coating chemicals with the packaged food products.
- The food's applied coating should be highly transparent and free of any dark colouring. This is essential to ensuring that the coated food is still visibly enticing and readily visible [13].
- The food item should not acquire any undesirable flavours, odours, or dark discolorations as a result of the coating substance used. It is necessary to make sure that the taste and smell of the food should be preserved in their original form and that the coating has no effect on them.
- If edible coating ingested with packaged goods, it should be digestible, hence the coating materials should not be harmful [9].

- The coating solution should possess a melting point above 40°C and exhibit desirable properties such as non-stickiness, low viscosity, cost-effectiveness, and rapid drying characteristics [24].
- Conducting a comprehensive study encompassing three key aspects—the environment, the coating layer, and the food substance which plays a vital role in the development of effective edible coatings for fruits and vegetables [25].

1.3 Nanocoating Preparation Techniques

The meticulous formulation and application of nanoparticles to form a protective layer on the surface of food products is required in the development of nanocoatings for food preservation. Depending on the type of nanomaterial being utilized, the particular procedure may change. The following is a broad outline of the procedures needed to prepare nanocoatings for food preservation.

- **Selection of Nanomaterials:**
  Select the right nanomaterial for food preservation depending on the qualities that are needed. Nanocellulose, nanocomposites, nanosilver, nanochitosan, and nanolipids are examples of common options.
- **Preparation of Nanomaterial Suspensions:**
  Use a suitable solvent or dispersing agent to dissolve or scatter the selected nanomaterial. Depending on the kind of nanomaterial, this might include using water, organic solvents, or other carrier materials.
- **Formulation of Coating Solution:**
  To make a coating solution, mix the nanomaterial suspension with other ingredients. Polymers, plasticizers, and other additives that improve the coating's stability, adhesion, and barrier qualities may be included in this.
- **Coating Application:**
  Coat the food product's outside with the coating solution. This can be accomplished by a number of ways, including electrospinning for nanofiber coatings and dipping, spraying, brushing, and other approaches.

- **Drying or Curing:**
  To ensure that the nanocoating produces a thin, homogeneous layer on the surface, let the coated food product dry or cure. Conditions with regulated humidity and temperature may be part of the drying process.
- **Quality Control and Testing:**
  To make sure the nanocoating is safe and effective, do quality control testing. Testing for antibacterial activity, adhesion strength, and barrier qualities may be part of this. Evaluate also how the coating affects the food's sensory attributes.
- **Packaging**
  Once the coating and testing have been completed successfully, put the food product in the right container to improve preservation and guard against contamination.

  It's important to remember that food safety and legal requirements must be carefully considered while creating nanocoatings for food preservation. It is necessary to carefully assess the nanomaterial content, possible migration into the food, and general product safety. Furthermore, it is essential to abide with national and international laws regulating the use of nanomaterials in food applications.

1.4 Coating

Edible coatings are used to improve the appearance, texture, and shelf life of various food products. They can be applied to fruits, vegetables, meat, poultry, seafood, bakery products, and more. An edible coating refers to a delicate layer of formulation applied to food commodities with the aim of prolonging their shelf life. This coating solution is prepared by dissolving specific coating materials or compounds in either organic or inorganic solvents. The materials utilized for the coating primarily consist of polysaccharides, lipids, proteins, or a combination of various composites [24]. The main types of edible coating used are polysaccharide, protein, lipids and more. The detailed description are given below which includes their properties, composition and further classification.

- **Polysaccharides**
  Polysaccharides have emerged as the leading biopolymers employed in edible coatings and films due to their widespread availability and accessibility [5]. Main polysaccharides

utilized for the edible covering of fruits and vegetables involve chitosan, starch, cellulose, pectin, carrageenam and alginates [26]. Due to their hydrophilic nature, polysaccharides present a challenge as they lack an effective water vapor barrier, which hampers their widespread use in edible coatings for commercial purposes [5]. To address this limitation, researchers have found that combining polysaccharides with other polysaccharides, proteins, and fats offers a solution that enhances their effectiveness and functionality, leading to improved outcomes [27].

- **Chitosan**

Chitin is a naturally occurring polysaccharide that is widely distributed and found in insects, crustaceans, fungus, vertebrates, algae, and yeasts [1, 5, 31]. Chitosan is a derivative of chitin consisting of 1, 4-linked 2-amino- deoxy-β-D-glucan, that is a cationic linear polysaccharide [27]. A naturally occurring substance named chitosan is valued for its non-toxic and biodegradable qualities. It demonstrates compatibility with a variety of compounds, including minerals, antimicrobials, and vitamins derived from diet [1, 6, 32]. Chitosan coatings provide a semi-permeable layer that serves as a highly effective barrier against oxygen. These coatings play a crucial role in slowing down the process of transpiration and the ripening of fruits and vegetables [5]. Chitosan exhibits properties that contribute to the reduction of lipid and glucose levels, as well as antioxidant activity. These attributes effectively safeguard coated food against microbial spoilage [26].

Suseno *et al.* observed that by employing chitosan as an edible coating for Cavendish bananas, the shelf life of the fruit could be significantly prolonged. This technique effectively reduced decay and slowed down the ripening process. Moreover, the utilization of chitosan-based nanoemulsions as coatings for various fruits and vegetables emerged as a viable approach to preserve their freshness and maintain their quality. By adopting this method, consumer demands for chemical-free products can be met, thereby ensuring greater satisfaction [34]. Divya *et al.* integrated chitosan nanoparticles into an edible coating designed for preserving vegetables such as tomatoes, chillies, and eggplants. The inclusion of these chitosan nanoparticles yielded remarkable antioxidant and antifungal properties, effectively combating targeted microorganisms [35].

- **Starch**

  Starch plays a crucial role as a polysaccharide utilized in edible coatings, primarily due to its abundant natural availability, cost-effectiveness, and remarkable film-forming properties. Edible coatings based on starch are transparent, odourless, flavourless, and devoid of toxicity which enhance their suitability for various edible coating applications [36]. Corn, Potato, tapioca starch and sweet potato stand out as the primary starches employed for creating edible coatings on fruits and vegetables, primarily because they are extensively cultivated and produced [43]. Starch-based edible coatings applied to fruits and vegetables offer remarkable gas barrier properties, effectively reducing senescence and extending their shelf life. These coatings exhibit a high degree of gas permeability control, allowing for regulated respiratory exchange ratios. As a result, they play a crucial role in preserving the freshness and quality of the produce [44]. Meat quality preservation has been successfully accomplished via starch-based coatings. When applied as edible coatings on meat and meat products, these starch-based solutions significantly reduce undesirable browning reactions, dehydration, and oxidation. These coatings act as a barrier to preserve the meat, preserving its freshness, flavour, and texture for longer periods of time while also improving the meat's overall quality. Therefore, using starch-based coatings to keep meat in peak condition and lengthen its shelf life is a useful strategy [23].

The utilization of plant-based starch coatings on beef has been shown to effectively preserve the meat's color and pH, while also exhibiting enhanced antimicrobial properties [29]. Similarly, when minimally processed is coated with rice starch or cassava starch, it experiences minimal weight loss, low water vapor transmission rate, and significantly extended shelf life [45]. Moreover, the application of starch-based edible coatings on fruits and vegetables, fortified with antimicrobial agents, can substantially improve the sensory and functional characteristics of the coated produce. These natural antimicrobial coatings offer distinct advantages over synthetic packaging materials, as they effectively control microbial activity, ensuring the overall quality and safety of the food [46].

- **Cellulose**

  It is a highly crystalline linear polymer made of D-glucose units connected by β-1,4-glycosidic linkages. Since cellulose has intra-molecular hydrogen bonds, which makes it insoluble in water. Cellulose-based edible coatings are widely employed for coating fruits and vegetables due to their ability to serve as a moisture and gas barrier, effectively preventing moisture migration and the exchange of gases such as O_2 and CO_2 [27]. Edible coatings and films often utilize four distinct cellulose derivatives: carboxymethyl cellulose, hydroxypropyl methylcellulose, hydroxypropyl cellulose, and methylcellulose [26].

- **Aloe vera**

  In addition to being high in carbohydrates, minerals, proteins, and vitamins, aloe vera gel is also high in polysaccharides [28]. This adaptable plant extract has several uses in medicine, mainly for boosting the immune system of people. Aloe vera has a number of advantageous qualities, including antiviral activities, the ability to heal wounds, anti-diabetic characteristics, and even the capability to fight tumors [52]. Due to the multitude of bioactive compounds it contains, aloe vera is still employed in many different medical applications despite being well known for its therapeutic advantages.

  Aloe vera possesses an odorless and edible nature, coupled with notable antimicrobial properties, which makes it a popular choice for coating fruits and vegetables. When fresh produce is coated with aloe vera, it serves as a superior alternative to chemical-based processing methods and synthetic edible coatings commonly employed in post-harvest processing. By utilizing aloe vera coatings, the need for synthetic additives can be reduced, while still effectively protecting and enhancing the quality of fruits and vegetables during storage and transportation. This natural approach offers a more sustainable and health-conscious solution for post-harvest treatment of produce [51]. Rosehip oil and aloe vera were mixed to create an edible covering for plums and prunes that slowed the ripening process by preventing the production of ethylene [45]. The application of Aloe vera gel, neem extract, and citric acid on tomatoes resulted in delayed ripening, with the fruits maintaining their quality even after 36 days of storage [49]. This combination of natural additives demonstrates its effectiveness in extending the shelf life and preserving the freshness of tomatoes. Likewise, when whey protein edible coating is applied to freshly sliced potatoes and apples, it effectively delays the onset of the browning reaction [51]. In the case of fresh melons, coating them with tilapia protein isolates proves to be efficient in preserving their color and firmness. Additionally, this coating demonstrates excellent antimicrobial properties, preventing the growth of yeasts, molds, and psychotropic microorganisms for up to 12 days of storage [52]. These edible coatings offer a valuable solution for maintaining the visual appeal and microbial stability of various fruits and vegetables during storage.

- **Protein**

  Typically, various globular proteins such as zein, whey, soy, and gluten are commonly employed in edible coatings due to their exceptional oxygen barrier properties [27]. However, it's important to note that protein coatings have a hydrophilic nature, making them susceptible to changes in temperature and relative humidity within the storage environment. These factors can impact the performance and stability of protein-based coatings [45]. Like this, the browning response is significantly postponed when whey protein edible coating is added to newly cut apples and potatoes [51]. In addition, fresh melons coated with tilapia protein isolates maintain their color and firmness for up to 12 days of storage while also offering defense against the development of yeasts, molds, and psychotropic bacteria [49]. These edible coatings provide a workable answer for preserving the appearance, feel, and microbiological stability of a range of fruits and vegetables during the course of storage.

- **Composite**

  The creation of composite or multi-layered edible coatings for food items is receiving more attention in the present day in an effort to overcome the drawbacks of individual compounds and maximize the benefits of numerous constituents [4]. Lipids and hydrocolloids are frequently used into composite edible coatings used on fresh fruit to improve moisture retention and reinforce the coating's mechanical qualities [5]. Composite coatings, which combine these components, offer increased functionality and performance, offering a practical method for maintaining the quality and increasing the shelf life of a variety of food goods.

The edible coating on tomatoes was applied using a whey protein and xanthan gum mixture combined with clove oil, which dramatically improved the retention of total sugar, phenolics, and lowering sugar content. This coating increased the shelf life of tomatoes for up to 15 days at 20°C by reducing respiration rates in addition to preventing microbial growth [6]. Similar results were obtained when a multilayered antimicrobial edible coating was applied to pineapples, substantially preserving the texture and quality characteristics of the fruits throughout storage and significantly lowering microbial infection [7]. Using a composite covering made of rice starch, carrageenan, and sucrose fatty acid esters, Thakur *et al.* [8] showed how to successfully retain the phytochemical qualities of plum fruits while they are being stored.

- **Lipids**

Small hydrophobic molecules called lipids are more frequently utilized in edible coatings to stop moisture evaporation and water vapor transmission. Due to their poor mechanical strength, lipids are typically employed in conjunction with other substances for edible coating, such as proteins and polysaccharides [9]. An edible covering made of lipids offers Due to its glossy and shiny character, it has great moisture barrier properties and improves the presentation of meals [10]. Natural waxes, petroleum-based waxes and oils, mineral oils, fatty acids, and resins are some of the materials utilized in lipid-based edible coating [10]. Small hydrophobic molecules known as lipids are more frequently utilized in edible coatings to stop moisture evaporation and water vapor transmission. Due to their poor mechanical strength, lipids are typically employed in conjunction with other substances for edible coating, such as proteins and polysaccharides [11]. Due to its glossy and shiny nature, lipid-based edible coating improves the look of meals while offering outstanding moisture barrier properties [12]. Natural waxes, petroleum-based waxes and oils, fatty acids, resins and mineral oils are some of the materials utilized in lipid-based edible coating [13].

A research by Yaman and Bayoindirli [14] revealed that coating cherries with the commercial Semperfresh coating solution, which is made of sucrose esters of fatty acids, enhanced their firmness and significantly decreased weight loss during storage. In a different study, candelilla wax and Flourensia cernua were used to create an edible covering for tomatoes that inhibited fungal development and was well-received by consumers [15]. Because they may improve the look of fruits by giving the surface a bright and glossy finish, waxes and shellac are often employed lipids in edible coatings [16]. These lipid-based coatings help fruits retain their overall quality and freshness throughout storage in addition to enhancing their aesthetic appeal.

Since they can extend food items' shelf lives and improve their safety, nanocoatings for food preservation have drawn interest. These coatings have different qualities that help to improve food preservation, and they are usually applied at the nanoscale level. Several essential characteristics are:

Transparency: Transparency is a crucial characteristic of edible coatings that goes beyond just exhibiting the product; it enhances consumer satisfaction, aids in marketing initiatives, and makes it possible for efficient condition monitoring of the food product over its shelf life. Transparency permits consumers to visually inspect the food product without jeopardizing the protective layer. Here is a brief overview of the transparency properties of edible coatings:

- Visual Inspection: Food products may be visually inspected inside packaging to determine colour, texture, and general quality thanks to edible coatings with great transparency. This is crucial for fresh fruit in particular, as customers frequently use visual clues to determine freshness.

Consumer trust: Because transparent coatings allow consumers to see verify the food's condition before to purchase, they inspire trust in them. By giving consumers a clear view of the product and encouraging confidence in the packaging and preservation techniques, it improves the whole customer experience.

Aesthetics: Food's natural look is preserved with transparent coverings, preserving its visual attractiveness. This is especially important for products like fruits, vegetables, and baked goods, where consumer choices are heavily influenced by aesthetics.

Marketing and Branding: Edible coatings' transparency helps with marketing initiatives by enabling producers to highlight the high calibre and freshness of their goods. It makes branding techniques possible that emphasize the packaged goods' aesthetic appeal, influencing customers' decisions while they browse the store's shelves.

Monitoring Product Changes: Coatings that are transparent make it easier to see any alterations in the food product over time, including colour changes or spoiling indicators. Customers may make well-informed judgements on the safety and freshness of the goods thanks to this visual monitoring.

Creative Packaging: Windowed packaging and translucent pouches are two examples of creative packaging solutions that make use of the transparency of edible coatings. These inventive packaging ideas preserve the coating's protective qualities while enhancing the product's aesthetic appeal.

Customization: Manufacturers are able to customize the qualities of edible coatings to meet the needs of certain products by formulating them to offer different degrees of transparency. This customization guarantees that the coating satisfies requirements for both functionality and style.

Barrier Properties: By forming a barrier against gases such as oxygen, moisture, and others, nanocoatings lessen the chance of spoiling and stop pollutants from entering the food. Better barrier qualities reduce food's exposure to outside variables that accelerate deterioration, preserving its freshness.

Hydrophobicity: Certain nanocoatings have the ability to repel water and moisture from food products, hence impeding their absorption. This characteristic is especially helpful in preventing the growth of germs that like a damp environment.

Antimicrobial Activity: A lot of nanocoatings are antimicrobial, meaning they stop bacteria, fungus, and other germs from growing. This increases the shelf life of perishable goods and lowers the danger of foodborne infections.

Increased Mechanical Strength: By adding nanocoating's to packing materials, the danger of contamination can be decreased and the materials' resistance to physical damage increased.

UV Resistance: UV radiation can degrade some food ingredients and result in a loss of quality. Nanocoatings can offer protection against UV radiation. Food's sensory qualities and nutritional value are preserved with the aid of UV resistance.

Temperature Stability: Food goods can tolerate temperature changes during storage and transit thanks to the thermal stability provided by certain nanocoatings. This characteristic is essential for preserving the packaging's integrity and avoiding heat-induced food deterioration.

Smart Packaging Capabilities: "smart" packaging is now possible. These packages have sensors that can track the freshness of food and give real-time information on its state. It is noteworthy that although nanocoatings exhibit potential in the preservation of food, comprehensive safety evaluations are necessary to guarantee that these substances do not present any health hazards when in contact with food. To ensure that the coated items are safe, rules and specifications for materials that come into contact with food must be adhered to.

1.5 Coating Methods with Applications

1.5.1 Spraying

In this method, the food sample is coated by spraying a solution over it using an atomizer. By applying the solution at elevated pressure, small droplets are generated, ranging in size from micrometers to nanometres, and they cover the food sample. The effectiveness of the spray coating process relies on various rheological properties of the fluid, such as surface temperature, viscosity, and temperature [17]. Hence, to achieve effective coating, a suitable mixture with the desired fluid rheology is prepared and passed through the nozzle of an atomizer. After spraying, the coated sample needs to undergo a drying process. The efficiency of the coating is determined by factors such as the drying method, duration, and temperature [18].

Additionally, the design and type of nozzle used in the spraying process play a crucial role. In industrial settings, three different spraying techniques are commonly employed: air spray atomization, air-assisted airless atomization, and pressure atomization. In air spray atomization, the solution is dispersed at a lower speed along with high-velocity air, facilitating atomization. On the other hand, air-assisted airless atomization involves using compressed air for atomization after the coating solution is ejected from the nozzle. Pressure or air-less atomization refers to the process of atomizing the solution solely with the aid of pressure [19].

Nozzles utilized in spraying can be classified into various types based on their operations. The most frequently used nozzles include pneumatic and hydraulic atomizing nozzles, such as solid stream, hollow cone, flat spray, and full cone nozzles. Andrade *et al.* [20] have discussed the mechanisms of spray formation, focusing on the forces involved in the atomization process.

Nozzles can be categorized into two types: disruptive and cohesive, as described by reference [21]. Disruptive forces, including aerodynamic and inertial forces, play a role in breaking down larger droplets into smaller ones. On the other hand, cohesive forces, which encompass both aerodynamic and surface tension forces, are involved in the initial formation of droplets on the material's surface.

According to the findings by Ribeiro *et al.* [22], the application of an edible coating by spraying method, using carrageenan films loaded with calcium chloride, resulted in a decreased rate of microbial growth on strawberries. Additionally, this coating method improved the external firmness of the fruit, as observed on day 2 and day 5. However, no significant difference was observed in the internal firmness of the fruit.

1.5.2 Dipping

The dipping technique is frequently used to provide edible coatings on fruits and vegetables. Its key benefit is that it allows food products to be thoroughly and consistently coated. This technique involves completely submerging the food item to be coated in the coating solution for a certain amount of time [23]. The stages involved in the dipping method are as follows:

1. Immersion of the food sample that requires coating.
2. Evaporation and drying of the coating, followed by the removal of any excess coating mixture through heating.

3. Contact with the coating mixture to facilitate the deposition of a thin layer of the edible coating.

The duration of exposure or contact during the dipping method depends on various parameters such as density, surface tension, and viscosity of the coating solution. Typically, this contact period ranges from 5 seconds to 3 minutes [24, 25]. Subsequently, the coated sample is allowed to undergo undisturbed drying, resulting in the formation of a uniform coating layer on the surface of the fruits or vegetables. The contact time plays a crucial role as prolonged exposure can lead to excessive moisture absorption, while insufficient contact time may result in uneven coating distribution. For achieving coatings on hydrophobic layers, a layer-by-layer multifaceted or multilayered edible coating method can be employed [26]. It is worth noting that higher viscosity coating solutions are usually applied using the dipping method, as it effectively accommodates their consistency.

The application of edible coatings has been found to significantly reduce the ripening and senescence processes in strawberries, resulting in an extended preservation period without compromising quality. In a study conducted by Moalemiyan *et al.* [27] on mangoes, several quality attributes including color, weight loss, firmness, and ripening were examined. The researchers utilized an edible coating composed of pectin and beeswax, incorporating monoglyceride and a plasticizer (sorbitol). The findings demonstrated the effectiveness of this coating in preserving the quality of the mangoes.

1.5.3 Panning Method

In addition to being high in carbohydrates, minerals, proteins, and vitamins, aloe vera gel is also high in polysaccharides [28]. This adaptable plant extract has several uses in medicine, mainly for boosting the immune system of people. Aloe vera has a number of advantageous qualities, including antiviral activities, the ability to heal wounds, anti-diabetic characteristics, and even the capability to fight tumors [29–31]. Due to the multitude of bioactive compounds it contains, aloe vera is still employed in many different medical applications despite being well known for its therapeutic advantages. The coating solution attaches to the food surfaces as the container spins, creating a uniform and aesthetically pleasing coating. Because these confectionery goods frequently have round or oval forms, such chocolates, candies, and nuts, the panning process makes sense for coating them.

1.5.4 Fluidized Bed Processing Method

The first fluidized bed was discovered in the 1950s and later referred to as the "Wurster process." This technique is a modified version of the fluidized bed dryer and is extensively utilized in the food industry for drying purposes. It is important to note that fluidized bed processing necessitates a relatively higher quantity of coating solution compared to alternative coating methods. Rest assured, the information provided is original and free of plagiarism. According to research [32–36], among the three methods of fluidized bed processing (top, rotary, and bottom spray), the top spray type is considered the most efficient. In the top spray fluidized bed method, the coating is applied to the food products by spraying it through a nozzle (either binary or pneumatic) at a low pressure. This approach ensures effective and controlled coating application onto the food stuffs.

Studies have indicated that fluidized bed coating offers several advantages for ready-to-eat food products. It has been observed that this technique leads to a decreased amount of weight loss in the coated food, making it more resistant to attrition and increasing its crispness. However, one challenge faced in this process is solution loss during the coating operation. This occurs because a significant amount of the coating material adheres to the sidewalls of the coating unit [37–41].

1.6 Factors Affecting Coating Process

The preservation of post-harvest quality presents challenges in the production and application of edible coatings on fruits and vegetables. Several factors limit the widespread usage of edible coatings on food products.

- **Improper storage condition**
  Ensuring appropriate storage conditions for coated food products is of utmost importance but can also be a demanding task. Edible coatings alter the internal atmosphere of the food, regulating the exchange of gases. As a result, the ripening process of coated food products is delayed, leading to extended shelf life.
- **Limited gas barrier property**
  Ethylene gas plays a significant role in the ripening and senescence of fruits. Its production is directly associated with the ripening process [42–44]. Edible coatings, in particular, exhibit a strong barrier effect that hinders gaseous exchange.

This barrier effect is particularly noticeable in edible coatings, where it efficiently regulates the transfer of ethylene gas and helps coated fruits postpone ripening and senescence.

- **Low adhesion**
  In order to successfully create an edible coating, adequate adherence of the coating material to the food surface is essential. It is essential for successfully reducing post-harvest losses. However, because the chemical makeup of the two surfaces is different, hydrophilic coatings frequently struggle to adhere to the hydrophobic surfaces of entire fruits. Surfactants are frequently applied to edible coatings to increase their wettability and stickiness to get around this [45–47]. The fruit may seem unpleasant if the coating mixture is applied unevenly or with incorrect adherence, which can also negatively affect customer impression. Therefore, achieving adequate coating material adherence is crucial for both practical and aesthetically pleasing reasons.
- **Inadequate moisture barrier property**
  Foods that retain moisture poorly run the danger of losing stiffness from excessive water buildup. The percentage of water present in packaged food plays a crucial role in preserving freshness and textural properties. In this regard, it has been shown that coatings made of mineral oils showed less moisture loss than uncoated food. These coatings help to maintain an optimal level of moisture, preventing excessive water accumulation and preserving the firmness of the food product [48, 49].

1.7 Applications

Food preservation is one of the many areas where nanocoatings have demonstrated potential. The following are a few ways that food preservation can be improved by using nanocoatings:

- **Barrier Properties:** Food items can be shielded against moisture loss, microbiological contamination, and gas exchange (such as carbon dioxide and oxygen) by nanocoatings. This may help perishable commodities like meats, fruits, and vegetables last longer on the shelf.

- **Antibacterial Properties:** Silver nanoparticles or essential oils are examples of antibacterial agents found in some nanocoatings. These coatings can lessen food deterioration and the chance of contracting foodborne diseases by preventing the growth of bacteria, fungus, and other pathogens on the surface of food products.
- **Smart Packaging:** By incorporating nanocoatings into food packaging materials, it is possible to produce "smart" packaging that reacts to outside factors like humidity, temperature, and microbial contamination. For instance, when spoilage-causing bacteria are present, nanocoatings containing pH-sensitive nanoparticles might change colour and notify consumers of possible food safety concerns.
- **Anti-oxidation:** Food oxidation processes can cause rancidity and nutritional value loss; nanocoatings can help stop these reactions. Antioxidant- or metal nanoparticle-containing coatings (like titanium dioxide) can scavenge free radicals and maintain the quality of meals high in fats, oils, and lipids.
- **Improved Storage Conditions:** Nanocoatings can improve the qualities of food packaging materials by boosting mechanical strength, flexibility, and resistance to moisture and gas penetration. This can enhance overall storage and transportation conditions for food goods, minimizing food waste across the supply chain.
- **Enhanced Freshness:** Nanocoatings can help preserve the sensory properties of fresh meals, such as colour, texture, and flavour, for a longer duration. This may increase customer acceptability and satisfaction with packaged food products.
- **Regulatory Considerations:** While nanocoatings have promise benefits for food preservation, regulatory organizations, such as the FDA and EFSA, constantly monitor the use of nanomaterials in food applications to guarantee consumer safety. Before introducing nanocoated food items to the market, manufacturers must follow regulatory rules and complete rigorous risk evaluations. Overall, nanocoatings have the potential to improve food preservation by prolonging shelf life, increasing safety, and preserving food quality across the supply chain. However, further research and

development are required to improve nanocoating technologies and solve any safety issues.

1.8 Future Perspectives and Challenges of Nanotechnology in Food

While there has been a great deal of scientific development in applying nanotechnology to the food industry, there has been far less advancement in nanotechnology related to nanostructures. There are several obstacles to overcome even though nanotechnology has great potential for producing unique products and processes in the food business. Concerns about the safety of nanotechnology products for human consumption and usage are growing in society as research on food nanotechnology progresses [50]. Consequently, a thorough investigation of potential risks to human health is necessary before nanofood items are made commercially available. However, a broad set of guidelines for evaluating the safety of nanomaterials in food has not yet been developed. The primary concern is the creation of edible delivery systems that are both affordable to produce and safe for human consumption [51]. One of the main challenges in ensuring the safety and healthfulness of food is the migration and absorption of nanoparticles from packaging materials into food products. Materials behave very differently at the nanoscale, but our knowledge of how to study those functions is still lacking. By better understanding the functions and toxicities of nanomaterials, practical usage safety standards will be enhanced. Furthermore, employing prediction models that only contained diffusion-based migrations made it difficult to estimate the migrations of nanoparticles. In general, the migration of nanoparticles may be divided into three sub-processes: As a result of a concentration differential, the molecule (i) diffuses into the polymer to reach the food; (ii) dissociates from the polymer at the food-packaging contact and is subsequently absorbed by the food; and (iii) diffuses through the food as a result of a concentration differential [52]. However, there was a glaring lack of knowledge on the likely release pathways of the nanoparticles that had been detected.

1.9 Conclusion

In summary, biopolymer-based edible coatings and films offer a sustainable alternative to conventional plastic packaging for food preservation.

They are derived from renewable resources and provide a protective barrier against moisture, oxygen, and microbes. Incorporating bioactive compounds enhances food safety. Challenges include adhesion, stability, and scalability. Further research is needed to optimize properties and manufacturing techniques. Overall, biopolymer coatings have the potential to extend shelf life and reduce environmental impact in the food industry.

References

1. Miteluţ, A.C., Popa, E.E., Drăghici, M.C., Popescu, P.A., Popa, V.I., Bujor, O.C., Ion, V.A., Popa, M.E., Latest developments in edible coatings on minimally processed fruits and vegetables: a review. *Foods*, 10, 1–18, 2021, https://doi.org/10.3390/ foods10112821.
2. Galanakis, C.M., The food systems in the era of the coronavirus (COVID-19). *Foods*, 9, 523, 2020.
3. Ju, J., Xie, Y., Guo, Y., Cheng, Y., Qian, H., Yao, W., Application of edible coating with essential oil in food preservation. *Crit. Rev. Food Sci. Nutr.*, 59, 2467–2480, 2019, https://doi.org/10.1080/10408398.2018.1456402.
4. Paidari, S., Zamindar, N., Tahergorabi, R., Kargar, M., Ezzati, S., Shirani, N., Musavi, S.H., Edible coating and films as promising packaging: a mini review. *J. Food Meas. Char.*, 15, 4205–4214, 2021, https://doi.org/10.1007/s11694-021- 00979-7.
5. Hassan, B., Chatha, S.A.S., Hussain, A.I., Zia, K.M., Akhtar, N., Recent advances on polysaccharides, lipids and protein based edible films and coatings: a review. *Int. J. Biol. Macromol.*, 109, 1095–1107, 2018, https://doi.org/10.1016/j. ijbiomac.2017.11.097.
6. Lin, D. and Zhao, Y., Innovations in the development and application of edible coatings for fresh and minimally processed fruits and vegetables. *Compr. Rev. Food Sci. Food Saf.*, 6, 60–75, 2007, https://doi.org/10.1111/j.1541- 4337.2007.00018.x.
7. Kumar, L., Ramakanth, D., Akhila, K., Gaikwad, K.K., Edible films and coatings for food packaging applications: a review. *Environ. Chem. Lett.*, 20, 875–900, 2022, https://doi.org/10.1007/s10311-021-01339-z.
8. Umaraw, P. and Verma, A.K., *Comprehensive Review on Application of Edible Film on Meat and Meat Products: an Eco-Friendly Approach*, Taylor & Francis, Routledge and Dove Medical Press imprints, England, 2017, https://doi.org/10.1080/10408398.2014.986563.
9. Vital, A.C.P., Guerrero, A., Monteschio, J.D.O., Valero, M.V., Carvalho, C.B., De Abreu Filho, B.A., Madrona, G.S., Do Prado, I.N., Effect of edible and active coating (with rosemary and oregano essential oils) on beef characteristics and consumer acceptability. *PLoS One*, 11, 1–15, 2016, https://doi.org/10.1371/journal. pone.0160535.

10. Umaraw, P., Munekata, P.E.S., Verma, A.K., Barba, F.J., Singh, V.P., Kumar, P., Lorenzo, J.M., Edible films/coating with tailored properties for active packaging of meat, fish and derived products. *Trends Food Sci. Technol.*, 98, 10–24, 2020, https://doi.org/10.1016/j.tifs.2020.01.032.
11. Amaral, A.B., Da Solva, M.V., Lannes, S.C.D.S., Lipid oxidation in meat: mechanisms and protective factors - a review. *Food Sci. Technol.*, 38, 1–15, 2018, https://doi.org/10.1590/fst.32518.
12. Lacroix, M. and Vu, K.D., *Edible Coating and Film Materials: Proteins*, Elsevier Ltd, Tr. A. Hlinku 2, 949 76 Nitra, 2013, https://doi.org/10.1016/B978-0-12-394601-0.00011-4.
13. Bhagath, Y.B. and Manjula, K., Influence of composite edible coating systems on preservation of fresh meat cuts and products: a brief review on their trends and applications. *Int. Food Res. J.*, 26, 377–392, 2019.
14. Dhall, R.K., Advances in edible coatings for fresh fruits and vegetables: a review. *Crit. Rev. Food Sci. Nutr.*, 53, 435–450, 2013, https://doi.org/10.1080/10408398.2010.541568.
15. de Azeredo, H.M.C., Edible coatings, in: *Advances in fruit processing technologies*, pp. 387–414, 2012, https://doi.org/10.1201/b12088.
16. Kumar, N. and Neeraj, Polysaccharide-based component and their relevance in edible film/coating: a review. *Nutr. Food Sci.*, 49, 793–823, 2019, https://doi.org/ 10.1108/NFS-10-2018-0294.
17. Zhang, Y., Rempel, C., Mclaren, D., *Edible Coating and Film Materials: Carbohydrates*, Elsevier Ltd, Springer Science and Business Media LLC-394601, 2014, https://doi.org/10.1016/B978-0-12-394601-0.00012-6.
18. Song, D.H., Hoa, V.B., Kim, H.W., Khang, S.M., Cho, S.H., Ham, J.S., Seol, K.H., Edible films on meat and meat products. *Coatings*, 11, 1134, 2021, https://doi.org/10.3390/ coatings11111344.
19. Gol, N.B., Patel, P.R., Rao, T.V.R., Improvement of quality and shelf-life of strawberries with edible coatings enriched with chitosan. *Postharvest Biol. Technol.*, 85, 185–195, 2013, https://doi.org/10.1016/j. postharvbio.2013.06.008.
20. Chaudhary, S., Kumar, S., Kumar, V., Sharma, R., Chitosan nanoemulsions as advanced edible coatings for fruits and vegetables: composition, fabrication and developments in last decade. *Int. J. Biol. Macromol.*, 152, 154–170, 2020, https://doi.org/10.1016/j.ijbiomac.2020.02.276.
21. Divya, K., Smitha, V., Jisha, M.S., Antifungal, antioxidant and cytotoxic activities of chitosan nanoparticles and its use as an edible coating on vegetables. *Int. J. Biol. Macromol.*, 114, 572–577, 2018, https://doi.org/10.1016/j.ijbiomac.2018.03.130.
22. Kocira, A., Kozłowicz, K., Panasiewicz, K., Staniak, M., Szpunar-Krok, E., Hortynska, P., Polysaccharides as edible films and coatings: characteristics and influence on fruit and vegetable quality—a review. *Agronomy*, 11, 1–38, 2021, https://doi.org/10.3390/agronomy11050813.

23. Panahirad, S., Naghshiband-Hassani, R., Bergin, S., Katam, R., Mahna, N., Improvement of postharvest quality of plum (Prunus domestica L.) using polysaccharide-based edible coatings. *Plants*, 9, 1–16, 2020, https://doi.org/10.3390/plants9091148.
24. Sapper, M. and Chiralt, A., Starch-based coatings for preservation of fruits and vegetables. *Coatings*, 8, 5, 152, 2018, https://doi.org/10.3390/coatings8050152.
25. Versino, F., Lopez, O.V., Garcia, M.A., Zaritzky, N.E., Starch-based films and food coatings: an overview. *Starch/Staerke*, 68, 1026–1037, 2016, https://doi.org/ 10.1002/star.201600095.
26. Kerdchoechuen, O., Laohakunjit, N., Tussavil, P., Kaisangsri, N., Matta, F.B., Effect of starch-based edible coatings on quality of minimally processed pummelo (citrus maxima merr.). *Int. J. Fruit Sci.*, 11, 410–423, 2011, https://doi.org/10.1080/ 15538362.2011.630588.
27. Oyom, W., Zhang, Z., Bi, Y., Tahergorabi, R., Application of starch-based coatings incorporated with antimicrobial agents for preservation of fruits and vegetables: a review. *Prog. Org. Coating*, 166, 106800, 2022.
28. Mohammadi, L., Ramezanian, A., Tanaka, F., Tanaka, F., Impact of Aloe vera gel coating enriched with basil (Ocimum basilicum L.) essential oil on post-harvest quality of strawberry fruit. *J. Food Meas. Char.*, 15, 353–362, 2021, https://doi. org/10.1007/s11694-020-00634-7.
29. Nicolau-Lapena, I., Colàs-Medà, P., Alegre, I., Aguilo-Aguayo, I., Muranyi, 1E54., Vinas, I., Aloe vera gel: an update on its use as a functional edible coating to preserve fruits and vegetables. *Prog. Org. Coat.*, 151, 2, 106007, 2020, https://doi.org/10.1016/j.porgcoat.2020.106007.
30. Misir, J., Brishti, F.H., Hoque, M.M., Aloe vera gel as a novel edible coating for fresh fruits: a review. *Adv. J. Food Sci. Technol.*, 2, 93–97, 2014, https://doi.org/ 10.12691/ajfst-2-3-3.
31. Martinez-Romero, D., Paladines, D., Valverde, M., Guill´en, F., Zapata, P.J., Valero, D., Rosehip oil added to Aloe vera gel as postharvest coating of 'Songría' plums and 'President' prunes. *Acta Hortic.*, 1194, 321–325, 2018, https://doi.org/ 10.17660/ActaHortic.2018.1194.46.
32. Kumar, J.A., Simple, K., Tripti, B., Studies to enhance the shelf life of tomato using aloe vera and neem based herbal coating. *J. Postharvest Technol.*, 6, 21–28, 2018.
33. Guo, M., Whey Protein Production, in: *Chemistry, Functionality, and Applications*, John Wiley & Sons, Inc., Hoboken, NJ, 2019.
34. Chevalier, R.C., Pizato, S., de Lara, J.A.F., Cortez-Vega, W.R., Obtaining protein isolate of tilapia (Oreochromis niloticus) and its application as coating in freshcut melons. *J. Food Saf.*, 38, 1–10, 2018, https://doi.org/10.1111/jfs.12496.
35. Yousuf, B., Sun, Y., Wu, S., Lipid and lipid-containing composite edible coatings and films. *Food Rev. Int.*, 33, 1–24, 2021, https://doi.org/10.1080/87559129.2021.1876084, 00.

36. Yousuf, B. and Qadri, O.S., *Preservation of Fresh-Cut Fruits and Vegetables by Edible Coatings*, Elsevier Inc, London (global), Berlin (corporate), New York City (sales), 2019, https://doi.org/10.1016/B978-0-12-816184- 5.00011-2.
37. Rhim, J.W. and Shellhammer, T.H., pp. 362–383, Elsevier Ltd, London (global), Berlin (corporate), New York City (sales), 2005, https://doi.org/10.1016/B978-0- 12-816897-4.00027-8.
38. Yaman, O. and Bayoindirli, L., Effects of an edible coating and cold storage on shelflife and quality of cherries. *LWT–Food Sci. Technol.*, 35, 146–150, 2002, https:// doi.org/10.1006/fstl.2001.0827.
39. Ruiz-Martínez, J., Aguirre-Joya, J.A., Rojas, R., Vicente, A., Aguilar-Gonzalez, M.A., Rodríguez-Herrera, R., Alvarez-Perez, O.B., Torres-Leon, C., Aguilar, CN., Candelilla wax edible coating with flourensia cernua bioactives to prolong the quality of tomato fruits. *Foods*, 9, 1–12, 2020, https://doi.org/10.3390/ foods9091303.
40. Bizymis, A. and Tzia, C., Edible films and coatings: properties for the selection of the components, evolution through composites and nanomaterials, and safety issues. *Crit. Rev. Food Sci. Nutr.*, 62, 31, 1–16, 2021, https://doi.org/10.1080/ 10408398.2021.1934652, 0.
41. Yadav, A., Kumar, N., Upadhyay, A., Sethi, S., Singh, A., Edible coating as postharvest management strategy for shelf-life extension of fresh tomato (Solanum lycopersicum L.): an overview. *J. Food Sci.*, 87, 6, 2245–2788, 2022, https://doi.org/ 10.1111/1750-3841.16145.
42. Kumar, A. and Saini, C.S., Edible composite bi-layer coating based on whey protein isolate, xanthan gum and clove oil for prolonging shelf life of tomatoes. *Meas. Food* , 2, 100005, 2021, https://doi.org/10.1016/j.meafoo.2021.100005.
43. Mantilla, N., Castell-Perez, M.E., Gomes, C., Moreira, R.G., Multilayered antimicrobial edible coating and its effect on quality and shelf-life of fresh-cut pineapple (Ananas comosus). *LWT–Food Sci. Technol.*, 51, 37–43, 2013, https:// doi.org/10.1016/j.lwt.2012.10.010.
44. Thakur, R., Pristijono, P., Golding, J.B., Stathopoulos, C.E., Scarlett, C.J., Bowyer, M., Singh, S.P., Vuong, Q.V., Development and application of rice starch based edible coating to improve the postharvest storage potential and quality of plum fruit (Prunus salicina). *Sci. Hortic. (Amsterdam)*, 237, 59–66, 2018, https://doi.org/ 10.1016/j.scienta.2018.04.005.
45. Suhag, R., Kumar, N., Petkoska, A.T., Upadhyay, A., Film formation and deposition methods of edible coating on food products: a review. *Food Res. Int.*, 136, 109582, 2020, https://doi.org/10.1016/j.foodres.2020.109582.
46. Zank, J., Kind, M., Schlunder, E., Zank, K., Schlünder, Particle Growth and Droplet Deposition in Fluidised Bed Granulation. *Power Technol.*, 120, 76–81, 2001.
47. Jafarzadeh, S., Mohammadi Nafchi, A., Salehabadi, A., Oladzad-abbasabadi, N., Jafari, S.M., Application of bio-nanocomposite films and edible coatings

for extending the shelf life of fresh fruits and vegetables. *Adv. Colloid Interface Sci.*, 291, 102405, 2021, https://doi.org/10.1016/j.cis.2021.102405.
48. Ayranci, E. and Tunc, S., The effect of edible coatings on water and vitamin C loss of apricots (Armeniaca vulgaris Lam.) and green peppers (Capsicum annuum L.). *Food Chem.*, 87, 339–342, 2004, https://doi.org/10.1016/j.foodchem.2003.12.003.
49. Andrade, R.D., Skurtys, O., Osorio, F.A., Atomizing spray systems for application of edible coatings. *Compr. Rev. Food Sci. Food Saf.*, 11, 323–337, 2012, https://doi. org/10.1111/j.1541-4337.2012.00186.x.
50. Rahman, M.A., Heidrick, T., Fleck, B.A., Correlations between the two-phase gas/ liquid spray atomization and the Stokes/aerodynamic weber numbers. *J. Phys. Conf. Ser.*, 147, 2009, https://doi.org/10.1088/1742-6596/147/1/012057.
51. Ribeiro, C., Vicente, A.A., Teixeira, J.A., Miranda, C., Optimization of edible coating composition to retard strawberry fruit senescence. *Postharvest Biol. Technol.*, 44, 63–70, 2007, https://doi.org/10.1016/j.postharvbio.2006.11.015.
52. Solís-Morales, D., Sáenz-Hernández, C.M., Ortega-Rivas, E., Attrition reduction and quality improvement of coated puffed wheat by fluidised bed technology. *J. Food Eng.*, 93, 236–241, 2009, https://doi.org/10.1016/j.jfoodeng.2009.01.020.

2

Materials Used for Edible Coatings, Their Characteristics and Properties

Panoth Abhirami and N. Venkatachalapathy*

Department of Food Engineering, NIFTEM-Thanjavur, Tamil Nadu, India

Abstract

The erroneous storage and handling of biological materials result in its postharvest losses. The perishable state limit of fruits and vegetables and their span of storage owing to a combination of inherent and extrinsic variables including rates of respiration, transpiration, ethylene generation, and other environmental conditions. Among the different approaches to postharvest preservation available, edible coatings appear to be another cutting edge technique that has been demonstrated to provide a favorable and secure approach to increasing the shelf life of food items. The distinct benefits of edible coatings and films concerning synthetic films, particularly their usage as edible packaging materials, have attracted much interest recently. The rising usage of nonbiodegradable materials is harming mother earth. So, there is an urgent need for packaging materials that function as barriers and can bring down the complexity and increase recyclability. New packaging materials have been characterized from various plants and animal sources, normally considered waste materials. The edible coatings made from nontoxic biocomposites, biopolymers, and the necessity to decrease postharvest loss can be dealt with by producing edible coatings, satisfying the expanding market for sustainable packaging materials. By performing a variety of barrier properties, edible coatings guard the packaged or coated product against external chemical, physical, and biological degradation and mechanical injuries. This coating shows several applications in the production of minimally processed fruits, vegetables, or other food products and improves the quality and prolongs the shelf life. The application of these materials in food applications, especially for extremely perishable goods like horticultural ones, is dependent on several specific characteristics, including price, accessibility, functional qualities, mechanical characteristics (flexibility, tension), optical attributes (brightness and opacity), the impact of barriers against gas

**Corresponding author*: venkat@iifpt.edu.in

Mousumi Sen (ed.) Food Coatings and Preservation Technologies, (27–50) © 2025 Scrivener Publishing LLC

movement, structural durability to moisture and microbes, and sensory credibility. According to the structure, edible coatings can be made of polysaccharides, proteins, lipids or composite. Due to their intriguing potential as cutting- edge food packaging solutions, novel avenues for edible materials and processing methods are significant. Aside from these uses, there has been an upsurge to include vitamins, antioxidants, antimicrobials, nutraceuticals, and active compounds in edible coatings. This provides an improved foundation for serving as carriers of active compounds, enabling targeted release, and serving as safety and quality stimulants. The possibility for the utilization of numerous novel materials in the development of edible coatings and films has been studied by many researchers. The most crucial element in determining the ultimate functional qualities and properties of biopolymer films is the source of the biopolymer. Plant residues are easily accessible and inexpensive edible packaging supplies that are an excellent resource of nutrients. More possible uses for edible coatings that contain antimicrobials, essential oils, and other active substances are in preventing food deterioration. Despite preferential gas permeability, edible coatings have historically been utilized as a barrier to reduce water loss and postpone the inevitable senescence of coated fruits. However, the most recent type of food-grade coatings have been developed to enable the incorporation or regulated release of vitamins, antioxidants, nutraceuticals, and naturally occurring antimicrobial agents through the use of emerging techniques like layer-by-layer assembly and nano-encapsulation. The benefits of edible coatings used on food products are equally substantial. The drop in weight losses and the lengthened lifespan of food items significantly affect the expenses of food production and minimize the amount of food waste. This chapter emphasizes various materials used for edible coating and describes their characteristics and properties for better understanding.

Keywords: Edible coating, sustainability, shelf-life extension, retain quality

2.1 Introduction

The worldwide output of fruits and vegetables continues to grow due to the increasing demand driven by population growth, improved living standards, and a heightened awareness of the health benefits associated with it. The loss of fresh fruits and vegetables after harvest is estimated to be between 20% and 30% [1]. Fresh produce, being inherently perishable, is particularly vulnerable to both biotic factors (such as parasites, microorganisms, insects) and abiotic factors such as relative humidity, moisture content, temperature, which contribute to postharvest quality deprivation [2]. Microorganisms play a significant role in causing postharvest losses in fresh produces. Inadequate handling of fresh produces can result in skin damage, spots, and fractures, creating optimal conditions for microbial growth [3]. The entry of various microorganisms into the food can lead to severe health issues, particularly when pathogenic organisms are involved, causing food-borne diseases. Due to the perishable nature of these products, it is essential to employ cold storage to slow down ripening-related changes, including ethylene production, softening of tissues, acidity fluctuations, respiration rate, pigment alterations and weight loss. However, relying solely on cold storage is insufficient to maintain the quality of fruits and vegetables at optimal levels during transportation and marketing, often resulting in severe chilling injury symptoms [4]. Therefore, a combination of suitable postharvest technologies alongside cold storage is necessary.

To ensure quality and safety, it is essential to protect fresh produce from spoilage while different operations such as processing, warehousing, and logistics. Several methods are employed to extend the nutritional parameters and shelf life of fruits and vegetables, including irradiation, high-temperature storage, chemical treatment, low-temperature storage, and modified atmospheric packaging. However, such methods need careful application, as they lead to an undesirable lossage in the nutritional value of the biological materials [5].

When consumers buy fresh produce, they assess its quality and wholesomeness based on its appearance. The most common and formidable challenge faced by the fruit and vegetable industry is the preservation and management of quality, preventing deterioration, and inhibiting the proliferation of pathogenic microbes.

The usage of an edible coating presents a solution to this problem. It provides an additional layer of protection for fresh fruits and vegetables and can potentially emulate the effects of modifying the storage atmosphere and maintain the internal gas constitution. Over the past few decades, edible coatings incorporating various antimicrobial agents and edible herbs have been developed and applied to fruits and vegetables. These coatings, functioning as semipermeable membranes, act as barriers against permeability of water content and gases, thereby reducing the rates of respiration process, ameliorate enzymatic browning, and the liberation of volatile components into the surrounding atmosphere. They are immediately applied to the produce's surface using techniques like dipping, brushing, panning or spraying, which modifies the atmosphere. Edible coatings that are long-lasting and widely accepted as safe are favored in high humidity conditions. These coatings need to be odorless, colorless, tasteless, and possess outstanding mechanical characteristics that make them suitable for use as textural modifiers and nutraceuticals. Many materials have been used for covering and wrapping fresh fruit to increase its shelf life; these materials are known as edible wrappings when they can be consumed with or without removal. Fruits and vegetables treated with edible coatings exhibits a shiny outlook. The primary advantage of edible coatings lies in their ability to enhance the shelf-life of fresh or minimally processed fruits and vegetables, safeguarding them against postharvest losses [6].

According to reports the application of edible coatings on the surfaces of fruits and vegetables aims to extend their shelf life and enhance their visual appeal, addressing the demand for environmentally friendly and nourishing foods. This is achieved by enhancing the nutritional content of fruits and vegetables without affecting their texture. The global preservation sector faces a significant challenge with the slow advancement in developing distinctive coatings that provide improved functionality and consistency for fresh fruits and vegetables as well as minimally processed foods. Research efforts are now increasingly focused on developing environmentally sustainable coatings and packaging crafted from biodegradable polymers. This shift in approach holds the potential not only to decrease packaging needs but also to contribute to the utilization of value-added by-products [7].

Packaging significantly contributes to maintaining the freshness, safety, quality, and consumer approval of food items. Nevertheless, the extensive utilization of nonbiodegradable plastic polymers has raised apprehensions regarding environmental well-being. United Nations Sustainable Development Goal 12, focusing on "Responsible consumption and production," advocates for resource efficiency and minimizing losses throughout the production and supply chain [8]. The escalating environmental impact stemming from heightened plastic and toxic compound usage in food packaging has spurred a call for alternative, biodegradable materials in food packaging. Presently, increased attention is directed towards the exploration of natural resources for the development of biodegradable edible coatings, given the persistent nondegradability of synthetic packaging materials. Research endeavors in this field primarily concentrate on formulating edible coatings and enhancing their characteristics to encompass key attributes akin to conventional synthetic materials. These sought-after features include elevated mechanical strength, lightweight composition, softness, water resistance, and transparency. Edible biopolymers, encompassing proteins, polysaccharides, and lipids, stand out as viable alternatives to traditional plastics in this quest for sustainable packaging solutions [9].

2.2 Classification of Edible Coatings

Edible coating refers to the process of applying a thin layer of a specially formulated solution to food items, with the goal of extending their shelf life. The formulation of this coating involves dissolving specific coating materials or compounds in either organic or inorganic solvents. These materials commonly include polysaccharides, lipids, proteins, or various composites. Different materials are used for the development of edible coating which are illustrated in Figure 2.1. Enhancing the physicochemical properties of the coating involves the introduction of a required amount of additives or plasticizer as ingredient. Antimicrobial agents, antioxidants, essential oils, vitamins, and plasticizers are among the active components utilized during the coating emulsion development. The solution undergoes thorough homogenization before being applied through dipping, fluidized bed, foaming, spraying, brushing, or panning methods.

2.2.1 Polysaccharides

Polysaccharides are polymers developed through the linkage of monomers via glycoside bonds, resulting in a long, continuous chain. The formation of polysaccharide films involves disrupting interactions within the polysaccharide chain and establishing new intermolecular connections. Chitosan, cellulose, pectin, alginate, kefiran, starch, carrageenan, and pullulan are commonly employed for this purpose due to their ready availability, renewability, nontoxic nature, and cost-effectiveness. Due to their structured arrangement, which includes a hydrogen network, polysaccharides serve as efficient barriers against oxygen. However, these substances exhibit a limited ability to block water vapor due to their hydrophilic nature [10]. To overcome this constraint, polysaccharides are utilized in

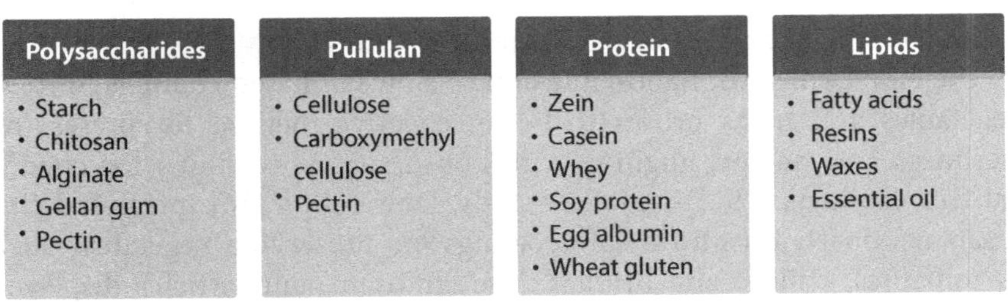

Figure 2.1 Various types of materials used for developing edible coating.

combination with other proteins, fats or polysaccharides [11]. Panahirad *et al.* [12] explored the utilization of pectin and carboxymethyl cellulose in the development of edible coating for diverse fresh produce. Pectin-based coatings applied to persimmon, plum, and peach apple exhibited notable success in minimizing microbial proliferation and deterioration. Additionally, the combination of moringa leaf extract and gum arabica extended the shelf life of guava, reducing spoilage and rhizopus infection while cold storage. The introduction of fatty acids into cellulose-coating formulations was observed to decrease water vapor permeability, with oleic acid demonstrating superior control over weight loss. In fruit coating applications, it is crucial to have films with minimal water vapor transmission rates. Therefore, plasticizers are added to starch-based biodegradable films to improve shelf life when compared to noncoated alternatives [13].

2.2.2 Chitosan

Chitin, a naturally occurring polysaccharide, is abundantly found in yeasts, fungi, vertebrates, crustaceans and insects. Chitosan, a derivative of chitin composed of 1,4-linked 2-amino-deoxy-β-D-glucan, is a cationic linear polysaccharide. Natural chitin can undergo deacetylation to produce chitosan, a polysaccharide employed in the development of edible coating. Chitosan finds significant use in coating fresh produces due to its antimicrobial and anti-oxidant properties. Recognized for being resistant to microbes, biocompatible, biodegradable and nontoxic chitosan is presently under extensive scientific scrutiny to explore its potential applications across various fields. Functioning as semi permeable films and coatings, chitosan can regulate internal structures by reducing transpiration rates and slowing down the ripening process in both foods and vegetables. Suseno and co-researchers [14] discovered that employing chitosan as an edible coating on Cavendish bananas prolonged their shelf life by ameliorating the spoilage and slowing down the ripening.

The employment of nano-emulsion of chitosan based coatings to fresh vegetables and fruits proves to be an effective method for preserving freshness and quality, aligning with consumer expectations for chemical-free products [15]. In another study, Divya *et al.* [16] integrated chitosan nanoparticles into edible coatings for preserving vegetables such as tomatoes, chilies, and brinjals. The chitosan nanoparticles displayed notable antifungal and antioxidant effects against specific microbes. When chitosan coatings were combined with supplementary techniques like UV radiation, they successfully prolonged the ripening process of fruits, such as sweet cherries, without compromising the quality [17]. Research on the application of chitosan coating in controlling postharvest softening involved the examination of gene expression and also enzymatic activities related to wall-modifying enzymes. This research enhanced our understanding of the significant role of chitosan in controlling postharvest tissue degradation in fruits [18].

2.2.3 Starch

Starch plays a crucial role as a polysaccharide in edible coatings, owing to its natural abundance, cost-effectiveness, and film-forming capabilities. Starch-based coatings are ideal for edible applications as they are tasteless, colorless, odorless, and nontoxic. Notably, tapioca starch, potato, corn and sweet potato are extensively utilized for coating fresh produces due to their widespread production and cultivation. Edible coatings derived from starch offer excellent gas barrier properties, effectively reducing senescence and extending the shelf life of fruits and vegetables. This type of coating exhibits a favorable gas permeability ratio, allowing control over the respiratory exchange ratio. Starch-based coatings also prove effective in preserving the qualities of meat by reducing browning reactions, dehydration, and oxidation when applied to meat and meat products [9].

2.2.4 Protein

Typically, various globular proteins such as zein, gluten, soy and whey are utilized in the production of edible coatings, showcasing impressive oxygen barrier properties. However, it's worth noting that protein coatings have a hydrophilic nature and are susceptible to changes in relative humidity and temperature in the storage environment [9]. Storage investigations on guava with flaxseed protein-based edible coatings demonstrated effective control of fruit browning. These coatings prolonged fruit decay, enhancing the shelf life of guava by an additional 6 days compared to untreated fruits [19]. Similarly, applying whey protein coatings to sliced potatoes and apples delayed the onset of enzymatic browning reactions [20]. In the case of fresh melons, edible coatings made from tilapia protein isolates proved efficient in preserving color and firmness while preventing the proliferation of psychrotropic microbes, yeasts and molds for up to 12 days of storage period [21]. The incorporation of thymol nano-emulsions into coatings made from quinoa protein and chitosan proved effective in addressing weight loss and preserving the structure of strawberries [22].

Hassani *et al.* [23] developed an edible coating by combining rice bran oil with whey protein. The resulting composite edible film, comprising rice bran oil and whey protein concentrate enriched with glycerol additives, effectively maintained the firmness and other organoleptic properties of kiwi fruit over a 28-day storage period. The inherent fragility of albumen proteins, due to their disordered structure upon denaturation, is mitigated by incorporating various plasticizers like glycerol. These plasticizers reduce intermolecular pressures within the protein chain, enhancing flexibility and translocation in the protein-polymer chains. The preservation of Lactobacillus acidophilus was achieved using a fluidized bed drying and electro spraying method, employing stearic acid and an egg albumen coating [24].

2.2.5 Lipids

Lipids, identified as compact hydrophobic compounds, are commonly employed in the development edible coating of foods due to their shining and glossy characteristics. Usually, lipids are mixed with other substances like polysaccharides and proteins in edible coatings to address their limited mechanical properties. The components utilized in lipid-based edible coatings encompass mineral oils, resins, oils, natural waxes, fatty acids and petroleum-based waxes [25]. Reports documented that the application of commercial Semperfresh™ coating solutions, comprising fatty acids and their sucrose esters, to cherries enhanced its structural properties including firmness and effectively minimized loss in weight during storage [26]. Similarly, the edible coating with Flourensia cernua and candelila wax on tomatoes not only resisted fungal proliferation but also garnered consumer endorsement. Shellac and waxes are extensively used as lipids in edible coatings to enhance the appearance of fruits by providing shine and glossiness [9].

Lipids are blended with various components extracted from natural resources and applied as coatings on a variety of fruits and vegetables. Key components utilized in lipid coatings include waxes, resins, oils or lipids. All these lipid components inherently possess hydrophobic properties. Pérez-Gago and Rhim [27] explored the development of composite films by combining lipids with polysaccharides and proteins in various configurations. Some commonly used lipid materials in this context include mineral oil, corn oil, candelilla wax,palm oil, carnauba wax, polyethylene wax oleic acid, paraffin wax. beeswax, wood rosin, lauric acid, and coumarone-indin resin.

2.2.6 Composites

In the present era, there is a current emphasis on composite or multilayered edible coatings for fresh produces, aiming to overcome the limitations of individual components and capitalize on their positive side. The practice of applying composite coatings to fruits and vegetables typically involves incorporating lipids and hydrocolloids, enhancing moisture retention and bolstering the mechanical properties of the coating. Various combinations suitable for the development of composite edible coatings include protein and lipid, as well as carbohydrate and lipids. The mechanical characteristics of the coating in these combinations are significantly influenced by the choice of carbohydrate or proteins. The application of a shellac-tannic acid edible coating extended the shelf-life of mangoes at ambient condition, preserving both firmness and minimizing weight loss. The study concludes that the inclusion of tannic acid enhanced the antifungal property of the coatings, resulting in an increased shelf life [28]. The use of composite edible coating on pineapples effectively maintained the textural and other quality parameters of the fruits while storage period, and this also significantly reducing microbial contamination [29].

2.2.7 Characteristics of Edible Coatings

An edible coating or edible is a material distinguished by a thickness less than 0.3 mm, formed by blending biopolymers and suitable additives dissolved in aqueous medium. Evaluating edible coatings is a pivotal strategy in confirming their suitability and effectiveness in maintaining the safety and quality of food products. This evaluation encompasses various aspects, including the coatings' capacity to function as effective barriers, mechanical characteristics, thickness, microstructural uniformity and its adherence to the fresh produces (illustrated in Figure 2.2). The assessment

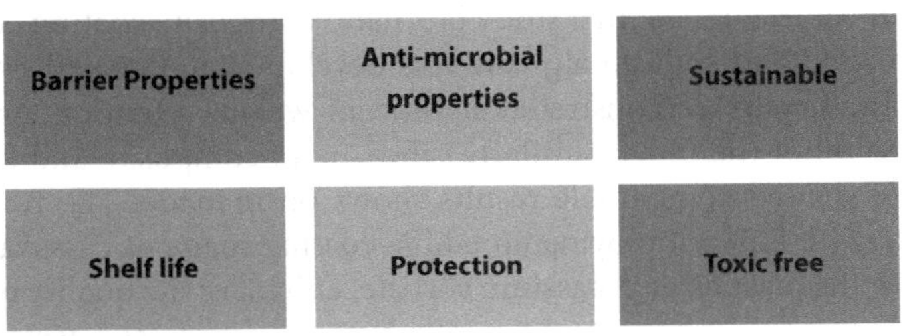

Figure 2.2 Characteristics of edible coating.

of barrier properties, such as hindrance towards oxygen and water vapor transmission is crucial for determining the coating's ability to safeguard against moisture loss and exposure to oxygen [30].

The mechanical characteristics, including tensile strength, flexibility and resistance to puncture, play a crucial role in assessing the coatings' ability to withstand handling and protect the fruits and vegetable effectively. Examining the microstructure, utilizing techniques like atomic force microscopy and scanning electron microscopy and gives significant insights into the internal composition of the coating. Additionally, evaluating thickness and uniformity guarantees consistent performance, while adhesion test measures the coating effectiveness in adhering to the food surface [31]. By characterizing the properties of edible coatings, manufacturers can customize formulations and various application techniques to address the unique requirements of different food products, thereby enhancing their storage life and quality. On-going research endeavors are fine-tuning these characterization techniques, aiming to optimize the effectiveness of edible coatings within the food industry.

2.2.8 Moisture and Gas Barrier Properties

Edible coating must act as barriers, preventing moisture and aroma loss, water uptake, and even oxygen penetration in food materials. This creates favorable storage conditions for these food products. The effectiveness of food-grade coatings to provide a layer of protection depends on a number of variables, including the kind of coating material used, the microstructural properties of coating, its thickness and other abiotic factors such as humidity and temperature also the requirement of particular food product that has to be coated [32]. All these parameters collectively contribute to barrier efficiency, a critical aspect for preserving the quality and safety of the food product. Edible coatings not only enhance texture and improve product appearance but also extend shelf life by establishing semi-permeable barriers. In a study bio-nano composite coating comprising nano-ZnO and sodium alginate was developed and coated on strawberries. The results demonstrated significant weight retention compared to those without the films. Similarly, silver nano composite and titanium packaging yielded comparable results shows on mangoes [1]. According to Garcia *et al.* [33], employing an edible coating made of cassava starch, along with the plasticizer potassium sorbate, enhances the quality of stored strawberries. This enhancement is attributed to increased water vapor resistance and a reduced respiration rate. In a similar vein, Ma *et al.* [28] utilized a combination of olive oil and gelatin to develop composite film through a microfluidic emulsification process. This research aimed to augment the hydrophobic characteristics and investigate whether the presence of oil affects the arrangement of droplets of lipid in suspension that form films. The findings suggested that olive oil and gelatin can be employed to develop coatings or films for various food products. Additionally, it was noted that the addition of olive oil diminishes the breaking strength and water vapor permeability of the film made of gelatin while imparting a glossy look.

2.2.9 Physicochemical Properties

Edible coatings are applied on the fresh produces to enhance their shelf life. The physicochemical properties of edible coating plays a crucial role in determining its effectiveness. Various physic-chemical properties such as appearance, transparency, solubility, permeability, adhesion properties are taken to be considered while selecting the materials for developing edible coating. The visual appeal of edible coatings is significant, as it can influence consumer perception. Generally, coatings provide glossiness to the fresh produces. Coatings are designed to enhance or maintain the natural color and appearance of the food product and it should not affect the mouth feel of the product. Depending on the application, edible coatings may need to be transparent or opaque. Transparency is important when maintaining the visual appeal of fruits and vegetables, while opacity may be preferred for certain confectionery items. Uniformity of the application of edible coating on the surface of fruits and vegetables directly correlate with the solubility of the material. Hydrophobic coatings can provide a moisture barrier to prevent food spoilage, while hydrophilic coatings may be used to retain moisture in certain products. It should have controlled permeability properties for moisture and gases. Good adhesion properties ensure that the edible coating sticks on to the surface of the fruits and vegetables [34].

2.2.10 Sensory and Organoleptic Properties

The key attributes of coated fruits and vegetables involve extending shelf life and improving texture and appearance. Understanding parameters related to wettability is crucial for edible coating development. Choi *et al.* [35] highlighted significant factors such as interfacial tension between the coating solution and the surface of the product which need to be coated, surface-free energy, and contact angle. These variables mostly impact the coating emulsion's adhesion to the product surface, which in turn influences the fruit's final coating thickness.

In a related background, Fama *et al.* [36] discovered that draining time, viscosity, surface tension of both the fruit and the liquid, density of biopolymer solutions, and the abrasions on the fruit surfaces influenced the mean liquid film thickness on coated apples. Preserving the aroma, fundamental flavors, gloss, color, firmness and other organoleptic attributes during coating application is imperative. Methods such as consumer and free-choice profiling panels or descriptive analysis are frequently employed to assess the sensory characteristics of fruits with coatings [37].

Nevertheless, there are instances where customers may decline samples because of synthetic coloring and a glossy appearance, particularly when coatings include lipids. It is crucial to acknowledge that alterations to the internal environment of coated fruits, leading to a metabolic slowdown, can also affect their mechanical properties. Compression studies using a Universal Testing Machine or Texture Profile Analyzer are frequently employed to assess these properties, with firmness or fracture resistance being the most commonly reported measures when evaluating the quality of preserved coated fruits [38].

2.2.11 Disperse System Forming Edible Coatings Packaging Materials

Biopolymeric substances utilized in the development of bio-packaging are integrated diversely based on their carbohydrate (polysaccharides), protein, or lipid properties. This leads to the formation of various systems such as dispersed emulsion-type (lipid-based) or colloidal systems (polysaccharide or protein-based). Emulsions are complex systems comprised of semi-liquid or liquid substances that do not naturally mix, like oil and water, which can be uniformly mix with emulsifiers. Emulsifier typically features both hydrophobic and hydrophilic zones, demonstrating an attraction for both nonpolar and polar sites. Hence, emulsions can be categorized into two types based on the ratios of their phases: water/oil or oil/water with the dispersed phase serving as the second constituent and being present in a lesser concentration within the emulsion. O/W systems are preferred, while developing edible coating due to their enhanced thermodynamic stability and their ability to dissolute antimicrobial constituents such as essential oil and other bioactive constituents such as phytosterols, quinones, carotenoids or fatty acids,. Colloid systems, which are made up of proteins or polysaccharides that have been dissolved in water, provide a thick matrix that shields active ingredients like antimicrobials and antioxidants and permits their regulated release inside the matrix. The dispersed systems need to be integrated or coated onto the product for drying (or vice versa), creating a firm matrix that functions as an edible coating or film. This process is highly dependent on the mode of application. Commonly used application techniques are Dipping, Spreading or Spraying.

2.2.12 Legislation, Standards for Edible Coatings

The substance employed in edible coating comes into direct interaction with fresh produces and must receive safety approval before commercial use. Hence, the material used for the development of edible coating needs to be designated as Generally Recognized as Safe by the USFDA prior to its use. Additionally using authorized natural plant extracts and essential oils may result in allergic responses and exhibit hazardous consequences based on its dosage. Hence, it is imperative to implement measures to routinely assess the toxicity and allergic potential of essential oils and extracts used in the production of edible coatings. Green packaging initiatives, supported by organizations such as the US Environmental Protection Agency, drive research towards the development of pectin-based edible coatings to enhance the shelf life and quality of packaged foods [39].

The materials utilized in the formation of edible coatings must adhere to food-grade standards, ensuring they are nontoxic and processed with appropriate hygiene practices. Development of edible coating needs to adhere to Good Manufacturing Practice guidelines. In accordance with the regulations set forth by the US FDA in 2006 and the European Directive in 1998, edible coatings are typically classified into categories such as food additives, food ingredients or substances, and food packaging materials. The list of acceptable additives may vary between countries in accordance with their respective regulations. The Directive regulation in Europe classifies Arabic gums, beeswax pectin, shellac, fatty acids, karaya gums, polysorbates and lecithin, and as food components. Meanwhile, polydextrose, castor oil, and cocoa butter, sucrose fatty acids are identified as food additives in the formulation of edible coating under the USFDA regulations [40].

To comprehend the type of chemicals existed in food industry, regulatory bodies such as the European Food Safety Authority and USFDA and have categorized them into three groups. The incorporation of antimicrobial, nanoparticles, antifungal, antioxidant, and anti-browning agents, into coating formulations falls under the term "food coating materials".

Food contact articles (FCA) are the finished products, which might be coatings or films, while food contact substances (FCS) are the materials used to make them. Food manufacturers need to adhere to the Codex Alimentarius standards and abide by rules set out by the country of transit as well as the region of origin when developing materials for packaging. Compliance with the newly formulated regulations, namely no. 1935/2004 and no. 2023/2006, is essential for meeting packaging requirements and maintaining Good Manufacturing Practice (GMP) standards for food coating materials (FCM). These regulations play a crucial role in ensuring the safety and quality of food contact articles (FCA) in the food packaging industry [41].

The materials used for developing edible coatings must meet stringent standards to ensure their safety and adherence to regulations. These materials are edible, and must compliance with hygienic practices and good manufacturing practices [42]. Coating materials deemed Generally Recognized as Safe (GRAS) or approved as food additives by the Food and Drug Administration (FDA) can be utilized for fresh fruits and vegetables. In European countries, the major ingredients for developing edible coatings are typically derived from listed food additives. However, it is essential to note that regulations governing approved coating polymers and food additives vary between countries, requiring careful consideration during the development of new edible coating materials. Biopolymers not recognized as GRAS should not be employed in edible coatings, as they may pose toxicity and allergenicity risks upon human consumption. In India, it is permissible to use either bee wax or carnauba wax as coating of fresh produces under the Prevention of Food Adulteration Act. This regulation ensures appropriate labeling and declaration of materials employed for developing edible coating. Standards established by the USFDA and European legislation are strictly followed while applying synthetic and chemically altered resins for the edible coating on fresh produces. Shellac, recognized as GRAS, is widely used resin for the development of edible coating and serves as food additive [1, 39].

2.3 Recent Advances in Edible Coating

2.3.1 Edible Coatings: As Functional Packages

According to the terminology given by Kris-Etherton *et al.* [43] a bioactive component is any ingredient that is present in food in minute levels also has the potential to affect health when consumed. Numerous bioactive compounds, including phytoestrogens and flavonoids which are known to have anti-inflammatory, anti-cholesterol, antimutagenic, apoptotic, anti-cancer, and antioxidant effects have been the focal point of substantial epidemiological research. As a result, a lot of research projects have been focused on incorporating a variety of these bioactive substances into edible coatings and films [44]. For example, a number of studies have evaluated the antioxidant capacity of edible films that include phenolic components derived from a variety of sources [45]. Nogueira *et al.* [46] established that the antioxidant property of blackberry powder, attributed to the presence of anthocyanin, remained intact when incorporated into arrowroot starch edible films. Additionally, the inclusion of the powder enhanced water solubility, tensile strength and elongation properties, while preserving the organoleptic properties of blackberries. Rodsamram and Sothrnvit [45] developed edible wrappings using coconut water and protein, harnessing antioxidant property from phenolic components in coconuts; the film possess a brown color which served as a UV light barrier. Assis *et al.* [47] developed edible films with improved solute transmission and antioxidant property by incorporating and encapsulated β-carotenes from carrot in a tapioca starch matrix. From these investigations, the antioxidant property was attributed to the presence of bioactive components, significantly retarding substrate oxidation and inhibiting reactions involving free radicals.

2.3.2 Edible Coatings: As Pathogen Inhibitors

Among the various attributes of edible coatings and films, the safeguarding of fresh produces through biological means stands out as very crucial, as it directly affects the shelf life of the product. Consequently, there is a need to inhibit or eradicate microorganisms, and their derivatives, capable of instigating or hastening the decay of food through the actions of their enzymes and metabolic by-products. For instance, in an experiment by Aloui *et al.* [48], *Wickerhamomyces anomalus* was inoculated into two distinct edible wrapping—one based on locust bean gum and other with alginate. The results of the investigation showed that the bacteria were more stable in the alginate matrix. Furthermore, when applied to oranges, the alginate edible film effectively prevented *Penicillium digitatum*, enhancing the quality of fruit for a period of 13 days. Another study by Parafati *et al.* [49] involved the inoculation of *Aureobasidium pullulans*, *Metschnikowia pulcherrima* and *Wickerhamomyces anomalus*, in mandarins coated with edible carob gum films. Results indicated a more pronounced inhibitory effect with *Metschnikowia pulcherrima* and against *Penicillium italicum* and *Penicillium digitatum* [50, 51].

2.3.3 Edible Coatings: Probiotic Carriers

Probiotics, categorized as bioactive compounds, offer distinct health advantages. As per the 2002 definition by the World Health Organization and Food and Agriculture Organization (WHO/FAO), probiotics are described as "viable microorganisms that, when consumed in sufficient quantities, provide a health advantage to the host." The probiotic properties of LAB belonging to the Lactobacillus genus have been extensively investigated. They play a crucial role in safeguarding the microbiota, preventing its deterioration, and inhibiting pathogenic microorganisms, particularly in the oral cavity and colon.

Due to this, the most comprehensive investigations into probiotic assessment have focused on incorporating microorganisms into edible films and coatings rather than their effectiveness in inhibiting external pathogens (arising from environmental contamination). The primary objective of preserving or ensuring the viability of probiotics is to enable them to endure the challenges posed by gastric juices and intestinal fluids upon ingestion. This resilience allows for periodic dosing, facilitating the realization of their probiotic effects [51]. As demonstrated by Gbassi *et al.* [52], they investigated the viability of *Lactobacillus plantarum* encapsulated in alginate and subsequently incorporated into an edible whey protein film under gastrointestinal conditions (pH 1.8 at 37 °C). The study revealed that the probiotics remained viable for up to 180 minutes. The significance of this research lies in the simplicity of the matrix, its effectiveness, and the cost efficiency associated with the use of alginate and whey protein.

2.3.4 Edible Coating: As Preservatives

At present, food industries faces the challenge of delivering fresh, high-quality food with health benefits to consumers. Yet, meeting these expectations proves to be a significant challenge as no food can endure under optimal conditions and maintain its physic-chemical, structural, functional attributes indefinitely. This is ascribed to the inherent decay resulting from chemical, physical alterations and biochemical reactions. The degradation of food is influenced by elements including water content, temperature, oxygen availability, pH and relative humidity [53]. Hence, in addition to the characteristics of the edible coatings and films it is crucial to preserve the qualities of the food, encompassing: (a) compliance with microbiological standards as mandated by regulatory bodies; (b) preservation of nutritional wholesomeness; and (c) safeguarding physic-chemical, organoleptic attributes, such as appearance, texture, aroma, and taste. This collective effort contributes to prolonging the storage life of the fresh produces. Research indicates that the maturation of fruits and vegetables, along with the proliferation of mold and microbes, can be postponed by employing edible films and coatings. This type of preservation method maintains distinct qualities such as vitamin C content, freshness, texture, and overall nutritional quality. Additionally, introducing diverse bioactive solutes into these films and coatings can confer novel biological activities, such as antioxidant properties. When applied to any dairy or other animal products, edible coatings or films aid in preserving bioactive components and sensory characteristics, enhancing antimicrobial and antioxidant activities, and prolonging shelf life. Table 2.1 illustrates various preservative effects of edible coatings. In the realm of bakery products, employing these coatings leads to decreased moisture content and an `extended shelf life.

Table 2.1 Preservative effect of edible coating.

Food application	Coating	Additives used	Obtained results	Reference
Figs	Chitosan	Canola oil, rosella extract, acetic acid, essential oil from cinnamon	Color changes was delayed, antioxidant property of the fruit is preserved	[54]
Papaya	Carrageenan	Citric acid and glycerol	Ripening process is delayed and shelf life is extended	[55]
Straw berries	Cassava starch	Extract from propolis	Vitamin C content was promoted	[56]
Fresh cut water melon	Pectin, alginate, Calcium lactate	Glycerol	Preserved the texture and the shelf life was extended	[57]
Tomato	Rice bran wax	Emulsifier	Delay the ripening process and extend the shelf life	[58]
Chicken breast	Chitosan and carrageenan	Mustard extract and glycerol	Reduction in *Campylobacter jejuni* and shelf life extension was observed	[59]

2.4 Conclusion and Future Scope

Consumers are becoming increasingly sophisticated, demonstrating a growing interest in pursuing wholesome foods and way of living. Maintaining equilibrium between supply and demand requires on-going research and advancements in edible coating methods. However, each developmental method comes with inherent flaws and demerits. A critical challenge for future research involves enhancing the shelf life of newly harvested and minimally processed foods without compromising their nutritional and sensory characteristics. Several investigations are currently in the early stages of exploration, focusing on creative and cost-efficient packaging solutions for fresh produce, prior to potential widespread industrial adoption. The exploration of innovative edible coating materials is anticipated to result from a comprehensive examination of biochemistry and its connections to physicochemical characteristics, antimicrobial properties, potential toxicity factors, and risk assessment. Mainly, progress is required in coating materials, particularly concerning mineral migration efficiency, cost-effectiveness ratios, and nutritional content, existing technological procedures, to guarantee and improve the storage life while maintaining the quality of coated fresh and minimally processed fruits.

Most research has concentrated on assessing the tensile strength of edible films, as maintaining consistent film qualities is vital for preserving enclosed food and ensuring an extended shelf life. However, edible films are prone to deterioration with age, and there is limited literature on how aging affects film properties. There is an urgent need to investigate approaches that control time-dependent changes in film characteristics. Although many studies have delved into the water and gas permeability of edible films or coatings, there is a scarcity of research on aroma permeability, oil permeability, and the permeability of additives. In the beverage industry, where a substantial amount of packaging materials is used, there has been limited exploration of the application of edible coating or films for beverage packaging.

To ensure the effectiveness of edible coatings or films, they need to align with existing packaging technologies. Serving as a vital barrier between food and its environment, packaging materials play a crucial role. Emerging food packaging technologies, such as smart packaging, intelligent food packaging and active food packaging present diverse interactions and functionalities. Active packaging enables specific interactions between food and its surroundings. Intelligent packaging observes the condition of the food and supplies data, while smart packaging surpasses mere monitoring by enabling customers to monitor and control food quality using indicators, sensors, and RFID systems.

Despite the potential advantages, there are substantial challenges in incorporating slender film electronic sensors while developing the edible films and integrating these devices into communication networks. These technologies are categorized as "green technologies," aiming to safeguard fresh produce while mitigating or eliminating the adverse impacts of food processing. Although the manufacturing of edible coatings is in its initial phases with limited production volumes, the current cost of edible coatings exceeds that of petroleum-origin films. The acceptance of edible films by customers is significantly affected by their pricing, highlighting the necessity of competitive pricing to attract consumers, ideally lower than or on par with the cost of petroleum-based polymers.

This chapter underscores the essential role of edible preservation technology, often referred to as "green technology," in the contemporary era. Compared to conventional food preservation technologies, these methods not only prevent food deterioration but also preserve the wholesomeness of the food. Utilizing bio-edible coatings emerges as a more efficient approach for improving the quality and safety of fresh produce in comparison to existing methods. These coatings, comprising an ingestible layer that encases the food, control the transmission of oxygen, water vapor carbon dioxide, leading to a prolonged shelf life without compromising quality and safety standards.

Many regulatory bodies in the food and pharmaceutical sectors, as well as agencies overseeing food safety, have not just sanctioned but also set safe thresholds for the application of edible coatings. As a result, the acceptance of edible coatings and films as practical substitutes for synthetic packaging is steadily growing. Despite the introduction of numerous new options in recent developments related to edible films and coatings, there is still a requirement for heightened focus to bridge the current gap and unleash the market potential of edible coatings.

References

1. Oduro, K.O.A., Edible coating, in: *Postharvest Technology-Recent Advances, New Perspectives and Applications*, 2021.
2. Sharma, P., Shehin, V.P., Kaur, N., Vyas, P., Application of Edible Coatings on Fresh and Minimally Processed Vegetables: A Review. *Int. J. Veg. Sci.*, 25, 3, 295–314, 2019, doi: 10.1080/19315260.2018.1510863.

3. Hauzoukim, Sagarika., S. and Mohanty, B., Functionality of protein-Based edible coating. *J. Entomol. Zool. Stud.*, *8*, 4, 1432–1440, 2020.
4. Strano, M.C., Restuccia, C., De Leo, R., Mangiameli, S., Bedin, E., Allegra, M., Pulvirenti, A., Efficacy of an antifungal edible coating for the quality maintenance of Tarocco orange fruit during cold storage, in: *Crop Protection*, vol. *148*, p. 105719, 2021.
5. Khezerlou, A., Zolfaghari, H., Forghani, S., Abedi-Firoozjah, R., Alizadeh Sani, M., Negahdari, B., McClements, D.J., Combining Non-Thermal Processing Techniques with Edible Coating Materials: An Innovative Approach to Food Preservation. *Coatings*, *13*, 5, 830, 2023, doi: 10.3390/coatings13050830.
6. Maringgal, B., Hashim, N., Tawakkal, I.S.M.A., Mohamed, M.T.M., Recent advance in edible coating and its effect on fresh/fresh-cut fruits quality, in: *Trends in Food Science & Technology*, vol. *96*, pp. 253–267, 2020.
7. Suhag, R., Kumar, N., Petkoska, A.T., Upadhyay, A., Film formation and deposition methods of edible coating on food products: A review, in: *Food Research International*, vol. *136*, p. 109582, 2020.
8. Subhasri, D., Nimbkar, S., Moses, J.A., Anandharamakrishnan, C., Biobased edible food coatings for food applications, in: *Advanced Applications of Biobased Materials*, pp. 263–302, Elsevier, 2023, doi: 10.1016/B978-0-323-91677-6.00011-8.
9. Priya, K., Thirunavookarasu, N., Chidanand, D.V., Recent advances in edible coating of food products and its legislations: A review. *J. Agric. Food Res.*, 12, 100623, 2023, doi: 10.1016/j.jafr.2023.100623.
10. Adiletta, G., Di Matteo, M., Petriccione, M., Multifunctional role of chitosan edible coatings on antioxidant systems in fruit crops: A review. *Int. J. Mol. Sci.*, *22*, 5, 2633, 2021, doi: 10.3390/ ijms22052633.
11. Zhang, Y., Rempel, C., Mclaren, D., Edible coating and film materials: Carbohydrates. *Innov. Food Packag.*, 305–323, 2014, doi: 10.1016/B978-0-12-394601-0.00012-6.
12. Panahirad, S., Dadpour, M., Peighambardoust, S.H., Soltanzadeh, M., Gullón, B., Alirezalu, K., Lorenzo, J.M., Applications of carboxymethyl cellulose-and pectin-based active edible coatings in preservation of fruits and vegetables: A review, in: *Trends in Food Science & Technology*, vol. *110*, pp. 663–673, 2021.
13. Pellá, M.C., Silva, O.A., Pellá, M.G., Beneton, A.G., Caetano, J., Simões, M.R., Dragunski, D.C., Effect of gelatin and casein additions on starch edible biodegradable films for fruit surface coating, in: *Food chemistry*, vol. *309*, p. 125764, 2020.
14. Suseno, N., Savitri, E., Sapei, L., Padmawijaya, K.S., Improving shelf-life of cavendish banana using chitosan edible coating, in: *Procedia chemistry*, vol. *9*, pp. 113–120, 2014.
15. Chaudhary, S., Kumar, S., Kumar, V., Sharma, R., Chitosan nanoemulsions as advanced edible coatings for fruits and vegetables: Composition, fabrication

and developments in last decade. *Int. J. Biol. Macromol.*, *152*, 154–170, 2020, doi: 10.1016/j.ijbiomac.2020.02.276.
16. Divya, K., Smitha, V., Jisha, M.S., Antifungal, antioxidant and cytotoxic activities of chitosan nanoparticles and its use as an edible coating on vegetables. *Int. J. Biol. Macromol.*, *114*, 572–577, 2018, doi: 10.1016/j.ijbiomac.2018.03.130.
17. Abdipour, M., Malekhossini, P.S., Hosseinifarahi, M., Radi, M., Integration of UV irradiation and chitosan coating: A powerful treatment for maintaining the postharvest quality of sweet cherry fruit, in: *Scientia Horticulturae*, vol. *264*, p. 109197, 2020.
18. Wang, K., Li, T., Chen, S., Li, Y., Rashid, A., The biochemical and molecular mechanisms of softening inhibition by chitosan coating in strawberry fruit (Fragaria x ananassa) during cold storage, in: *Scientia Horticulturae*, vol. *271*, p. 109483, 2020.
19. Sharma, M. and Saini, C.S., Efficacy of flaxseed protein-based edible coatings on the quality of whole guava (Psidium guajava) during storage. *Food Sci. Appl. Biotechnol.*, *4*, 2, 93–104, 2021, doi: 10.30721/fsab2021.v4.i2.132.
20. Guo, M. (Ed.), *Whey protein production, chemistry, functionality, and applications*, John Wiley & Sons, 2019.
21. Chevalier, R.C., Pizato, S., de Lara, J.A.F., Cortez-Vega, W.R., Obtaining protein isolate of tilapia (Oreochromis niloticus) and its application as coating in fresh-cut melons. *J. Food Saf.*, 38, 5, e12496, 2018, doi: 10.1111/jfs.12496.
22. Robledo, N., López, L., Bunger, A., Tapia, C., Abugoch, L., Effects of antimicrobial edible coating of thymol nanoemulsion/quinoa protein/chitosan on the safety, sensorial properties, and quality of refrigerated strawberries (Fragaria× ananassa) under commercial storage environment, in: *Food and Bioprocess Technology*, vol. *11*, pp. 1566–1574, 2018.
23. Hassani, F., Garousi, F., Javanmard, M., Edible coating based on whey protein concentrate-rice bran oil to maintain the physical and chemical properties of the kiwifruit Actinidia deliciosa. *Trakia J. Sci.*, *10*, 1, 26–34, 2012.
24. Pitigraisorn, P., Srichaisupakit, K., Wongpadungkiat, N., Wongsasulak, S., Encapsulation of Lactobacillus acidophilus in moist-heat-resistant multilayered microcapsules. *J. Food Eng.*, *192*, 11–18, 2017, doi: 10.1016/j.jfoodeng.2016.07.022.
25. Yousuf, B. and Qadri, O.S., Preservation of fresh-cut fruits and vegetables by edible coatings, in: *Fresh-cut fruits and vegetables*, pp. 225–242, Academic Press, 2020, doi: 10.1016/B978-0-12-816184-5.00011-2.
26. Yaman, O. and Bayoındırlı, L., Effects of an edible coating and cold storage on shelf-life and quality of cherries. *LWT Food Sci. Technol.*, *35*, 2, 146–150, 2002, doi: 10.1006/fstl.2001.0827.
27. Pérez-Gago, M.B. and Rhim, J.W., Edible coating and film materials: Lipid bilayers and lipid emulsions, in: *Innovations in food packaging*, pp. 325–350, Academic Press, 2014, doi: 10.1016/B978-0-12-394601-0.00013-8.

28. Ma, J., Zhou, Z., Li, K., Li, K., Liu, L., Zhang, W., Zhang, H., Novel edible coating based on shellac and tannic acid for prolonging postharvest shelf life and improving overall quality of mango. *Food Chem.*, *354*, 129510, 2021, doi: 10.1016/j.foodchem.2021.129510.
29. Mantilla, N., Castell-Perez, M.E., Gomes, C., Moreira, R.G., Multilayered antimicrobial edible coating and its effect on quality and shelf-life of fresh-cut pineapple (Ananas comosus). *LWT Food Sci. Technol.*, 51, 1, 37–43, 2013, doi: 10.1016/j.lwt.2012.10.010.
30. Tayeb, A.H., Tajvidi, M., Bousfield, D., Enhancing the oxygen barrier properties of nanocellulose at high humidity: numerical and experimental assessment. *Sustain. Chem.*, *1*, 3, 14, 2020, doi: 10.3390/suschem1030014.
31. Falsafi, S.R., Rostamabadi, H., Assadpour, E., Jafari, S.M., Morphology and microstructural analysis of bioactive-loaded micro/nanocarriers via microscopy techniques; CLSM/SEM/TEM/AFM. *Adv. Colloid Interface Sci.*, *280*, 102166, 2020, doi: 10.1016/j.cis.2020.102166.
32. Odetayo, T., Tesfay, S., Ngobese, N.Z., Nanotechnology-enhanced edible coating application on climacteric fruits. *Food Sci. Nutr.*, *10*, 7, 2149–2167, 2022, doi: 10.1002/fsn3.2557.
33. Garcia, L.C., Pereira, L.M., de Luca Sarantópoulos, C., II, Hubinger, M.D., Selection of an edible starch coating for minimally processed strawberry. *Food Bioprocess Technol.*, *3*, 834–842, 2010, doi: 10.1007/s11947-009-0313-9.
34. Armghan Khalid, M., Niaz, B., Saeed, F., Afzaal, M., Islam, F., Hussain, M., Al-Farga, A., Edible coatings for enhancing safety and quality attributes of fresh produce: A comprehensive review. *Int. J. Food Prop.*, *25*, 1, 1817–1847, 2022.
35. Choi, W.Y., Park, H.J., Ahn, D.J., Lee, J., Lee, C.Y., Wettability of chitosan coating solution on 'Fuji'apple skin. *J. Food Sci.*, *67*, 7, 2668–2672, 2002, doi: 10.1111/j.1365-2621.2002.tb08796.x.
36. Famá, L., Goyanes, S., Gerschenson, L., Influence of storage time at room temperature on the physicochemical properties of cassava starch films. *Carbohydr. Polym.*, *70*, 3, 265–273, 2007, doi: 10.1016/j.carbpol.2007.04.003.
37. Han, C., Lederer, C., McDaniel, M., Zhao, Y., Sensory evaluation of fresh strawberries (Fragaria ananassa) coated with chitosan-based edible coatings. *J. Food Sci.*, *70*, 3, S172–S178, 2005, doi: 10.1111/j.1365-2621.2005.tb07153.x.
38. Patil, V., Shams, R., Dash, K.K., Techno-functional characteristics, and potential applications of edible coatings: A comprehensive review. *J. Agric. Food Res.*, 14, 100886, 2023, doi: 10.1016/j.jafr.2023.100886.
39. Kocira, A., Kozłowicz, K., Panasiewicz, K., Staniak, M., Szpunar-Krok, E., Hortyńska, P., Polysaccharides as edible films and coatings: Characteristics and influence on fruit and vegetable quality—A review. *Agronomy*, *11*, 5, 813, 2021, doi: 10.3390/agronomy11050813.
40. Vargas, M., Pastor, C., Chiralt, A., McClements, D.J., Gonzalez-Martinez, C., Recent advances in edible coatings for fresh and minimally

processed fruits. *Crit. Rev. Food Sci. Nutr.*, *48*, 6, 496–511, 2008, doi: 10.1080/10408390701537344.

41. Umaraw, P., Munekata, P.E., Verma, A.K., Barba, F.J., Singh, V.P., Kumar, P., Lorenzo, J.M., Edible films/coating with tailored properties for active packaging of meat, fish and derived products. *Trends Food Sci. Technol.*, *98*, 10–24, 2020, doi: /10.1016/j.tifs.2020.01.032.
42. Aayush, K., McClements, D.J., Sharma, S., Sharma, R., Singh, G.P., Sharma, K., Oberoi, K., Innovations in the development and application of edible coatings for fresh and minimally processed Apple. *Food Control*, *141*, 109188, 2022, doi: 10.1016/j.foodcont.2022.109188.
43. Kris-Etherton, P.M., Hecker, K.D., Bonanome, A., Coval, S.M., Binkoski, A.E., Hilpert, K.F., Etherton, T.D., Bioactive compounds in foods: their role in the prevention of cardiovascular disease and cancer. *Am. J. Med.*, *113*, 9, 71–88, 2002, doi: 10.1016/S0002-9343(01)00995-0.
44. Ma, W., Tang, C.H., Yin, S.W., Yang, X.Q., Wang, Q., Liu, F., Wei, Z.H., Characterization of gelatin-based edible films incorporated with olive oil. *Food Res. Int.*, *49*, 1, 572–579, 2012, doi: 10.1016/j.foodres.2012.07.037.
45. Rodsamran, P. and Sothornvit, R., Bioactive coconut protein concentrate films incorporated with antioxidant extract of mature coconut water. *Food Hydrocolloids*, *79*, 243–252, 2018, doi: 10.1016/j.foodhyd.2017.12.037.
46. Nogueira, G.F., Fakhouri, F.M., de Oliveira, R.A., Effect of incorporation of blackberry particles on the physicochemical properties of edible films of arrowroot starch. *Drying Technol.*, *37*, 4, 448–457, 2019, doi: 10.1080/07373937.2018.1441153.
47. Assis, R.Q., Pagno, C.H., Costa, T.M.H., Flôres, S.H., Rios, A.D.O., Synthesis of biodegradable films based on cassava starch containing free and nanoencapsulated β-carotene. *Packag. Technol. Sci.*, *31*, 3, 157–166, 2018, doi: 10.1002/pts.2364.
48. Aloui, H., Licciardello, F., Khwaldia, K., Hamdi, M., Restuccia, C., Physical properties and antifungal activity of bioactive films containing *Wickerhamomyces anomalus* killer yeast and their application for preservation of oranges and control of postharvest green mold caused by *Penicillium digitatum*. *Int. J. Food Microbiol.*, *200*, 22–30, 2015, doi: 10.1016/j.ijfoodmicro.2015.01.015.
49. Parafati, L., Vitale, A., Restuccia, C., Cirvilleri, G., The effect of locust bean gum (LBG)-based edible coatings carrying biocontrol yeasts against Penicillium digitatum and Penicillium italicum causal agents of postharvest decay of mandarin fruit. *Food Microbiol.*, *58*, 87–94, 2016, doi: 10.1016/j.fm.2016.03.014.
50. Gbassi, G.K., Vandamme, T., Ennahar, S., Marchioni, E., Microencapsulation of Lactobacillus plantarum spp in an alginate matrix coated with whey proteins. *Int. J. Food Microbiol.*, *129*, 1, 103–105, 2009, doi: 10.1016/j.ijfoodmicro.2008.11.012.

51. Holzapfel, W. (Ed.), *Advances in fermented foods and beverages: improving quality, technologies and health benefits*, Elsevier, 2014.
52. Gbassi, G.K., Vandamme, T., Ennahar, S., Marchioni, E., Microencapsulation of Lactobacillus plantarum spp in an alginate matrix coated with whey proteins. *Int. J. Food Microbiol.*, *129*, 1, 103–105, 2009, doi: 10.1016/j.ijfoodmicro.2008.11.012.
53. Kong, F. and Singh, R.P., Chemical deterioration and physical instability of foods and beverages, in: *The stability and shelf life of food*, pp. 43–76, Woodhead Publishing, 2016, doi: 10.1016/B978-0-08-100435-7.00002-2.
54. Saavedra, S.C., Ventura-Aguilar, R., II, Bautista-Baños, S., Barrera-Necha, L.L., Biodegradable chitosan coating for improving quality and controlling Alternaria alternata growth in figs. *World J. Adv. Res. Rev.*, 7, 2, 115–125, 2020, doi: 10.30574/wjarr.2020.7.2.0246.
55. Hamzah, H.M., Osman, A., Tan, C.P., Ghazali, F.M., Carrageenan as an alternative coating for papaya (Carica papaya L. cv. Eksotika). *Postharvest Biol. Technol.*, *75*, 142–146, 2013, doi: 10.1016/j.postharvbio.2012.08.012.
56. Thomas, A.B., Nassur, R. D. C. M. R., Boas, A.C.V., Lima, L.C.D.O., Cassava starch edible coating incorporated with propolis on bioactive compounds in strawberries. *Ciência e Agrotecnologia*, *40*, 87–96, 2016, doi: 10.1590/S1413-70542016000100008.
57. Sipahi, R.E., Castell-Perez, M.E., Moreira, R.G., Gomes, C., Castillo, A., Improved multilayered antimicrobial alginate-based edible coating extends the shelf life of fresh-cut watermelon (Citrullus lanatus). *LWT Food Sci. Technol.*, *51*, 1, 9–15, 2013, doi: 10.1016/j.lwt.2012.11.013.
58. Abhirami, P., Modupalli, N., Natarajan, V., Novel postharvest intervention using rice bran wax edible coating for shelf-life enhancement of Solanum lycopersicum fruit. *J. Food Process. Preserv.*, *44*, 12, e14989, 2020, doi: 10.1111/jfpp.14989.
59. Olaimat, A.N., Fang, Y., Holley, R.A., Inhibition of Campylobacter jejuni on fresh chicken breasts by κ-carrageenan/chitosan-based coatings containing allyl isothiocyanate or deodorized oriental mustard extract. *Int. J. Food Microbiol.*, *187*, 77–82, 2014, doi: 10.1016/j.ijfoodmicro.2014.07.003.

3

Edible and Biodegradable Polymeric Materials for Food Packaging or Coatings

D. Kavya Gupta, M. H. M. Phanisree, M. Penchalaraju and A. Surendra Babu*

Department of Food Science and Technology, School of Agricultural Sciences, Malla Reddy University, Hyderabad, Telangana, India

Abstract

The global population is estimated to increase by 3.1 billion between 2020 and 2100; hence, the demand for food production is also expected to rise from 59% to 95%. Controlling food waste and loss is inevitable to promote food security and achieve a better balance between supply and demand for food. Hence, fresh agriculture produces and processed food products should be preserved using appropriate packing materials. The packaging material helps to extend the shelf life of the food and enhance the quality while it is being transported, stored, and distributed. Plastic continues to be the preferred material for food packaging and after which a significant amount of it is released into the environment, which is matter of concern as its degradation can take up to a thousand years. Innovations and developments in biodegradable food packaging have become indispensable in the food packaging industry. In recent years, edible and biodegradable packaging or coating have drawn a lot of interest as an alternative to plastic packaging. Biopolymers of plant, animal, and microbial origin, such as polysaccharides, proteins, cellulose, and lipids, have been successfully used to develop environmentally friendly biodegradable food packaging materials. These biopolymer-based films can be decomposed into CH_4, CO_2, H_2O, and inorganic compounds by the action of microorganisms in the soil after being used. They are designed to act as a potential carrier for bioactive compounds. Biopolymers were classified into three different types based on their origin and method of synthesis. The first group of biopolymers includes those directly derived from biomass (carbohydrates and proteins). The second group of polymers, such as polylactic acid, were produced from biobased monomers using the traditional polymerization process, and the third category of polymers represents that were synthesized by microorganisms.

**Corresponding author*: surendrababu.postdoc@gmail.com

Mousumi Sen (ed.) Food Coatings and Preservation Technologies, (51–84) © 2025 Scrivener Publishing LLC

In this context, edible films developed using biopolymers have drawn much interest as an alternative to petroleum-based packaging systems. Edible packaging is often made from polysaccharides, such as starch, lipid, and protein, and their composites have several advantages over synthetic polymers. Composite edible films and coating have enhanced physical, barrier, and barrier properties. Plasticizers, crosslinking agents, and emulsifiers are included to improve the properties of composite films. Edible packaging is eco-friendly, can be consumed safely as a component of food products and extends the shelf life of fresh goods. Edible films/packaging systems are generally loaded with plasticiser, such as momo-, di-, or oligosaccharides, such as glucose, fructose, sucrose, polyols, etc., for enhanced flexibility and durability. Further, antioxidants are also included in edible packaging materials to decrease the oxidative deterioration of foods. Recent trends in edible coating include the application of antimicrobial agents, anti-browning agents, nutraceuticals, colors, and flavoring agents to increase the quality of food products. In addition, edible coatings are used as a carrier of nutrients to enhance the nutritional value of the finished product. Potential role of edible films/packaging materials as smart packaging an emerging trend, which needs exploration and extending it to industrial level will be a revolution in packaging sector. This chapter aims to provide a comprehensive understanding of the current state and critical developments in biodegradable polymer packaging systems for food applications.

Keywords: Biopolymer, nanocrystal, starch, cellulose, SEM, food packaging, biofilm, PVC

3.1 Introduction

With an increased global population, the food requirement is expected to raise between 59% and 98% by 2050 [1]. Consequently, food production and food safety are most vital for ensuring safe and efficient delivery of food to the needy [2]. Center for Disease Control and Prevention reported that more than 70 million people every year suffer from food-borne illness, which results in economic crisis and deaths. In this scenario, food packaging has become crucial for enabling food safety throughout the food supply chain providing protection from external environment, contamination, spillage, microorganisms, and physical damage. Moreover, packaging reduces the food wastage and maintains the nutritional value of the food [2]. Packaging developed with various materials, such as glass, paper, plastic, biodegradable polymers, metals, etc., are employed in enclosing the food products.

Polymers are classified as natural or synthetic. Natural polymers or also known as biopolymers are derived from plants or animals and includes carbohydrate polymers, nucleic acids and proteins, etc. [3], while synthetic polymers are obtained by chemical modification of other polymers even natural polymers. Thermoplastic polymers and thermoset polymers are two types synthetic polymers classified basing on the heat resistance. Thermoplastic are polymers do not experience any chemical change when exposed to heat and can be remoulded. Some examples are polyvinyl chloride, polyethylene, polytetrafluroethylene or Teflon, polypropylene, and polystyrene. On the other hand, thermoset plastics can be melted, moulded into required shapes permanently on cooling. They include bakelite, polyepoxide, vulcanized rubber [4].

The term "plastic" is originated from Greek word "plastikos" that indicates fit for moulding, representing its flexibility during manufacturing or processing that enables them to be cast, pressed into various shapes including plates, tubes, film, bottles, boxes, etc. [5]. There is a rapid increase in the growth of global plastic industry due to its wide applications and cheaper rates since 1950s and the production rate of plastic is increasing by 4% every year [6]. Plastic materials used in packaging are synthetic organic polymers employed in variety of applications such as water bottles, food packaging, medical, and electronic goods [7]. Plastic was once thought to be innocuous and inert, nevertheless prolonged disposal of plastic into the environment have resulted in several associated problems. Plastic waste is now considered as the major environmental pollution particularly aquatic environments where plastic undergoes prolonged breakdown, pose a serious threat to wildlife and possess limited options for plastic elimination [8, 9]. Plastic waste generation for the year 2015 across the globe was reported to be 6.3 billion tonnes. Out of which 9% of the total waste was recycled, 12% waste was incinerated, and 79% waste was gathered in landfills or environment. If this trend persists, around 12 billion metric tonnes of plastic waste will be transferred to landfills and natural environment by 2050 [10].

3.1.1 Effect of Plastic Wastes on the Environment

Plastic waste generation is associated with the human population. The demand for plastic and plastic based products has increased with an increase in the global population. The indiscriminate disposal of plastic waste to the surroundings may cause environmental pollution which is evident in a number of ways, such as the destruction of natural environmental beauty, entanglement and loss of aquatic life, blockage of sewage systems in towns and cities, particularly in the developing nations, which provide a favorable condition for the breeding of mosquitoes and other disease-causing vectors and further, generation of fouls smell. In addition, there will be a reduction in water percolation and regular aeration of agriculture soils, which reduces the soil fertility and productivity of crops in such lands [11, 12]. Incineration of plastic debris releases greenhouse gases which leads to global warming. Landfilling of plastic waste especially in large quantities is not a good alternative as the toxic chemicals found in plastic including heavy metals, phthalates, bisphenol-A, phenanthrene, polychlorinated biphenyl ethers, brominated flame retardants, nonylphenol, dichlorodiphenyldichloroethylene (DDE) and antimony trioxide may pass into the groundwater (Table 3.1). These harmful substances flow into rivers, seas, oceans, etc. casing threat to aquatic life [13].

3.1.2 Effect of Plastic Waste on Animals

Animal food supplements intended for human consumption may cause adverse effects if the animals become consume hazardous chemicals found in plastic and components derived from plastic. There have been reports of enormous marine mammals being threatened by the massive levels of plastic debris that are finding their way into the rivers and oceans [23]. Microplastics which implies a particle less than five millimetres (0.2 inches) in diameter. Fish and other aquatic life may be poisoned by the consumption of microplastics, which may indeed cause oxidative damage, growth retardation, reduced food intake, and aberrant behavior [6]. As a result, millions of animals are exterminated by plastics annually including fish, birds, and other marine species. When the animal intestine was loaded with plastics, it decreases appetite. Animals, such as seals, turtles, whales, etc., are choked by abandoned fishing gear or cast-off six-pack rings [24]. Nevertheless, nanoplastics would penetrate the tissues, generating reactive oxygen species (ROS), terminating lipid metabolism, in addition causes defects at molecular level. For example, fishes when intake nanopolystyrene particles via aquatic food chain demonstrated a considerable difference in cholesterol level in liver and muscles, serum triglyceride to cholesterol ratio, body weight, and several metabolic parameters [6].

3.1.3 Effect of Plastic Waste on Public Health

Long-term storage of liquid foods in PET bottles may leach antimony trioxide (carcinogenic agent) as the later together with antimony acetate and antimony glycolate is the most widely used catalyst in the manufacturing of PET resin at a concentration of 200 to 300 ppm. Trace amounts of antimony may migrate into the beverages from the packaging material when exposed to hot environment for a long period of time [5]. In addition, consumption of food or water packed in poly vinyl chloride (PVC)

Table 3.1 Different types of toxic additives used in plastic manufacturing and their potential hazards.

Additives	Plastic types	Potential hazards on animals/ human health	References
Bisphenol A	Polyvinyl chloride (PVC), Polycarbonate (PC)	Ovarian disorders, mimics oestrogen	[14, 15]
Phthalates	Polystyrene (PS) Polyvinyl chloride (PVC)	Affects sperm motility, testosterone function and fetal development	[15, 16]
Persistent organic pollutants (POPs)	All plastics	Neurological, reproductive disorders and hepatotoxicity	[15, 17]
Dioxins	All plastics	Cancer and interferes with testosterone	[15, 18]
Polycyclic aromatic hydrocarbons (PAHs)	All plastics	Reproductive and developmental toxicity, Risk of cardiovascular disease	[15, 19]
Polychlorinated diphenyls (PCBs)	All plastics	Affects the function of thyroid hormone	[15, 20]
Styrene monomers	Polystyrene	Carcinogen and DNA damage	[15, 21]
Nonylphenol	PVC	Mimics oestrogen	[15, 22]

or contaminated with PVC, inhaling emissions from PVC manufacturing units, incinerating PVC or physical contact with PVC-based materials on skin can have harmful effect to humans [25]. Exposure or inhalation of these toxic substances may lead to several health complications such as weakened immune system, liver and kidney disorders, sensory dysfunction, may also cause cancer, neuroreproductive disorder, and age-related toxicity [26].

Henceforth, in recent times, considerable efforts have been taken to protect the environment and public health by exploring natural renewable eco-friendly and biodegradable substances as packaging materials. One such materials are biopolymers, a type of polymers that are derived from natural sources, such as alginate, agar-agar, carrageenan, which originated from seaweeds, chitosan found in shells of certain organisms, insects and microbes including algae, fungi, and yeast [27]. Biopolymers are made up of the monomeric units, such as amino acids, sugars, nucleotides. Some examples of biopolymers are cellulose, starch, proteins, chitin, DNA, and RNA [28].

3.2 Classification of Biobased Polymers

The main factors influencing interest in biopolymers made from natural resources like plants are the need for sustainable alternatives to the limited supply of petrochemical products, rising public concern over ecological issues, and increased regulation. Biopolymers can perform just as well as synthetic polymers, but they have several environmental benefits, including virtually no harmful emissions, sustainable and renewable raw materials, and extensive spread. Biopolymers are classified based on origin, chemical structure, properties, and their potential industrial applications [29].

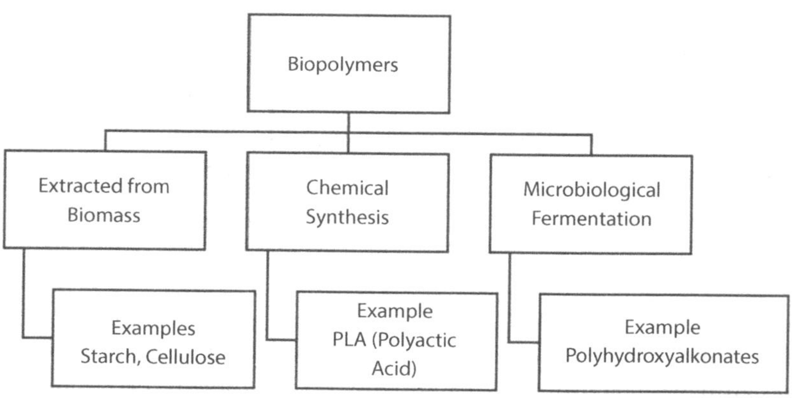

Figure 3.1 Classification of biopolymers.

Biopolymers are polymers generally produced from natural and biological material or biosynthesized living organisms, such as yeast, mould, and bacteria (Figure 3.1).

3.2.1 Carbohydrate-Based Biopolymer

Carbohydrate polymers like starch extracted from various sources have been studied widely for the synthesis of bioplastics. It is a low-cost, abundant and biodegradable natural biopolymer. Hence, extensive research has been done to develop packaging materials that can replace synthetic or petrochemical alternates. Starch is composed of amylose and amylopectin chains. Amylose is a linear crystalline polymer while amylopectin is a branched chain and amorphous polysaccharide. Amylose and amylopectin accounts to about 10% to 20% and 80% to 90%, respectively, in starch depending on the type of starch [30, 31]. The most common sources of industrial starches for the preparation of biopolymers are potato, tapioca, maize, wheat, and rice. Starch is used as a thermoplastic material [32]. It is produced by mixing starch and plasticizer with enough water and heating in an extruder at high temperature, pressure, and shear force. Addition of glycerol as a plasticizer is crucial to obtain a rubbery texture with minimum brittleness [33]. Tapioca starch was studied extensively to produce biopolymer compared to the other starches. Jaramillo *et al.* [34] have demonstrated the application of cassava starch for the production of biofilm. Moisture content of the cassava starch was found to be 0.27%, which is significantly lesser than potato starch, maize starch, and rice starch. The cassava-based biofilm showed an improved water vapor permeability and elasticity. In another study blending of starch with hemicellulose, polycaprolactone, and zein have reported water resistance ability [35]. Addition of citric acid in thermoplastic starch showed an improved thermal stability, reduced tensile strength, and enhanced water resistance [36].

On the other hand, cellulose, which is a most abundant polysaccharide, is also explored for the synthesis of biopolymer. It is an unbranched polymer of glucose. Cellulose derivatives such as carboxy cellulose, hydroxy methyl cellulose, methyl cellulose, hydroxy propyl cellulose, and cellulose nanofibers are natural and renewable polymers which are gaining considerable attention. Further, these polymers exhibit outstanding biocompatibility, mechanical properties, thermal properties, biodegradability, water absorbance, antimicrobial properties, and crystallinity. Yaradoddi *et al.* [37] have developed a carboxy methyl cellulose (CMC)-based biofilm extracted from agricultural waste using glycerol as a plasticizer agent.

The biofilm presented an excellent mechanical property, water permeability, and biodegradability, making it a good material for packing.

3.2.2 Protein-Based Biopolymer

Several plant and animal proteins, including gluten, whey protein, casein, zein, and gelatine, are extensively investigated to produce biopolymers because of their biodegradability and sustainability [38, 39]. They can be easily processed and formulated into films. These protein based-polymers possess great capacity for film formation with high nutritional value and biodegradability. They exhibit better mechanical properties than lipid and polysaccharide-based films. The quality of the biopolymers can be improved by modification and various treatments, such as (i) cross-linking using chemicals like $CaCl_2$, glutaraldehyde, 1-ethyl-3-(3-dimethyl-aminopropyl) carbodiimide/N-hydroxysuccinimide (ED/NHS) [40, 41]; (ii) blending plasticizer/polymer with the protein film for enhancing flexibility and tensile strength [42]; (iii) addition of natural antioxidants such as vitamin E to enhance the shelf life and stability [43]; (iv) incorporation of antimicrobial agents including nisin (peptide from *Lactococcus lactis*) to improve shelf life and protect foods from microbial spoilage; (v) blending with other polysaccharides including cellulose, chitosan, starch, and alginate to enhance mechanical and barrier properties [44]; (vi) surface modification of films to improve the film properties; (vii) insertion of functional groups into the protein structure to enable entanglement of proteins side chains [45].

The biofilms prepared from soy protein isolate are eco-friendly, biodegradable and can be an alternative to synthetic packaging materials which presented good mechanical, barrier properties and enhanced the shelf life of packaged foodstuffs [46]. Zein is a protein isolated from maize is nontoxic and edible, which is widely used to develop biopolymers for packaging applications. These films exhibit excellent barrier properties against moisture, oxygen, light, and microorganism [47]. While gelatin-based biopolymers are flexible, transparent and exhibits good gas barrier properties and can be used in fruit and vegetable, meat, and confectionary packaging [48].

3.2.3 Polylactic Acid

Polylactic acid (PLA) is produced from bioderived monomers and polyhydroxyalaknoates are produced by microorganisms. PLA is a biopolymer made from renewable feedstock that shows great potential for use in various biomedical applications. PLA is an exceptional material for these applications due to its remarkable biocompatibility, biodegradability, excellent mechanical strength, and effortless processability. PLA is considered one of the most promising biopolymers due to its ability to be produced from nontoxic materials. Generally, PLA biopolymers are produced by chemical synthesis [49]. PLA and poly hydroxyalkanoates (PHA) are linear and branched molecules of repeating cells of monomers. LAB (lactic acid bacteria) is capable to produce PLA and PHA monomers. PLA can be employed for preparation of bioplastic and is useful for production of loose fill packaging, drinking cups, salad cups, overwrap, food packaging, and compost bags. PLA can also be used in textile industry, construction material for frameworks and fishing nets. PLA offers a great opportunity to replace less environmentally friendly materials in the packaging industry because of its favorable physical properties (transparency, high elastic modulus, high melting temperature), which frequently improve function while lowering environmental impact [50, 51].

3.2.4 PHA

Microorganisms synthesize polyhydroxy alkanoates, which resemble synthetic polymers and are entirely biodegradable. PHAs, or poly hydroxyalkanoates, are synthetic polymers that can be processed similarly, and they are thought to be an alternative to plastics made from petroleum. Fifty percent of the entire expense of producing commercialized PHAs is attributed to the presence of bacteria and significant amounts of organic carbon sources in the culture medium. The PHA's commercial application can only increase in the event that production costs are lowered. Many studies have indicated that microalgae, which are photoautotrophs that get their energy from light and CO_2, may be able to produce PHAs at a lower cost. They also only need a small amount of nutrients [29]. Polyhydroxyalkanoic acid (PHA) is the building block of thermoplastic polymers. The PHA is produced via microbial fermentation. Polyhydroxyalkanotes, which are classified as both biodegradable and biocompatible polyesters while maintaining physical properties similar to most petroleum-derived plastics, are esters of 3-, 4-, 5-, and 6-hydroxyalkanoic acids produced by an assortment of bacterial species. PHA can be divided into three types based on the length of comprising monomers: short chain PHAs (scl-PHA) containing up to C5 monomers, medium chain PHAs (mcl-PHA) with C6-C14, and long chain PHAs (lcl-PHA) with more than C14 monomers [52]. Polyhydroxyalkanoates (PHAs), one type of bioplastic, are thought to be the best option for use in food packaging. Hydrophobic, biodegradable, nontoxic, thermoplastic, piezoelectric, and impermeable to gases and/or water are just a few of the promising qualities of PHAs [53].

3.3 Advanced Methods for the Synthesis of Biopolymers for Packaging Applications

The synthesis of biopolymers involves several steps, from deriving the biopolymers to the development of the final biopolymer material for packaging applications. They are derived naturally or synthetically [54]. The natural sources include plants, animals, microorganisms, and marine (Figure 3.2). Biopolymers can be made into a wide range of products, such as packaging films, coatings, trays, cups, cutlery items, etc. [55]. The primary applications for biopolymer packaging are as coatings in industrial processes, and their conventional coating techniques, including spray fluidization, falling and pan coatings, spraying, dipping, and brushing, are employed [56]. The production of biopolymer film involves the same old techniques of casting, extrusion, or blow molding. To enhance the mechanical, thermal, and barrier properties, biopolymers are often mixed or combined with other biopolymers, nanoparticles, or active agents [57]. The selection of raw materials, the synthesis method, and the characteristic features of the biopolymer depend on the end-use application.

The raw material is selected in such a way that it meets the specific characteristics needed to tailor the end-use application. Depending on the raw material chosen, the procedure to extract the biopolymer varies.

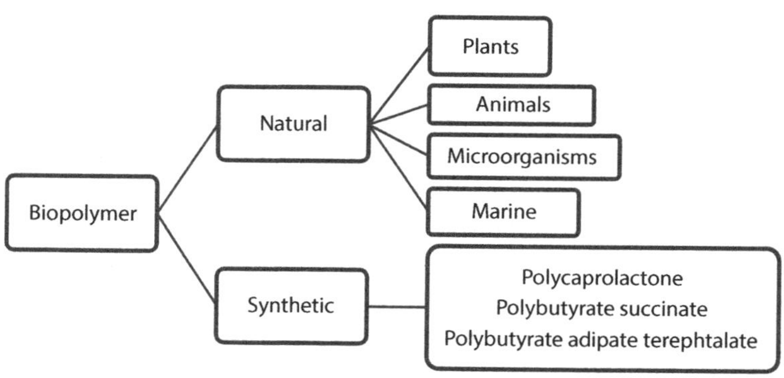

Figure 3.2 Classification of biopolymers based on origin (source: Jariyasakoolroj *et al.* [54]).

3.3.1 Cellulose

Cellulose can be synthesis naturally and synthetically (Table 3.2). The natural sources include plant, bacteria, and agricultural waste. To form a cellulose film, the cellulose is converted to its derivative form. The cellulose structured is altered by substituting the hydroxyl group with acetate or methyl group. These derivatives are categorized into two groups—cellulose esters and cellulose ether [58]. Depending on the source, the extraction method of cellulose varies. The table below discuss a few green approaches for the extraction of cellulose.

Table 3.2 Extraction of cellulose from different sources.

Source	Extraction method
Sugarcane bagasse	The fibres are extracted from dried sugarcane bagasse by subjecting it to a 13 bar steam explosion at 195°C. These fibers undergo a 1-hour treatment with 20 μg of xylanase (1:10 fiber to liquid ratio) at 50°C with constant stirring using. The dried bagasse is then bleached for 1 hour at 70°C with 0.7% sodium chlorite ($NaClO_4$) by maintain the pH at 4 using acetic acid. For everyone hour, sodium chlorite and acetic acid are added until the fibers turn white. After filtering, the product is washed to a neutral pH and dried for a day at 55°C.
Rice husk	The rice husk is boiled in 4% wt NaOH solution for 2 hours. The filtered product is then bleached with buffering solution containing acetic acid and aqueous chlorite (1.7% by wt) at 100–130°C for 4 hours. The mixture is then cooled and clarified. The clarified product undergoes acid hydrolysis by 10M of H_2SO_4 at 50°C for 40 minutes under continuous agitation. The mixture is then centrifuged for 10 minutes at G-force 12298 at 10°C.
Palm Kernel cake	The palm kernel cake first undergoes a hot-water treatment. The cake is boiled in water at 160–180°C for 30 minutes. The cake is then washed with ethanol and water to remove the acids and dried. The dried cake now undergoes a liquid phase oxidation. In this process, the cake is treated with hydrogen peroxide until the lignin are removed. The addition of cold water terminates the liquid phase oxidation. The filtered residue obtained is the cellulose.

(Continued)

Table 3.2 Extraction of cellulose from different sources. (*Continued*)

Source	Extraction method
Coconut waste	The coconut mesocarp is first treated with NaOH for 2 hours at 80°C. The residue is then water washed and bleached with $NaClO_2$. The mixture is treated with HNO_3 solution at 70°C. the residue undergoes a alkaline treatment with NaOH and $Na_2B_4O_7$ at 28°C. The mixture is then treated with H_2O_2 at 45°C. The obtained residue is again treated with NaOH at 45°C followed by a acid treatment with acetic acid and nitric acid (10:1) at 120°C. The residue then undergoes acid hydrolysis with conc. Sulphuric acid at 45°C. The obtained cellulose is water washed until the pH is neutral.
Bamboo	The bamboo undergoes size reduction to obtain bamboo fiber. The bamboo is first grounded and made into a mixture of chips and powder which is then dried in an oven for 72 hours at 70°C. The produced bamboo fibres are mixed with a toluene-ethanol mixture (2:1) and heated. This step ensures the dewaxing of bamboo fibres. The de-waxed bamboo fiber is treated with hydrogen peroxide and acetic acid and heated at 130°C for 2 hours. The filtered residue washed to obtain a neutral pH and is dried at 70°C. The obtained residue is called as residue is delignified bamboo fiber (DLBF). This DLBF is treated with NaOH under agitation at 150rpm and heated for 2 hours at 80°C followed by stirring for 8 hours. The obtained residue is washed to a neutral pH and dried at 85°C for 48 hours.

Source: Khan *et al.* [58].

3.3.2 Chitin (CT) and Chitosan (CS)

Chitin is the next abundant biopolymer to cellulose. It is constituting a 2-acetamido-2-deoxy-β-D-glucose monomer [59]. It is mostly composed of the cell walls of certain fungus and the exoskeletons of marine insects and other invertebrates. Chitosan is a cationic biopolymer which is produced by, a deacetylating chitin. It is composed of β-(1-4)-2-amino-2-deoxy-D-glucopyranose [60, 61].

Extraction of CT: The fundamental procedure for extracting chitin entails pulverizing, drying, and washing the raw material to create a fine powder.

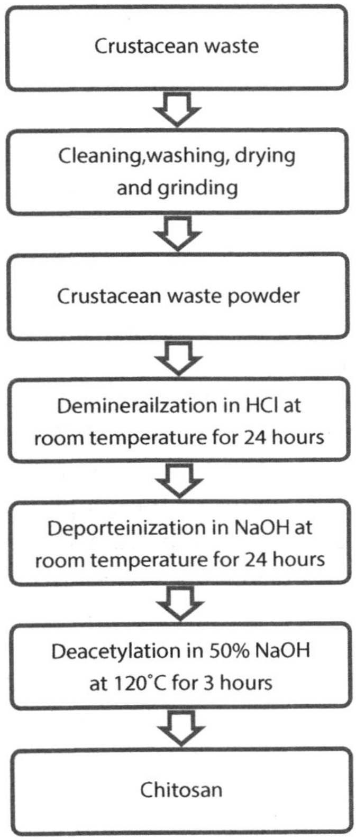

Figure 3.3 Flow sheet for production of chitosan (source: Hameed *et al.* [62]).

This fine powder undergoes through decolorization, deproteinization and demineralization. For both chemical and biological approaches, these steps are the same (Figure 3.3) [59, 62].

Conversion of CT to CS: The conversion of CT to CS is achieved by alkaline method and enzymatic method. The enzymatic method uses Chitin deacetylase for the conversion of CT to CS. In alkaline method, the chitin powder is treated with a 40% to 50% NaOH solution resulting in the hydrolysis of acetyl groups. The formed N-acetyl-D-glucosamine units are transformed into D-glucosamine units with free amine groups [59].

3.4 Properties of Biopolymers

3.4.1 Structural Properties

When starch is combined with chitosan and gelatin using the solution casting method, it is observed that the biopolymer has a smooth surface with no visible pores, a high density, and minimal moisture ingression when tested with Fourier transform infrared spectroscopy (FTIR) and Scanning electron microscope (SEM). Using electrospinning method, Poly vinyl alcohol (PVA) and oxidized starch-based biopolymer were exhibited a reduced average diameter with change in concentration and weight ratio. At a weight ratio below 1:3, the fibres were observed to be irregular and scattered with shuttle shape bead under SEM [55]. Starch when plasticized with glycerol and urea showed good structural stability. The characteristic bands for the individual granules of thermoplastic starch by FTIR spectrum presented a stronger hydrogen bond between the starch chains. The presence of urea (15–25%) resulted a breakdown of the starch crystalline structure causing increased amorphous nature ensuring the stable structure of thermos plastic starch at 25°C, 50% RH for 30 days [63]. The surfaces of chitosan-roselle films with ratios of 80:20, 70:30, and 60:40 was smoother than those with ratios of 100:0 and 90:10. A larger chitosan/roselle ratio results in a smoother surface structure. The greater heat resistance of these mix films was attributed to the chitosan-roselle films' smoother surface [64]. Fourier transform infrared (FTIR) and differential scanning calorimetry (DSC) analysis were used to characterize cod gelatin films produced using 1-ethyl-3-(3-dimethylaminopropyl) carbodiimide (EDC) or transglutaminase (TGase) and it was compared film produced from pig gelatin and observations revealed that the development of the biofilm from pig gelatin featured water to water, amide to amide and water to amide hydrogen bonds and indicated amide I band at peak 1663 cm^{-1} and amide II band at peak 1537 cm^{-1} had more number of hydrogen bonds from pig gelatin biofilm by increasing the strength of helical structure than cod gelatin biofilm [65].

3.4.2 Mechanical Properties

Mechanical properties play a vital role in protecting the food during handling, storage, and processing. The structure and design of the biopolymer matrix mainly depends on mechanical property such as tensile strength, elongation at break and elastic modulus [57]. Young's modulus (YM), elongation at break (EAB), and tensile strength (TS) of the synthesized biopolymers were examined for arrowroot starch-based biopolymer prepared by employing solution casting process and glycerol as plasticiser (15, 30 and 45% w/w). compared to other plasticized films 15% glycerol film showed higher tensile strength with increase in elongation at break (EAB) from 2.41% to 57.33% and tensile strength and modulus of plasticized arrowroot starch films minimized from 9.34 to 1.95 MPa and 620.79 to 36.08 MPa, respectively [57]. Using 0% (CS-00), 30% (CS-30), 60% (CS-60), and 90% (CS-90) glycerol with chitosan, four moulds were made. Tensile strength and percentage elongation at break were measured on the samples. A peak tensile strength is reached at TS 85 MPa and EAB 30.8% of CS-G30 hence, at the condition of increase in the glycerol content, the tensile strength dropped and the film's thickness and elongation at break was increased [67]. The mechanical properties were modified by using pectin and alginate by casting process with the solutions of 2.5%, 5%, and 7.5% (w/w) pectin and alginate powders for each polymer at pH ranges of 9 to 10 and pH 3 to 3.5. Results have showed that the pH 3 to 3.5 have improved the polymer organization with reduced intermolecular spacing in between polymer chains by gelation of charged polymers [68]. Another study stated that the starch incorporated chitosan film with glycerol as additive had affected the mechanical properties like tensile strength that has been increased up to 2.24 times than native starch when combined with glycerol as additive. The chitosan and cassava starch added with gallic acid and glycerol. The tensile strength increased from 0.51 to 0.83 MPa [66]. The tensile strength increased from 1.12 to 10.03 MPa (7.96-fold increment) and the elongation at break increased from 67.00% to 90.77% (0.35-fold increment) when 15% w/w CNP (size range, 20–50 nm) was added to the starch films (starch/CNP films). The 15% w/w starch/CNP film's opacity improved from 8.07 to 14.67 as a result of the CNP's presence in the starch matrix, which prevents light from passing through [69]. Films developed with rice starch and PVA with ratio 2:8 observed high tensile strength and in combination of rice starch, PVA and sugar with ratio 1:8:1 showed high TS along with good elongation at break [69].

3.4.3 Thermal Properties

Modifying the starch molecular structure by incorporating borax was used as an intermolecular bond reinforcer. Studies stated that addition of borax enhanced the thermal conductivity of biopolymers. This phenomenon is associated with the crosslinking effect, which facilitates the primary biopolymer chain's integration, resulting in a stronger linked network that promotes heat transfer. A 1.40% borax have significantly increased the thermal conductivity to its maximum value. On treating chitosan nanoparticles with different concentrations of 5%, 10%, 15%, 20%, 25%, 30%, 35% w/w of solid starch by solvent casting method, the mechanical, thermal, and barrier properties of tapioca starch films were examined [70]. Thermogravimetric analysis (TG) results indicate that, in comparison to unmodified chitosan, the addition of micro- and nanometric cellulose of each polymorphic form decreased the thermal stability of composites. It is interesting to note that films containing CNC II (nanocrystalline cellulose II) had greater thermal stability than films containing CNC I (nanocrystalline cellulose I). The CNC II film exhibited greater thermal stability compared to the equivalent cellulose I (C II) composite. This difference could likely be attributed to the less effective bonding between the C II particles and the matrix [71].

3.4.4 Barrier Properties

Packaging materials should exhibit transparency, strong mechanical performance, and an excellent barrier to oxygen and water vapor. The nature of the penetration, processing techniques, structure and morphology of polymers are the main factors that influence the barrier properties of biopolymers. Enhancements in polymer mixing, multi-layer coextrusion, chain configuration, crystallization/orientation, nanotechnology, and coating can all help increase the barrier properties of biodegradable polymers [60]. It has been noted that adding organically modified layered nanoclays to biopolymers by reinforcement significantly increases their characteristics. Low nanoclay concentrations (less than 10 weight percent) are known to significantly improve rigidity (elastic modulus), thermal and dimensional stability, and barrier qualities to gases and vapors without sacrificing other qualities like toughness and transparency. Only polyhydroxy butyrate (PHB) and PHB nanocomposites made by melt mixing had oxygen permeability values that were lower than those of pure polyethylene terephthalate (PET), according to the permeability data. Out of all the biodegradable materials, the PHB nanocomposite showed the lowest oxygen permeability value, approaching the value of the PET nanocomposite. A 5% of scattered nano clays increased the oxygen barrier significantly for all the biodegradable materials, making them more suitable for packaging oxygen-sensitive products [72]. The oxygen permeability coefficient (OPC), carbon dioxide permeability coefficient (CO_2 PC), and water vapor permeability coefficient

(WVPC) respectively measure how much O_2, CO_2, or water vapor permeates a packaging material per unit area and time (kg m m^{-2} s^{-1} Pa^{-1}). These metrics are used to quantify the oxygen, carbon dioxide, and water vapor barrier properties of a film [73]. One popular way to determine crosslinking is to evaluate the degree of swelling. In aqueous conditions, neither neat (uncoated) nor beeswax-coated PLA films expand. Yet, in order to establish a connection between the crosslinking density and barrier qualities, swelling tests on crosslinked chitosan coated PLA were carried out. The effect of crosslinking agent on swelling behavior and water vapor transmission rate (WVTR) was investigated by Aris *et al.* [74]. A 1% concentration of sodium tripolyphosphate (cross linking agent) have improved the barrier properties non significantly. Nevertheless, higher concentrations of cross-linking agent have showed a remarkable increase in barrier properties. The ideal sodium tripolyphosphate concentration was reported to be 10% (w/v) [74]. This study examines the permeability and hydrophilicity of starch biopolymer, which was modified by adding nanocrystalline cellulose (NCC) (1–15%) and ethylene vinyl alcohol (EVOH) (0–10%). Casting was used to create starch-EVOH nanocomposites in various compositions. Scanning electron microscopy was used to confirm that NCC was produced by acid hydrolysis. The pure starch nanocomposite with 8% NCC had the lowest observed moisture absorption. The sample with 7.8% EVOH and 13% NCC and the starch nanocomposite with 8% NCC have the lowest permeability to water vapor. Additionally, the composite containing the most EVOH had the lowest solubility. The biofilm's ability to absorb moisture is decreased when NCC and EVOH are used [75]. A 2% cassava starch was used to produce biopolymeric films, different concentrations of beeswax (BW) were introduced to the polymeric network as a hydrophobic agent (0%, 5%, and 10%). When comparing the behavior of the films with and without beeswax (BW) for fruit shelf life, the contact angle of the systems including BW revealed the hydrophobic behavior of the films, concurring with the water vapor permeability (WVP) analysis [76] decrease was noted in the film' s oxygen permeability from 7.38 × 10–3 to 3.59 × 10^{-3} cm^3/m day Pa (0.51-fold increment) and water vapor permeability from 1.1 × 10–11 to 0.63 × 10–11 g/Pa h m (4-fold increment). The improved mechanical and barrier qualities of the starch/CNP films created in these studies demonstrated their intriguing potential for use as food packaging materials [69].

3.5 Tests for Determining the Biodegradation of Biopolymer

The term "polymer biodegradation" describes the process by which microorganisms break down polymer carbon to produce energy through the generation of CO_2 (and CH4) (catabolism) and microbial biomass (anabolism) [77]. The generation of CO_2 or the mineralization of materials can be used to gauge the rate of biodegradation [60]. The tests are either performed in the laboratory or in field. There are different analytical methods used to measure the biodegradability of polymer, which can be categorized into physical, thermal, chromatographic, spectroscopic, and respirometric (Figure 3.4) [78].

In respirometric tests, the degree of biodegradation is measured by the utilization of oxygen. In aerobic biodegradation the of substances undergo oxidation to their mineral constituents, such as CO_2, H_2O, sulfur dioxide (SO_2), phosphorous pentoxide (P_2O_5). Thus, the biological oxygen demand (BOD), which is a measurement of the amount of oxygen used during incubation, indicates the level of biodegradation. Further, the measurement of TOD and COD are also used to check the biodegradability [79].

The tests used to check the biodegradability of plastics can be incorporated to understand the biodegradability of biopolymers. International standard organizations, such as American Society for Testing and Materials (ASTM), European Standard, and International Organization

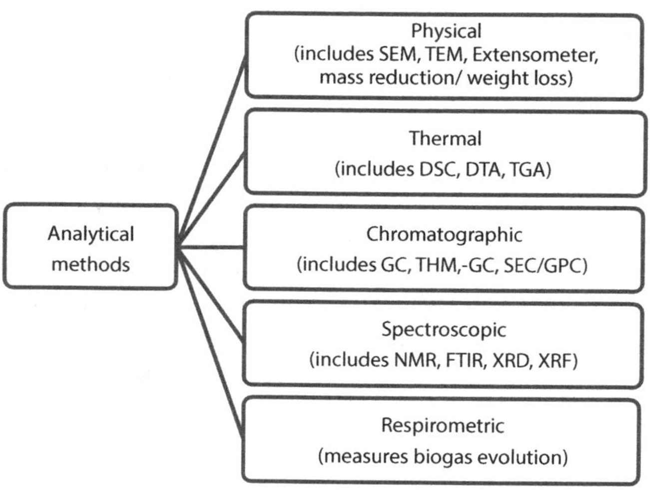

Figure 3.4 Analytical methods for determining the biodegradation of biopolymer [78].

for Standardization (ISO) have proposed various tests methods or guidelines to check the disintegration and biodegradability of material and to standardize the labelling requirements of a packaging material. Selection of the tests depends on the factors, such as material type, testing conditions (aerobic or anaerobic), test duration, and for regulatory compliance. The biodegradability of a packaging material can be understood more by combining multiple test methods. The biodegradation assays can be categorized depending on the disposal environment in soil, compost, and aquatic system [80].

3.5.1 Assay in Soil

There are various soil assays used for measuring the biodegradation of a polymer. In these tests, the specimen is biodegraded in the soil which is solely prepared for the test purpose. The soil either collected from the fields or forest which is fertile and free from the pollutants [80]. ASTM D5988-18 and ISO 17556:2019 are the standard test methods which use the soil technique for biodegradation. Both the test methods use natural and fertile soil collected from fields and forests having a pH of 6 to 8 [80].

3.5.2 Assay in Compost

Composting is an aerobic solid waste management technique in which biodegradable materials are biologically broken down into humus—a valuable source of nutrients for boosting soil productivity and agricultural yield—under the supervision of microorganisms [80]. Test standards like ASTM D5338, ISO 14855-1, ISO 14855-2, ISO 20200, and ISO 17556 provides methodologies to determine the biodegradability in a compost environment. These test methods discuss a testing environment where the parameter such as temperature, humidity, aeration are monitored and controlled and use of a municipal waste as an inoculum.

3.5.3 Assays in Aquatic System

The raising concern of plastic making into water as micro and nano plastic have led to the assays in aquatic system. Like assays in soil and compost, the aquatic testing also incorporates testing environment that mimics the real aquatic system. The inoculum is collected from different aquatic sources and the testing parameters such as temperature, light, nutrients are monitored and controlled. The degradation of the polymer is analysed through analytical methods. Standard test methods such as ISO 18830:2016, ISO 19679:2016, ISO 14853:2016, ISO 23977-1:2020, ISO 23977-2:2020, ISO 15314:2018, ISO 22766:2020, ISO 62:2008, ASTM D6691-17, ASTM D7473/D7473M-21, ASTM D7991-15, ASTM D570-98(2018), and ASTM D5229/D5229M-20 provide an approach for the testing the biodegradability of polymers in aquatic environment [80].

3.5.4 ASTM Standards

ASTM D6400-23—It is a specification for labelling the plastic developed which are aerobically composted in industrial and municipal facilities. The specification set forth the conditions under which goods and materials, including plastic packaging, must be labelled as "compostable in aerobic industrial and municipal composting facilities." This test is in accordance with ISO 17088 test protocol. ASTM D6868—This specification is for used for labelling the end items that have plastics and polymers as coatings or additives with paper and other substrates developed to aerobically composted in industrial and municipal facilities [80].

3.5.5 ISO Standards

There are various ISO standards to determine the biodegradability of a plastics (Table 3.3). ISO 14855 consists of two test methods—ISO 14855-1 (pilot-scale test) and ISO 14855-2 (laboratory-scale test) [81]. ISO 14855-2 standard describes a gravimetric method for measuring the amount of carbon dioxide evolved to determine the final aerobic biodegradability of plastic materials under controlled composting conditions. The technique modifies the temperature, aeration, and humidity of the composting tank to produce the best possible rate of biodegradation. The degradation rate is periodically estimated by measuring the mass of the evolved carbon dioxide in an absorption column that is loaded with soda talc and soda lime using an electronic balance [81].

3.5.6 EN 13432

This standard is entitled as "Packaging—Requirements for packaging recoverable through composting and biodegradation—Test scheme and evaluation criteria for the final acceptance of packaging." The standard outlines the criteria and processes for figuring out whether packaging and packaging materials are compostable and anaerobic treatable. It covers the

Table 3.3 ISO standards for the biodegradation of plastics

ISO standard no.	Test title	Content/ environment	Remarks
14851	Determination of the ultimate aerobic biodegradability of plastic materials in an aqueous medium	Aqueous	Biodegradation is checked by measuring the oxygen demand in a closed respirometer
14852	Determination of the ultimate aerobic biodegradability of plastic materials in an aqueous medium	Aqueous	Biodegradation is checked by measuring the carbon dioxide produced
14855-1	Determination of ultimate aerobic biodegradability of plastic materials under controlled composting conditions	Compost	Biodegradation is checked by measuring the carbon dioxide produced
14855-2	Determination of the ultimate aerobic biodegradability of plastic materials under controlled composting conditions	compost	This is laboratory-scale test. Biodegradation is checked by gravimetrically measuring the carbon dioxide produced
16929	Plastics - Determination of the degree of disintegration of plastic materials under defined composting conditions in a pilot-scale test	Disintegration	This test is not suitable for the determination of aerobic biodegradability of a test material.
20200	Plastics—Determination of the degree of disintegration of plastic materials under simulated composting conditions in a laboratory-scale test	Disintegration	This test is being conducted on a laboratory scale in an environment that is similar to an industrial compost pile.

(*Continued*)

Table 3.3 ISO standards for the biodegradation of plastics. (*Continued*)

ISO standard no.	Test title	Content/ environment	Remarks
17556	Plastics—Determination of the ultimate aerobic biodegradability in soil by measuring the oxygen demand in a respirometer or the amount of carbon dioxide evolved	Soil	
14853	Plastics—Determination of the ultimate anaerobic biodegradation of plastic materials in an aqueous system—Method by measurement of biogas production	Anaerobic	Biodegradation is checked by measuring the biogas produced
15985	Plastics—Determination of the ultimate anaerobic biodegradation and disintegration under high-solids anaerobic-digestion conditions— Method by analysis of released biogas	Anaerobic	Biodegradation is checked by measuring the biogas produced
22403	Plastics-Assessment of the intrinsic biodegradability of materials exposed to marine inocula under mesophilic aerobic laboratory conditions	Aerobic	No preliminary environmental exposure or pre-treatment is required. Further the standard to not applicable to the claim "marine biodegradable" for a biodegradable polymer

Source: Funabashi *et al.* [80].

concepts of biodegradability, disintegration (during biological treatment), impact on process of biological treatment and impact on the final compost's quality.

3.6 Biopolymers and Their Potential Application as Edible, Nonedible Packaging, and Coating

3.6.1 Starch

Starch acts a building block of the biopolymer structure and the addition of plasticizer has reported to improve the flexibility and processability of thermoplastic starch [82]. Studies have showed that the biodegradable films were developed and reinforced with wheat and corn husks using starch as a matrix. The incorporation of hulls was found to improve modulus of the starch matrix, its tensile strength, and impact strength [83]. Coating of walnuts with rice starch in combination with chitosan and emulsifier (sodium caseinate) using dipping technology enhanced its shelf life. Cassava starch, used as a coating material for various fruits such as papaya reduced the fruit maturation period at different concentrations (2%, 3% and 4%). The shelf life of white mushrooms was improved by coating with nano-SiO_2—potato starch film and the further the mechanical qualities, water resistance, thermal resistance, and UV aging of the film were enhanced along with preservation [84]. Shelf life of strawberries was extended when stored at 4°C and 85% relative humidity (RH) when coated with yam starch and glycerol [85].

3.6.2 Chitosan

The major sources of chitosan include crustacean shells and exoskeletons. It is a deacetylated derivative of chitin. Chitosan can alternatively be made from the cell walls of some fungus without relying on chemicals that destroys the environment [86]. Chibio Biotech, a Chinese company, makes chitosan from fungi like *Aspergillus* and mushrooms for food and non-food applications. Another firm that makes high purity food grade chitosan from mushrooms is Chinova Bioworks, Canada. Further Meron biopolymers a company from India produces chitosan from shells of crabs and shrimps. Chitosan forms a protective film on the food's surface that resembles cellophane, shielding it from microbiological attack [87]. The chitosan layer acts as a barrier between food and outer environment mainly obstructing the oxygen entry into the food preventing the availability of oxygen to aerobic bacteria [88]. Chitosan in combination with polyvinyl alcohol (PVA) exhibits superior mechanical qualities along with increased water and oxygen barrier capabilities [89]. It was discovered that the chitosan grafted salicylic acid films were effective in preventing chilling injury to cucumbers, preserving their quality, and extending their shelf life [90]. Recently, chitosan layers containing different preservatives were integrated into PET/PP films, and employed as a food packaging material. The films exhibited antibacterial activity against the foodborne bacteria *B. subtilis* and *E. coli* [91]. Biopolymers made using chitosan film and zinc oxide had showed moisture barrier ability and mechanical strength [92]. Tensile strength (TS) and elongation at break (EB) values of chitosan films were similar to those of cellophane, LDPE, and high-density polyethylene (HDPE) [93]. It was shown that the chitosan films coated with silver nanoparticles exhibited increased antibacterial activity against both Gram-negative bacteria, such as *P. aeruginosa* and *E. coli* and Gram-positive bacteria including *S. aureus* and *B. subtilis.* Another study demonstrated an improved antibacterial activity and mechanical strength of chitosan films embedded with silver nanoparticles [94]. Chitosan nanoparticles (CNP) enhanced the mechanical characteristics and water sorption of starch films by acting as a natural filler. The starch/CNP nanocomposite films increased the shelf life of perishable foods by preventing the growth of *S. aureas, B. cereus, E. coli,* and *S. typhimurium* [95]. In another study Chitosan–gelatin edible films extended the shelf life and reduced microbiological deterioration of perishable foods [96].

3.6.3 Pectin

Pectin is a polymer which occurs naturally from plants for cell growth and expansion. Biopolymers can be prepared from pectin plant source to develop bioplastics for the usage of food packaging [97]. Fruits with pectin coatings have a different environment surrounding them, which alters the fruit's internal oxygen levels and prevents physiological deterioration. It has been investigated that adding nanoparticles to pectin-based films can enhance their functional and physical characteristics. Pectin plays a major role in active food packaging, bio composites made of a biopolymer matrix containing metal or metal oxide nanoparticles are becoming more and more significant [97]. Films made of pectin are edible, non-toxic, and have a high degree of biocompatibility. However, combining pectin films with other biopolymers and fillers improves its high-water permeability and poor mechanical strength [98].

Dissolving high ester pectin powder (6% w/v) in distilled water at room temperature for 20 minutes at 500 rpm until fully dissolved, coating dispersion was prepared by combining 60% of pectin solution with 40% of prepared nano emulsions and stirring it for 10 minutes at 500 rpm. On concluding the microbiological quality properties of chilled chicken fillets were enhanced by nano emulsions/pectin coatings made with essential oils of garlic and cinnamon. This resulted in a delay in the appearance of microbial spoilage, which in turn reduced total volatile nitrogen (TVN) and (thiobarbituric acid) TBA values and postponed the deterioration of the texture of the chicken fillets, extending their shelf life by up to 12 days [99].

TiO2 nanoparticles were combined with a solution of lemon waste pectin and sweet potato starch to create biodegradable starch-pectin titanium oxide nanoparticles (TiO2-NPs) blend edible films. The impact of adding varying amounts of TiO2 nanoparticles (0.5–4% wt) to edible films based on starch and pectin resulted in notable reductions in the films' moisture content, solubility, and moisture absorption [100].

3.7 Future Aspects of Biopolymers in Packaging Applications

Plastic packaging materials have been the principal source of food packaging over the years. However, there exists several challenges and limitations associated with plastic usage with respect to environmental pollution. Subsequently, the application of biopolymer for food packaging applications has increased considerably in the recent times. Biopolymers/bioplastics including starch, cellulose, soy protein, PLA and PBAT have got a great attention to produce packaging material because of their environmentally friendly, non-toxic, biodegradability, and renewability. Biopolymers in their native form exhibit poor mechanical and barriers properties, low thermal properties and are expensive compared to the synthetic polymers. These undesirable properties can be over combed by reinforcing it with active agents and nanofillers. Nevertheless, the improved properties are not comparable to the synthetic polymers henceforth there is a need of novel technologies to match the physicochemical, functional, and mechanical properties of synthetic polymers in future. Extensive research should be carried out on the toxicity and allergic responses of plant-based protein biopolymers particularly gluten, soy and casein and carbohydrate polymer such as starch may trigger allergic reactions,

In vitro and *in vivo* experimentation should be planned to address the biosafety concerns. Further research is required to assess the pharmacokinetics of biofilms incorporated with active substances which includes digestion, bio-accessibility, biochemical transformation, cell absorption, excretion and so on. These factors should be considered by the manufacturers for successful commercialization of packaging films made of biopolymers.

3.8 Concluding Remarks

The biopolymers are acquinted mainly to ameliorate the environmental impact of non-degradable polymers derived from fossil fuels. However, the most efficient method of reducing the harmful effects of plastics on the environment is still to follow the basic (3R) reduce, reuse, and recycle strategy. A biodegradable and bioactive material made of lactic acid building blocks is called as polylactic acid, or polylactide (PLA). PHAs are thermoplastic polymers made of linear hydroxyalkanoic acid. Similar to synthetic polymers, polyhydroxyalkanoates are made entirely of biological matter and are produced by microorganisms. Biopolymers are considered to be more environmentally friendly than existing petroleum-based polymers. They come from renewable plant sources or are biodegradable. Nevertheless, because of their high production costs and restricted availability, their use is still quite low even though their properties are similar to those of petroleum-based polymers. Furthermore, a variety of procedures including plantation, irrigation, pest-control, procedures for harvest, separation, fermentation, polymerize, and chemical processing are needed to produce final products when producing biopolymers from plant sources. As a result, the environmental impact of biopolymers is almost the same as that of conventional plastic types that are sold commercially. The production technology of biopolymers requires advanced knowledge, which ultimately limits their use.

References

1. Elferink, M. and Schierhorn, F., Global demand for food is rising. Can we meet it. *Harv. Bus. Rev.*, 7, 04, 2016.
2. Zhang, L., Qin, K., Li, Y., Wu, L., Public-Public Collaboration for Food Safety Risk Management: Essence, Modes, and Identification of Key Influencing Factors Using DANP. *Front. Public Health*, 10, 944090, 2022.

3. Harsulkar, A.A., Sreenivas, S.A., Mandade, R.J., Wakada, R.B., Polymers in mucoadhesive drug delivery system-A review. *Int. J. Drug Dev. Res.*, 2, 3, 61–67, 2011.
4. Ugoeze, K.C., Nwachukwu, N., Nwodo, C.C., Excipient functionality of a novel hydrophilic biopolymer derived from Ipomoea batatas tubers. *Indo Am. J. Pharm. Res.*, 7, 7, 360–368, 2017.
5. Ugoeze, K.C., Amogu, E.O., Oluigbo, K.E., Nwachukwu, N., Environmental and public health impacts of plastic wastes due to healthcare and food products packages: A Review. *J. Environ. Public Health*, 5, 1, 1–31, 2021.
6. Li, S., Ding, F., Flury, M., Wang, Z., Xu, L., Li, S., Wang, J., Macro-and microplastic accumulation in soil after 32 years of plastic film mulching. *Environ. Pollut.*, 300, 118945, 2022.
7. Proshad, R., Kormoker, T., Islam, M.S., Haque, M.A., Rahman, M.M., Mithu, M.M.R., Toxic effects of plastic on human health and environment: A consequences of health risk assessment in Bangladesh. *Int. J. Health*, 6, 1, 1–5, 2018.
8. Rochman, C.M., Hoh, E., Kurobe, T., Teh, S.J., Ingested plastic transfers hazardous chemicals to fish and induces hepatic stress. *Sci. Rep.*, 3, 1, 1–7, 2013.
9. Wilcox, C., Van Sebille, E., Hardesty, B.D., Threat of plastic pollution to seabirds is global, pervasive, and increasing. *Proc. Natl. Acad. Sci.*, 112, 38, 11899–11904, 2015.
10. Banning single-use plastic: lessons and experiences from countries, in: UN Environment Programme report, 2018.
11. Lithner, D., Damberg, J., Dave, G., Larsson, A., Leachates from plastic consumer products–screening for toxicity with Daphnia magna. *Chemosphere*, 74, 9, 1195–1200, 2009.
12. Adane, L. and Muleta, D., Survey on the usage of plastic bags, their disposal and adverse impacts on environment: A case study in Jimma City, Southwestern Ethiopia. *J. Toxicol. Environ. Health Sci.*, 3, 8, 234–248, 2011.
13. Lohmann, R., Correction to critical review of low-density polyethylene's partitioning and diffusion coefficients for trace organic contaminants and implications for its use as a passive sampler. *Environ. Sci. Technol.*, 49, 6, 3985–3985, 2015.
14. Eskenazi, B., Warner, M., Samuels, S., Young, J., Gerthoux, P.M., Needham, L., Mocarelli, .., P. Serum dioxin concentrations and risk of uterine leiomyoma in the Seveso Women's Health Study. *Am. J. Epidemiol.*, 166, 1, 79–87, 2007.
15. Halden, R.U., Plastics and health risks. *Annu. Rev. Public Health*, 31, 179–194, 2010.
16. Heudorf, U., Mersch-Sundermann, V., Angerer, J., Phthalates: toxicology and exposure. *Int. J. Hyg. Environ. Health*, 210, 5, 623–634, 2007.
17. Du, Y., Xu, X., Liu, Q., Bai, L., Hang, K., Wang, D., Identification of organic pollutants with potential ecological and health risks in aquatic environments: Progress and challenges. *Sci. Total Environ.*, 806, 150691, 2022.

18. Verma, R., Vinoda, K.S., Papireddy, M., Gowda, A.N.S., Toxic pollutants from plastic waste-a review. *Procedia Environ. Sci.*, 35, 701–708, 2016.
19. Holme, J.A., Brinchmann, B.C., Refsnes, M., Låg, M., Ovrevik, J., Potential role of polycyclic aromatic hydrocarbons as mediators of cardiovascular effects from combustion particles. *Environmen. Health*, 18, 1–18, 2019.
20. Ribas-Fito, N., Sala, M., Kogevinas, M., Sunyer, J., Polychlorinated biphenyls (PCBs) and neurological development in children: a systematic review. *J. Epidemiol Community Health*, 55, 8, 537, 2001.
21. Cohen, J.T., Carlson, G., Charnley, G., Coggon, D., Delzell, E., Graham, J.D., Thompson, .., K. A comprehensive evaluation of the potential health risks associated with occupational and environmental exposure to styrene. *J. Toxicol. Environ. Health Part B*, 5, 1-2, 1–263, 2002.
22. Wang, P.W., Huang, Y.F., Wang, C.H., Fang, L.J., Chen, M.L., Prenatal to preschool exposure of nonylphenol and bisphenol A exposure and neurodevelopment in young children, *Pediatr. Neonatol.*, 65, 1, 76–84, 2023.
23. Karleskint, G., Turner, R., Small, J., *Introduction to marine biology*, Cengage Learning, Boston, Massachusetts, United States, 2012.
24. Alabi, O.A., Ologbonjaye, K., I., Awosolu, O., Alalade, O.E., Public and environmental health effects of plastic wastes disposal: a review. *J. Toxicol. Risk Assess*, 5, 021, 1–13, 2019.
25. Lewandowski, K. and Skórczewska, K., A brief review of poly (vinyl chloride) (PVC) recycling. *Polym.*, 14, 15, 3035, 2022.
26. Kudzin, M.H., Piwowarska, D., Festinger, N., Chruściel, J.J., Risks Associated with the Presence of Polyvinyl Chloride in the Environment and Methods for Its Disposal and Utilization. *Mater.*, 17, 1, 173, 2023.
27. Atiwesh, G., Mikhael, A., Parrish, C.C., Banoub, J., Le, T.A.T., Environmental impact of bioplastic use: A review. *Heliyon*, 7, 9, 2422, 2021.
28. Hassan, M.E.S., Bai, J., Dou, D.Q., Biopolymers; definition, classification, and applications. *Egypt. J. Chem.*, 62, 9, 1725–1737, 2019.
29. Ranganadhareddy, A., Production of Polyhydroxyalkanoates from Microalgae-A Review. *J. Biochem. Technol.*, 13, 2, 1–6, 2022.
30. Weber, C.J., Haugaard, V., Festersen, R., Bertelsen, G., Production and applications of biobased packaging materials for the food industry. *Food Addit. Contam.*, 19, S1, 172–177, 2002.
31. Adeyeye, S.A.O., Babu, A.S., Guruprasath, N., Ganesh, P.S., Starch Nanocrystal and its Food Packaging Applications. *Curr. Res. Nutr. Food Sci.*, 11, 1, 01–21, 2023.
32. Mangaraj, S., Mohanty, S., Swain, S., Yadav, A., Development and characterization of commercial biodegradable film from PLA and corn starch for fresh produce packaging. *J. Package Technol. Res.*, 3, 127–140, 2019.
33. Carvalho, A.J.F. Starch: Major sources, properties and applications as thermoplastic materials. In Monomers, Polymers and Composites from Renewable Resources; Elsevier: Amsterdam, The Netherlands, pp. 321–342, 2008.

34. Medina Jaramillo, C., Gonzalez Seligra, P., Goyanes, S., Bernal, C., Famá, L., Biofilms based on cassava starch containing extract of yerba mate as antioxidant and plasticizer. *Starch-Stärke*, 67, 9-10, 780–789, 2015.
35. Gaspar, M., Benkő, Z., Dogossy, G., Reczey, K., Czigany, T., Reducing water absorption in compostable starch-based plastics. *Polym. Degrad. Stab.*, 90, 3, 563–569, 2005.
36. Jiugao, Y., Ning, W., Xiaofei, M., The effects of citric acid on the properties of thermoplastic starch plasticized by glycerol. *Starch-Stärke*, 57, 10, 494–504, 2005.
37. Yaradoddi, J.S., Banapurmath, N.R., Ganachari, S.V., Soudagar, M.E.M., Mubarak, N.M., Hallad, S., Fayaz, H., Biodegradable carboxymethyl cellulose based material for sustainable packaging application. *Sci. Rep.*, 10, 1, 21960, 2020.
38. Byun, Y., Kim, Y.T., In: Innovations in Food Packaging, Hn J.H., ed. 2nd edition, Academic Press, San Diego, 369–390, 2014.
39. Jabeen, N., Majid, I., Nayik, G.A., Bioplastics and food packaging: A review. *Cogent Food Agric.*, 1, 1, 1117749, 2015.
40. Mohamed, M.H., Udoetok, I.A., Wilson, L.D., Animal biopolymer-plant biomass composites: Synergism and improved sorption efficiency. *J. Compos. Sci.*, 4, 1, 15, 2020.
41. Pushp, P. and Gupta, M.K., Cardiac tissue engineering: A role for natural biomaterials, in: *Bioactive Natural Products for Pharmaceutical Applications*, pp. 617–641, 2021.
42. Fahmy, T., Elhendawi, H., Elsharkawy, W.B., Reicha, F.M., AC conductivity and dielectric relaxation of chitosan/poly (vinyl alcohol) biopolymer polyblend. *Bull. Mater. Sci.*, 43, 1–10, 2020.
43. Shams, B., Ebrahimi, N.G., Khodaiyan, F., Development of antibacterial nanocomposite: whey protein-gelatin-nanoclay films with orange peel extract and tripolyphosphate as potential food packaging. *Adv. Polym. Technol.*, 4, 1–9, 2019.
44. Wu, C., Zhu, Y., Wu, T., Wang, L., Yuan, Y., I., Chen, J., Pang, J., Enhanced functional properties of biopolymer film incorporated with curcurmin-loaded mesoporous silica nanoparticles for food packaging. *Food Chem.*, 288, 139–145, 2019.
45. Bhatia, S., Al-Harrasi, A., Al-Azri, M.S., Ullah, S., Makeen, H.A., Meraya, A.M., Anwer, M.K., Gallic Acid Crosslinked Gelatin and Casein Based Composite Films for Food Packaging Applications. *Polym.*, 14, 19, 4065, 2022.
46. Zubair, M. and Ullah, A., Recent advances in protein derived bionanocomposites for food packaging applications. *Crit. Rev. Food Sci. Nutr.*, 60, 3, 406–434, 2020.
47. Kasaai, M.R., Zein and zein-based nano-materials for food and nutrition applications: A review. *Trends Food Sci. Technol.*, *79*, 184–197, 2018.

48. Lu, Y., Luo, Q., Chu, Y., Tao, N., Deng, S., Wang, L., Li, L., Application of gelatin in food packaging: A review. *Polym.*, 14, 3, 436, 2022.
49. Kaseem, M., Rehman, Z.U., Hossain, S., Singh, A.K., Dikici, B., A review on synthesis, properties, and applications of polylactic acid/silica composites. *Polym.*, 13, 18, 3036, 2021, https://doi.org/10.3390/polym13183036.
50. Ubeda, Aznar, M., Nerín, C., Determination of volatile compounds and their sensory impact in a biopolymer based on polylactic acid (PLA) and polyester. *Food Chem.*, 294, 171–178, 2019, https://doi.org/10.1016/j.foodchem.2019.05.069.
51. Mehmood, A., Raina, N., Phakeenuya, V., Wonganu, B., Cheenkachorn, K., The current status and market trend of polylactic acid as biopolymer: Awareness and needs for sustainable development. *Mater. Today: Proc.*, 72, 6, 3049–3055, 2023, https://doi.org/10.1016/j.matpr.2022.08.387.
52. Aydemir, D. and Gardner, D.J., Biopolymer blends of polyhydroxybutyrate and polylactic acid reinforced with cellulose nanofibrils. *Carbohydr. Polym.*, 250, 116867, 2020, https://doi.org/10.1016/J.CARBPOL.2020.116867.
53. Kavitha, G., Rengasamy, R., Inbakandan, D., Polyhydroxybutyrate production from marine source and its application, International. *J. Biol. Macromol.*, 111, 108, 2018, https://doi.org/10.1016/j.ijbiomac.2017.12.155.
54. Jariyasakoolroj, P., Leelaphiwat, P., Harnkarnsujarit, N., Advances in research and development of bioplastic for food packaging. *J. Sci. Food Agric.*, 100, 14, 5032–5045, 2020, https://doi.org/10.1002/jsfa.9497.
55. Yadav, A., Mangaraj, S., Singh, R., Das, K., Kumar, N., Arora, S., Biopolymers as packaging material in food and allied industry. *Int. J. Chem. Stud.* 6, 2, 2411–2418, 2018.
56. Popović, S.Z., Lazić, V.L., Hromiš, N.M., Šuput, D.Z., Bulut, S.N., Biopolymer Packaging Materials for Food Shelf-Life Prolongation, in: *Biopolymers for Food Design*, pp. 223–277, Elsevier Inc., Cambridge, Massachusetts, United States, 2018, https://doi.org/10.1016/B978-0-12-811449-0.00008-6.
57. Perera, K.Y., Jaiswal, A.K., Jaiswal, S., Biopolymer-Based Sustainable Food Packaging Materials: Challenges, Solutions, and Applications, *Foods*, 12, 12, 2422, MDPI, Basel, Switzerland, 2023, https://doi.org/10.3390/foods12122422.
58. Khan, R., Jolly, R., Fatima, T., Shakir, M., Extraction processes for deriving cellulose: A comprehensive review on green approaches, *Polym. Adv. Technol.*, 33, 7, 2069–2090, John Wiley and Sons Ltd, Hoboken, New Jersey, United States, 2022, https://doi.org/10.1002/pat.5678.
59. Priyadarshi, R. and Rhim, J.W., Chitosan-based biodegradable functional films for food packaging applications. *Innov. Food Sci. Emerg. Technol.*, 62, 2020.
60. Agarwal, A., Shaida, B., Rastogi, M., Singh, N.B., Food Packaging Materials with Special Reference to Biopolymers-Properties and Applications, *Chem. Afr.*, 6, 1, 117–144, Springer Science and Business Media Deutschland GmbH, Berlin, Germany, 2023, https://doi.org/10.1007/s42250-022-00446-w.

61. Ali, G., Sharma, M., Salama, E.S., Ling, Z., Li, X., Applications of chitin and chitosan as natural biopolymer: potential sources, pretreatments, and degradation pathways, *Biomass Convers. Biorefin.*, Springer Science and Business Media Deutschland GmbH, Berlin, Germany, 2022, https://doi.org/10.1007/s13399-022-02684-x.
62. Hameed, A.Z., Raj, S.A., Kandasamy, J., Baghdadi, M.A., Shahzad, M.A., Chitosan: A Sustainable Material for Multifarious Applications. *Polym.*, 14, 2335, 2022, https://doi.org/10.3390/polym14122335.
63. Paluch, M., Ostrowska, J., Tyński, P., Sadurski, W., Konkol, M., Structural and Thermal Properties of Starch Plasticized with Glycerol/Urea Mixture. *J. Polym. Environ.*, 30, 2, 728–740, 2022, https://doi.org/10.1007/s10924-021-02235-x.
64. Nainggolan, I., Nasution, T., I., Ahmad, K.R., *Thermal Properties of Chitosan-roselle Films.* scitepress.org, Setubal, Portugal, 2020.
65. Staroszczyk, H., Pielichowska, J., Sztuka, K., Stangret, J., Kołodziejska, I., Molecular and structural characteristics of cod gelatin films modified with EDC and TGase. *Food Chem.*, 130, 2, 335–343, 2012, https://doi.org/10.1016/j.foodchem.2011.07.047.
66. Jamal, T., Sapuan, S., Abdan, K., Harussani, M.M., Sherwani, S., Mechanical properties of arrowroot starch-based biopolymers plasticized with glycerol. *Sci. Rep.*, 11, 303–305, 2021.
67. Gadgey, K.K. and Sharma, G.S., Investigation of mechanical properties of chitosan based films prepared from narmada riverside crab shells. *Int. J. Mech. Eng.*, 11, 6, 2020, https://doi.org/10.34218/ijmet.11.6.2020.003.
68. van Rooyen, B., de Wit, M., Osthoff, G., van Niekerk, J., Hugo, A., Effect of pH on the Mechanical Properties of Single-Biopolymer Mucilage (Opuntia ficus-indica), Pectin and Alginate Films: Development and Mechanical Characterisation. *Polymers*, 15, 24, 4640, 2023, https://doi.org/10.3390/polym15244640.
69. Shapii, R.A., Othman, S.H., Basha, R.K., Naim, M.N., Mechanical, thermal, and barrier properties of starch films incorporated with chitosan nanoparticles. *Nanotechnol. Rev.*, 11, 1, 1464–1477, 2022, https://doi.org/10.1515/ntrev-2022-0094.
70. Franco-Bacca, A.P., Cervantes-Alvarez, F., Macías, J.D., Castro-Betancur, J.A., Pérez-Blanco, R.J., Giraldo Osorio, O.H., Arias Duque, N.P., Rodríguez-Gattorno, G., Alvarado-Gil, J.J., Heat transfer in cassava starch biopolymers: Effect of the addition of borax. *Polym.*, 13, 23, 4106, 2021, https://doi.org/10.3390/polym13234106.
71. Grząbka-Zasadzińska, A., Amietszajew, T., Borysiak, S., Thermal and mechanical properties of chitosan nanocomposites with cellulose modified in ionic liquids. *J. Therm. Anal.*, 130, 1, 143–154, 2017, https://doi.org/10.1007/s10973-017-6295-3.
72. Sanchez-Garcia, M.D., Lopez-Rubio, A., Lagaron, J.M., Natural micro and nanobiocomposites with enhanced barrier properties and novel

functionalities for food biopackaging applications. *Trends Food Sci. Technol.*, 21, 528–536, 2010, https://doi.org/10.1016/j.tifs.2010.07.008.

73. Mangaraj, S., Yadav, A., Bal, L.M., Dash, S.K., Mahanti, N.K., Application of Biodegradable Polymers in Food Packaging Industry: A Comprehensive Review. *J. Package Technol. Res.*, 3, 1, 77–96, 2019, https://doi.org/10.1007/s41783-018-0049-y.
74. Aris, Z.F.M., Bavishi, V., Sharma, R., Nagarajan, R., Barrier properties and abrasion resistance of biopolymer-based coatings on biodegradable poly(-lactic acid) films. *Polym. Eng. Sci.*, 59, 9, 1874–1881, 2019, https://doi.org/10.1002/pen.25187.
75. Amidi-Fazli, F. and Amidi-Fazli, N., Barrier properties of starch-ethylene vinyl alcohol nanocomposites. *Int. J. Food Eng.*, 9, 2, 249–252, 2015.
76. Oliveira, V.R.L., Monteiro, M.K.S., Santos, F.K.G., Leite, R.H.L., Aroucha, E.M.M., Effect of drying temperature in biopolymeric films of cassava starch and its effect on wettability, water vapor barrier and mechanical properties, *Mater. Sci. Forum*, 930, 270–275, Trans Tech Publications Ltd, Seestrasse 24c, CH-8806 Baech, Switzerland, 2018.
77. Sander, M., Weber, M., Lott, C., Zumstein, M., Künkel, A., Battagliarin, G., *Polymer Biodegradability 2.0: A Holistic View on Polymer Biodegradation in Natural and Engineered Environments*, pp. 65–110, 2023, https://doi.org/10.1007/12_2023_163.
78. Baidurah, S., Methods of Analyses for Biodegradable Polymers: A Review. *Polym.*, 14, MDPI, 2022, https://doi.org/10.3390/polym14224928.
79. Van Der Zee, M., Methods for evaluating the biodegradability of environmentally degradable polymers, in: *Handbook of Biodegradable Polymers*, Springer, Cham, Switzerland, pp. 1–21, 2020, https://doi.org/10.1515/9781501511967-001.
80. Pires, J.R.A., Souza, V.G.L., Fuciños, P., Pastrana, L., Fernando, A.L., Methodologies to Assess the Biodegradability of Bio-Based Polymers—Current Knowledge and Existing Gaps. *Polym.*, 14, 1359, 2022, https://doi.org/10.3390/polym14071359.
81. Funabashi, M., Ninomiya, F., Kunioka, M., Biodegradability evaluation of polymers by ISO 14855-2. *Int. J. Mol. Sci.*, 10, 3635–3654, 2009, https://doi.org/10.3390/ijms10083635.
82. Zakuan, N.D., Jumaidin, R., Selamat, M.Z., dan Pembuatan, F.T.K.M., Biopolymer from starch. *Proc. Inst. Mech. Eng.*, 13, 323–324, 2020.
83. Gadhave, R.V., Das, A., Mahanwar, P.A., Gadekar, P.T., Starch Based Bio-Plastics: The Future of Sustainable Packaging. *Open J. Polym. Chem.*, 08, 02, 21–33, 2018, https://doi.org/10.4236/ojpchem.2018.82003.
84. Zhang, R., Wang, X., Cheng, M., Preparation and characterization of potato starch film with various size of Nano-SiO2. *Polymers*, 10, 10, 2018, https://doi.org/10.3390/POLYM10101172.
85. García-Guzmán, L., Cabrera-Barjas, G., Soria-Hernández, C.G., Castaño, J., Guadarrama-Lezama, A.Y., Rodríguez Llamazares, S., Progress in

Starch-Based Materials for Food Packaging Applications. *Polysaccharides*, 3, 1, 136–177, Multidisciplinary Digital Publishing Institute (MDPI), 2022, https://doi.org/10.3390/polysaccharides3010007.
86. Sun, X., Wang, Z., Kadouh, H., Zhou, K., The antimicrobial, mechanical, physical and structural properties of chitosan-gallic acid films. *LWT Food Sci. Technol.*, 57, 1, 83–89, 2014.
87. Kumari, P., Barman, K., Patel, V.B., Siddiqui, M.W., Kole, B., Reducing post-harvest pericarp browning and preserving health promoting compounds of litchi fruit by combination treatment of salicylic acid and chitosan. *Sci. Hortic.*, 197, 555–563, 2015.
88. Dotto, G.L., Vieira, M.L.G., Pinto, L.A.A., Use of chitosan solutions for the microbiological shelf life extension of papaya fruits during storage at room temperature. *LWT Food Sci. Technol.*, 64, 126–130, 2015.
89. Giannakas, A., Vlacha, M., Salmas, C., Leontiou, A., Katapodis, P., Stamatis, H., Barkoula, N.M., Ladavos, A., Preparation, characterization, mechanical, barrier and antimicrobial properties of chitosan/PVOH/clay nanocomposites. *Carbohydr. Polym.*, 140, 408–415, 2016, https://doi.org/10.1016/j.carbpol.2015.12.072.
90. Zhang, Y., Zhang, M., Yang, H., Postharvest chitosan-g-salicylic acid application alleviates chilling injury and preserves cucumber fruit quality during cold storage. *Food Chem.*, 174, 558–563, 2015.
91. Lei, J., Yang, L., Zhan, Y., Wang, Y., Ye, T., Li, Y., Li, Plasma treated polyethylene terephthalate/polypropylene films assembled with chitosan and various preservatives for antimicrobial food packaging. *Colloids Surf., B*, 114, 60–66, 2014.
92. Tripathi, S., Mehrotra, G.K., Dutta, P.K., Chitosan-Silver Oxide Nanocomposite Film: Preparation and Antimicrobial Activity. *Bull. Mater. Sci.*, 34, 29–35, 2011, https://doi.org/10.1007/s12034-011-0032-5.
93. Butler, B.L., Vergano, P.J., Testin, R.F., Bunn, J.M., Wiles, J.L., Mechanical and barrier properties of edible chitosan films as affected by composition and storage. *J. Food Sci.*, 61, 5, 953–956, 1996.
94. Vimala, K., Yallapu, M.M., Varaprasad, K., Reddy, N.N., Ravindra, S., Naidu, N.S., Raju, K.M., Fabrication of curcumin encapsulated chitosan-PVA silver nanocomposite films for improved antimicrobial activity. *J. Biomater. Nanobiotechnol.*, 2, 01, 55, 2011.
95. Shapi'i, R.A., Othman, S.H., Nordin, N., Basha, R.K., Naim, M.N., Antimicrobial properties of starch films incorporated with chitosan nanoparticles: *In vitro* and *in vivo* evaluation. *Carbohydr. Polym.*, 230, 115602, 2020.
96. Poverenov, E., Rutenberg, R., Danino, S., Horev, B., Rodov, V., Gelatin-chitosan composite films and edible coatings to enhance the quality of food products: Layer-by-layer vs. blended formulations. *Food Bioproc. Tech.*, 7, 3319–3327, 2014.
97. Mellinas, C., Ramos, M., Jiménez, A., Garrigós, M.C., Recent trends in the use of pectin from agro-waste residues as a natural-based biopolymer for

food packaging applications. *Mater.*, 13, 673, 2020, https://doi.org/10.3390/ma13030673.

98. Prakash, O., Swami, S., Shiran, K., Mansuri, S.M., Kaukab, S., *Possible improvement of Biopolymer based Packaging Film*, 2020.
99. Abdou, E.S., Galhoum, G.F., Mohamed, E.N., Curcumin loaded nanoemulsions /pectin coatings for refrigerated chicken fillets. *Food Hydrocoll.*, 83, 445–453, 2018, https://doi.org/10.1016/j.foodhyd.2018.05.026.
100. Dash, K.K., Ali, N.A., Das, D., Mohanta, D., Thorough evaluation of sweet potato starch and lemon-waste pectin based-edible films with nano-titania inclusions for food packaging applications. *Int. J. Biol. Macromol.*, 139, 449–458, 2019, https://doi.org/10.1016/j.ijbiomac.2019.07.193.

4

Active and Intelligent Antimicrobial Coating Systems

Dineshkumar T., Sakthivel Sangeetha, Pragalyaashree M. M.* and Freeda Blessie R.

Division of Food Processing Technology, School of Agricultural Sciences, Karunya Institute of Technology and Sciences, Coimbatore, Tamil Nadu, India

Abstract

In today's world, the effective control of microbial contamination is of utmost importance across various industries. Active and intelligent antimicrobial coating systems have emerged as ground-breaking technologies that actively combat the growth and spread of microorganisms while incorporating intelligent features for real-time monitoring and adaptive responses. These advanced surface treatments have far-reaching implications, revolutionizing food preservation, healthcare practices, consumer product safety, and more. The significance of active and intelligent antimicrobial coating systems lies in their ability to address critical challenges associated with microbial contamination, food safety, and healthcare-associated infections. Microbial contamination poses significant risks to public health, leading to foodborne illnesses and healthcare-associated infections. By employing antimicrobial coatings, these risks can be mitigated, enhancing public health and safety. In the food industry, active and intelligent antimicrobial coatings play a vital role in reducing food spoilage and waste. They extend the shelf life of perishable foods by inhibiting microbial growth and preserving product quality. Additionally, these coatings help maintain the sensory attributes, taste, and appearance of food products, ensuring higher quality for consumers. In healthcare settings, antimicrobial coatings on medical devices, equipment, and surfaces contribute to infection control efforts. By reducing the transmission of pathogens, these coatings help minimize the occurrence of healthcare-associated infections, which can be life-threatening for vulnerable patients. The active and intelligent antimicrobial coating systems utilize a range of techniques, including physical deposition, chemical immobilization, sol-gel coating, electrochemical

**Corresponding author*: pragalyaashree@karunya.edu

Mousumi Sen (ed.) Food Coatings and Preservation Technologies, (85–126) © 2025 Scrivener Publishing LLC

deposition, plasma treatment, nanotechnology, encapsulation, photochemical coating, and smart polymers. Each of these methods offers unique features and benefits, depending on the specific application and desired performance of the antimicrobial coating. The selection of the appropriate method depends on factors such as the target microorganisms, the substrate material, the intended use, and the desired lifespan of the antimicrobial effect. In the food industry, these coatings are applied to food contact surfaces, packaging materials, and processed foods. They prevent microbial contamination, extend shelf life, and ensure food safety, reducing foodborne illnesses and food waste. The integration of intelligent features allows real-time monitoring of storage conditions, ensuring optimal preservation. In addition, the use of intelligent coatings with biofilm detection capabilities enhances infection control efforts in hospital settings. In consumer products, antimicrobial coatings are used in everyday items like utensils, textiles, and personal care products. They provide an added layer of protection, promoting hygiene and reducing microbial contamination. Consumers benefit from safer and more hygienic products in their daily lives. As technology continues to advance, the future of active and intelligent antimicrobial coating systems is promising. Ongoing research focuses on optimizing coating effectiveness, expanding the range of antimicrobial agents used, and enhancing intelligent features for greater adaptability. The development of eco-friendly and sustainable materials for these coatings aligns with the growing consumer demand for environmentally conscious products. Furthermore, advancements in nanotechnology may lead to even more efficient antimicrobial coatings with increased surface area and targeted action. In the healthcare industry, active and intelligent antimicrobial coatings are expected to play a pivotal role in infection control, supporting patient safety, and reducing healthcare expenses associated with treating infections. Overall, the active and intelligent antimicrobial coating systems have ushered in a new era of microbial control, offering versatile solutions with profound implications across various industries. By actively inhibiting microbial growth, providing controlled release of antimicrobial agents, and incorporating intelligent features, these coatings empower industries to enhance food safety, improve infection control practices, and promote sustainability. With their multifaceted benefits and ongoing research and development, active and intelligent antimicrobial coating systems hold immense potential for a healthier, safer, and more sustainable future. As their adoption continues to grow, these coatings will undoubtedly remain at the forefront of microbial control strategies, contributing to a world where microbial contamination is effectively managed, and public health is safeguarded.

***Keywords*:** Real-time monitoring, immobilization, smart polymers, oxygen scavenging, foodborne illnesses, food safety, sustainability

4.1 Introduction

In this modern era, the prevalence of foodborne illnesses has become a pressing issue that troubles not only consumers, but also the food safety authorities and food industry. As people strive to find nourishment that is both wholesome and beneficial to their well-being, the demand for natural and healthy food has surged. This has led to a growing need for safe, high-quality food that is minimally processed, conveniently prepared, and readily available for consumption [1]. The convergence of emerging social trends and advancements in food processing techniques has fuelled a growing desire for food products that have an extended shelf life. In response to this demand, there has been a significant focus on the development of innovative packaging materials that offer enhanced preservation capabilities. These materials include active, intelligent, and even edible systems, which play a crucial role in ensuring the longevity and quality of food products [2]. Food products, being inherently perishable, are susceptible to degradation from various environmental factors, including bacterial and fungal contamination. Consequently, processed food necessitates safeguarding against spoilage throughout its storage, handling, as well as dissemination processes to enhance its product longevity as well as preserve characteristics. It is crucial to note that certain microorganisms, particularly pathogens, have the potential to pose significant health risks to consumers, particularly when food is mishandled or distributed under improper conditions. Therefore, ensuring proper handling and storage practices is essential to protect consumer well-being [3].

Microorganisms commonly associated with spoilage and pathogenic risks in packaged foods comprise *Salmonella, Bacillus cereus, Campylobacter, Clostridium perfringens, Aspergillus niger, Saccharomyces cerevisiae*, and *Escherichia coli 0157*. These microorganisms are influenced by a range of extrinsic factors, related to storage conditions, such as, time, temperature and relative humidity as well as intrinsic factors, including oxygen and pH [1]. In addition, unfavourable reactions may transpire to alter the fresh food's flavor, texture, color, and odor. Although the food industry has long employed conventional preservation methods like acidification or salting, heat treatment or to inhibit the proliferation of harmful microorganisms in food, these methods frequently lead to undesirable reductions in their nutritional attributes. Nevertheless, recent advancements in preservation methods have shown to be greater efficacy in safeguarding food while preserving nutritional attributes and its sensory characteristics [3]. Likewise, the safety of artificial additives used in food has also drawn the attention of consumers. In order to increase food quality and shelf life, this has consequently led to the discovery of natural antimicrobials that can prevent the growth of bacteria and fungi in food products. it is driving up demand for natural products as substitutes for preservatives in food. The sources of these natural antimicrobials are diverse and include bacteria, fungi, algae, plants, and animals [4].

Consequently, in order to impede the proliferation of unwanted microorganisms in food, antimicrobials can be introduced directly into the formulation of the product, applied as a coating on its integrated, or surface into the packaging material [1]. This is especially relevant in coating as well as film for antimicrobial system, which has piqued a curiosity in food sector as well as research community. These materials have promptly to derived on organic matter as well as undergo simple processing on making coating by utilization of films, allowing regulator over oxidation processes, moisture permeation also gas exchange on targeted products, along with mitigating microbial growth. Additionally, these biopolymer films have the capacity to accommodate nutrients as well as additives, gradually dispensing on food, acting as a fundamental basis on packaging [5]. Through the integration of antimicrobial additives, these edible coatings can expand their capabilities, effectively protecting food items against microbial spoilage. This process aids in extending the shelf life of the products while enhancing safety [6].

4.2 Antimicrobial Agents

Generally, diverse antimicrobial agents can be integrated into packaging systems, encompassing mineral acids, isothiocyanates, phenolic compounds, organic acids, and inorganic compounds. These agents are classified as natural or synthetic. Despite regulatory approval for chemical antimicrobial agents like propionates sodium benzoates, nitrites, chlorides, sorbic acid, potassium sorbates, sulfites, nisin, triclosan, tartaric acid and others, many of these compounds continue to pose potential nutritional or health risks to consumers. Consequently, natural antimicrobial agents are gaining prominence and attention due to their potential to extend food shelf life and effectively control or prevent the growth of microorganisms [7].

4.2.1 Natural Antimicrobial Agents for Food Preservation

The natural based antimicrobial substance has been explored by the researchers as well as the potential substitutes for chemical ones in order to preserve the quality of food. These natural compounds have demonstrated their effectiveness in combating foodborne pathogens and food spoilage. Natural antimicrobial agents can be categorized into various groups based on their source of production. These groups primarily consist of phytochemicals, like procurement of plants material as well as essential oils, antimicrobial agents obtained from animals, antimicrobial substances derived from antagonistic microorganisms and bacterial cell metabolism [8].

a) **Plant Derived Compounds:** The bounty of natural plant-derived compounds can be accessed from a variety of sources, including vegetables, fruits, leaves, seeds, as well as spices and herbs. In olden days, essential oils (EOs) and plant-based extracts have been added to food to improve flavor and extend shelf life by reducing microbial contamination and preventing rancidity. Actually, these compounds have the ability to decrease or inhibit the growth of pathogenic microorganisms because of higher levels in its bioactive compounds, which are primarily isoflavones, phenolic compounds, ketonic compounds, aldehydes, terpenes, acidic compounds, and aliphatic alcohols. The effectiveness of plant-derived compounds as antimicrobials relies significantly on factors such as the, size of the inoculum, type of microorganism, extraction method, culture medium and the approach used to determine antimicrobial activity [4]. Depending on their chemical formulation, plant extracts and essential oils (EOs) can exhibit antimicrobial effects through various mechanisms of action. These mechanisms encompass, disruption of the cytoplasmic membrane, alterations in cell membrane permeability, phospholipid composition, modifications in fatty acid and release of cellular components, changes in RNA and DNA synthesis, and interference with protein translocation.

   Over recent decades, numerous studies have investigated the prospective application used in organic compounds along with different extract from plants normally act as natural substance of antimicrobial system in a wide range of foods [9].

b) **Animal Derived Compounds:** The animal derived antimicrobial substance from nature source are protein and enzyme found in egg and milks include lactoferrin, lactoperoxidase, lysozyme as well as certain polysaccharides like chitosan (CH), primarily extracted from crustaceans and shrimp shells. Lysozyme has been commercially integrated into semi hard fromage for combat secondary fermentation due to Clostridium tyrobutyricum [10]. Various methods have been suggested to enhance lysozyme's antimicrobial potency contrast to anti-Gram-negative bacterial, which has aiming to expand their uses in food sectors. These different strategies are employed to alter or modify or proteins or other biomolecules for various purposes, such as improving solubility, enhancing stability, modifying their biological activity or increasing bioavailability [11]. The lactoperoxidase system (LPS), abundant in bovine milk, demonstrates substantial multidirectional antimicrobial effectiveness in contradiction of various pathogenic microorganisms in food like *Staphylococcus aureus, Salmonella, Campylobacter* as well as *Listeria monocytogenes.* Activation of LPS has been widely employed to prevent the degradation of fresh milk via combating Unwanted microbial presence throughout procurement, processing, as well as transportation. Specifically, the areas lacking adequate refrigeration.

Lactoferrin, an antimicrobial glycoprotein, exhibits effectiveness against various foodborne microorganisms including *E. coli, Carnobacterium, Klebsiella,* and *L. monocytogenes* [12]. Apart from peptides, chitosan extracted from insects, arachnids, and decapods exoskeletons displays a broad-spectrum antibacterial activity against a range of bacteria based on Gram staining (*Shigella dysenteriae, Listeria monocytogenes, Bacillus cereus, Escherichia coli, Staphylococcus aureus* and *Salmonella),* including disease-causing fungi *Alternaria species, Aspergillus flavus, Cladosporium species* and *Penicillium species.* CH's antimicrobial effect varies based on factors such as molecular composition, mass, microbial variety, acidity, as well as deacetylation level [13]. The CH exerts its antimicrobial action normally happen due to their low molecular weight, may penetrate the bacterial cell wall after interaction. They are primarily originating in the exterior barrier of endotoxins. These polymeric cations might then associate with the DNA compound carrying a negative charge, inhibiting transcription on DNA, affecting RNA as well as protein synthesis, and accordingly impeding enzymes facilitating the proliferation of the intended microbial species. Despite its efficacy against numerous foodborne microorganisms, CH's mainly preserving the food quality remains imposed by the substance's dissolvability characteristics in specific conditions of pH [8].

c) **Microbial Derived Compounds:** Natural preservatives extracted from proposed use of beneficial microorganisms and their antimicrobial by-products to hinder or suppress the growth of harmful microorganisms in food systems. This serves to enrich the food security as well as extend its lifespan. Lactic acid bacteria are commonly recognized as preserving microflora among microbial antagonists [14]. LAB (Lactic acid bacteria) is not only generally recognized as safe (GRAS) but also exhibit the ability to compete with deteriorating food due to nutrient conflict or the secretion on microbial inhibitors. The survival of LAB during food storage is primarily influenced by factors such as the characteristics of the food composition as well as ambient factors, including ionic strength, thermal intensity and pH [15]. While research on protective cultures has predominantly concentrated to Lactic acid bacteria and its microbial peptides. Various studies are highlighted positive indications of yeast as beneficial organisms. Yeasts possess a significant suppressive potential as well as rapidly colonize fruit wounds, making them effective in controlling microbiological spoilage and maintaining postharvest qualities in vegetable and fruit products [16].

4.2.2 Methods of Coating

a) **Dipping:** While considered one of the oldest commercial methods, the dipping technique remains pertinent today. It involves immersing fresh food produce into a coating solution to ensure thorough saturation of the food surface.

The excess coating is then removed from the food surface by draining out the coating solution [17]. The parameters (viscosity, surface tension, density) of the coating solution determine the duration of exposure/contact, which typically ranges from 5 to 3 minutes. After that, the coated sample is left alone so that the coating can dry. This technique achieves a consistent coating layer over the food sample. Because prolonged exposure causes more moisture to be adsorbed, the duration of the contact is important. The food is unevenly coated as a result of the shorter duration of contact. Higher viscosity coating solutions are typically applied solely by dipping.

Prior to coating, the coating solution is mixed with the relevant additives, such as antioxidants or antimicrobials, in the proper ratios. Sometimes the coating is applied as foam, which when the food is rolled, leaves a uniform layer on the surface [18].

b) **Layer by Layer Method:** The layer-by-layer technique allows for more precise control over the functionality and properties of the coating by alternating the deposition of oppositely charged polyelectrolytes. Several layers of the films are produced using this method, which may enhance the edible coating's effectiveness [19].

c) **Vacuum Impregnation Technique:** The dipping method has been further improved by the vacuum impregnation technique. The environment of a vacuum during dipping makes a difference. That is, the fresh food is immersed in an airtight vacuum application rather than being dipped in a typical dipping tank. While submerged in the coating solution at atmospheric pressure, the food material is exposed to atmospheric restoration [19].

d) **Spraying Method:** This method involves using an atomizer to spray the coating solution over the food sample. A higher pressure causes the solution to spray onto the food sample, covering it in tiny droplets that range in size from mm to nm. High pressure spraying of less viscous coating solutions is more appropriate for the spraying method. The formation of polymeric coating through the use of a spraying system is influenced by temperature and drying time. A fluid's rheological characteristics, such as surface temperature, viscosity, and temperature, shows the effect of spray coating [20]. As a result, a coating mixture that has the right fluid rheology is prepared and forced through the atomizer's gun. Once the sample has been sprayed, it must be given time to dry. The efficiency of coating is determined at this stage by the drying technique, duration, and temperature. Furthermore, a major factor in spraying is the type and design of nozzle. Air-assisted airless atomization, pressure atomization, and air spray atomization are the three spraying techniques that have been successfully implemented in industries. The application of the spraying technique has the benefit of increasing the liquid coating's surface area through droplet formation [17].

e) **Foaming and Dripping Method:** Traditional coating methods include the foaming and dripping techniques. These techniques are currently becoming less and less common among academics and professionals working in the food industry. When using the dripping technique, brushes are used to apply the coating directly to the food surface; when using the foam application method, a foaming agent is mixed into the coating. Subsequently, the applicator tank is filled with compressed air. A lot of tumbling is used to break up the foam for even dispersal [19].

f) **Panning Method:** The panning method of coating involves the use of a large rotating container resembling a pan. This method involves putting the food to be coated inside the pan and then sprinkling or splattering the coating solution over it. To apply coating solution to the food items, a spray gun is installed inside the coating chamber. The pan is continuously rotated at a specific speed during the coating cycle to ensure that the coating material is distributed evenly throughout [20]. After coating is finished, the product is placed in the drying cycle, which involves forcing hot air through it to finish drying it completely. Using this technique, coating foods with round or oval shapes can be accomplished successfully [21].

g) **Fluidized Bed Processing Method:** The initial fluidized bed, known as the Wurster process, emerged in the 1950s as a modified version of the fluidized bed dryer, widely applied in food industries for drying purposes. Fluidized bed processing generally necessitates a substantial quantity of coating solution compared to alternative coating methods. Among the three fluidized bed methods (bottom spray, rotary, and top,), the top spray variant stands out as the most efficient [22]. However, this technique encounters solution loss as a significant portion of the coating adheres to the coating unit's sidewalls. There are instances of premature evaporation, potentially resulting in uneven coating on the food surface. Nevertheless, the advantages of utilizing a fluidized bed include prevention of cluster formation and reduced processing time [23].

4.2.3 Attributes of Antimicrobial Coatings

Antimicrobial based edible coating as well as film has proven to be effective in safeguarding food from spoilage and reducing the growth of pathogens. This is achieved by regulating the controlled release of embedded antimicrobial agents onto the food surface, thereby protecting it [24]. The choice of appropriate antimicrobial agents entails evaluating their efficacy against specific microbes, and its potential relationships to a polymer coating as well as the food packaging. Such associations have an ability to alter both the antimicrobial function as well as attributes of the films, and therefore significantly contribute to development in antimicrobial films as well as coatings [3]. Several physical tests have been conducted to evaluate the barrier as well as mechanical attributes on film and coating on edible products.

- Quasi-static tension is performed on consumable coatings into measure the physical properties like the stretch resistance, elongation at failure and elastic modulus [25].
- The WVP (water vapor permeability) of these coatings are obtained by using the ASTM E-96 static technique [26].

A variety of microorganisms and testing techniques, such as film contamination test, microbial colony counting, as well as disk diffusion method, can be used to evaluate the antimicrobial properties of edible coatings. The effectiveness of antimicrobial films has been assessed *in vitro* using these techniques. On the other hand, the number of native or vaccinated microbial population counts during food storage are used to assess the efficacy of coatings when they are applied to food [5].

4.2.4 Incorporation of Antimicrobials in Food Systems

Microbial growth in foods can be managed using antimicrobials, which are introduced in various forms. These can either be directly added to the food or released gradually from packaging materials [13].

4.2.4.1 Direct Applications

- Antimicrobial compounds derived from natural sources have been utilized directly in food systems, commonly in powder or liquid form. Their antimicrobial efficacy has been primarily studied in model food systems, particularly those derived from plant sources. Nevertheless, only a select few natural antimicrobials have been practically employed in the food industry, as their utilization as preservatives is frequently restricted by the potent aroma and taste they impart to these food items. Furthermore, the limited solubility of natural antimicrobials within complex food matrices poses an additional constraint on their application [27].

4.2.4.2 Indirect Applications

- **Edible Films and Coatings:** applying thin layer of material on the surface of the food products is known as edible coatings. By acting as barriers against moisture, oxygen, and potentially surface-contaminating microbes, edible films also enhance the quality of food [28]. furthermore, edible coatings and films assist in lessening the environmental issues raised by traditional plastic packaging. the application of edible coatings and films from diverse edible surfaces, including meat, vegetable as well as fruit commodities, in order to provide antimicrobial properties [29].
- **Nanomaterials:** The nanoparticles could be applied directly, as a coating, or as packaging in various food systems to prevent the growth of foodborne pathogens, thereby improving antimicrobial stability. It has been demonstrated that adding nanomaterials to food packaging improves the quality of fresh produce, baked goods, and confections by shielding food from lipids, gases, moisture, odors and off-flavors [13].

4.2.5 Applications of Antimicrobial Coatings

In general, antimicrobial coatings have been used extensively in fish, meat, fruits, and vegetables because of their chemical properties and rapid rates of degradation [3].

a) **Vegetables and Fruits:** Fresh vegetables and fruits that have undergone minimal processing are prone to spoilage and can be readily affected by microbial contamination (processing, transportation, retail presentation and storage) and postharvest physiological changes. Because it is difficult for them to maintain their freshness for extended periods of time, minimally processed vegetables and foods. Utilizing antimicrobial substances within the packaging of vegetables and fruits might serve as a tactic to manage microbial impact, prolong the duration of product shelf life, and enhance overall product quality [7]. The following Table 4.1 shows the application of antimicrobial coating on vegetables and fruits.
b) **Meat:** Meat provides an optimal environment for spoilage microorganism proliferation. Within packaged meats, microbial growth fosters the development of pathogens and causes unfavourable sensory changes as time elapses. Both white and red meats possess high potential for bacterial growth owing to their higher water activity [40]. Consequently, antimicrobial packaging is employed to safeguard meat from spoilage microorganisms by utilizing particular chemical agents or compounds within the packaging material and/or packaging environment. In meat items, assessing the effectiveness of antimicrobial packaging involves monitoring relevant microbial counts or quality indicators that indirectly relate to microbial proliferation [41]. As the majority of antimicrobial packaging methods depend on active antimicrobial agents migrating from the packaging material into the food, the active of this migration play a pivotal role in effectively suppressing microbial activity. Hence, optimizing antimicrobial packaging involves directing the active agent's migratory kinetics toward specific pathogenic or spoilage organisms [42].

Table 4.1 Application of antimicrobial coating on vegetables and fruits.

Fruit based products			
Product	**Antimicrobial agent**	**Inference**	**Reference**
Strawberry	Silver nanoparticles incorporated to low density polyethylene	Fruit quality and storage life were both enhanced by nano-silver packaging.	[30]
Grapes	Essential oils (alginate/ chitosan beads)	Beads made of alginate/ chitosan and emitting clove essential oils effectively preserved the quality of the produce.	[31]
Papaya	Essential oils (lemon grass, oregano and cinnamon)	Microorganism growth was found to have decreased; at the conclusion of storage, the cinnamon sachet had the largest reduction in microorganisms.	[32]
Sliced peach	Essential oils (cinnamon leaves)	Antibacterial-enriched pectin film exhibited optimal performance at a cinnamon leaves' concentration of 36.11 g/L, thereby inhibiting the growth of bacteria.	[33]
Mango	Essential oils (lemon grass and oregano)	The physicochemical characteristics of the produce were unaffected by the presence of essential oils. Active sachets containing essential oils inhibited the growth of fungi under test, but lemongrass was more successful.	[34]

(Continued)

Table 4.1 Application of antimicrobial coating on vegetables and fruits. (*Continued*)

Vegetable based products			
Product	**Antimicrobial agent**	**Inference**	**Reference**
Chillies	Essential oils (lemon grass)	The developed film proved to be effective in preserving chilies, and lemongrass essential oil was found to be effective in inhibiting microbial growth.	[35]
Sliced tomatoes	Essential oils (cinnamon bark oil, sweet basil oil and cloveleaf oil,)	Blended films containing cinnamon bark oil demonstrated the best performance when compared to films without essential oils.	[36]
Carrots	Essential oils (thyme and oregano)	Compared to the water treatment on carrots, the oregano solution showed a noticeably lower initial total viable count level. The sensory panel concluded that essential oil treatments were appropriate for carrots at all stages of storage.	[37]
Broccoli	carboxymethyl-cellulose and Chitosan	The coatings improve the retention of ascorbic acid and total chlorophyll while reducing stem hardening, browning, and yellowing of the product.	[38]
Cabbage	Malic, lactic, and acetic acid	Treatment with 2 % malic, 1 % lactic, and 2 % acetic acids significantly decreased the pathogens inoculated on cabbage.	[39]

Additionally, it's crucial to minimize the impact of active antimicrobial packaging on the sensory and visual aspects of the packed product for consumers, ensuring appropriate utilization of antimicrobial packaging in meat preservation [43]. The following Table 4.2 shows the application of antimicrobial coating on meat.

c) **Fish:** Fresh fish is inherently susceptible to quick spoilage owing to its biological structure. The primary factor contributing to its deterioration is the presence and activity of spoilage seafood microorganisms. These microorganisms prompt the degradation of vital components such as essential protein functionality, fat-soluble vitamins, and fatty acids. Furthermore, their activity leads to the generation of biogenic amines and the development of off-putting odors in the fish [1]. The following Table 4.3 shows the application of antimicrobial coating on fish.

4.3 Antimicrobial Agent for Active and Intelligent Package of Food

A realm in food packaging technology is present a constant state of evolution, adapting to the increasing challenges presented by modern society [54]. A contemporary emphasis has arisen, focusing on the advancement of smart and responsive antimicrobial coatings for packing supplies [55]. Incorporating or applying antimicrobial agents within or on food packaging serves to mitigate, inhibit, or delay the proliferation of microorganisms, thus elongating the lag phase and curtailing microbial growth [56]. Active packaging refers to altering the conditions of packaged food by introducing active substances like antimicrobials and antioxidants, with the aim of extending shelf life or enhancing the safety and sensory attributes of foods. Intelligent packaging encompasses indicators that manifest observable changes related to alterations in the freshness, quality, or safety of foods. These indicators typically respond to shifts in pH, gas concentrations, temperature, or humidity [55].

4.4 Active Coating

Active food packaging encompasses using materials that offer enhanced functionalities, like, biocatalytic properties, antioxidant, or antimicrobial.

Table 4.2 Application of antimicrobial coating on meat.

Meat based products			
Product	**Antimicrobial agent**	**Inference**	**Reference**
Meat pieces	Acetic acid and oregano essential oil	The growth and proliferation of pathogens (*S. aureus*) were inhibited by the combination of organic acids and essential oils.	[44]
Broiler chicken wings	fumaric acid, lactic acid, and chlorine dioxide	The samples that were treated with lactic acid demonstrated the highest level of effectiveness in reducing *mesophilic bacteria* and *E. coli.*	[45]
Fresh sausage	Essential oils (marjoram and oregano)	A bacteriostatic effect was produced by the addition of marjoram and oregano essential oil.	[46]
Meat balls	Lovage extracts, rosemary, and cranberry (0.2 %)	The rosemary extract demonstrated the most effectiveness in extending the product's shelf life to 13.3 days.	[47]
Fresh minced beef	Thymol	The higher thymol concentrations (750 mg/Kg) had better effects on product quality and prolong the shelf life by up to seven days.	[48]

Table 4.3 Application of antimicrobial coating on fish.

Fish based products			
Product	**Antimicrobial agent**	**Inference**	**Reference**
Peeled shrimps	Thymol	The products shelf life has increased to 14 days when compared to the control.	[49]
Crab meat	Sodium acetate	The products shelf life has increased to 12 days when dipped in a 2% solution for 2 minutes, compared to the control.	[50]
Fried mullet fish fillets	Essential oils (marjoram and thyme)	Strong effects were observed against Enterobacteriaceae when combination of marjoram and thyme.	[51]
Fish burgers	Essential oils (Rosemary extract)	The use of vacuum packaging and rosemary extract inhibit the growth of biochemical effect and microorganisms.	[52]
Rainbow trout fillets	Essential oils (oregano)	Using oregano (0.2%) in combination with vacuum packaging significantly extended the shelf life of trout fillets to 11–12 days compared to the control, which was packaged in air and had a shelf life of 5 days.	[53]

Achieving this involves integrating active compounds into commonly used packaging materials' matrix or applying coatings with similar functionalities via surface modification [57].

4.4.1 Method of Active Coating

a) **Controlled Release:** Controlled release coatings in active packaging involve embedding active agents within polymeric matrices to exert specific functions in packaged goods, like antimicrobial or antioxidant properties. Two main approaches for integrating active ingredients within polymer matrices include solution deposition and forming [58]. Solution casting dissolves the intended polymer and active compound simultaneously in a solvent, forming functional plastic films upon solvent evaporation. Conversely, extrusion involves melting the active compound with polymeric material to create blended films. However, challenges exist for practical application. The majority of polymers that are edible tend to Disintegrate at reduced temperatures when in contact with volatile organic compounds (VOCs), which restricts the steadiness and efficacy of various bioactive substances. Solution casting is explored more with biodegradable polymers that dissolve at lower temperatures, like poly (lactic acid) and cellulose derivatives. Although extrusion and blending are industrially scalable, they face limitations due to the thermal instability of many active compounds, which can degrade during heat transfer in operations [59]. Attaining regulated release requires intricate chemical and physical processes, encompassing the interaction of the bioactive ingredients integrated into the packaging structure, as well as the structure and permeability. These factors can influence the release rate, with lower affinity or porosity leading to slower releases. Multilayer systems can act as barriers to control compound release rates [60].

Temperature also influences compound diffusion, potentially advantageous for antimicrobial packaging, as microorganisms reproduce faster at higher temperatures [59]. Incorporating active compounds into plastic matrices, even at low concentrations, can significantly impact material properties crucial for processing and production.

These include tensile, mechanical, thermal, and gas barrier properties, undermining the goal of enhancing material performance in active packaging. Hence, the interest lies in thin coatings over bulk modifications to avoid affecting critical physical and chemical properties, thus preserving commercial applicability. Achieving active functionality without compromising essential material properties remains a challenge in the development of effective active packaging technologies [61].

b) **Surface Immobilization:** The materials used for smart packaging can be created by fixating bioactive compounds on the surface of materials packaged in mass, in addition to utilizing integrating methods. Although packaging that are active in nature are less commonly investigated, Paperboard can be modified to immobilize biologically active substances through modifying the composition of cellulose to participate readily in reactions like carboxylic, epoxy or aldehydes acids [62]. In contrast, polymer materials often need surface functionalization for active agent immobilization. Surface activation techniques include techniques involving physical properties. like corona discharge, flame, plasma as well as UV radiation, and wet methods involving corrosive liquids like piranha solution, sodium hydroxide, sulfuric acid, among others. These methods introduce oxygenated species like carbonyl, hydroxyl, and carboxylic acid groups onto the polymer surface. Industrial methods like flame, corona, and plasma treatments are common for their scalability and ability to enhance surface energy, wettability, and remove contaminants from polymer films without liquid reagents [63]. After undergoing functionalization, substances can be fixed in place using either chemical or nonchemical approaches. Nonchemical interaction is dependent including electrically charged or attraction connections, whereas covalent immobilization involves the direct coupling of active compounds onto the polymer surface via covalent bonds, preventing migration to the packaged product matrix. Chemical bonding linking the polymer substrate containing bioactive compounds involves forming amide, ether, ester, or thioether bonds, utilizing functional groups like amine, hydroxyl, thiol also carboxylic acid.

Covalent immobilization offers the advantage of preventing active compound migration, essential for commercial viability. Crosslinkers may be used to form permanent or "zero-length" covalent bonds between the substrate and active compound [64]. Surface immobilization technologies are versatile, allowing for a range of bioactive agents to be attached after incorporating reactive units on the surface of the polymer, such as enzymes or peptides. These methods present opportunities for enhancing active packaging materials, providing controlled release functionalities without compromising the packaging's properties or the product's safety and quality [65].

c) **Layer-by-Layer Method:** Layer-by-layer (LbL) method is a flexible technique for modifying surfaces, which involves sequential deposition of charged polymers carrying opposite polarity upon a firm substrate. To activate a polymeric substrate, surface activation methods or the covalent attachment of a polyelectrolyte onto its surface can be employed to facilitate the deposition of these layers. These polyelectrolytes, which encompass various polymers like proteins, polysaccharides, or synthetic variants, can be modified with Enhancing their ability to be deposited by incorporating reactive units [66]. Settling (deposition) happens by immersing the substrate in Electrolytic solutions or by dispersing solutions onto the underlying surface demonstrating capability for scalable efficient coating processes. Optimization of pH levels in solutions enhances charge presence, facilitating deposition through electrostatic interactions or building covalent bonds between layers using crosslinkers.

In theory, an infinite layers of electrolyte polymer can be added, a saturation point may halt further deposition. Molecular weight of polyelectrolytes exhibits a positive correlation with system thickness and stability, but the bonding type and susceptibility to hydrolysis in diverse environments also influence stability. LbL deposition serves as a platform to develop active packaging films by incorporating functional additives interspersed among the layers or as part of the composition of specific electrolyte polymers [67] experimental scale LbL application device hint at the potential scalability for mass production of active packaging coatings. This technique offers adaptability and scalability, enabling the creation of robust and multifunctional active packaging coatings. Its flexibility in incorporating various active agents positions LbL assembly as a promising method for the enhancement of packaging materials in the food industry [57].

d) **Photografting:** Photografting involves grafting polymer chains onto a surface using UV (ultraviolet radiation) light exposure in the presence of photoinitiators and monomers in the 315–400 nm range. UV radiation creates radicals on the polymeric surface, initiating polymerization of monomers through photoinitiators, which generate unpaired electrons upon UV exposure. Photografting occurs via three procedure include Abstraction of hydrogen, transfer of electrons, and breaking apart [68]. Abstraction of hydrogen involves UV radiation removing hydrogen from a substrate, leading to unpaired electrons on the surface. Electron transfer occurs through reduction of the photoinitiator's carbonyl group in an acceptor-donor system. The cleavage process includes the light-initiating agent breaking into free radicals, which kick-start polymerization upon exposure to UV light, producing free radicals initiating polymerization. widely used light-initiating agent include anthraquinone, benzophenone, phenyl azide as well as thioxantone. Additionally, macromolecular and naturally derived photoinitiators like curcumin have been studied [69]. Photografting can create functional packaging films through the direct inclusion of active compounds during photografting otherwise immobilizing them After attaching functionalized polymer chain such as acrylic acid. This method offers versatile ways to create active packaging coatings, offering flexibility in incorporating various active compounds into packaging materials [70].

4.4.2 Application of Active Coating

a) **Antimicrobial:** Active packaging technologies, especially those focused on antimicrobial packaging, have undergone extensive research to manage the proliferation of harmful and spoilage-causing microorganisms within packaged items. Diverse antimicrobial agents' organic acids, essential oil, biopolymers, enzymes, and peptides have been integrated into these packaging materials using various coating methodologies. One method to achieve controlled release involves a liquid formulation that includes essential oils, adhesion promotors, and fixatives. These components regulate the gradual release of essential oils when applied to standard packaging materials. Investigations have showcased the antifungal capabilities of coatings containing cinnamon essential oil, effectively curbing the growth of *Penicillium species* and *Aspergillus*. Similarly, controlled release coatings featuring thyme and oregano essential oils have demonstrated antimicrobial effects against *Listeria monocytogenes, Salmonella typhimurium,* and *Escherichia coli* [71]. The sol-gel technique, which involves condensing hydroxylated monomers and polymers, has found application in controlled release coatings. For example, A coating created through sol-gel encapsulation containing pimaricin was used on plasma-treated PLA films, effectively preventing. The growth of mold on cheese that has been kept at temperature of 4°C. Additionally, this technique has demonstrated promise in applying antimicrobial photocatalysts such as TiO2 [72].

Methods for targeted emission coatings have been combined with surface fixation methods to create coatings that are germ-resistant. In one instance, polydiacetylene liposomes containing cinnamaldehyde were covalently linked to surfaces, showcasing antimicrobial efficacy against *Bacillus cereus* and *Escherichia coli*. Coating PLA with a blend of 2,4-hexadienoic acid, chitosan, ethyl lauroyl arginate, as well as lactic acid sodium salt exhibited notable antimicrobial activity against diverse microorganisms, including *Listeria innocua, Listeria monocytogenes,* and *Salmonella Tiphymurium*. Another study utilized spray coating on corona-treated PLA with solutions containing lauric arginate ester, resulting in significant reductions in *S. Typhimurium* and *L. monocytogenes* present in processed meat samples that are encased [73]. Antimicrobial agents have been covalently attached to solid supports, focusing on cationic polymers and antimicrobial enzymes. Cationic pullulan films displayed significant reductions in *Escherichia coli* and *Staphylococcus aureus*, along with improved mechanical and thermal properties. Covalently binding the antimicrobial enzyme lysozyme to EVOH films resulted in decreased levels of *Listeria monocytogenes* in bacterial suspension. The functionalization of flexible polymers with quaternary phosphonium compounds demonstrated antimicrobial efficacy against *Escherichia coli*. Furthermore, atmospheric plasma treatment of multi-layered films led to the inhibition of *Candida albicans, Bacillus subtilis, Listeria monocytogenes, Staphylococcus aureus,* and *Escherichia coli* [74].

The layer-by-layer deposition technique has been investigated extensively across studies, predominantly aiming to enhance barrier properties in packaging films and integrate antimicrobial elements. The application of chitosan and κ-carrageenan onto aminated PET films notably enhanced gas barrier properties. Additionally, alginate, serving as another polyanion, exhibited potential by minimally affecting vapor transmission rate when applied to polyethylene terephthalate (PET) films treated with amines. Employing sequential layering of κ-carrageenan and the muramidase onto polyethylene terephthalate films with amine functionalization showed enhancements in both oxygen and water vapor permeability. Though antimicrobial active packaging has primarily centered on curtailing microbial growth, there's potential in leveraging food processing equipment coating technologies designed to prevent microbial adhesion. Adopting these methods might offer an extra safeguard against potential contamination [75].

b) **Antioxidant:** Oxidative degradation significantly affects the shelf life of diverse food and consumer products, resulting in quality decline during their transit and storage. This degradation is linked to oxidative reactions that lead to issues such as lipid rancidity, color fading, and vitamin breakdown. To safeguard oxidation-sensitive packaged goods, preservation techniques involve the direct introduction of antioxidants and the development of appropriate packaging solutions. A specific approach is packaging with antioxidant properties, which integrates antioxidants into packaging materials to bolster food preservation.

Presently, commercially accessible technology incorporating antioxidants in packaging encompass oxygen absorptive agents, commonly found in labels or sachets [76]. Current research initiatives are directed towards advancing Free radical-scavenging coatings, designed for active packaging uses, include both movable and non-movable cases. Coatings of migratory antioxidants are crafted to regulate the gradual dispensing of antioxidants throughout the product's shelf life. Previous research in this domain primarily concentrated by combining polymer composite incorporating antioxidants, followed by the method of extrusion., succeeded by extrusion. Despite that the incorporation of embedding free radical scavengers within wrapping components might impact the overall properties of the bulk material, restricting their potential applications.

Encouraging advancements in designing active coatings for the controlled release of antioxidants are currently underway. For instance, creating packaging infused with antioxidants by employing a atomized coating method involving application of methanol-dissolved citrus oil onto polyethylene terephthalate containers, showcasing heightened overall antioxidant activity [77]. Similarly, directly coated rosemary extract onto LDPE plastic wrap, resulting in a Substantial postponement in the onset of external fat deterioration in wrapped chicken burgers. Moreover, studies have investigated the application of metal oxide coatings on packaging materials to curtail oxygen diffusion and prolong product shelf life [78]. Additionally, nonmigratory antioxidant coatings can be employed by immobilizing functional groups covalently onto the surface of packaging materials. For instance, the immobilization of Gallic acid onto chitosan generated an active packaging material that notably diminished the oxidation of ground peanuts. Furthermore, photo-induced grafting polymer formation has been employed to create coating capable of chelating metal ions, thus extending the delay in Preventing fat deterioration in emulsified solutions by isolating metals in transition [79]. Apart from possible regulatory benefits associated with "transparent labelling", fixed antioxidant layers provide quality benefits for products compared to migratory antioxidant packaging coatings, as they not affect the sensory attributes of the packaged product. Nonetheless, there are still technical challenges in implementing nonmigratory active coatings for commercial use [80].

c) **Biocatalytic:** The collaboration between solid support materials and enzymes presents vast potential for biocatalytic coatings in active packaging. Enzymes, possessing heightened substrate specificity as proteins, find widespread applications across the diagnostic sectors pharmaceutical, and food. During food production, enzymes are instrumental in breaking down undesired components, enhancing product quality, and manufacturing ingredients. Incorporating these embedding biological catalysts in packaging allows using them as functional layers, enabling "in package processing" to enhance quality or prolong shelf life during storage and transit [81]. Numerous methods involved in functional packaging, leveraging biological catalysts are designed to control the pathogenic microorganisms and/or proliferation of spoilage, frequently utilizing lysozyme as a prominent enzyme for microbe-inhibiting packaging. The enzymatic components have been incorporated within active packaging layers using mixing, employing noncovalent interactions for regulated release and chemical fixation. A specific focus within enzyme-enhanced packaging layers is the reduction of gaseous element (O_2) to enhance food preservation duration. Successful endeavours integrating glucose oxidase into multi-layered packaging materials have shown remarkable outcomes, achieving initial activity retention of up to 97% even after exposure to elevated temperatures during production. Recent advancements have revealed the potential for enzyme modification through hydrophobic ion pairing, enabling solubility and sustained activity in solvents utilized in ink formulations. This innovation paves the way for the development of biocatalytic active packaging coatings utilizing conventional printing technology [82]. A significant portion of research in enzyme-enhanced packaging layers revolves around chemical fixation, enabling enzymatic coating has associate with packaging product excluding incorporating to the structure of food. Nonmigratory biocatalytic coatings present potential regulatory advantages, aligning with the increasing demand for "clean label" products. sequential layering enables for increased biocatalyst incorporation in biocatalytic coatings in contrast with conventional fixation methods. Additionally, photografting, often coupled with other surface modification techniques, has been utilized for immobilizing diverse enzymes, showcasing potential applications in active packaging coatings.

Novel advancements like biopatterning and electrospun nanofibers present fresh avenues for developing biocatalytic active packaging coatings. Electrospinning enables the integration of enzymes into polymer nanofibers, maintaining substantial enzyme activity even after prolonged dry storage. Biopatterning, which establishes precise patterns of immobilized biomolecules at micron or submicron levels, holds promise in active packaging coatings. This technique allows for targeted activity in specific regions within encasement structure, minimization of resource inefficiency in alteration and managing expensive coating cost [83]. Developing biocatalytic coatings necessitates a comprehensive approach that encompasses the packaged product, coating matrix, and enzyme. The effectiveness of the coating method relies on the stability of the bound enzyme. Attention should be devoted to reducing diffusional constraints of substrates and products, alongside retaining activity to enhance thermostability and pH stability. The integration incorporating enzymes into bioactive layers for active packaging brings about diverse prospects and complexities, prompting ongoing research aimed at refining these technologies for real-world implementation [84].

4.5 Intelligent Packaging

In recent years, Intelligent Packaging (IP) has emerged as a technology observing the state of packaged foods in transit as well as storage. The term "intelligent" signifies rationality, sound judgment, and the capability for processing and data storage. Effective communication, involving the storage, acquisition, sharing of information and processing is a prerequisite for making informed decisions—an area where IP plays a pivotal role. Essentially, Intelligent Packaging refers to a packaging system that performs advanced functionalities including sensing, observing, tracing, Capturing, implementing scientific method as well as broadcasting into aid judgment processes. Its aims include prolonging freshness, improving security, advancing excellence, dispensing knowledge, as well as signalling potential issues. The goal of IP is to transform packaging into an innovative and useful communication tool. Intelligent Packaging is commonly linked with Smart package devices, which are tags or affordable labels affixed to primary or secondary packaging (like trays, pouches, shipping containers or bottles). These devices enable seamless interaction across the distribution network, enabling adequate response for improve the excellence and security of food [85]. There are two categories of intelligent packaging systems are available:

- Packaged material signals like gas indicators, time-temperature indicators and biosensors Observing the external surroundings and complication alerts when needed.
- Information conduits like radio frequency identification tags as well as barcodes aimed at inventory management including transmitting data.

These devices vary not only in their physical composition but also in the quantity and nature of the data they carry, and how they capture and distribute that data. The usual intelligent packaging systems, numerous intelligent packaging contraption is strategically deployed at various points along the supply chain [86].

4.5.1 Method of Intelligent Packaging

a) **Gas Indicators:** The gaseous mixture within a void on the packaging materials undergoes fluctuation due to various factors such as the environmental elements, packaging properties, or food's activity. Activities like spoilage microorganism gas generation, fresh produce respiration, or leaks can all cause alterations in the gas composition inside the package or gas transmission through package materials. Gas indicators, often in the form of package printed films or labels, serve to track these changes, offering a means to monitor safety and food product quality. Among gas indicators, oxygen indicators are most prevalent in food packaging applications because oxygen in the air can lead to color shifts, oxidative rancidity, and microbial spoilage. Several oxygen indicators are crafted to reveal color changes in cases of tampering or package leaks [87].

Gas indicators for various gases like carbon dioxide, water vapor, hydrogen sulfide, ethanol, among others, also hold significance. Anticipated incorporation involving gas indicators within Radio Frequency Identification Data Tags or barcode is poised to permit transmission of gas indicator indications visually and electronically. Progress in smart ink and printing technology will further facilitate automatic reading of gas indicators distantly via optic platforms [88].

b) **Time-Temperature Indicators:** Temperature serves as a key environmental factor affecting both chemical changes and physical changes as well as microbial growth in food items. Time-temperature indicators (TTIs) are compact labels affixed to shipping containers or individual product packages. They visually signal temperature histories during distribution and storage, particularly vital for indicating temperature variations in chilled or frozen foods and estimating remaining shelf life for perishable items. These labels exhibit distinct temperature-dependent changes, such as dye diffusion or intensified color along a linear path. Commercially, three main types of TTIs exist: partial history indicators, critical temperature indicators, and full history indicators [89]. Lifelines Technologies in the early 1990s introduced a concept involving laser optical scanning of a TTI/barcode label for retrieving both temperature history and product data simultaneously. Despite the concept's potential, it was not widely adopted due to costly scanners and insufficient support within the supply chain infrastructure. Today, advancements in affordable scanning and wireless technologies have enabled companies to innovate advanced TTI systems for monitoring perishable food quality [90]. For instance, Bioett (Lund, Sweden) developed a TTI/barcode system enabling data retrieval via display on computer monitors, handheld scanners, and downloading into databases for interpretation. Similarly, German based industry launched power cell (Batteries) driven radio frequency identification data tags or time-temperature indicators employing batteries made from a bendable base material and featuring thin-film printing. In the 'Nard, (California), efforts are also underway to develop a TI/RFID tags powered by a battery.

Differing from conventional TTIs that rely on dispersion or chemical processes, the time-temperature indicator (TTI) with RFID technology label uses a semiconductor chip to detect and consolidate temperature changes over a period, determining a product's shelf life [91].

c) **Biosensors:** The landscape of food-borne infections is continuously evolving, with established pathogens being managed while new ones emerge. There's an imperative need for accurate, swift, real-time sensing technologies to analyze pollutants in situ, detect and identify pathogens, and monitor the quality metrics on food products. Typically, the biosensor has portable diagnostic system designed to detect, record, and transmit information associated with biological processes. This smart apparatus consists of two main components: a transducer that transforms bio signals into a quantifiable electrical reaction and a bioreceptor that identifies the specific analyte. biological recognition element, biotic substance such as hormone, enzyme, antigen, or nucleic acid or microbe, partners with a transducer that varies based on the parameters being measured (acoustic, optical, electrochemical, etc.). Key attributes of a biosensor encompass sensitivity, specificity, simplicity, portability, and reliability. Currently, in the market biosensors designed via IP is currently unavailable, in spite of some trial models have under development. In particular, the united states of america has working on the biosensors known as a food monitoring system designed to detect microorganisms in food packaging [92].

Within this system, a barcode segment is connected to an antibody targeting a particular pathogen; The existence of bacterial contamination results in the development of a dark stripe, causing the barcode to become unintelligible during the process of scan. When these platforms are envisaged to sensing significant infection, as this may lack to detect lower pathogen which stimulus to illness [93].

d) **Radio Frequency Identification Data Tags (RFID):** The RFID tag stands as a sophisticated data carrier for automated traceability and product identification. While previously used to track high-value products, RFID is an extensive deployment on packaged products in present times. Typically, a radio frequency identification data setup comprises the radio-frequency transmitter to collect information from a radio frequency identification data tags, which are transferred to a central server for analysis. Inside the RFID tag is a small microchip that is connected to a compact antenna. Two categories encompass these tags: inactive tags drawing devoid of a battery, power from the radio-frequency transmitter, as well as active tags with its individual batteries to power the circuitry of the microchip and send indications into radio-frequency transmitter. Active tags, being too costly possess capture radius exceeding 30.48 meter. Whereas passive tags have low cost, which possess capture radius read up to 4.57 meter. Various factors, such as frequency, reader power, and metal interference, influence the actual reading range. Distinct from barcodes, RFID tags offer several unique advantages. They operate without requiring line-of-sight, transmitting data through nonmetallic materials via radio waves. They boast substantial data storage capacities, reaching the maximum capacity up to one megabyte for higher end radio frequency identification data tags enabling storage of diverse information like humidity, temperature, cooking instructions and nutritional details. Radio frequency identification data tags in addition offer integration possibilities to the biosensors or time-temperature indicators to carry microbiological data and time-temperature history [94]. Major corporations, retailers, researchers, and government entities have shown keen interest in utilizing radio frequency identification data tags for diverse uses including Procurement network, inventory management, Integrity operation, and livestock feedstuff tracking. Mandates from retailers like Wal-Mart and Metro Group requiring RFID tags on shipping crates and pallets have accelerated technology adoption and infrastructure development [86].

e) **Barcodes:** Barcodes represent the most widely used and cost-effective application of information storage mediums. These systems developed in the year of 1970s period, which revolutionized stock reordering, inventory control, and checkout processes in grocery stores, becoming an integral part of their operations [95]. These barcodes rely on standardized languages known as “symbologies” to communicate with scanners and printers. Despite numerous symbologies developed over the years, less than 20 are currently in use. The traditional linear Universal Product Code barcode, with its limited storage capacity of 12 digits, primarily contains manufacturer and item identification information, lacking space for additional data encoding. To accommodate the expanding need into further information to a restricted area, a novel generation on barcode encoding systems known as RSS (Reduced Space Symbology), has emerged. Additionally, the RSS Expanded Barcode, offering up to 74 alphanumeric characters, serves variable measure products like meat and seafood, enabling encoding of extra details such as batch/lot number, packing date, and package weight. Advancements in barcode technology have seen the rise of two-dimensional barcodes, notably the PDF (Product Data File) 417 symbol, capable of storing up to 1.1 kilobytes of data in the space of a UPC barcode. These 2D barcodes allow encoding of diverse information beyond the capacity of linear barcodes, including cooking instructions, nutritional facts, graphics design, and manufacturer websites.

The benefit of portable data is immediate accessibility without relying on an external database. Enhancing versatility, the Uniform Code Council introduced Composite Symbology, combining a 2D barcode like PDF 417 with a linear barcode like UPC. The evolution of wireless handheld barcode scanners has expanded barcode applications. For instance, in healthcare settings, barcodes ensure the accurate administration of medication to patients [86].

4.5.2 Applications of Intelligent Packaging

a) **Enhancing Food Safety and Biosecurity:** Intelligent packaging, especially when aligned with science-driven principles, serves as a valuable tool for facilitating real-time data access, monitoring product conditions, and enabling prompt decision-making. These attributes are fundamental components of any effective biosecurity strategy or food safety. The utilization of intelligent packaging in augmenting traceability systems and Hazard Analysis Critical Control Points (HACCP) is explored in encompassing product flow tracking, Traceability and recordkeeping throughout the production process and supply chain, is widely acknowledged as pivotal in enhancing food safety and biosecurity [96]. The existing traceability systems vary in their scope, depth, and precision—terms denoting the recorded data, the extent of tracking forward or backward along the supply chain, and the accuracy in pinpointing product location, respectively. Given the challenge of precise tracing for all information, only a limited set of variables is typically tracked. However, these variables fail to provide specific technical insights into an integrity as well as security of the food products. Intelligent packaging has a potential to integrate with current traceability systems, establishing more efficient communication channels. RFID and Barcodes tags can facilitate electronic data storage and sharing, particularly when paired through exterior devices accomplished of swiftly assessing superiority parameters as well as overseeing food security. For instance, strategic placement of rapid microbial detection devices, water activity meters, pH meters or nondestructive quality measurement instruments along the supply chain allows for data exchange with write/ read RFID or Barcodes tags [97]. However, broadening and deepening a traceability system alone aren't adequate to enhance food safety; it requires science-based food models and user-friendly software to fully leverage the additional layer of food-related data available. In recent years, Hazard Analysis Critical Control Points (HACCP) has gained international recognition as a system for managing risks associated with food safety. The key aspect of this system is the sharing of information among locations and multiple devices. Information gathered from locations and multiple devices, particularly when they remain interconnected, is further valuable than data attained within the location alone or individual unit. While it is widely acknowledged that microbial detection alone is insufficient to guarantee food safety, when complemented with chemical and physical measurements, it allows for the rectification of safety issues and timely identification [86].

b) **Improving Food Excellence and Ease:** The present utilization on Intelligent Packaging are predominantly concentrated on employing Time-Temperature Indicator labels for temperature monitoring, though numerous alternative applications are viable. As an illustration, intelligent Packaging can be implemented in culinary apparatus, such as the intelligent microwave oven system [98]. The unique aspect of the existing system is its use of exchanging information to enhance ease and the quality of food. Barcode scanning is also beneficial for individuals with language comprehension or visual impairments difficulties.

Additionally, the internet connection enables convenient access to packaged food details, including the website, manufacturers, product information, allergen information, and recipes. Designing smart microwave appliance, necessitates applying technical expertise in developing the information strata framework and information processing. such information strata must include details about the food, packaging and microwave oven. However, the data layers can be intricate due to the various ways in which data can enter the system. This comprehensive approach enhances the usability and functionality of the intelligent microwave oven system, catering to a wide range of consumer preferences and needs [86].

4.6 Market Potential for Antimicrobial, Active, and Intelligent Coating

Antimicrobial coatings hold great promise in the food industry, with a wide range of benefits including improved food safety, extended shelf life, and enhanced consumer confidence [99]. As technology continues to advance, the industry is expected to witness a surge in innovative applications and increased adoption of these coatings across different sectors of the food industry [100]. A worldwide antimicrobial coating sector has anticipated to experience substantial growth, with anticipated worth of 8300 million U.S dollars by 2025 and a compound annual growth rate (CAGR) of 12.8%. This growth is primarily driven by the rising demand for new antimicrobial products, reflecting the industry's recognition of the need for effective antimicrobial solutions. Due to consumer demand for natural, secure, and longer-lasting food products, there is a substantial market opportunity for active food coating packaging (Imran). As a result, several coating technologies that improve the function of packaging materials have been developed, such as layer-by-layer deposition, immobilization, and embedding. With an emphasis on techniques like film formation and coating deposition, the edible packaging market is anticipated to expand at a compound annual growth rate of 6.81% between 2017 and 2023. The industry is moving toward sustainable barrier coatings and film technologies in response to sustainability mandates. These developments point to a robust market potential for packaging containing active food coatings, driven by industry innovation and consumer demand. As of 2024, the estimated market value for intelligent coated packaging is $26.77 billion. This form of packaging, equipped with attributes like radio frequency identification and time-temperature indicators is poised for substantial growth, especially within the food sector [101]. The intentional engagement of these packaging materials with food and the environment poses worries regarding substance migration and food safety, matters that have been attended to through recent EU regulations. Although these barriers, the demand for intelligent coated package is anticipated to grow in the future due to the rising need for novel packaging strategies [102].

4.7 Conclusion

The significant potential of antimicrobial coatings in the food industry, emphasizing benefits such as enhanced food safety, prolonged shelf life, and increased consumer confidence. It highlights the projected substantial growth of the global antimicrobial coatings market, reflecting industry recognition of the necessity for effective antimicrobial solutions. Furthermore, it emphasizes the incorporation of antimicrobial agents into packaging materials, which has demonstrated a reduction in microbial growth in food products, ultimately enhancing safety and quality.

Evaluating the efficacy of antimicrobial agents against specific microorganisms and their interactions with film-forming polymers is deemed crucial in developing coatings that effectively protect food items from spoilage and pathogen growth. The integration of active compounds into packaging materials or applying coatings has proven to enhance the preservation capabilities of food products, further contributing to improved safety and quality.

The pivotal role of intelligent packaging in meeting evolving consumer needs for safe, high-quality, and durable food products. Intelligent packaging aids real-time data access, monitors product conditions, and enables swift decision making, ensuring consumer well-being and food product integrity throughout the production process and supply chain. In overall as technology continues to advance the industry is expected to witness a surge in innovative applications and increased adoption of these coatings across different sectors of the food industry.

References

1. Lucera, A., Costa, C., Conte,A., Del Nobile, M.A., Food applications of natural antimicrobial compounds. *Front. Microbiol.*, 3, 287, 2012.
2. Mellinas, C., Valdés, A., Ramos, M., Burgos, N., Garrigos, M.C., Jiménez, A., Active edible films: Current state and future trends. *J. Appl. Polym. Sci.*, 133, 1–15, 2016.
3. Valdes, A., Ramos, M., Beltrán, A., Jiménez, A., Garrigós, M.C., State of the art of antimicrobial edible coatings for food packaging applications. *Coatings*, 7, 56, 2017.
4. Tajkarimi, M., Ibrahim, S.A., Cliver, D., Antimicrobial herb and spice compounds in food. *Food Control*, 21, 1199, 2010.
5. Campos, C., Gerschenson, L., Flores, S., Development of edible films and coatings with antimicrobial activity. *Food Bioprocess Technol.*, 4, 849–875, 2011.
6. Tavassoli-Kafrani, E., Shekarchizadeh, H., Masoudpour-Behabadi, M., Development of edible films and coatings from alginates and carrageenans. *Carbohydr. Polym.*, 137, 360, 2016.
7. Fadiji, T., Rashvand, M., Daramola, M., O., Iwarere, S.A., A Review on Antimicrobial Packaging for Extending the Shelf Life of Food. *Processes*, 11, 590, 2023.
8. Aloui, H. and Khwaldia, K., Natural antimicrobial edible coatings for microbial safety and food quality enhancement. *Compr. Rev. Food Sci. Food Saf.*, 15, 1080, 2016.
9. Amensour, M., Bouhdid, S., Fernández-López, J., Idaomar, M., Senhaji, N.S., Abrini, J., Antibacterial activity of extracts of Myrtus communis against food-borne pathogenic and spoilage bacteria. *Int. J. Food Prop.*, 13, 1215, 2010.
10. Pellegrino, L. and Tirelli, A., A sensitive HPLC method to detect hen's egg white lysozyme in milk and dairy products. *Int. Dairy J.*, 10, 435, 2000.
11. Masschalck, B. and Michiels, C.W., Antimicrobial properties of lysozyme in relation to foodborne vegetative bacteria. *Crit. Rev. Microbiol.*, 29, 191, 2003.
12. de los Alimentos, I., Risk analysis of the exacerbation of foodborne pathogens in raw milk activated with the lactoperoxidase system. *Rev. Salud Anim.*, 29, 176, 2007.

13. Gyawali, R. and Ibrahim, S.A., Natural products as antimicrobial agents. *Food Control*, 46, 412, 2014.
14. Ghanbari, M., Jami, M., Domig, K.J., Kneifel, W., Seafood biopreservation by lactic acid bacteria–a review. *LWT Food Sci. Technol.*, 54, 315, 2013.
15. da Cruz, A.G., Buriti, F.C.A., de Souza, C.H.B., Faria, J.A.F., Saad, S.M., II, Probiotic cheese: health benefits, technological and stability aspects. *Trends Food Sci. Technol.*, 20, 344, 2009.
16. El-Tarabily, K.A. and Sivasithamparam, K., Potential of yeasts as biocontrol agents of soil-borne fungal plant pathogens and as plant growth promoters. *Mycoscience*, 47, 25, 2006.
17. Ju, J., Xie, Y., Guo, Y., Cheng, Y., Qian, H., Yao, W., Application of edible coating with essential oil in food preservation. *Crit. Rev. Food Sci. Nutr.*, 59, 2467, 2019.
18. Khodaei, D., Hamidi-Esfahani, Z., Rahmati, E., Effect of edible coatings on the shelf-life of fresh strawberries: A comparative study using TOPSIS-Shannon entropy method. *NFS J.*, 23, 17, 2021.
19. Oduro, K.O.-A., Edible coating, in: *Postharvest Technology-Recent Advances, New Perspectives and Applications*, 2021.
20. Suhag, R., Kumar, N., Petkoska, A.T., Upadhyay, A., Film formation and deposition methods of edible coating on food products: A review. *Food Res. Int.*, 136, 109582, 2020.
21. Kumar, L., Ramakanth, D., Akhila, K., Gaikwad, K.K., Edible films and coatings for food packaging applications: A review. *Environ. Chem. Lett.*, 1, 2214–7993, 2022.
22. Solís-Morales, D., Sáenz-Hernández, C., Ortega-Rivas, E., Attrition reduction and quality improvement of coated puffed wheat by fluidised bed technology. *J. Food Eng.*, 93, 236, 2009.
23. Zank, J. and , Kind Particle growth and droplet deposition in fluidised bed granulation. *Powder Technol.*, 120, 76, 2001.
24. Min, S. and Krochta, J.M., Inhibition of Penicillium commune by edible whey protein films incorporating lactoferrin, lacto-ferrin hydrolysate, and lactoperoxidase systems. *J. Food Sci.*, 70, M87, 2005.
25. Lim, G.-O., Jang, S.-A., Song, K.B., Physical and antimicrobial properties of Gelidium corneum/nano-clay composite film containing grapefruit seed extract or thymol. *J. Food Eng.*, 98, 415, 2010.
26. Aider, M., Chitosan application for active bio-based films production and potential in the food industry. *LWT Food Sci. Technol.*, 43, 837, 2010.
27. Soković, M., Glamočlija, J., Marin, P.D., Brkić, D., Van Griensven, L.J., Antibacterial effects of the essential oils of commonly consumed medicinal herbs using an in vitro model. *Molecules*, 15, 7532, 2010.
28. Jang, S.A., Shin, Y.J., Song, K.B., Effect of rapeseed protein–gelatin film containing grapefruit seed extract on 'Maehyang'strawberry quality. *Int. J. Food Sci. Technol.*, 46, 620, 2011.

29. Ye, M., Neetoo, H., Chen, H., Control of Listeria monocytogenes on ham steaks by antimicrobials incorporated into chitosan-coated plastic films. *Food Microbiol.*, 25, 260, 2008.
30. Motlagh, N.V., Aliabadi, M., Rahmani, E., Ghorbanpour, S., The effect of nano-silver packaging on quality maintenance of fresh strawberry. *Int. J. Nutr. Food Eng.*, 14, 123, 2020.
31. Sangsuwan, J. and Sutthasupa, S., Effect of chitosan and alginate beads incorporated with lavender, clove essential oils, and vanillin against Botrytis cinerea and their application in fresh table grapes packaging system. *Packag. Technol. Sci.*, 32, 595, 2019.
32. Espitia, P.J.P., Soares, N.F.F., Coimbra, J.S.R., de Andrade, N.J., Cruz, R.S., Medeiros, E.A.A., Zinc oxide nanoparticles: synthesis, antimicrobial activity and food packaging applications. *Food Bioprocess Technol.*, 5, 1447, 2012.
33. Ayala-Zavala, J.F., Silva-Espinoza, B., Cruz-Valenzuela, M., Leyva, J., Ortega-Ramírez, L., Carrazco-Lugo, D., Pérez-Carlón, J., Melgarejo-Flores, B., González-Aguilar, G., Miranda, M., Pectin–cinnamon leaf oil coatings add antioxidant and antibacterial properties to fresh-cut peach. *Flavour Fragance J.*, 28, 39, 2013.
34. Medeiros, E.A.A., Soares, N., d., F.F., Polito, T.O.S., Sousa, M.M., Silva, D.F.P., Antimicrobial sachets post-harvest mango fruits. *Rev. Bras. de Fruticultura*, 33, 363, 2011.
35. Perdana, M., II, Ruamcharoen, J., Panphon, S., Leelakriangsak, M., Antimicrobial activity and physical properties of starch/chitosan film incorporated with lemongrass essential oil and its application. *LWT*, 141, 110934, 2021.
36. Wattananawinrat, K., Threepopnatkul, P., Kulsetthanchalee, C., Morphological and thermal properties of LDPE/EVA blended films and development of antimicrobial activity in food packaging film. *Energy Procedia*, 56, 1, 2014.
37. Gutierrez, J., Barry-Ryan, C., Bourke, P., Antimicrobial activity of plant essential oils using food model media: efficacy, synergistic potential and interactions with food components. *Food Microbiol.*, 26, 142, 2009.
38. Ansorena, M.R., Marcovich, N.E., Roura, S., II, Impact of edible coatings and mild heat shocks on quality of minimally processed broccoli (Brassica oleracea L.) during refrigerated storage. *Postharvest Biol. Technol.*, 59, 53, 2011.
39. Bae, Y.-M., Choi, N.-Y., Heu, S., Kang, D.-H., Lee, S.-Y., Inhibitory effects of organic acids combined with modified atmosphere packaging on foodborne pathogens on cabbage. *J. Korean Soc. Appl. Biol. Chem.*, 54, 993, 2011.
40. Bader, R., Becila, S., Ruiz, P., Djeghim, F., Sanah, I., Boudjellal, A., Gatellier, P., Portanguen, S., Talon, R., Leroy, S., Physicochemical and microbiological characteristics of El-Guedid from meat of different animal species. *Meat Sci.*, 171, 108277, 2021.
41. Talebi, F., Misaghi, A., Khanjari, A., Kamkar, A., Gandomi, H., Rezaeigolestani, M., Incorporation of spice essential oils into poly-lactic acid film matrix with

the aim of extending microbiological and sensorial shelf life of ground beef. *LWT*, 96, 482, 2018.
42. Rawdkuen, S., Punbusayakul, N., Lee, D., Antimicrobial packaging for meat products, in: *Antimicrobial food packaging*, p. 229, Elsevier, Netherlands, 2016.
43. Fang, Z., Zhao, Y., Warner, R.D., Johnson, S.K., Active and intelligent packaging in meat industry. *Trends Food Sci. Technol.*, 61, 60, 2017.
44. Hecer, C. and Guldas, M., Effects of lactic acid, fumaric acid and chlorine dioxide on shelf-life of broiler wings during storage. *Afr. J. Microbiol. Res.*, 23, 3880, 2011.
45. Souza, E.L.D., Barros, J.C.D., Conceição, M.L.D., Gomes Neto, N.J., Costa, A.C.V.D., Combined application of Origanum vulgare L. essential oil and acetic acid for controlling the growth of Staphylococcus aureus in foods. *Braz. J. Microbiol.*, 40, 387, 2009.
46. Busatta, C., Mossi, A.J., Rodrigues, M.R.A., Cansian, R.L., Oliveira, J.V.D., Evaluation of Origanum vulgare essential oil as antimicrobial agent in sausage. *Braz. J. Microbiol.*, 38, 610, 2007.
47. Jalosinska, M. and Wilczak, J., Influence of plant extracts on the microbiological shelf life of meat products. *Pol. J. Food Nutr. Sci.*, 59, 303–308, 2009.
48. Del Nobile, M.A., Conte, A., Cannarsi, M., Sinigaglia, M., Strategies for prolonging the shelf life of minced beef patties. *J. Food Saf.*, 29, 14, 2009.
49. Mastromatteo, M., Danza, A., Conte, A., Muratore, G., Del Nobile, M.A., Shelf life of ready to use peeled shrimps as affected by thymol essential oil and modified atmosphere packaging. *Int. J. Food Microbiol.*, 144, 250, 2010.
50. Lohalaksanadech, S. and Sujarit, C., Shelf life extension of refrigerated storage of blue swimming crab meat (Portunus pelagicus) by dipping in sodium acetate, in: *37th Congress on Science and Technology of Thailand*, p. 1, 2011.
51. Yasin, N.M. and Abou-Taleb, M., Antioxidant and antimicrobial effects of marjoram and thyme in coated refrigerated semi fried mullet fish fillets. *World J. Dairy Food Sci.*, 2, 1, 01–09, 2007.
52. Uçak, I., Özogul, Y., Durmuş, M., The effects of rosemary extract combination with vacuum packing on the quality changes of Atlantic mackerel fish burgers. *Int. J. Food Sci. Technol.*, 46, 1157, 2011.
53. Mexis, S., Chouliara, E., Kontominas, M., Combined effect of an oxygen absorber and oregano essential oil on shelf life extension of rainbow trout fillets stored at 4 C. *Food Microbiol.*, 26, 598, 2009.
54. Realini, C.E. and Marcos, B., Active and intelligent packaging systems for a modern society. *Meat Sci.*, 98, 404, 2014.
55. Yin, W., Qiu, C., Ji, H., Li, X., Sang, S., McClements, D.J., Jiao, A., Wang, J., Jin, Z., Recent advances in biomolecule-based films and coatings for active and smart food packaging applications. *Food Biosci.*, 52, 2212–4292, 2023.
56. Perera, K.Y., Sharma, S., Pradhan, D., Jaiswal, A.K., Jaiswal, S., Seaweed polysaccharide in food contact materials (active packaging, intelligent packaging, edible films, and coatings). *Foods*, 10, 2088, 2021.

57. Bastarrachea, L.J., Wong, D.E., Roman, M.J., Lin, Z., Goddard, J.M., Active packaging coatings. *Coatings*, 5, 771, 2015.
58. Han, J.H. and Floros, J.D., Casting antimicrobial packaging films and measuring their physical properties and antimicrobial activity. *J. Plast. Film Sheeting*, 13, 287, 1997.
59. Bastarrachea, L., Dhawan, S., Sablani, S.S., Engineering properties of polymeric-based antimicrobial films for food packaging: a review. *Food Eng. Rev.*, 3, 79, 2011.
60. Ahvenainen, R., *Novel food packaging techniques*, Elsevier, Netherlands, 2003.
61. Ratner, B.D., Surface modification of polymers: chemical, biological and surface analytical challenges. *Biosens. Bioelectron.*, 10, 797, 1995.
62. Kong, F. and Hu, Y.F., Biomolecule immobilization techniques for bioactive paper fabrication. *Anal. Bioanal. Chem.*, 403, 7, 2012.
63. Barish, J.A. and Goddard, J.M., Topographical and chemical characterization of polymer surfaces modified by physical and chemical processes. *J. Appl. Polym. Sci.*, 120, 2863, 2011.
64. Goddard, J.M. and Hotchkiss, J., Polymer surface modification for the attachment of bioactive compounds. *Prog. Polym. Sci.*, 32, 698, 2007.
65. Farkaš, P. and Bystrický, S., Chemical conjugation of biomacromolecules: A mini-review. *Chem. Pap.*, 64, 683, 2010.
66. Decher, G., Fuzzy nanoassemblies: toward layered polymeric multicomposites. *Science*, 277, 1232, 1997.
67. Bastarrachea, L.J., Peleg, M., McLandsborough, L.A., Goddard, J.M., Inactivation of Listeria monocytogenes on a polyethylene surface modified by layer-by-layer deposition of the antimicrobial N-halamine. *J. Food Eng.*, 117, 52, 2013.
68. Carlini, C. and Angiolini, L., Polymers as free radical photoinitiators, in: *Synthesis and Photosynthesis*, p. 127, 1995.
69. Dunkirk, S., Gregg, S., Duran, L., Monfils, J., Haapala, J., Marcy, J., Clapper, D., Amos, R., Guire, P., Photochemical coatings for the prevention of bacterial colonization. *J. Biomater. Appl.*, 6, 131, 1991.
70. Zhao, J., Lalevée, J., Lu, H., MacQueen, R., Kable, S.H., Schmidt, T.W., Stenzel, M.H., Xiao, P., A new role of curcumin: as a multicolor photoinitiator for polymer fabrication under household UV to red LED bulbs. *Polym. Chem.*, 6, 5053, 2015.
71. Minelli, M., De Angelis, M.G., Doghieri, F., Rocchetti, M., Montenero, A., Barrier properties of organic–inorganic hybrid coatings based on polyvinyl alcohol with improved water resistance. *Polym. Eng. Sci.*, 50, 144, 2010.
72. Zhu, Y., Buonocore, G., Lavorgna, M., Photocatalytic Activity of PLA/TIO 2 Nanocomposites and TIO 2-Active Multilayered Hybrid Coatings. *Ital. J. Food Sci.*, 24, 102–106, 2012.
73. Auxier, J.A., Schilke, K.F., McGuire, J., Activity retention after nisin entrapment in a polyethylene oxide brush layer. *J. Food Prot.*, 77, 1624, 2014.

74. Muriel-Galet, V., Talbert, J.N., Hernandez-Munoz, P., Gavara, R., Goddard, J., Covalent immobilization of lysozyme on ethylene vinyl alcohol films for nonmigrating antimicrobial packaging applications. *J. Agric. Food. Chem.*, 61, 6720, 2013.
75. Medeiros, B.G., d., S., Pinheiro, A.C., Teixeira, J.A., Vicente, A.A., Carneiro-da-Cunha, M.G., Polysaccharide/protein nanomultilayer coatings: construction, characterization and evaluation of their effect on 'Rocha'pear (Pyrus communis L.) shelf-life. *Food Bioprocess Technol.*, 5, 2435, 2012.
76. López-Rubio, A., Lagarón, J.M., Ocio, M.J., Active polymer packaging of non-meat food products, in: *Smart packaging technologies for fast moving consumer goods*, p. 19, 2008.
77. Contini, C., Álvarez, R., O'sullivan, M., Dowling, D.P., Gargan, S.O., Monahan, F.J., Effect of an active packaging with citrus extract on lipid oxidation and sensory quality of cooked turkey meat. *Meat Sci.*, 96, 1171, 2014.
78. Lee, C.H., An, D.S., Lee, S.C., Park, H.J., Lee, D.S., A coating for use as an antimicrobial and antioxidative packaging material incorporating nisin and α-tocopherol. *J. Food Eng.*, 62, 323, 2004.
79. Tovar, L., Salafranca, J., Sánchez, C., Nerín, C., Migration studies to assess the safety in use of a new antioxidant active packaging. *J. Agric. Food. Chem.*, 53, 5270, 2005.
80. Roman, M.J., Decker, E.A., Goddard, J.M., Performance of nonmigratory iron chelating active packaging materials in viscous model food systems. *J. Food Sci.*, 80, E1965, 2015.
81. Fernández, A., Cava, D., Ocio, M.J., Lagarón, J.M., Perspectives for biocatalysts in food packaging. *Trends Food Sci. Technol.*, 19, 198, 2008.
82. Soares, N. and Hotchkiss, J., Naringinase immobilization in packaging films for reducing naringin concentration in grapefruit juice. *J. Food Sci.*, 63, 61, 1998.
83. Krenkova, J., Lacher, N.A., Svec, F., Highly efficient enzyme reactors containing trypsin and endoproteinase LysC immobilized on porous polymer monolith coupled to MS suitable for analysis of antibodies. *Anal. Chem.*, 81, 2004, 2009.
84. Larsen, E.K.U., Mikkelsen, M.B.L., Larsen, N.B., Facile photoimmobilization of proteins onto low-binding PEG-coated polymer surfaces. *Biomacromolecules*, 15, 894, 2014.
85. Suppakul, P., Miltz, J., Sonneveld, K., Bigger, S.W., Active packaging technologies with an emphasis on antimicrobial packaging and its applications. *J. Food Sci.*, 68, 408, 2003.
86. Yam, K.L., Takhistov, P.T., Miltz, J., Intelligent packaging: concepts and applications. *J. Food Sci.*, 70, R1, 2005.
87. Smiddy, M., Fitzgerald, M., Kerry, J., Papkovsky, D., O'sullivan, C., Guilbault, G., Use of oxygen sensors to non-destructively measure the oxygen content in modified atmosphere and vacuum packed beef: impact of oxygen content on lipid oxidation. *Meat Sci.*, 61, 285, 2002.

88. Hong, S.-I. and Park, W.-S., Use of color indicators as an active packaging system for evaluating kimchi fermentation. *J. Food Eng.*, 46, 67, 2000.
89. Singh, R., Scientific principles of shelf life evaluation, in: *Shelf life evaluation of foods*, p. 3, Springer, United States, 1994.
90. Bhushan, N. and Gummaraju, K., A petri net based simulation approach for evaluating benefits of time temperature indicator and wireless technologies in perishable goods retail management, in: *FOODSIM'2002 The Second International Conference on Simulation and Modeling in the Food and Bio-Industry*, 2002, http://www. infosys. com/technology/tl_pervasive_infrastructure. asp.
91. Giaglis, G., Minis, I., Tatarakis, A., Zeimpekis, V., Real time decision support systems in urban distribution: opportunities afforded by mobile and wireless technologies, in: *3rd International ECR Research Symposium*, 2003.
92. Ayala, C. and Park, D., New bar code will help monitor food safety. *Louisiana Agric.*, 43, 8, 2000.
93. Bodenhamer, W.T., Jackowski, G., Davies, E., Surface binding of an immunoglobulin to a flexible polymer using a water soluble varnish matrix, 2004, ed: Google Patents.
94. Nambi, S., Nyalamadugu, S., Wentworth, S.M., Chin, B., Radio frequency identification sensors. *Citeseer*, 2003.
95. Manthou, V. and Vlachopoulou, M., Bar-code technology for inventory and marketing management systems: A model for its development and implementation. *Int. J. Prod. Econ.*, 71, 157, 2001.
96. Golan, E.H., Krissoff, B., Kuchler, F., Calvin, L., Nelson, K.E., Price, G.K., *Traceability in the US food supply: economic theory and industry studies*, Economic Research Service of the U.S. Department of Agriculture, Washington, 2004.
97. Bhagwat, A.A., Rapid detection of Salmonella from vegetable rinse-water using real-time PCR. *Food Microbiol.*, 21, 73, 2004.
98. Yam, K.L., Intelligent packaging for the future smart kitchen. *Packag. Technol. Sci.Int. J.*, 13, 83, 2000.
99. Sorrentino, A., Gorrasi, G., Vittoria, V., Potential perspectives of bionanocomposites for food packaging applications. *Trends Food Sci. Technol.*, 18, 84, 2007.
100. Ahmad, V., Khan, M.S., Jamal, Q.M.S., Alzohairy, M.A., Al Karaawi, M.A., Siddiqui, M.U., Antimicrobial potential of bacteriocins: in therapy, agriculture and food preservation. *Int. J. Antimicrob. Agents*, 49, 1, 2017.
101. Schaefer, D. and Cheung, W.M., Smart packaging: opportunities and challenges. *Procedia CIRP*, 72, 1022, 2018.
102. Restuccia, D., Spizzirri, U.G., Parisi, O., II, Cirillo, G., Curcio, M., Iemma, F., Puoci, F., Vinci, G., Picci, N., New EU regulation aspects and global market of active and intelligent packaging for food industry applications. *Food Control*, 21, 1425, 2010.

5

Edible Coatings—Medical Foods and Nutraceuticals

Sakthivel Sangeetha, Dineshkumar T., Pragalyaashree M. M.* and Freeda Blessie R.

Division of Food Processing Technology, School of Agricultural Sciences, Karunya Institute of Technology and Sciences, Coimbatore, India

Abstract

Medical foods and nutraceuticals are specialized products designed to meet specific dietary and health requirements for individuals with medical conditions or those seeking additional health benefits beyond basic nutrition. Coating these products plays a vital role in enhancing their overall effectiveness, stability, and consumer acceptance. The importance of coating in the context of medical foods and nutraceuticals, highlighting its significance in taste masking, patient compliance, stability, targeted delivery, and improved consumer experience. Coating formulations can be designed to improve the solubility of poorly water-soluble compounds, thereby facilitating their absorption in the body. This increased bioavailability ensures that patients receive the intended therapeutic benefits, leading to more consistent and reliable treatment outcomes. Taste acceptance is another crucial factor influencing patient compliance with medical foods and nutraceuticals. The inherent bitterness or unpleasant taste of some bioactive compounds can discourage patient adherence to treatment regimens. Coating technologies can be employed to mask these undesirable tastes, making the products more palatable and increasing patient satisfaction. Flavour-masking coatings, in particular, have shown promising results in improving patient acceptance and adherence to medical food and nutraceutical therapies. Moreover, Long-term treatment often requires multiple doses of medical foods or nutraceuticals, which can become burdensome for patients. By using coated formulations with extended release profiles, dosing frequency can be reduced, enhancing patient convenience and adherence to treatment plans. There are various coating techniques, such as film coating, enteric coating, microencapsulation, and matrix encapsulation are in

**Corresponding author*: pragalyaashree@karunya.edu

Mousumi Sen (ed.) Food Coatings and Preservation Technologies, (127–166) © 2025 Scrivener Publishing LLC

vogue. These Coating techniques contribute to improved dietary management and overall health outcomes for individuals with specific medical conditions and those seeking additional health benefits. Additionally, the key coating materials, such as hydroxypropyl methylcellulose, ethylcellulose, gelatin, and lipids, are examined for their suitability in protecting active ingredients and ensuring safety and regulatory compliance. While coating technologies offer numerous advantages, regulatory considerations and safety evaluations are essential to ensure the quality and efficacy of coated medical foods and nutraceuticals. Regulatory bodies play a critical role in assessing the safety, effectiveness, and compliance of coated products, providing consumer confidence and promoting responsible innovation in the industry. Medical foods are products that are carefully formulated to meet the dietary requirements of particular medical conditions that have unique nutritional needs. In contrast, nutraceuticals refer to bioactive compounds or dietary supplements that offer potential health benefits beyond the essential nutrients they provide. Both medical foods and nutraceuticals offer a unique approach to personalized healthcare, leveraging the therapeutic potential of specific nutrients and bioactive compounds. In conclusion, innovative coating technologies present a promising avenue to enhance the delivery, stability, and efficacy of medical foods and nutraceuticals. By overcoming challenges related to stability, bioavailability, and patient acceptance, coatings contribute to better treatment outcomes and patient experiences. Collaborations between researchers, manufacturers, healthcare professionals, and regulatory bodies are crucial to further advance coating technologies and maximize the potential of medical foods and nutraceuticals in personalized healthcare. As this field continues to evolve, the integration of coating technologies with medical foods and nutraceuticals is poised to revolutionize the landscape of modern healthcare, offering patients safer, more effective, and convenient therapeutic options for a wide range of medical conditions.

***Keywords*:** Nutraceuticals, bioavailability, enteric coating, gelatin, dietary management

5.1 Introduction on Edible Coatings

Generally, edible coatings consist of a thin substance that either wraps around the food item like a film or is made separately and applied onto or between its parts. Edible coatings can be directly added to the surface of the food product to provide extra protection and maintain its stability and quality. The specific characteristics of the food product, such as its water content, along with any changes in those characteristics that occur during storage and production, determine the necessary criteria for edible coatings. In practical terms, coatings that regulate the movement of the molecular components of the food product from the interior to the exterior of the packaging can help slow down undesirable reactions that cause negative changes in the food items. The efficiency of edible coatings relies on their mass transport characteristics. The primary function of edible coatings is to safeguard the superior quality of a food product. As coatings are ingested along with the product, their properties and composition must adhere to the standards set for food products. This necessitates the implementation of new regulations to reduce the impact on environment and ensure their safety [1].

A perfect coating is described as one that can prolong the shelf life of fresh vegetables and fruits preventing anaerobiosis and minimizing decay without compromising their quality. In the past, edible coatings were employed primarily to minimize water loss. However, advancements in the formulation of edible coatings with a broader range of permeability properties have expanded their potential for use in preserving fresh produce. Coatings derived from fruits offer improved nutritional benefits to products, thereby boosting their market worth. Biodegradable and edible coatings need to fulfil various specific functional criteria, such as solubility in lipids or water, non-toxicity, desirable appearance and colour, moisture resistance, barrier properties against solutes or gases, mechanical properties, and more. The impact of coatings on vegetables and fruits is heavily influenced by factors such as alkalinity, variety, temperature, condition of the produce and coating thickness. The necessary functional attributes of the coating are determined by the moisture content range of the product matrix and the specific decline in quality of the product undergoes [2].

5.1.1 Pros and Cons of Edible Coatings

Below are some of the advantages of edible coatings as described [3]:

- Enhancing the visual appeal of coated food products by providing a glossy appearance.
- Reducing the respiration rate and ethylene gas emission, thereby delaying the senescence phase of fruits and vegetables.
- Eliminating the need for non-biodegradable and hazardous packaging.
- Preventing the loss of volatile compounds present in food products.
- Maintaining fruit firmness and mitigating weight loss to some extent.

- Contributing to the extension of the shelf life of fruits and vegetables.
- Being free from toxic and reactive chemicals that could have harmful effects on humans.
- Minimizing moisture loss by acting as a barrier.
- Providing antimicrobial protection to coated food products through the incorporation of certain active ingredients.
- Exerting control over some physical damages.

However, there are certain drawbacks associated with edible coatings, which include:

- Compromised food quality due to modifications in the coating affecting gaseous exchanges during packaging.
- Undesirable changes in flavor, weight loss, and texture in the coated food product caused by increased coating thickness.
- Sensory-related alterations observed when incorporating antimicrobial compounds.
- Allergic reactions in consumers resulting from the raw materials used in coating preparation.
- Some coating methods incurring significant costs, ultimately leading to an increase in commodity prices.

5.1.2 The Primary Advantages of Edible Coatings/Films

5.1.2.1 Moisture Barrier

Edible films prevent the food material from experiencing loss of aroma, oxygen penetration, water uptake, or moisture loss, thereby creating favourable storage conditions for these food products. Furthermore, they enhance the product's visual appeal and texture and while prolonging its shelf life through the creation of barriers that allow for limited permeability [4].

5.1.2.2 Oxygen Scavengers

The existence of oxygen can significantly harm certain fresh food items that are packaged. Certain edible films have been identified to incorporate humidity control mechanisms and oxygen scavengers, playing a crucial role in diminishing gases that contribute to the deterioration of vegetables and fruits [5].

5.1.2.3 *Ethylene Scavenger*

Managing ethylene during storage is crucial for prolonging the freshness of produce. The effectiveness of chitosan-TiO2 nanocomposites was assessed in controlling ethylene levels on tomatoes and found a reduction in concentration of ethylene [6].

5.1.2.4 *Antimicrobial Properties*

Improper packaging stands out as a major cause of contamination for vegetables and fruits. To reduce the spoilage and microbial growth of fresh produce, it is possible to utilize an active packaging system that contains antimicrobial agents. The antimicrobials present in the edible coatings improve the safety and prolong the shelf life of vegetables and fruits [7]. Among the antimicrobial substances utilized are microbial bacteriocins such as those derived from certain polypeptides like lysozymes, lactic acid bacteria, and organic acids, like lactic acid and citric acid [8].

5.1.2.5 *Antibrowning and Antioxidant Properties*

The discoloration and browning observed in minimally-processed fruits and vegetables are a result of enzymatic reactions catalyzed by the polyphenol oxidase (PPO) enzyme. This enzyme converts polyphenolic compounds into dark pigments when exposed to oxygen, leading to the discoloration of phenolic compounds. The incorporation of antibrowning substances into edible coatings can effectively regulate the activity of polyphenol oxidase (PPO) while simultaneously creating a robust barrier against oxygen. The commonly employed antibrowning substances include carboxylic acids (such as oxalic and citric acid), ascorbic acid, resorcinol's, phenolic acids and compounds containing thiols (such as glutathione and cysteine), These substances facilitate the conversion of o-quinones, which are produced by the activity of PPO enzymes, back into their original phenolic substrates [9].

5.1.2.6 *Texture Modifiers for Inhibition of Physical Damages*

The activity of pectolytic enzymes, which causes a decrease in fruit firmness, can be suppressed to maintain firmness. Edible coatings with texture enhancers can reduce softening during storage [10]. Calcium salts are effective in managing the softening of fresh-cut fruits [11]. They can also decrease pectin solubilization and preserve firmness [12]. However, calcium treatments can increase cell wall degrading enzyme activity [13].

5.1.2.7 *Nutraceuticals for Preservation of Nutritional Quality*

Nutraceuticals are substances that improve the nutritional content of products that are lacking in vitamins, micronutrients, bioactive compounds, and minerals are examples of nutraceuticals that can be included in the formulation of active coatings. This enhances the nutritional value of certain vegetables and fruits, particularly when these micronutrients are found in limited amounts [14].

5.2 Edible Coating on Tablets

A tablet is a solid form of medication that consists of a combination of active ingredients and excipients. Typically, these ingredients are in powdered form and are compressed or compacted to form a stable and solid dosage form. Capsule dosage form is highly popular and widely preferred around the globe. The majority of drug molecules can be prepared in tablet form, and the process of manufacturing drugs in tablets is straightforward and highly adaptable. Coating involves applying a dry outer layer of coating material to the surface of a dosage form in order to achieve specific advantages or benefits. Coating can be applied to various types of oral solid dosage forms, including multiparticulates, capsules, drug crystal, and tablets [15]. Tablet coating involves applying a coating composition to a moving bed of tablets while utilizing heated air to facilitate solvent evaporation.

The purposes of tablet coating include [15]:

a) Minimizing the impact on the release pattern as much as possible and avoiding marked changes in appearance.
b) Modifying release with specific requirements and release mechanisms tailored to the characteristics within the digestive tract.
c) Customizing modified release with specific requirements and release mechanisms to match the characteristics of the body in the digestive tract.

d) Providing insulation through colour coating.
e) Incorporating another drug or additional adjuvants in the coating to prevent chemical incompatibilities or achieve sequential drug release.
f) Improving pharmaceutical aesthetics through the use of unique colours and contrasting printing.

5.2.1 Objectives of Tablet Coating

The objective is to conceal the taste, color, or unpleasant smell of the tablet in order to enhance patient adherence. The purpose is to provide physical and/or chemical safeguarding to the medication, shielding it from the external environment (especially, moisture, light, and air) to enhance its stability. The aim is to extend the duration of the drug's effectiveness and improve the ease of swallowing larger dosage forms. The purpose is to slow down the evaporation of volatile components and adjust and/or regulate the pace at which the drug is released, as seen in products with sustained-release, delayed release (enteric coated), and repeat-action formulations. To combine drugs that are not compatible into a single form of medication, while also enhancing the overall strength of the medication. Enhancing the visual appeal of the product and aiding in its identification by the pharmacist, patient, often through the use of color and manufacturer. Concealing variations in the visual aspects of raw materials across different batches, while also enhancing the overall resilience of the product [16].

5.3 Edible Coating on Nutraceuticals

Numerous scientists have made efforts to include fatty acids, vitamins, and minerals in edible coating formulations with the aim of boosting the nutritional content of certain vegetables and fruits that have low levels of micronutrients. Thorough examination of the nutrient concentration incorporated into the coatings is crucial as it is essential to understand how it affects their fundamental functionality, specifically their mechanical and barrier properties [17].

5.4 Edible Coating Methods

5.4.1 Dipping

The dipping technique involves submerging the fresh food produce into the coating solution to ensure thorough saturation of the food material's surface. Subsequently, the excess coating on the food surface is removed by draining out the coating solution. The duration of exposure or contact is determined by the characteristics of the coating solution, such as viscosity, density, and surface tension. Typically, this period ranges from 5 seconds to 3 minutes. Subsequently, the coated sample is allowed to dry undisturbed in order to facilitate the drying process of the applied layer. By employing this method, the vegetables or fruits sample achieves a consistent and even coating layer. The duration of contact is of utmost importance as prolonged exposure results in the absorption of a greater amount of moisture. Conversely, a shorter duration of contact results in an uneven coating on the food. Table 5.1 shows the dipping technique for various foods [3].

5.4.2 Layer by Layer Method

The layer-by-layer technique relies on the sequential application of oppositely charged polyelectrolytes, leading to improved regulation of the coating's functionality and properties. By employing this approach, multiple layers of films are generated, thereby enhancing the efficacy of the edible coating [18].

5.4.3 Vacuum Impregnation Technique

The vacuum impregnation technique represents a progression from the dipping method, distinguished by the presence of a vacuum environment during the immersion of fruits. In other words, rather than immersing the food material in a conventional dipping tank, the fresh food is placed in a sealed vacuum apparatus. While the food material is submerged in the coating solution at atmospheric pressure, it undergoes a process of atmospheric restoration [18].

5.4.4 Spraying Method

The spraying method is better suited for coating solutions with lower viscosity, as they can be effectively applied at elevated pressures through spraying. The drying temperature and time have an impact on the development

Table 5.1 Dipping technique for various foods [3].

Products	Coating agents	Additives	Advantages
Tomato	Basil mucilage	Cumin essential oil	Reduced loss of weight and extended lifespan.
	Shellac	Aloe Vera	Decelerated the aging process.
	Aloe Vera	Garlic oil, Sodium alginate	Prolonged the shelf life duration while lowering the production of β-carotene, ethylene, and lycopene.
	Chitosan	Aloe Vera	Prolonged duration of shelf life and postponed the ripening process.
Strawberry	Cactus mucilage	Glycerol	Prolonged the shelf life duration.
	Chitosan	Lemon essential oil	Improvement of the fermentation procedure.
	Wheat Gluten / Soy Protein	Thyme essential oil	Reduced losses after harvesting.
Plums	Aloe Vera	Rose hip oil	Enhanced qualities after harvesting.
Pork Patty	Pectin	Green tea leaf extract	lower the rate at which lipids oxidize.
Green Chillies	Shellac	Sodium alginate	Prolonged duration of shelf life.
Kiwi Fruit	Cactus Pear	Tween 20 / Glycerol	preserved the pectin, ascorbic acid, and firmness of the fruit.
Nectarine	Aloe Vera	Tween 20	Delays the process of ripening.

of a polymeric coating when utilizing a spraying system. One benefit of using the spraying technique is that it enhances the surface area of the liquid coating by creating droplets and evenly distributing them across the surface of the fruit [18]. The effectiveness of spray coating relies on the rheological characteristics of the fluid, including temperature, surface temperature, and viscosity. Following the spraying process, it is necessary to allow the coated sample some time for drying to occur. During this stage, the efficiency of the coating is determined by the chosen drying method, as well as the duration and temperature of the drying process. Furthermore, the type of nozzle and design also have a substantial impact on the spraying process. In industrial applications, three distinct methods of spraying have been achieved, specifically pressure atomization, air spray atomization, and air-assisted airless atomization [3].

5.4.5 Foaming and Dripping Method

The dripping and foaming methods are recognized as conventional techniques for coating. Using the dripping technique, brushes are employed to apply the coating directly to the fruit surface. In contrast, the foam application involves adding a foaming agent to the coating. Next, compressed air is introduced into the applicator tank, creating a blowing effect. This results in a vigorous tumbling action that breaks down the foam, ensuring its even distribution [18].

5.4.6 Panning Method

The panning method, a process for applying edible coating, involves the use of a rotating container resembling a large pan. In this technique, the food that needs to be coated is placed within the pan, and the coating solution is either sputtered or sprinkled onto it. This technique is particularly effective for coating foods that have oval or round shapes [3].

5.4.7 Fluidized Bed Processing Method

The Wurster process, coined in the 1950s, refers to the discovery of the initial fluidized bed. It is an adapted form of the fluidized bed dryer, commonly employed in the food industry for the drying process. Compared to other coating methods, fluidized bed processing necessitates a significant quantity of coating solution. The top spray method is considered the most efficient among the three types of fluidized bed processes (bottom, top, and rotary spray) [19]. During the coating process, this technique experiences solution loss as a significant portion of the coating adheres to the side walls of the coating unit [20]. Occasionally, there is premature evaporation that may result in an uneven coating on the food surface. One benefit of utilizing a fluidized bed is the reduced processing time and absence of cluster formation [3].

5.4.8 Traditional Method in Edible Coating Process

5.4.8.1 Coating with Sugar

The original development of tablet coating stemmed from the utilization of sugar to both conceal the taste and enhance the visual appeal of the core tablet [21]. Colored or colorless sugar layers can be applied to compressed tablets. The coating is soluble in water and rapidly dissolves upon ingestion. The sugar coating serves as a protective shield for the enclosed medication, safeguarding it from the surroundings and preventing any unpleasant odor or taste. In addition, the sugar coating improves the visual appeal of the compressed tablet and allows for the inclusion of manufacturing information through imprinting. The sugar coating offers a variety of benefits, including masking of taste, insulation, colouring, modified release and smoothing of the tablet core. One drawback of sugar coating is the expertise and time needed for the coating process, which in turn leads to increased weight, shipping, and size, expenses [16, 22].

a. Sealing (Waterproofing): To prevent moisture from entering the tablet core, a seal coat is administered on top of the tablet [21]. The seal coat creates a barrier against moisture and increases the hardness of the tablet surface, aiming to reduce attritional effects. Tablets with extremely fast disintegration rates could potentially initiate the process of disintegration during the initial stage of sugar coating. Typically, the sealants are film or polymers formers that are insoluble in water and are applied using a solution made from organic solvents. The amount of material applied for the sealing coat will mainly rely on the porosity of the tablet. Tablets with high porosity have a tendency to absorb the initial solution application, hindering its uniform spread across the entire batch of tablets. Therefore, it may be necessary to apply one or more additional layers of resin solution to ensure effective sealing of the tablet cores. Sealants commonly employed include zinc, shellac, polyvinyl acetate phthalate, cellulose acetate phthalate (CAP), Hydroxypropyl ethylcellulose, Hydroxypropylcellulose, and other similar materials [22].

b. Sub coating: The tablet core is typically coated with significant amounts of sugar coating, resulting in a weight increase of approximately 50% to 100% [15]. The purpose of this step is to smooth the edges and enhance the weight of the tablet [16, 21]. The application of a sub coating leads to a quick accumulation that smooths out the edges of the tablet. The sub coating marks the beginning of the sugar coating process and facilitates the rapid buildup required to round off the edges of the tablet. Additionally, it serves as the base for the subsequent coloring layers and smoothing [22].

c. Smoothing/syrup coating: To conceal and address the flaws on the tablet surface resulting from the sub-coating process [15]. This addresses the imperfections from the sub-coating and enlarges the tablet to achieve the dimensions that are predetermined [15, 16]. The initial syrup coating includes certain powdered substances that are suspended within and are commonly referred to as grossing syrups. To facilitate consistent coating in subsequent stages, diluted colorants can be included to create a tinted base. Syrup solutions, incorporating the dye, are employed until the desired size and color are attained [15]. The flaws on the tablet surface are concealed, and the predefined size is attained. This stage demands a high level of expertise [21]. The smoothing or grossing process is specifically designed to even out and eliminate any surface irregularities that may have occurred during the sub coating phase. Furthermore, it enlarges the tablet to a predetermined size. Typically, smoothing can be achieved by administering a straightforward syrup solution with a sugar solid content of approximately 60% to 70%. If necessary, this syrup commonly includes starch, pigments, acacia, opacifier or a gelatin. In instances where there are inconsistencies in the coating, a small amount of color suspension can be applied to give a hint of the desired color [22].

d. Finishing: The last step involves applying a few transparent layers of syrup coating [15]. The successful completion of a sugar coating process often hinges on this stage, which entails the repeated application of a syrup solution (with a sugar solid content of 60–70%) that includes the necessary coloring agent. To achieve the desired color in sugar coating, primarily soluble dyes were employed, as these dyes have the ability to migrate to the surface during the drying process. In contemporary times, insoluble certified lakes have largely supplanted soluble dyes in the coating of pharmaceutical tablets. Utilizing a pre-dispersed suspension of an opacified lake is the most effective method for color coating [22].

e. Polishing: The desired shine or gloss is achieved during this concluding stage [15, 16]. Coating pans that are clean and have a lining made of canvas [15]. To achieve the desired shine, powdered wax (either carnauba or beeswax) is applied [21]. Polishing results in the desired glossy appearance. To achieve a final elegant appearance, sugar-coated tablets require polishing. To polish tablets, a combination of waxes such as carnauba wax, beeswax, hard paraffin waxs, or candelilla wax is applied to them in a polishing pan [22]. Warm solution of wax or powdered waxes is applied by a suitable solvent in volatile form [15].

5.4.8.2 Film Coating

The equipment and process parameters for sugar coating and film coating are identical. Due to its time-consuming nature, sugar coating has been supplanted by film coating technology. Tablet coating process includes spraying onto a rotating bed of tablets. Polymer selection depends on the desired target site of release and delivery rate [15]. Two methods of film coating are:

- **Pan-Pour method:** It is identical to the pan-pour sugar coating process. The method is comparatively slow and heavily relies on the technique of the operator and expertise. Due to the issue of localized over-wetting, aqueous-based film coating is not appropriate [15].
- **Pan-Spray method:** Implementing an automated spraying system [15].

5.4.8.3 Materials Used in Film Coating

Film formers, which can be colorants, either enteric or nonenteric, solvents, opaquant-extenders, plasticizers, and other components found in coating solutions [22].

The desirable characteristics for film coating materials can be outlined as follows [22]:

- Ability to dissolve in the chosen solvent for coating preparation
- Specific solubility criteria for the intended purpose, such as being water-soluble, having gradual water solubility, or exhibiting pH-dependent solubility
- Capability to yield a visually appealing final product

- Strong stability when exposed to heat, light, moisture, air, and the substrate being coated
- Absence of inherent color, taste, or odor
- Excellent compatibility with other additives in the coating solution
- Non-toxic and devoid of pharmacological activity
- High resistance to cracking
- Film-forming properties that avoid bridging or filling of debossed tablets
- Compatibility with the printing process

5.4.8.4 Enteric Coating

An enteric coating functions as a barrier that regulates the specific location within the digestive system where oral medication is absorbed. The term "enteric" pertains to the small intestine; thus, enteric coatings serve the purpose of inhibiting the release of medication until it reaches the small intestine. The enteric coated polymers maintain their non-ionized state under low pH conditions, which keeps them insoluble. However, when the pH in the gastrointestinal tract (GIT) rises, the functional groups that are acidic can undergo ionization, causing the polymer to either dissolve or swell in the intestinal fluid [16, 21]. The following are the polymers utilized for enteric coating [22].

5.4.8.5 Enteric Sugar Coating

In this case, the sealing layer is customized to incorporate an adequate amount of one of the enteric polymers, meeting the criteria for enteric disintegration testing. The subsequent coating and sub coating, the steps are performed in the same manner as traditional coating using sugar [22].

5.4.8.6 Enteric Film Coating

Film coating processes allow for the formation of a direct film using enteric polymers. To achieve an effective enteric effect, it is necessary to use an adequate amount of enteric polymer. The combination of enteric coating with polysaccharides that undergo enzymatic degradation in the colon, such as galactomannan and cyclodextrin, is possible [22].

5.4.8.7 *Controlled Release Coating*

Controlled release coating utilizes polymers, such as water-insoluble ethyl cellulose and modified acrylates [22].

5.4.9 Modern Method in Edible Coating Process

5.4.9.1 *Compression Coating*

These types of tablets consist of components such as an outer coating and an inner core coating. The core is a compact and porous tablet that is positioned on a single turret. To prepare the final tablet, a larger die cavity is utilized in alternate turrets. Initially, the coating material is filled halfway, followed by the automatic transfer of the core tablet. Then, the remaining space is filled with the coating material, and final applying compression force. Frequently, the coating is soluble in water and readily breaks down upon ingestion [15]. A specialized tablet machine is necessary for applying this type of coating. While compression coating is not commonly employed, it offers advantages in certain scenarios where the tablet core cannot withstand water or organic solvents, but still requires coating for delayed release, taste masking, enteric properties, or to prevent incompatibility between ingredients [22].

5.4.9.2 *Electrostatic Coating*

In the process of electrostatic powder coating, we apply powder by utilizing a region with a strong electric field and a high concentration of free ions. As they traverse this region, the particles become charged. Pauthenier's equation guides the process of electrically charging powder particles within the corona discharge electric field. The strength of charging is primarily influenced by field intensity, the size and shape of powder particles, and the duration of time the particle is exposed to the charging area [15]. The application of coatings to conductive substrates through electrostatic coating is a highly efficient method. The substrate is subjected to a powerful electrostatic charge. The charged substrate is coated by spraying a coating material that contains conductive ionic species with opposite charges. The effectiveness of achieving a complete and uniform coating, particularly in corners, is constrained when applying this method to pharmaceuticals which are relatively nonconductive substrates [22].

5.4.9.3 *Electrostatic Dry Coating*

The introduction of electrostatic dry powder coating in a pan coater device marked the first development of a method for applying this coating technique to tablets. The capsules produced using the optimized dry powder coating method exhibit a coating, smooth surface, a release profile, and uniform that aligns with that of the tablet cores [15].

5.4.9.4 *Dip Coating*

The process of applying the coating involves submerging the tablet in a liquid coating solution, followed by drying the wet tablets using traditional coating pans. The sequence of drying and dipping steps can be iterated multiple times to attain coating effectively [16]. The tablet cores are coated by immersing them into the coating liquid. The moisture-laden tablets are dried using a traditional method within a coating pan. To achieve the intended coating, it is possible to repeat the process of drying and dipping multiple times. The reliability, speed and versatility of spray-coating techniques are not present in this process. Although specialized equipment has been created for dip-coating tablets, there has been no successful commercial pharmaceutical implementation of this technique [22].

5.4.9.5 *Press Coating*

Compression is employed to create a coating around a pre-existing core, primarily to isolate chemically incompatible substances [16].

5.4.9.6 *Vacuum Film Coating*

A novel coating technique called vacuum film coating utilizes a baffles pan which are designed specifically. The pan is equipped with a hot water jacket and can be tightly sealed to establish a vacuum system. After the tablets are positioned within the sealed pan, nitrogen is introduced to displace the air present in the pan until the required level of vacuum is reached. Subsequently, the coating solution is administered using a spray system that is airless. The heated pan induces evaporation, and the vacuum system eliminates the obtained vapor. Due to the absence of heated air with high-velocity, the energy demand is minimal, resulting in efficiency of coating is high. This coating system allows for the effective use of organic solvents with safety and minimal environmental issues [22].

5.5 Edible Coating Materials

Compound films composed of different substances in varying proportions exhibit superior functionality performance compared to coatings with single-component. Hydrocolloids, such as carbohydrates, lipids, and proteins, are the typical primary materials utilized in the production of consumable coatings [23].

5.5.1 Hydrocolloid Materials

They typically consist of lengthy polymer chains and belong to a category of materials that exhibit hydrophilic properties. This implies that they readily dissolve in water and can be sourced from animals, plants, and microorganisms [24]. They typically contain a significant quantity of hydroxyl (-OH) groups. Recent research has indicated that these are predominantly used as the primary ingredient in solutions that create films, with the purpose of controlling or enhancing color, affecting the flavor, achieving the desired texture and improving durability in food products. All types of hydrocolloids can fully or partially disperse in water, primarily aiming to increase the viscosity of the aqueous phase. This thickening effect is achieved through the presence of a gelling agent [25]. Because of their stabilizing properties, they serve as emulsifying agents. Hydrocolloids are generally categorized into two groups:

1) Hydrocolloids based on polysaccharides and
2) Hydrocolloids based on proteins.

5.5.1.1 Hydrocolloid Based on Polysaccharides

Commonly used polysaccharides for coating edible food items include and carrageenan (used for cold set gels), gellan gum, alginate, starch, chitosan, cellulose, and others. Due to their natural ability to form gels, strong adhesive, and high viscosity properties during coating production, as well as their effectiveness in providing delicacy, tightness, and solidity to the final product, these substances are extensively employed in the food industry. These coatings have the potential to slow down the ripening process and significantly increase the shelf life or longevity of edible food items, particularly for climacteric fruits that continue to ripen without inducing anaerobiosis, even after being harvested [23].

a. Cellulose and derivatives based coatings: Cellulose, a natural polymer found in both plants and animals, consists of repeating units of D-glucose connected by ß-1, 4 glycosidic bonds. The coatings and films based on cellulose derivatives exhibit enhanced film-forming capabilities. Typically, these coatings or films are characterized by their flexibility, lack of taste, low energy, absence of odor, resistance to oil and fat, transparency, hydrophilicity, moisture resistance, and moderate oxygen permeability. Among cellulose derivatives, MC (Methylcellulose) is the least hydrophilic and demonstrates exceptional water resistance. HPMC (Hydroxypropyl methylcellulose) and MC (Methylcellulose) have been widely employed to prevent oil absorption in deep-fried food items. MC (Methylcellulose) has been utilized as a coating and film material on confectionery products, acting as a barrier to the passage of lipids [26].

b. Chitin and chitosan-based coatings: Chitin, a polysaccharide that occurs naturally, is present in insects, invertebrates, diatoms, marine, fungi and algae. Chitosan is primarily derived from chitin through deacetylation, which involves treating chitin with a concentration of alkali solution. Chitosan is a polymer made up of N-acetyl-D-glucosamine units [27]. It is a natural and safe substance, free from allergens, that is biocompatible and offers various health benefits [28]. Chitosan coatings have found effective application in the food industry primarily because of their structural characteristics that enable the creation of a seamless coating layer on food products. Chitosan coatings and films exhibit favourable barrier properties against carbon dioxide (CO2) and oxygen (O2) and display exceptional antimicrobial qualities. The primary application of chitosan coatings is as an antimicrobial coating for bell peppers, cucumbers, and strawberries. It is also used as a gas barrier for pears, apples, plums, and peaches. There are certain limitations when using chitosan as an antioxidant and antifungal agent. These limitations include the need to determine the ideal concentration of chitosan coating solution to effectively fungal growth and inhibit oxidation, while avoiding any negative impact on the sensory characteristics of the end product [26].

c. Starch-based coatings: Starch is a naturally occurring carbohydrate polymer that consists of residues namely anhydroglucose. The majority of starches are composed of amylopectin, which is a branched chain polymer of glucose, and amylose, which is a linear chain polymer [29]. Starch-based coatings and films are commonly used due to their lack of odor or taste, transparency, and their ability to effectively block O_2 and CO_2 [30, 31]. Nevertheless, starch-based coatings and films inadequate water vapour barrier and exhibit water solubility characteristics because of their hydrophilic nature [32]. Starch is commonly employed in the manufacturing of food products, as it possesses desirable mechanical properties in most cases [33]. Corn starch, which contains a high content of amylose, provides a significant supply for the synthesis of edible coatings and films. The typical corn starch composition consists of approximately amylopectin is 75% and amylose is 25% [34].

d. Pectin-based coatings: Pectin is a additive that is crucially found in citrus, serving as a valuable by product. In general, pectin consists of straight chains of α-1, 4-galacturonic acids (homogalacturonan) with additional branches of α-1, 4-galacturonic acid to α-1,2-rhamnose (rhamnogalacturonan) chains. These branches, which contain sugars that are neutral, are connected through residues of rhamnose [26].

e. Pullulan-based coatings: Pullulan can act as a thickening agent that has the ability to create films that are semipermeable. Glutathione is known for its potent reducing properties, while chitooligosaccharide exhibits antibacterial activity. Utilizing pullulan-based coatings and films alongside chitooligosaccharides and glutathione is an effective method to extend the shelf life of different types of fruits [26].

f. Alginate-based films and coatings: Alginate, a polysaccharide obtained from brown seaweeds that is widely recognized for its significant role as a valuable component in biopolymer coatings and films. This is primarily due to its unique colloidal properties, which include its capacity for stabilizing emulsions, thickening, creating films, and forming gels [26].

5.5.1.2 Hydrocolloid Based on Proteins

Protein-based hydrocolloids typically exhibit comparable characteristics to films based on polysaccharides. These qualities include resistance to oils, excellent oxygen resistance, and appropriate odor resistance. Typically, their hydrophilic nature results in reduced resistance to water vapour. These coatings are derived from both animals and plants. Protein-based coatings, being naturally hydrophilic, do not offer significant resistance to water vapour. However, they possess favorable mechanical and organoleptic properties [23]. The various types of proteins, including soy protein, whey protein, gelatin, wheat gluten, casein, and corn zein have been employed to create edible coatings or films [26].

a. corn zein-based films and coatings: Corn is a significant provider of zein, a type of protein namely prolamin that can be effectively dissolved in ethanol solutions containing 70% to 80% concentration [35, 36]. Zein possesses inherent hydrophobic properties. The hydrophobic nature of the substance can be attributed to its high concentration of non-polar amino acids [37]. Zein, with its favorable film-forming properties, is utilized in the production of biodegradable coatings and films. The biofilm and coating made from zein exhibit the presence of disulfide and hydrogen bonding [38]. The process of creating coatings and films involves drying a solution of ethanol in zein that is mixed with water. To enhance flexibility, plasticizers are employed in the production of zein coatings and films, as they have a tendency to be brittle in their natural state. Compared to other films, zein coatings or films exhibit superior properties of blocking moisture. To enhance the barrier characteristics of water vapor of zein coatings and films, cross-linking or fatty acids agents are utilized [26].

b. Gelatin-based coatings and films: Producing films using sources of gelatin as the main biopolymer is highly desirable due to their affordability and easy accessibility. Gelatin-based coatings or films demonstrate excellent barrier properties, transparency, and mechanical strength, can be fabricated through either casting or extrusion methods [39–41]. Gelatin is derived from the hydrophobic fibrous protein through hydrolysis. Naturally existing collagen is used as the primary constituent of skin, connective tissues and bones. In gelatin, amino acids are arranged in a precise and organized manner. Gelatin is characterized by its significant concentration of the hydroxyproline, amino acids proline, and glycine. Additionally, gelatin contains both double and single unfolded chains with hydrophilic properties. During the process of gelation, the chains experience a conformational unfolding and strive to revert back to the triple-helix structure characteristic of collagen. The formulation for gelatin coatings and films typically consists of 40% to 70% water, 10% to 30% plasticizer (such as sorbitol or glycerine), and 20% to 30% gelatin. Gelatin is commonly employed in the drug company and for encapsulating oil-based food products [26].

c. Wheat gluten-based films and coatings: Wheat flour contain wheat gluten in which hydrophobic protein is present, is classified as a globular protein consisting of a mixture of molecules made of polypeptide. The elasticity and cohesiveness of gluten contribute to the ease of forming a film during the process. Glutenin and gliadin, commonly known as glutelin and prolamine, are the two components of wheat gluten, which is derived from wheat flour protein [26]. The level of purity in wheat gluten significantly influences both the mechanical properties and visual appearance of the film. Films made from higher purity gluten exhibit enhanced clarity and stability [42]. Films from wheat gluten are produced using spray-drying demonstrate superior stability compared to films from gluten are made from flash-drying with larger particle sizes [43]. The strength of gluten films can be enhanced by the application of a cross-linking agent, leading to improved tensile properties [42, 44]. The higher water content in wheat gluten enhances the interaction between the protein matrix and carbon dioxide, resulting in excellent sorption values at high relative humidity levels [45].

d. Whey protein-based films and coatings: Coatings and films are suitable for consumption are derived from polysaccharides, proteins, lipids, or combinations of these additives. Materials which is protein-based are considered particularly attractive among these options due to their added nutritional value [46]. Proteins from whey are commonly utilized in the production of sports nutrition products and infant formulas, demonstrating a high degree of effectiveness in forming films. As a result, edible substances based on whey protein have been a highly regarded research area in recent decades. Protein films of whey exhibit exceptional gas barrier properties and excellent mechanical resistance, particularly under low humidity conditions. Additionally, these coatings demonstrate exceptional barrier properties against oil and aroma compounds [47].

5.5.2 Composite Coatings

In order to eliminate or overcome the drawbacks of using single-component edible films, we propose the concept of combining the unique characteristics of various separate coating materials in desired proportions is termed as composite coatings [23]. In recent times, there has been a significant emphasis on multicomponent or composite in the field of edible films components. The main objective of this focus is to explore the fundamental advantages of each component while concurrently minimizing their limitations [47, 48]. The primary goal of producing coatings and heterogeneous composite films is to improve the mechanical properties or permeability. Films can be utilized in various forms such as dispersion, emulsion, suspension, or arranged in consecutive layers, commonly known as bi-layer films [38]. A significant disadvantage of bi-layer coatings and films is that their preparation method necessitates four stages, which include two drying stages and two casting stages. Despite of having strong barriers against water vapour, as a result laminated films are less used in the food business [49].

5.5.3 Lipid Coatings

Typically, coatings contain lipids are the preferred choice for preserving fruits and vegetables, as they give them a shiny and glossy appearance. Coating material which is lipid based primarily consist of mineral oil, carnauba wax, vegetable oil, and beeswax [23].

These coatings have certain drawbacks [23].

1) Inadequate structural integrity.
2) Inadequate structural stiffness
3) Rigidness
4) They are commonly resistant to penetration

Lipids provide effective protection against the movement of moisture [50]. When lipids are combined with polysaccharides and proteins, they create coatings that possess enhanced barrier properties and mechanical strength. Apart from that, these films based on composite might have increased susceptibility to moisture compared to films made solely from lipids [51]. Compounds containing lipids, such as surfactants, acetylated monoglycerides, and herbal wax, are used to form a protective coating or film. Beeswax and paraffin represent the most elementary forms of compounds containing lipids. Their low polarity makes coatings and films containing lipids are highly efficient in impeding the transfer of moisture, resulting in their recognized effectiveness. Due to their hydrophobic properties, lipid based coatings or films are generally more fragile and have a greater thickness compared to usual [52]. With an increasing concentration of the hydrophobic phase, here is a decrease in the permeability of water vapor [53]. Nevertheless, there have been reports indicating that films or coatings containing lipids can negatively impact the visual appeal and gloss of the food products they are applied to [52, 53].

a. Waxes and paraffin-based films and coatings: The process of fraction distillation is employed to extract paraffin wax from crude petroleum. It is comprised of complex combinations of hydrocarbons that are generated by catalytically polymerizing ethylene. Typically, coatings and films containing paraffin wax are employed for preserving uncooked vegetables, uncooked fruits and cheese. The leaves of palm tree *(Copoernica Cerifera)* are used in extraction to acquire carnauba wax. The source of honeybee and candelilla plant is used to extract beeswax and candelilla wax [54]. These waxes are utilized to create coatings and film that serve the purpose of preventing gasoline and moisture penetration, while also enhancing the visual appeal of various food items. When these coatings are administered in a substantial thickness, they should be discarded before consumption (such as certain cheeses); however, when used in thin layers, they are deemed edible. These type waxes are effective compounds that create a strong barrier against moisture, making them safe to consume [55].

b. Acetoglyceride-based films and coatings: The combination of acetylated glycerol monostearate with acetic anhydride results in the formation of 1-stearodacetylglycerol. The molten state of the acetylated monoglyceride can be effortlessly transformed into a flexible, wax-like solid form [56]. The majority of lipids can be elongated to a significant extent, reaching up to 102% of their original length, without experiencing significant cracking. GMS (glycerol monostearate) can be stretched to a length that is 800% of its original size [57]. Coatings and films of this kind exhibit reduced vapour transmission compared to coatings and films based on glycan. Typically, coatings and films are utilized on poultry and meat portions to prevent the loss of moisture while in storage [58].

c. Shellac resins based films and coatings: lac resin is a complex combination of cyclic alcohols polymers, derived from the exudation of the lac insect. Gums dissolve readily in basic alcoholic solutions. Gums do not have general recognition as safe (GRAS); they are only accepted as an ancillary substance in food-grade coatings. The primary application of the coatings and films is in the pharmaceutics, with minimal research conducted on their utilization in food products [54]. The coatings and films are derived from these gums are widely used for fruit preservation. Typically, these coatings are developed to add extra shine to the surface of food when being acquisition by the buyer. Coatings and films create an additional obstacle that gases must traverse to pass through. Lac coated citrus fruits typically exhibit reduced internal oxygen levels, improved internal carbon dioxide levels, and higher ethanol content [59]. Elevated ethanol concentrations reduce the flavour of the food product [60–62].

5.6 Edible-Coated Products

The coating materials used for fresh foods must be created with approved substances that comply with regulations, as they are a part of the edible section of the food and are produced using proper manufacturing and hygienic practices. Table 5.2 shows the examples of edible coating for various products [2].

Table 5.2 Example of edible coating for various products [18, 63].

Coating agent	Product	Advantages
Chitosan	Carrot	Preserve sensory attributes, extend the longevity of the product, and minimize the risk of spoilage.
	Banana	Retard alterations in pH, reduction in weight, and acidity. Safeguard against microbial deterioration. Preserve sensory attributes throughout the storage period.
	Apricot	Prolong the duration of shelf life while preserving both total phenolic content and antioxidant activity.
Arabic Gum	Apple	Prevents microbial deterioration and maintains its quality characteristics while being stored. microbial growth is reduced in comparison to the control.
	Cucumber	Reduce the loss of weight. Maintain the nutritive value and additional sensory aspects. Extend the shelf life of product.

(*Continued*)

Table 5.2 Example of edible coating for various products [18, 63]. (*Continued*)

Coating agent	Product	Advantages
Pectin	Sliced Persimmon	Suppress the proliferation of microorganisms, slow down the browning process, and prolong the shelf life.
	Blackberries	Prolong the life of storage. Stop the fungus from decomposing. minimize variations in phenolic content, acidity, and pH while storing.
Guar Gum	Tomato	Minimize water loss, maintain firmness, and preserve other desirable qualities. Slow down alterations in acidity and soluble solids throughout storage.
Xanthan Gum	Table grapes	Maintain the quality characteristics throughout storage, with a decrease in both respiration rate and weight loss.
Pullulan	Pepper	Preserve the nutrients, colour, and firmness while minimizing weight reduction and extend the product shelf life.
Starch	Pumpkin	Enhance the quality characteristics by reducing acidity and mass loss, while preventing the growth of mold, bacteria, and yeast.
Cassava Starch	Pineapple	Efficient in minimizing respiration rate and weight loss, leading to an enhancement in overall quality characteristics.
Alginate	Plum	Prevent the reduction of weight, decelerate the softening process, and inhibit the impact of ethylene.
Aloe Vera Gel	Cherry	Antioxidant potential and phenolic content can be increased by reducing the rate of respiration and also retains the quality and firmness of the product.

(*Continued*)

Table 5.2 Example of edible coating for various products [18, 63]. (*Continued*)

Coating agent	Product	Advantages
Zein	Pumpkin	Growth of micro-organisms is supressed thereby preserving the physiochemical characteristics. Extend the duration of the shelf life.
	Apple	A zein coating containing propylene glycol acts as a barrier to gases which slow down the process of browning and minimize the loss of weight.
	Apricot	Zein coating helps to prevent from acidity loss, microbial growth and moisture loss. It also has ability to increase the shelf life.
	Pears	It prolongs the shelf life and has good gas barrier properties which minimizes the weight loss and slow down the rate of respiration.
Soy Protein	Egg Plant	Prolong the storage life by maintaining firmness, colour and also helps to regulate enzymatic browning process.
	Sliced Melon	Reduction in Vitamin C, Loss of weight and Decrease in rate of respiration. Protect against microbial deterioration and also preserves sensory attributes.
Casein	Carrots	Maintain their firmness and tactile attributes. increase the duration of the shelf life
Egg White	Tomato	The gelatin coating infused with blueberry juice safeguards the quality characteristics. prolong the shelf life and prevent microbial deterioration.
Whey Protein Isolates	Apple	Prolong the shelf life by upholding the colour and firmness. Decrease in rate of respiration thereby preventing from loss of weight.

(*Continued*)

Table 5.2 Example of edible coating for various products [18, 63]. (*Continued*)

Coating agent	Product	Advantages
Gelatin	Strawberry	The incorporation of grapefruit seed extract in a gelatin coating extends the shelf life of the fruit, preserving its sensory characteristics and preventing from spoilage of microbes.
Wheat Protein		Preserve both sensory attributes and firmness, prolong the storage period, and decrease of decay in fungal activity.
Carnauba Wax	Bitter Guard	Reduce weight loss, respiration rate, and microbiological spoiling. prolong the life of storage.
	Jujube Fruit	Prevent microbial deterioration. decrease in respiration rate, water loss, and extended shelf life. maintain sensory attributes throughout storage.
	Sweet Potato	Maintain the nutrition value and sensory characteristics. Postpone growth of microbes, colour change, and loss of mass.
	Avocado	Retain firmness, colour, and reduce mass loss without altering flavour. prolong the shelf life while postponing the ripening process.
	Tomato	Diminish alterations in colour, soluble solids, and pH. Prolong the sensory attributes. Breakdown of chlorophyll is delayed.
	Nectarines	Postpone the growth of fungi. Prolong the postharvest period. Fruit should become more vibrant and firm.

(*Continued*)

Table 5.2 Example of edible coating for various products [18, 63]. (*Continued*)

Coating agent	Product	Advantages
Bees Wax	Tomato	Diminish loss of firmness and weight. Prevent the degradation of microbes. When storing, preserve flavour, colour, and aroma.
	Mandarin	Minimize the loss of weight and rate of respiration, prevent the deterioration of microbes, and extend the shelf life.
	Pears	Preserve the acidity, phenolic content, soluble solids, and pH. Extend the duration of the shelf life.
Candelilla Wax	Avocado	The addition of ellagic acid to candelilla wax coating extends the avocado's shelf life while improving its sensory appeal and nutritional value.
	Strawberries	Strawberry shelf life was increased by candelilla wax containing the *Bacillus subtillis* HFC103 strain, which inhibited the growth of *rhizopus stolonfier.*
Cumin Essential Oil		Maintain colour integrity, postpone the occurrence of fungal deterioration and hue shifts. prolong the shelf life.
Salvia Officinalis Essential Oil		Guard against decay caused by fungi. Retain colour and tactile qualities. Increase the duration of the shelf life.
Cinnamon Essential Oil	Apple	Increase the potential for antioxidants. Minimize the growth of micro-organisms. Preserve firmness, colour, and acidity while storing.

5.7 Properties of Edible Coating

Figure 5.1 shows the various properties of edible coating which are described below.

5.7.1 Mechanical Properties

Modulus of elasticity and Pulling strength are the key attributes for analysing different characteristics of such coatings, but these parameters are not definitive for the films. Several other attributes of food coatings and films are associated, either explicitly or implicitly, In relation to these factors and Attributes. In conclusion, we can assert that these properties impact the quality of the product [64].

5.7.2 Optical Properties

Optical properties impact certain crucial aspects, such as the visual appeal of food. The primary optical features of these coatings include translucency, sheen, and colour. These attributes are surface properties that are determined by human vision. The lustrous finish of the food is determined by the light reflection intensity at the interaction zone of the air and the coating, which directly influences the overall reflection [23].

5.7.3 Thickness

The appropriate thickness of the coating directly impacts the organic traits and the extended shelf life of the coatings on food by influencing absorption and translucency. The quantity of light dispersed by the solution of the

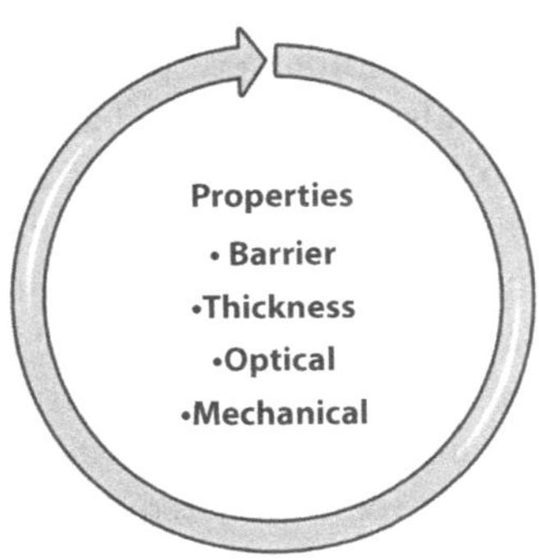

Figure 5.1 Properties of edible coating.

coating. Manages the film thickness and additionally oversees the effectiveness of these film thickness [64].

5.7.4 Barrier Properties

This characteristic shields the food from the aroma-bearing substances, permeation of gases, water and oil. These are the primary factors for maintaining the quality of storage food products during and ensuring extended shelf life. They also determine the inherent worth of the coating. It also determines the capability quality of the coating [23].

5.8 Stability of Edible Coating

After the reap, vegetables and fruits persist in respiring and utilizing the oxygen present in the yield. The edible coating does not replace O_2 as quickly, resulting in the buildup of CO_2 within the yield. The vegetable and fruit will gradually transform to partial oxygen-free respiration, which requires 1-3% less O_2. In low-oxygen conditions, the synthesis of ethylene is hindered, and the biological loss of water is reduced. As a result, the vegetables and fruits maintain their firmness, freshness, and nutritional value for a prolonged duration which is nearly doubling their shelf life. The degree of variation in the internal environment (CO_2 and O_2) and the degree of reduction in loss of weight will depend on the vegetables and fruits inherent hindrance as well as the type and thickness of coating [2].

In order to preserve sensory qualities (appearance, flavor, and scant), the edible coating and films are intended to serve as the main food packaging material for foods containing edible ingredients. Edible films are coated on vegetables and fruits exhibit prolonged shelf life and delayed maturation stages [17].

Coatings that are edible are employed to prolong the shelf life of vegetables and fruits. These can also be consumed as part of the product without introducing undesirable traits to the food [65]. Films or coatings that are edible extend the longevity of vegetables and fruits while also being eco-friendly. Recently, innovative coatings and films that are edible have emerged, incorporating a variety of herbs that are antimicrobials and edible to preserve the freshness of vegetables and fruits [66]. Coatings that are edible are additionally aided in the preservation of rigidity and dampness, thereby preventing their loss. They regulate the process of ripening, growth, and rate of respiration. Coatings that are edible help to reduce the proliferation of microbes and inhibit enzymatic browning in various vegetables and fruits, such as cucumbers, cherries, tomatoes and many more [67]. Films and coatings that are edible contribute to extending the shelf life of vegetables and fruits by minimizing dehydration, transport of solute, exchange of gases, and alleviating biological conditions. Coatings that are edible show significant promise in managing the discoloration, microbe proliferation of fruits and vegetables, unpleasant taste, reduce the activities of browning, and in prolonging their shelf life [2].

5.9 Bioavailability of Edible Coating

Assessing the bioavailability is a common method used to evaluate the drug absorption administration through various route. Bioavailability measures the amount of an active element that is absorbed into the body when it is included in a specific form of dosage (e.g., tablet) and administered through a specific route (e.g., orally). Bioavailability assessment involves measuring its rate, proportion of the drugs incorporated into the dose format it reaches the circulatory system or size. Furthermore, the magnitude-based evaluation of bioavailability can be either relative or absolute in nature. Additionally, a variety of physiological factors, including bile salt, liver metabolism, gut wall metabolism, food effects, gastric emptying, gastrointestinal pH, and absorption mechanism, may have an impact on oral bioavailability. Moreover, oral drug absorption may be impacted by formulation elements like coating type, formulation excipients, or particle size [68, 69]. Especially the pH of gastrointestinal fluids can be impacted the oral medicine of bioavailability due to their composition and natural factor. The pH levels exhibit mutually intra and inter -discrete variability depending on various aspects, and the pH levels within distinct regions vary the tract of gastrointestinal [70, 71]. In many pharmacological dosage forms are consumed orally which can be coated with enteric coatings including microcapsules, drug delivery system, granules, microparticles, tablets, nanoparticles, pellets and the characteristics of the active ingredient in biopharmaceutical particularly its bioavailability can be influenced in various ways [72]. As per the Biopharmaceutics Classification System, managing bioavailability revolves around two crucial factors such as drug molecules solubility and permeability of the intestines [73].

The release rate of the drug from these systems relies on various factors, including the coating leachable levels on components, density of coating, variation in oncotic pressure over the membrane and dissolving of medicine [74]. The application of simple osmosis system regulates the distribution of high-water soluble medications with short to medium biologic intervals and substantial variations in bioavailability through oral administration. Furthermore, it has been discovered that simple osmotic systems are excellent for increasing the bioavailability of medications with inadequate oral absorbance, because of its controlled distribution of drugs [75]. These systems have undergone a number of modifications in recent years to advance the osmotic bursting systems, sandwiched-type osmotic tablets, floating osmotic pumps, liquid-filled osmotic pumps, controlled porosity osmotic pumps, and multiple-chamber push-pull osmotic pumps to modify the drug's absorption properties from the devices in order to improve therapeutic effect and increase oral absorption [76].

5.10 Delivery System of Edible Coating

Typically, the small bowel is regarded as the main location for uptake of drug, making it a favored segment of the alimentary canal for aiming medications using Assorted sustained release approaches. Targeted delivery of drugs to the colon not only enhances the availability of the drug at the desired location but also reduces the necessary dosage and potential side effects. Recent research has highlighted the growing significance of Pharmaceutical delivery mechanisms specifically designed to target the colon, which are being utilized for the treatment of localized Intestinal conditions such as colonic cancer, IBD (inflammatory bowel disease), and regional enteritis, as well as for the systemic administration of peptide and protein therapeutics. This occurs due to the susceptibility of protein and peptide therapeutics to degradation or inactivation within the stomach gastric acidic milieu or as a result of activity of enzymes secreted by the pancreas in the small bowel [77]. Targeting medications to the colon can also be advantageous when a postponement in uptake of drug is sought for medicinal reasons. This is especially pertinent for addressing conditions that manifest their most severe symptoms during the morning hours, such as nighttime asthma, chest pain, or joint inflammation. Several strategies, including the development of inactive drug forms, compositions sensitive to pH changes, temporally regulated compositions, and bacterial-responsive mechanisms, have been devised for the purpose of delivering drugs specifically to the colon [78].

5.11 Evaluation of Edible Coating

Evaluation parameters for coated tablets such as:

5.11.1 Weight Variation

The mean weight of 20 capsules is determined by weighing each capsule independently. The independent capsule weight is evaluated to mean weight [79].

5.11.2 Thickness

Independent capsule thickness can be determined by using a micrometer. Which enabling precise measurements and offering insights into the differences among to the capsule [79].

5.11.3 Firmness

Capsule firmness can be calculated by using capsule firmness testing instrument (Pfizer and Monsanto type). The first method capsule is positioned among the jaws and the impact force, that breaks the capsule has been noted [80]. In other method Monsanto type is ascertained by consuming of 6 capsules from each batch [79].

5.11.4 Friability

The certain amount of capsule is weighted and deposited in the device, in that the capsule are subjected to repetitive shock while falling six inches every time the device turns. Whereas the device revolves sliding 6 inches at each moment [79]. Once the process duration reaches 240 seconds or the number of cycles reaches 100, the capsule is subjected to weighing, and their weight is then contrasted to the weight taken initially. Afterwards the capsule friability is determined by the amount of losses from friction [80].

5.11.5 Disintegration Test

The disintegration test is carried out utilizing a disintegration testing apparatus that includes a rack containing six plastic tubes that are open at both ends, each with a 10-mesh screen covering the bottom. The basket is placed into a container of appropriate liquid maintained at 37°C, ideally within a

1L glass or beaker. The duration for disintegration was evaluated using the Electro lab-manufactured USP disintegration apparatus in 0.1N HCl for a period of 2 hours, subsequent to that, the sample was subjected to pH of 6.8 phosphate buffer with a for a duration of one hour while maintaining at 37 ± 2°C temperature [79, 80].

5.11.6 *In Vitro* Dissolution Study

The assessment for dissolution was conducted utilizing apparatus 2 (USP dissolution tester) employing the paddle method, with a medium consisting of (0.067M) Phosphate buffer at 37°C temperature, 6.8 pH, 50 RPM rotation speeds, 1000 ml volume, and sampling scheduled at intervals of 10, 20, 30, and 45 minutes. The solutions' absorbance was assessed at a 248nm UV wavelength. Inject approximately 10 μl of the liquid used solely for the purpose of the blank, preparation of reference and introduce the sample into the chromatograph. Capture the chromatograms and integrated peak areas for the desired compound peak.

5.11.7 Drug Content Studies

The amount of drug present in the tablets was assessed by selecting tablets (ten) in random from each of the three formulated gastro – resistant tablets and then pulverizing them using a pestle and mortar. A 250 mg quantity of azithromycin was measured, and it was methanol soluble, potentially with dilution. Afterward, the absorbance was determined at a 215 nm wavelength employing a UV 150-02 Dual-Beam Spectrometer manufactured in Japan (by Shimadzu Corp) [80].

5.12 Market Potential of Edible Coating in Food

The increasing demand for eco-friendly packaging solutions is driving the potential market for edible coatings on food products, including medical foods and nutraceuticals, is driven by the increasing demand for sustainable solutions of package [81]. The anticipated growth of the market globally for coatings and films that are edible from 2022 to 2028 is projected to result in a value of 4.87 billion USD by 2033, with a compound annual growth rate of 7.7% (CAGR) [82].

Coatings present a hopeful substitute for materials used for packaging conventionally and have the capacity to support eco-friendly food manufacturing [81]. Nevertheless, they encounter difficulties including complying with regulatory requirements, securing consumer endorsement, and shifting concepts in research to business applications on a broader scale [83]. In spite of these obstacles, films and coatings that are edible are recognized as feasible substitutes for conventional techniques for package, offering the potential to extend the shelf life and enhance the taste of various food products [81].

5.13 Conclusion

To conclude, the intricate production of medical foods and nutraceuticals demands meticulous attention to diverse factors such as, targeted delivery, stability, and taste. The enhancement of product efficacy and consumer approval is significantly influenced by coating, utilizing a diverse methods and materials employed to achieve the intended results. Starting from sugar coating to film coating, the coating process encompasses several stages, such as sealing, polishing, and finishing, each presenting distinct advantages and difficulties. As the demand for specialized dietary products continues to increase, it is imperative to stay current with the latest advancements in coating technology. This is necessary to guarantee the production of products that are both safe and effective, while also being visually appealing, and able to meet the diverse needs of patients and consumers across the globe.

References

1. Kokoszka, S., Lenart, A., Edible coatings-formation, characteristics and use-a review. *Pol. J. Food Nutr. Sci.*, 57, 399, 2007.
2. Dhall, R., Advances in edible coatings for fresh fruits and vegetables: a review. *Crit. Rev. Food Sci. Nutr.*, 53, 435, 2013.
3. Priya, K., Thirunavookarasu, N., Chidanand, D., Recent advances in edible coating of food products and its legislations: A review. *J. Agric. Food Res.*, 12, 100623, 2023.
4. Emamifar, A. and Bavaisi, S., Nanocomposite coating based on sodium alginate and nano-ZnO for extending the storage life of fresh strawberries (Fragaria× ananassa Duch.). *J. Food Meas. Charact.*, 14, 1012, 2020.
5. Resende, N., Gonçalves, G., Reis, K., Tonoli, G., Boas, E., Chitosan/cellulose nanofibril nanocomposite and its effect on quality of coated strawberries. *J. Food Qual.*, 2018, 1–13, 2018.

6. Kaewklin, P., Siripatrawan, U., Suwanagul, A., Lee, Y.S., Active packaging from chitosan-titanium dioxide nanocomposite film for prolonging storage life of tomato fruit. *Int. J. Biol. Macromol.*, 112, 523, 2018.
7. Jafarzadeh, S., Nafchi, A.M., Salehabadi, A., Oladzad-Abbasabadi, N., Jafari, S.M., Application of bio-nanocomposite films and edible coatings for extending the shelf life of fresh fruits and vegetables. *Adv. Colloid Interface Sci.*, 291, 102405, 2021.
8. de Jesús Salas-Méndez, E., Vicente, A., Pinheiro, A.C., Ballesteros, L.F., Silva, P., Rodríguez-García, R., Hernández-Castillo, F.D., Díaz-Jiménez, M., d., L.V., Flores-López, M.L., Villarreal-Quintanilla, J.A., Application of edible nanolaminate coatings with antimicrobial extract of Flourensia cernua to extend the shelf-life of tomato (Solanum lycopersicum L.) fruit. *Postharvest Biol. Technol.*, 150, 19, 2019.
9. El Hosry, L., Auezova, L., Sakr, A., Hajj-Moussa, E., Browning susceptibility of white wine and antioxidant effect of glutathione. *Int. J. Food Sci. Technol.*, 44, 2459, 2009.
10. Oms-Oliu, G., Rojas-Graü, M.A., González, L.A., Varela, P., Soliva-Fortuny, R., Hernando, M., II, Munuera, I.P., Fiszman, S., Martín-Belloso, O., Recent approaches using chemical treatments to preserve quality of fresh-cut fruit: A review. *Postharvest Biol. Technol.*, 57, 139, 2010.
11. Stow, J., Effect of calcium ions on apple fruit softening during storage and ripening. *Postharvest Biol. Technol.*, 3, 1, 1993.
12. Ortiz, A., Graell, J., Lara, I., Cell wall-modifying enzymes and firmness loss in ripening 'Golden Reinders' apples: A comparison between calcium dips and ULO storage. *Food Chem.*, 128, 1072, 2011.
13. Kittemann, D., Neuwald, D., Streif, J., Influence of calcium on fruit firmness and cell wall degrading enzyme activity in'Elstar'apples during storage, in: *VI International Postharvest Symposium*, vol. *877*, p. 1037, 2009.
14. Basaglia, R.R., Pizato, S., Santiago, N.G., de Almeida, M.M.M., Pinedo, R.A., Cortez-Vega, W.R., Effect of edible chitosan and cinnamon essential oil coatings on the shelf life of minimally processed pineapple (Smooth cayenne). *Food Biosci.*, 41, 100966, 2021.
15. Himaja, V., Sai, K., Karthikeyan, R., Srinivasa, B., A comprehensive review on tablet coating. *Austin Pharmacol. Pharm.*, 1, 1, 2016.
16. Arora, R., Rathore, K.S., Bharakatiya, M., An overview on tablet coating. *Asian J. Pharm. Res. Dev.*, 7, 89, 2019.
17. Miteluţ, A.C., Popa, E.E., Drăghici, M.C., Popescu, P.A., Popa, V., II, Bujor, O.-C., Ion, V.A., Popa, M.E., Latest developments in edible coatings on minimally processed fruits and vegetables: A review. *Foods*, 10, 2821, 2021.
18. Oduro, K.O.-A., Edible coating, in: *Postharvest Technology-Recent Advances, New Perspectives and Applications*, 2021.
19. Solís-Morales, D., Sáenz-Hernández, C., Ortega-Rivas, E., Attrition reduction and quality improvement of coated puffed wheat by fluidised bed technology. *J. Food Eng.*, 93, 236, 2009.

20. Zank, J., Kind, M., Schlünder, E.-U., Particle growth and droplet deposition in fluidised bed granulation. *Powder Technol.*, 120, 76, 2001.
21. Bagade, O., Pujari, R., Nemlekar, N., As, P.K., On, M.A., Tablet Coating and Its Outcome with Complementary Sprouting Technology. *Res. J. Pharm. Biol. Chem. Sci. Apprais.*, 5, 298, 2014.
22. Hemchand, P., Doye Avinash, R., Tukaram, M.S., Mahavir, B., Recent advances in different aspects of tablet coating. *Asian J. Pharm. Res. Dev.*, 16, 89, 2017.
23. Singh, A., Pramanik, J., Gururani, P., Different Materials Used For Edible Coating, Their Characteristics and Properties. *Indian J. Pure Appl. Biosci.*, 8, 70, 2020.
24. Phillips, G.O. and Williams, P.A., *Handbook of hydrocolloids*, Elsevier, Netherlands, 2009.
25. Baldwin, E.A., Nisperos-Carriedo, M.O., Baker, R.A., Use of edible coatings to preserve quality of lightly (and slightly) processed products. *Crit. Rev. Food Sci. Nutr.*, 35, 509, 1995.
26. Hassan, B., Chatha, S.A.S., Hussain, A., II, Zia, K.M., Akhtar, N., Recent advances on polysaccharides, lipids and protein based edible films and coatings: A review. *Int. J. Biol. Macromol.*, 109, 1095, 2018.
27. Maghsoudlou, A., Maghsoudlou, Y., Khomeiri, M., Ghorbani, M., Evaluation of anti-fungal activity of chitosan and its effect on the moisture absorption and organoleptic characteristics of pistachio nuts. *Res. Innov. Food Sci. Technol.*, 1, 87, 2012.
28. Devlieghere, F., Vermeulen, A., Debevere, J., Chitosan: antimicrobial activity, interactions with food components and applicability as a coating on fruit and vegetables. *Food Microbiol.*, 21, 703, 2004.
29. Rodríguez, L., Lopez, D.J., Preston, T., Peters, K., New Cocoyam (Xanthosoma sagittifolium) leaves as partial replacement for soya bean meal in sugar cane juice diets for growing pigs. *Livest. Res. Rural Dev.*, 18, 1–14, 2006.
30. Jiang, Z., Neetoo, H., Chen, H., Efficacy of freezing, frozen storage and edible antimicrobial coatings used in combination for control of Listeria monocytogenes on roasted turkey stored at chiller temperatures. *Food Microbiol.*, 28, 1394, 2011.
31. Neetoo, H., Ye, M., Chen, H., Bioactive alginate coatings to control Listeria monocytogenes on cold-smoked salmon slices and fillets. *Int. J. Food Microbiol.*, 136, 326, 2010.
32. Campos, C.A., Gerschenson, L.N., Flores, S.K., Development of edible films and coatings with antimicrobial activity. *Food Bioprocess Technol.*, 4, 849, 2011.
33. Xu, Y., Qiu, Y., Schnable, J.C., Functional modeling of plant growth dynamics. *Plant Phenome J.*, 1, 1, 2018.
34. Daniel, J.R. and Whistler, R.L., Principal changes in starches during food processing, in: *Chemical changes in food during processing*, p. 305, Springer, Heidelberg, New York, 1985.

35. Dickey, L., Kurantz, M., Johnston, D., McAloon, A., Moreau, R., Grinding and cooking dry-fractionated corn germ to optimize aqueous enzymatic oil extraction. *Ind. Crops Prod.*, 32, 36, 2010.
36. Dickey, L., Parris, N., Craig, J., Kurantz, M., Ethanolic extraction of zein from maize. *Ind. Crops Prod.*, 13, 67, 2001.
37. Shukla, R. and Cheryan, M., Zein: the industrial protein from corn. *Ind. Crops Prod.*, 13, 171, 2001.
38. Guilbert, S., Gontard, N., Cuq, B., Technology and applications of edible protective films. *Packag. Technol. Sci.*, 8, 339, 1995.
39. Hanani, Z.N., McNamara, J., Roos, Y., Kerry, J., Effect of plasticizer content on the functional properties of extruded gelatin-based composite films. *Food Hydrocoll.*, 31, 264, 2013.
40. Molinaro, S., Cruz-Romero, M., Sensidoni, A., Morris, M., Lagazio, C., Kerry, J.P., Combination of high-pressure treatment, mild heating and holding time effects as a means of improving the barrier properties of gelatin-based packaging films using response surface modeling. *Innov. Food Sci. Emerg. Technol.*, 30, 15, 2015.
41. Wang, L.Z., Liu, L., Holmes, J., Kerry, J.F., Kerry, J.P., Assessment of film-forming potential and properties of protein and polysaccharide-based biopolymer films. *Int. J. Food Sci. Technol.*, 42, 1128, 2007.
42. Gennadios, A. and Weller, C., Tensile strength increase of wheat gluten films, in: *American Society of Agricultural Engineers. Meeting (USA)*, 1992.
43. Heralp, T., Gnanasambandam, R., McGuire, B., Hachmeister, K., Degradable wheat gluten films: preparation, properties and applications. *J. Food Sci.*, 60, 1147, 1995.
44. Koelsch, C. and Labuza, T.P., Functional, physical and morphological properties of methyl cellulose and fatty acid-based edible barriers. *LWT Food Sci. Technol.*, 25, 404, 1992.
45. Pochat-Bohatier, C., Sanchez, J., Gontard, N., Influence of relative humidity on carbon dioxide sorption in wheat gluten films. *J. Food Eng.*, 77, 983, 2006.
46. Cao, N., Fu, Y., He, J., Preparation and physical properties of soy protein isolate and gelatin composite films. *Food Hydrocoll.*, 21, 1153, 2007.
47. Kurek, M., Galus, S., Debeaufort, F., Surface, mechanical and barrier properties of bio-based composite films based on chitosan and whey protein. *Food Packag. Shelf Life*, 1, 56, 2014.
48. Galus, S., Lenart, A., Voilley, A., Debeaufort, F., Effect of oxidized potato starch on the physicochemical properties of soy protein isolate-based edible films. *Food Technol. Biotechnol.*, 51, 403, 2013.
49. Debeaufort, F. and Voilley, A., Effect of surfactants and drying rate on barrier properties of emulsified edible films. *Int. J. Food Sci. Technol.*, 30, 183, 1995.
50. Debeaufort, F. and Voilley, A., Lipid-based edible films and coatings, in: *Edible films and coatings for food applications*, vol. 135, 2009.

51. Bravin, B., Peressini, D., Sensidoni, A., Influence of emulsifier type and content on functional properties of polysaccharide lipid-based edible films. *J. Agric. Food. Chem.*, 52, 6448, 2004.
52. Perez-Gago, M., Rojas, C., DelRio, M., Effect of lipid type and amount of edible hydroxypropyl methylcellulose-lipid composite coatings used to protect postharvest quality of mandarins cv. fortune. *J. Food Sci.*, 67, 2903, 2002.
53. Valencia-Chamorro, S.A., Pérez-Gago, M.B., Del Río, M.A., Palou, L., Effect of antifungal hydroxypropyl methylcellulose-lipid edible composite coatings on penicillium decay development and postharvest quality of cold-stored "ortanique" mandarins. *J. Food Sci.*, 75, S418, 2010.
54. Hernandez, E., Edible coatings from lipids and resins, in: *Edible coatings and films to improve food quality*, vol. 1, p. 279, 1994.
55. Bourtoom, T., Edible films and coatings: characteristics and properties. *Int. Food Res. J.*, 15, 237, 2008.
56. Feuge, R., Vicknair, E., Lovegren, N., Modification of vegetable oils. XIII. Some additional properties of acetostearin products. *J. Am. Oil Chem. Soc.*, 30, 283, 1953.
57. Jackson, F. and Lutton, E., The polymorphism of 1-stearyl-and 1-palmityldiacetin,-dibutyrin,-dicaproin and 1-stearyldipropionin. *J. Am. Chem. Soc.*, 74, 4827, 1952.
58. Kester, J.J. and Fennema, O., Edible films and coatings: a review. *Food Technol. (USA)*, 40, 47–59, 1986.
59. Hagenmaier, R.D. and Baker, R.A., Reduction in gas exchange of citrus fruit by wax coatings. *J. Agric. Food. Chem.*, 41, 283, 1993.
60. Ahmad, M. and Khan, I., Effect of waxing and cellophane lining on chemical quality indices of citrus fruit. *Plant Foods Hum. Nutr.*, 37, 47, 1987.
61. Cohen, E., Shalom, Y., Rosenberger, I., Postharvest Ethanol Buildup and Off-flavor inMurcott'Tangerine Fruits. *J. Am. Soc. Hortic. Sci.*, 115, 775, 1990.
62. Ke, D. and Kader, A.A., Tolerance ofValencia'oranges to controlled atmospheres, as determined by physiological responses and quality attributes. *J. Am. Soc. Hortic. Sci.*, 115, 779, 1990.
63. Armghan Khalid, M., Niaz, B., Saeed, F., Afzaal, M., Islam, F., Hussain, M., Mahwish, Muhammad Salman Khalid, H., Siddeeg, A., Al-Farga, A., Edible coatings for enhancing safety and quality attributes of fresh produce: A comprehensive review. *Int. J. Food Prop.*, 25, 1817, 2022.
64. Paul, S.K., *Edible films and coatings for fruits and vegetables*, Springer, Heidelberg, New York, 2020.
65. Baldwin, E., Nisperos, M., Chen, X., Hagenmaier, R., Improving storage life of cut apple and potato with edible coating. *Postharvest Biol. Technol.*, 9, 151, 1996.
66. Momin, M.C., Jamir, A.R., Ankalagi, N., Henny, T., Devi, O.B., Edible coatings in fruits and vegetables: A brief review. *Pharma Innov. J.*, 10, 71, 2021.

67. Kumar, S. and Bhatnagar, T., Studies to enhance the shelf life of fruits using Aloe vera based herbal coatings: A review. *Int. J. Agric. Food Sci. Technol.*, 5, 211, 2014.
68. Martinez, M.N. and Amidon, G.L., A mechanistic approach to understanding the factors affecting drug absorption: a review of fundamentals. *J. Clin. Pharmacol.*, 42, 620, 2002.
69. Song, N.-N., Zhang, S.-Y., Liu, C.-X., Overview of factors affecting oral drug absorption. *Asian J. Drug Metab. Pharmacokinet*, 4, 167, 2004.
70. Abuhelwa, A.Y., Foster, D.J., Upton, R.N., A quantitative review and meta-models of the variability and factors affecting oral drug absorption—part I: gastrointestinal pH. *AAPS J.*, 18, 1309, 2016.
71. Abuhelwa, A.Y., Williams, D.B., Upton, R.N., Foster, D.J., Food, gastrointestinal pH, and models of oral drug absorption. *Eur. J. Pharm. Biopharm.*, 112, 234, 2017.
72. Maderuelo, C., Lanao, J.M., Zarzuelo, A., Enteric coating of oral solid dosage forms as a tool to improve drug bioavailability. *Eur. J. Pharm. Sci.*, 138, 105019, 2019.
73. Keen, J.M., McGinity, J.W., Williams III, R.O., Enhancing bioavailability through thermal processing. *Int. J. Pharm.*, 450, 185, 2013.
74. Verma, R., Mishra, B., Garg, S., Osmotically controlled oral drug delivery. *Drug Dev. Ind. Pharm.*, 26, 695, 2000.
75. Sareen, R., Jain, N., Kumar, D., n insight to osmotic drug delivery. *Curr. Drug Deliv.*, 9, 285, 2012.
76. Gupta, B.P., Thakur, N., Jain, N.P., Banweer, J., Jain, S., Osmotically controlled drug delivery system with associated drugs. *J. Pharm. Pharm. Sci.*, 13, 571, 2010.
77. Yang, L., Chu, J.S., Fix, J.A., Colon-specific drug delivery: new approaches and *in vitro/in vivo* evaluation. *Int. J. Pharm.*, 235, 1, 2002.
78. Ravi, V. and SiddaramaiahPramod Kumar, T., Influence of natural polymer coating on novel colon targeting drug delivery system. *J. Mater. Sci.: Mater. Med.*, 19, 2131, 2008.
79. Pole, S., Maurya, S., Hasnale, P., Rathod, N., Bendale, S., Khutle, N., A detail understanding of enteric coated tablet: manufacturing and evaluation. *Eur. J. Pharm. Med. Res.*, 3, 135, 2016.
80. Patil, A., Payghan, S., Disouza, J., Formulation and evaluation of enteric coated tablets of azithromycin dehydrate. *Int. J. Chem. Tech. Res.*, 3, 1479, 2011.
81. Iversen, L.J.L., Rovina, K., Vonnie, J.M., Matanjun, P., Erna, K.H., Aqilah, N.M.N., Felicia, W.X.L., Funk, A.A., The emergence of edible and food-application coatings for food packaging: a review. *Molecules*, 27, 5604, 2022.
82. Rossi-Márquez, G., Dávalos-Saucedo, C.A., Di Pierro, P., *Edible Films and Coatings Applied in the Food Industry*. vol. 13, p. 670, MDPI, 2023.
83. Janjarasskul, T. and Krochta, J.M., Edible packaging materials. *Annu. Rev. Food Sci. Technol.*, 1, 415, 2010.

6

Edible Packaging: Extension of Shelf Life and Improvement of Food Quality

Ritika[1,2] and Rizwana[2]*

[1]Department of Home Science, University of Delhi, New Delhi, India
[2]Department of Food Technology, Bhaskaracharya College of Applied Sciences, University of Delhi, New Delhi, India

Abstract

Consumers' desire to purchase healthy food products has increased, with fewer ingredients being accepted and fresher options being preferred. In order to preserve their natural and fresh appearance for an extended period as possible while also making them safe, this has led to the creation of many cutting-edge approaches. For the commodity to receive proper (mechanical and functional) protection, packaging has been a critical component in various preservation methods. The need to minimize plastic and food waste creation has become more widely recognized in society due to our present ecological predicament. Researchers are creating environmentally friendly active packaging technologies, such as edible packaging manufactured from leftover industrial materials, to address environmental concerns and lessen plastic waste. Increasing food quality and safety, extending shelf life, and lowering plastic and food waste are the goals of these developments. Edible coatings have come to light as an alternative that offers several benefits in preserving food quality. This is especially significant in light of the increased fresh and minimally processed foods market. Edible coatings and packaging have been suggested as an alternative or addition to traditional packaging, as synthetic packaging materials lead to environmental pollution. Edible coatings have thus been viewed as sustainable packaging for food that can substitute for conventional plastic packaging, enhancing the shelf life of the product and taking into account the food industry's aim to prevent food spoilage along the food chain. Edible coatings are thin films of biopolymers that are directly applied to the surface of food products to enhance their physicochemical properties. Edible packaging is a type of active packaging. It is crucial in packaged meals because

**Corresponding author*: rizwana.haleem@bcas.du.ac.in

Mousumi Sen (ed.) Food Coatings and Preservation Technologies, (167–210) © 2025 Scrivener Publishing LLC

it favorably interacts with the food and performs technological functions. One type of edible packaging is an edible coating placed directly on the food product, while the other type is premade film wrapped over the food product. There are two ways to make edible films: wet (casting) and dry (extrusion). When placing edible coatings on the surface of food products, techniques like dipping, spraying, fluidized beds, and panning are utilized. Typically, these coatings contain other substances, such as nanoparticles or essential oils. In order to extend shelf life, these materials must have acceptable sensory properties and operate as a barrier against gases, including water vapor, moisture loss, development of microbes, and oxidation. They ought to be inexpensively made, safe, and generated using basic technologies. Additionally, they can serve as an efficient transporter for additives like antioxidants, flavors, colors, nutrients, or anti-microbials. The utility of edible materials as a supplement to packaged foods has recently been the subject of various research. Therefore, the chapter's objective is to combine the most recent information available to provide a comprehensive understanding of the formulation method and the many approaches to enhancing the properties of the coatings applied to food products.

Keywords: Edible packaging, wet casting, dry extrusion, shelf life, applications

6.1 Introduction

The world of packaging in the food industry has undergone a significant transformation with the emergence of edible packaging. Unlike traditional packaging materials that contribute to environmental pollution, edible packaging offers a sustainable solution by utilizing materials that are safe for consumption [1]. Edible packaging can be defined as a category of materials derived from various sources such as proteins, polysaccharides, and lipids, which are safe for consumption and serve as both a container and a component of the food product. Edible packaging refers to materials designed to protect and contain food, but with the twist that they can be consumed alongside the food or biodegrade quickly and safely [2]. It is a revolutionary concept aimed at tackling the immense plastic waste problem generated by conventional food packaging. The concept of edible packaging predates written history, with early examples dating back to ancient civilizations. In the 12th century, Chinese farmers utilized beeswax to coat citrus fruits, extending shelf life and enhancing visual appeal [3]. Similarly, 16th-century Japan saw mochi dough employed as a natural wrapper for rice cakes, a tradition that continues to delight palates today [4]. Across continents, Native Americans relied on readily available leaves and corn husks to preserve meat and fish, showcasing the ingenuity and resourcefulness of pre-industrial societies [5]. The 20th century witnessed a dramatic shift in food packaging with the invention of cellophane and other plastic-based materials. These offered unparalleled convenience and durability, revolutionizing the industry. However, their environmental impact became increasingly evident, leading to concerns about plastic pollution and the accumulation of waste in landfills and oceans [6]. By the 1970s, growing environmental consciousness ignited a renewed interest in biodegradable packaging alternatives, reigniting exploration into edible packaging solutions. The 21st century has witnessed a blossoming of innovation in the field of edible packaging. Seaweed-based films, derived from readily available and renewable resources, have risen to prominence, effectively wrapping fruits, vegetables, and even encapsulating water in edible bubbles [7]. Starch-based coatings, applied to fruits, extend shelf life and offer a subtle sweetness, further highlighting the potential for functional integration and waste reduction [8]. Moreover, edible cutlery crafted from wheat bran or rice bran provides a sustainable alternative to plastic utensils, further minimizing the environmental footprint of food consumption [9]. In the current paradigm of sustainability, edible packaging emerges as a captivating contender, poised to revolutionize the food industry. Transcending the traditional role of mere protection, this innovative approach offers a plethora of intriguing possibilities, including the potential to enhance food quality, reduce waste generation, and even deliver essential nutrients. However, effectively harnessing the full spectrum of benefits necessitates a meticulous examination of the underlying properties that define this novel technology. Therefore, this chapter embarks on a critical exploration of the key properties of edible packaging, meticulously dissecting their multifaceted roles in ensuring food safety and quality. By demystifying these characteristics, we aim to illuminate the intricate interplay between functionality and sustainability, paving the way for a comprehensive understanding of this revolutionary technology. This analysis will not only delve into the essential barrier properties—moisture, gas, and oil barriers—but also delve into the mechanical aspects such as strength and flexibility, and functional properties like antimicrobial activity and nutrient fortification potential. Each facet will be meticulously examined, highlighting both its contributions and the inherent challenges associated with optimization [10].

In the food industry, packaging plays a far more significant role than simply containing and protecting products. It acts as a silent partner, influencing everything from food safety and quality to consumer behavior and brand identity [11]. Understanding the importance of packaging in this context is crucial for stakeholders across the entire food chain, from producers and processors to retailers and consumers. The importance of packaging in the preservation and presentation of food cannot be overstated. It serves as a protective layer, safeguarding products from external factors that could compromise their quality, such as moisture, oxygen, light, and microorganisms.

1. Ensuring Food Safety and Quality:

- Protection: Packaging safeguards food from physical damage, contamination, and spoilage during transportation, storage, and distribution. This includes preventing exposure to light, moisture, oxygen, and pests, all of which can compromise food quality and safety [12].
- Hygiene and Sanitation: Packaging plays a vital role in maintaining hygiene and sanitation throughout the food supply chain. Proper packaging materials and processes can minimize the risk of cross-contamination with harmful bacteria and pathogens [13].
- Extended Shelf Life: Certain packaging technologies, such as modified atmosphere packaging and vacuum packaging, can significantly extend the shelf life of food products by controlling the environment within the package. This reduces food waste and improves food security [14].

6.2 Facilitating Convenience and Functionality

- Portion Control and Convenience: Packaging allows for convenient portion control, catering to individual needs and reducing food waste [15]. Additionally, features like resealable closures and easy-open packaging enhance convenience for consumers.
- Information and Transparency: Packaging provides a platform for essential information about the food product, including ingredients, nutritional value, allergens, and production and expiry dates. This transparency empowers consumers to make informed choices [16].
- Marketing and Branding: Packaging serves as a powerful marketing tool, influencing brand perception and consumer purchase decisions. Attractive and informative packaging

designs can differentiate products, build brand loyalty, and drive sales [17].

6.3 Addressing Environmental Concerns

- Sustainability: The food industry is increasingly focusing on sustainable packaging practices to minimize environmental impact. This includes using biodegradable and compostable materials, reducing packaging waste, and adopting recycling initiatives [18].
- Food Waste Reduction: Packaging plays a crucial role in reducing food waste throughout the supply chain [19]. By protecting food from spoilage and damage, packaging optimizes shelf life and minimizes losses.

In today's globalized world, the food and consumer goods industries rely heavily on packaging. While packaging plays a crucial role in protecting products, ensuring hygiene, and facilitating convenience, its environmental impact is increasingly becoming a concern of critical importance. The current reliance on conventional, non-sustainable packaging solutions poses a significant threat to our planet and raises a pressing need for immediate action [20, 21].

The Alarming Scale of the Problem:

- Plastic Pollution: Conventional packaging, particularly plastic-based materials, contributes significantly to plastic pollution. An estimated 8 million metric tons of plastic enter the oceans every year, harming marine life and ecosystems. Landfills are also overflowing with plastic waste, taking centuries to decompose and leaching harmful chemicals into the environment [22–24].
- Resource Depletion: Producing traditional packaging materials uses valuable resources like trees, water, and fossil fuels. Deforestation contributes to climate change, and water scarcity remains a global challenge [25].
- Climate Change Impact: The production and disposal of non-sustainable packaging generates greenhouse gas emissions, accelerating climate change and its devastating consequences [26, 27].

The Imperative for Sustainable Alternatives:

To address these growing environmental concerns, transitioning to sustainable packaging solutions is no longer an option, but a necessity. This necessitates a shift towards:

- Biodegradable and compostable materials: Utilizing materials derived from renewable resources like plant-based polymers, seaweed, or mycelium allows for decomposition without harming the environment [28].
- Reusable and refillable packaging: Encouraging and facilitating the reuse of packaging minimizes waste and extends the lifespan of materials.
- Reduced packaging: Adopting minimalist packaging designs and optimizing material usage further decreases environmental impact [29].
- Improved recycling infrastructure: Efficient recycling systems and consumer awareness are crucial for closing the loop on packaging materials [30].

Beyond Environmental Benefits:

Transitioning to sustainable packaging offers not just environmental benefits, but also economic advantages:

- Cost savings: Long-term, utilizing renewable resources and reducing waste can lead to cost efficiencies in production and disposal [31].
- Brand reputation: Consumers are increasingly conscious of environmental concerns and value companies that prioritize sustainability. Building a brand image rooted in eco-friendly practices can attract and retain customers.
- Innovation and market opportunities: Investing in sustainable packaging solutions fosters innovation and opens up new market opportunities for businesses catering to the growing demand for environmentally responsible products [32].

The transition to sustainable packaging is not without challenges. Higher initial costs, concerns about performance and shelf life, and limited availability of certain materials are some of the hurdles that need to be addressed [33]. However, ongoing research and development, collaboration between stakeholders across the supply chain, and government policies promoting sustainable practices can pave the way for a smoother transition. Edible packaging emerges as a solution to this predicament, aligning with the global push towards sustainable and eco-friendly practices [34]. The primary objective of this chapter is to provide a comprehensive understanding of edible packaging, its scientific underpinnings, mechanisms for extending shelf life, its impact on food quality, and the challenges and solutions associated with its implementation. This chapter delves into the realm of edible packaging, exploring its historical development, current applications, and the pivotal role it plays in extending the shelf life of food products while simultaneously enhancing their overall quality.

6.3.1 The Science Behind Edible Packaging

As mountains of plastic waste continue to pollute our oceans and landfills, the need for sustainable alternatives has never been more pressing. Edible packaging offers a glimmer of hope, providing a solution that not only protects food but also nourishes the planet. Table 6.1 provides the glimpse of the various natural edible material available for packaging. The

Table 6.1 Various edible materials used for packaging.

Edible material	Source	Potential benefits	Limitations
Proteins	Gelatin, Collagen	Structural integrity Flexibility Biodegradability	Limited barrier properties Sensitivity to moisture
Polysaccharides	Starch, cellulose, chitosan, alginate	Barrier properties (moisture control) Biodegradability Antimicrobial properties	Limited mechanical strength
Lipids	Waxes, liposomes	Moisture resistance Controlled release of bioactive compounds Compatibility with lipophilic substances	Limited barrier properties Potential for oxidation

rise of edible packaging aligns with a growing consumer demand for eco-conscious products and a circular economy where waste is minimized.

6.4 Proteins

Origin: Proteins used in edible packaging are often derived from animal or plant sources. Gelatin, a widely used protein in edible films, is typically extracted from the collagen found in animal connective tissues, such as bones and skin. Collagen, another protein, is sourced similarly from animal tissues [35].
Chemical Structure: Gelatin is a mixture of peptides and proteins produced by partial hydrolysis of collagen. It is composed of amino acids, primarily glycine, proline, and hydroxyproline. Collagen, the precursor to gelatin, is a fibrous protein with a triple helical structure, providing it with strength and stability [36].
Stability: Gelatin-based films have good film-forming properties and flexibility. However, they exhibit sensitivity to moisture, limiting their barrier properties. The stability of gelatin-based films is influenced by humidity levels, and they may undergo degradation over time, affecting their mechanical properties. The ability of proteins to form films and coatings makes them suitable for encapsulating and protecting food products. Polysaccharides, on the other hand, exhibit excellent moisture barrier properties, preventing dehydration of the packaged items [37].

6.5 Polysaccharides

Origin: Polysaccharides used in edible packaging, such as starch and chitosan, are often derived from plant and marine sources. Starch is primarily extracted from cereals like corn, wheat, and potatoes, while chitosan is derived from chitin, a natural polymer found in the exoskeletons of crustaceans like shrimp and crab [38].
Chemical Structure: Starch is a carbohydrate consisting of glucose units linked by glycosidic bonds. It has an amylose-amylopectin structure. Chitosan is derived from chitin through deacetylation and is composed of N-acetylglucosamine units. The chemical structure of chitosan provides it with antimicrobial properties [39].
Stability: Starch-based films offer good barrier properties against water vapor but may have limitations in mechanical strength. Chitosan, with its antimicrobial activity, enhances the stability of edible films by inhibiting

microbial growth. However, its stability can be influenced by pH and temperature variations [40].

6.6 Lipids

Origin: Lipids used in edible packaging, including waxes and liposomes, are sourced from various plant and animal fats. Waxes can be derived from natural sources like beeswax, carnauba wax, or soy wax. Liposomes, on the other hand, are artificial lipid vesicles created from phospholipids, often sourced from soy or sunflower lecithin [41].
Chemical Structure: Waxes are composed of long-chain fatty acids and esters, providing them with water-repellent properties. Liposomes are spherical structures composed of a lipid bilayer, mimicking cell membranes. The lipid bilayer is typically made up of phospholipids, which consist of hydrophilic head groups and hydrophobic tails [42].
Stability: Waxes contribute to the moisture resistance of edible films, creating a protective barrier. However, their barrier properties may be limited concerning gases like oxygen. Liposomes, with their versatile structure, can encapsulate bioactive compounds and release them slowly, enhancing stability. However, lipid oxidation is a potential concern that can affect the stability of liposomes. Lipid-based edible packaging materials not only provide a protective layer but can also be engineered to release bioactive compounds, adding an extra layer of functionality [43].

6.6.1 Manufacturing Process

Wet casting and dry extrusion are two distinct methods employed in the manufacturing of edible packaging, each offering unique advantages and applications within the field. These processes play a crucial role in creating edible films and coatings that contribute to sustainable packaging solutions [44]. Figure 6.1 describes the manufacturing process of wet casting along with characteristics and the applications. Figure 6.2 explains the dry extrusion.

The growing quest for sustainable solutions has cast a spotlight on edible packaging, offering an intriguing alternative to traditional plastics. This innovative approach goes beyond mere protection, potentially contributing to food quality, reducing waste, and even delivering nutrients. Yet, harnessing the full potential of edible packaging requires a nuanced understanding of its diverse properties. Just like traditional packaging, edible films and coatings must strike a delicate balance between various

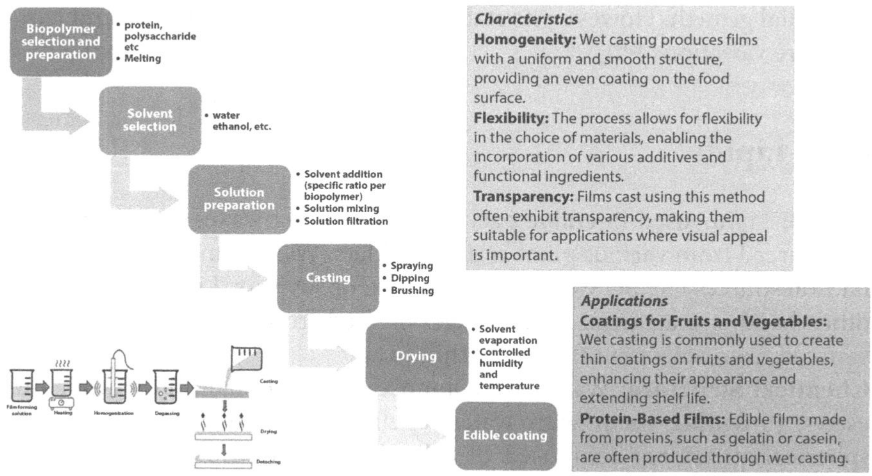

Figure 6.1 Wet casting of edible packaging films [45].

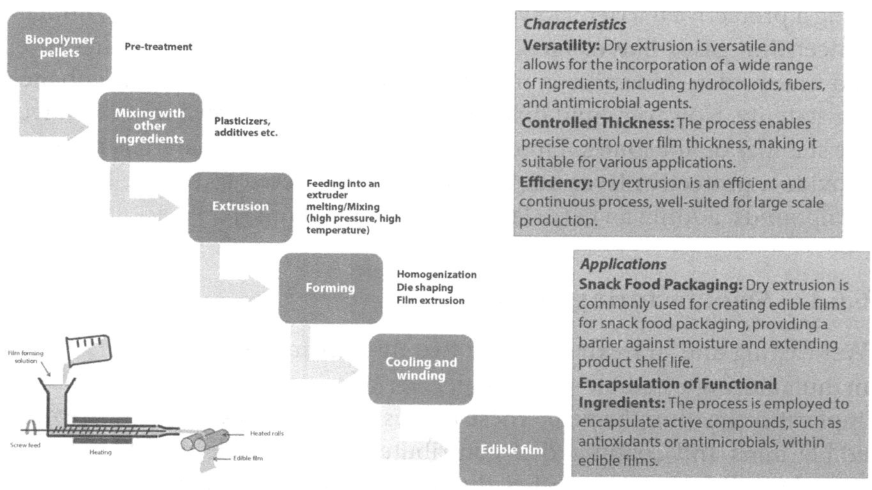

Figure 6.2 Dry extrusion of edible packaging films [46].

characteristics. Table 6.2 delve into the key properties of edible packaging, exploring their roles in safeguarding food while considering the challenges and opportunities that lie ahead.

Edible packaging materials possess a set of distinctive properties that render them suitable for diverse applications within the food industry.

Table 6.2 Comparison between the wet casting and dry extrusion technique.

Feature	Wet casting	Dry extrusion
Solvent usage	Required	Not required
Processing temperature	Lower	Higher
Control over film properties	Higher (thickness, uniformity)	Lower
Scalability	Limited	High
Incorporation of ingredients	Better suited for sensitive ingredients	Limited by high temperatures
Production cost	Generally lower	Generally higher
Suitable biopolymers	Wide range (proteins, polysaccharides, lipids)	Mainly starches, protein isolates
Energy consumption	Lower	Higher
Waste generation	Can generate solvent waste	Minimal waste
Environmental impact	Can be environmentally friendly with solvent selection	Lower dependence on solvents, but high energy use
Advantages	Precise control, suitable for diverse materials, good for sensitive ingredients	Scalable, high throughput, no solvents
Disadvantages	Solvent use, limited scalability, slower process	High temperatures, lower control over properties, limited material suitability

These inherent characteristics contribute to the efficacy, safety, and sustainability of edible packaging solutions. Key properties of note include biodegradability, as edible packaging materials commonly exhibit the capacity to undergo natural decomposition, aligning with sustainability imperatives. Moreover, edibility is a defining attribute, ensuring that these materials are fit for direct consumption without posing health risks. These properties contribute to the effectiveness, safety, and sustainability of edible packaging solutions. Here are some key properties of edible packaging:

1. **Biodegradability:** Edible packaging materials are often biodegradable, meaning they can break down naturally into harmless substances over time. This property aligns with sustainability goals and reduces environmental impact compared to traditional non-biodegradable packaging.
2. **Edibility:** The primary distinguishing feature is that edible packaging is safe for consumption. This property allows for direct contact with food items without concerns about the transfer of harmful substances.
3. **Film-Forming Ability:** Edible packaging materials have the capability to form thin films or coatings on the surface of food products. This property enables the creation of protective layers that can extend shelf life, preserve freshness, and enhance food safety.
4. **Barrier Properties:** Edible packaging provides barrier properties against external factors such as oxygen, moisture, and light. Effective barrier properties contribute to the preservation of the enclosed food by preventing oxidation, moisture-induced spoilage, and degradation of sensitive components.
5. **Flexibility:** Edible films and coatings are often flexible, allowing them to conform to the shape of various food products. This property is crucial for creating coatings on irregularly shaped items and ensuring that the packaging does not affect the overall texture or appearance of the food.
6. **Taste Neutrality:** Edible packaging materials are designed to be taste-neutral, meaning they do not impart undesirable flavors or aromas to the enclosed food. This property ensures that the packaging does not compromise the sensory attributes of the food product.
7. **Texture:** Depending on the application, edible packaging materials can have varying textures, ranging from thin and transparent films to thicker and more robust coatings. The texture is often tailored to suit the specific requirements of the food product.
8. **Solubility:** Some edible packaging materials are soluble in water or other liquids. This solubility can be advantageous for certain applications, such as single-use sachets

or water-soluble coatings that dissolve upon contact with moisture.

9. **Adaptability to Food Types:** Edible packaging can be customized and adapted to different types of food. This adaptability ensures that the packaging material complements the specific characteristics and requirements of various food items.
10. **Safety and Regulatory Compliance:** Edible packaging materials must meet safety standards and regulatory requirements for contact with food. This property ensures that the materials used are safe for consumption and do not introduce harmful substances.
11. **Incorporation of Functional Ingredients:** Edible packaging can be enriched with functional ingredients such as antimicrobial agents, antioxidants, or vitamins. This property allows for the creation of packaging solutions that actively contribute to food safety and quality.
12. **Temperature Stability:** Edible packaging materials exhibit stability across a range of temperatures, ensuring that they maintain their structural integrity and protective properties during storage, transportation, and handling.
13. **Transparency:** Transparency is a desirable property for certain edible films, as it allows consumers to see the food product inside the packaging. This property is particularly important for items where visual appeal is a significant factor.
14. **Printability:** Edible packaging materials can often be printed on, allowing for the inclusion of branding, nutritional information, or other relevant details. This property enhances the marketability of packaged food products.

6.7 Mechanisms of Shelf Life Extension by Edible Packaging

The extension of shelf life in edible packaging involves the implementation of various mechanisms that contribute to preserving the quality and safety of the enclosed food products. Shelf life extension in the context of edible packaging represents a multifaceted endeavor at the intersection of food science and packaging technology. These mechanisms collectively address factors such as oxidation, moisture absorption, microbial contamination, and external environmental influences. The following elucidates the key mechanisms employed for shelf life extension in the context of edible packaging:

6.7.1 Barrier Properties

One of the foundational mechanisms contributing to shelf life extension is the incorporation of materials with robust barrier properties. The packaging materials serve as barriers against oxygen and moisture, two key elements that can significantly impact the quality of food products. Effective oxygen barriers prevent or minimize the ingress of oxygen, mitigating oxidative reactions that can lead to the degradation of fats, flavors, and other sensitive compounds. Moisture barriers, on the other hand, play a crucial role in controlling moisture absorption, which is vital for preventing spoilage, maintaining texture, and inhibiting the growth of microorganisms [47].

6.7.2 Antimicrobial Properties

The integration of antimicrobial properties into edible packaging materials represents an innovative strategy to combat microbial contamination and extend shelf life. This involves the incorporation of natural antimicrobial agents, such as essential oils or plant extracts, into the packaging. These agents act as inhibitors against the growth of bacteria and fungi, providing an added layer of protection against spoilage and microbial-induced degradation [48].

6.7.3 pH Regulation

Maintaining the optimal pH of packaged food products is paramount for ensuring their stability and longevity. Edible packaging can contribute to pH stabilization, preventing undesirable changes that may occur due to acidification or alkalization. The preservation of the original pH helps in retaining the taste, texture, and color of the food, thereby contributing to an extended shelf life [49].

6.7.4 Temperature Control

Temperature fluctuations can have a profound impact on the quality of food products. Edible packaging materials are designed to provide insulation, helping to maintain the desired storage temperature for the enclosed food. This is particularly crucial for preserving freshness, preventing accelerated degradation, and extending the overall shelf life of the product [50].

6.7.5 Light Protection

Protection against external environmental factors, such as light exposure, is another critical mechanism in shelf life extension. Edible packaging materials can shield the enclosed food from ultraviolet (UV) light, which has the potential to cause photochemical reactions leading to nutrient degradation, color changes, and flavor alterations [51].

6.7.6 Active Packaging

Active packaging involves the incorporation of components that actively interact with the food to extend its shelf life. Examples include oxygen scavengers, moisture absorbers, and ethylene absorbers, which help control environmental conditions within the packaging. These active components contribute to the preservation of the enclosed food by minimizing factors that could lead to spoilage [52].

6.7.7 Edible Coatings

The application of edible coatings directly to the surface of food products represents an additional mechanism for shelf life extension. These coatings, often derived from natural polymers, create a protective barrier that can include antimicrobial agents or antioxidants. Edible coatings help in preserving the quality and safety of the food, contributing to an extended shelf life [53].

6.7.8 Controlled Release of Additives

To enhance the efficacy of preservation, edible packaging can be designed to release additives gradually. This controlled release of preservatives or antioxidants provides sustained protection against factors that contribute to spoilage over an extended period, thus contributing to a longer shelf life for the food product.

6.7.9 Adaptability to Food Types

The customization and adaptability of edible packaging solutions to different food types play a crucial role in ensuring the effectiveness of preservation strategies. Understanding the unique characteristics of each food product help in designing packaging that optimally preserves its quality. Tailored solutions contribute to a more targeted and efficient approach to shelf life extension [54].

6.7.10 Sensory Preservation

Preserving the sensory attributes of the food is integral to the success of shelf life extension in edible packaging. This involves minimizing the impact of external factors on taste, aroma, and texture. The preservation of these sensory qualities contributes not only to an extended shelf life but also to an enhanced consumer experience [55].

6.8 Enhancement of Food Quality

Edible packaging, being an integral part of the food product itself, can have a direct impact on the sensory attributes of taste, aroma, and color. The choice of edible packaging materials and their interactions with the enclosed food play a significant role in shaping the overall consumer experience. Table 6.3 shows how edible packaging influences each of these sensory attributes

6.8.1 Texture Retention in Packaged Foods

Texture retention in edible packaging constitutes a paramount aspect of food preservation, emphasizing the maintenance of the desired textural qualities of enclosed food throughout its shelf life [56]. Edible packaging materials play a pivotal role in this endeavor by preventing moisture absorption, thereby preserving the crispness or crunchiness of certain snacks. Effective barriers against oxygen are crucial to impede oxidative reactions that may compromise the structural integrity of foods. The stability of edible packaging across varying temperatures is vital to ensuring that the original texture of the food is maintained during transportation, storage, and display [57]. Additionally, edible packaging must provide physical protection against external factors, safeguarding against potential damage or alteration of the food's textural attributes. The choice of packaging materials, their thickness, and composition all contribute to the successful retention of texture, with customization being key to addressing the diverse needs of different food types. Ultimately, texture retention in edible packaging represents a delicate balance between preservation and

Table 6.3 Influence of edible packaging on sensory attributes.

Sensory attribute	Potential benefits	Potential challenges	Influencing factors
Taste	Flavor masking of undesirable elements	Off-flavors from packaging material	Material selection
	Flavor enhancement through added ingredients	Migration of flavors between food and packaging	Processing techniques
	Controlled release of encapsulated flavors/spices	Masking desirable flavors	Food characteristics
Aroma	Aroma preservation by acting as a barrier	Aroma absorption by packaging material	Storage conditions (temperature, humidity)
	Aroma enhancement through controlled release	Off-odors from packaging material	Food aroma profile
	Masking undesirable aromas	Complex interactions with food aroma	Material-food compatibility
Color	Color preservation by blocking light	Migration of color pigments from packaging Color mismatch between packaging and food	Material selection Food properties (e.g., fat content, pH)

protection, enhancing the overall quality and consumer experience of packaged foods. Ongoing research endeavors seek to further refine and innovate in this area, addressing the dynamic challenges inherent in maintaining texture over extended periods [58].

6.8.2 Nutrient Retention and Preservation

While edible packaging presents a promising pathway towards a greener future, its impact on nutrient content and preservation within food products is not straightforward. On one hand, exciting opportunities exist. Edible packaging can be fortified with vitamins and minerals, directly contributing to the product's nutritional value [59]. It can also act as a barrier against light, oxygen, and moisture, potentially slowing down nutrient degradation compared to conventional options. Additionally, incorporating natural antioxidants within the packaging can further protect nutrients and extend shelf life. However, challenges remain. Unwanted nutrient migration between the packaging and food can alter the nutritional profile and compromise safety. Some edible materials might not be suitable for preserving certain light- or heat-sensitive nutrients, and processing techniques used to create the packaging can inadvertently degrade nutrients. Careful consideration of material selection, processing methods, food characteristics, and storage conditions is crucial to maximize the benefits and minimize the drawbacks. Ongoing research is exploring novel materials with superior barrier properties and improved nutrient compatibility, while optimizing processing techniques and addressing regulatory hurdles. Ultimately, by balancing these opportunities and challenges, edible packaging can revolutionize the food industry, offering both sustainability and nutrient preservation benefits for consumers and the environment [60–62].

6.9 Consumer Acceptance and Perception of Edible Packaging

Consumer acceptance and perception of edible packaging represent critical aspects in the successful adoption of this innovative approach within the packaging industry. The degree to which consumers embrace edible packaging is influenced by several factors, encompassing sensory experience, environmental considerations, and overall perceptions of convenience and safety [63]. From a sensory perspective, consumers evaluate edible packaging based on its impact on the taste, aroma, and overall eating experience of the enclosed food. The seamless integration of edible packaging with the food product, without compromising its flavor or texture, is a key determinant of consumer acceptance. Additionally, transparency and clear communication about the edible packaging's composition contribute to building trust and addressing potential concerns related to taste and safety. Environmental considerations play a significant role in shaping consumer attitudes toward edible packaging. With an increasing emphasis on sustainability, consumers often view edible packaging favorably when it aligns with eco-friendly practices. Biodegradability, compostability, and the use of renewable resources are key attributes that positively influence consumer perceptions, reflecting a growing demand for packaging solutions that minimize environmental impact. Convenience is another crucial factor influencing consumer acceptance. The ease of use, storage, and disposal of edible packaging materials contributes to their overall appeal [64]. Consumers appreciate packaging solutions that enhance convenience without compromising the functionality or shelf life of the enclosed product. Figure 6.3 gives a brief about the consumer acceptance and perception about edible packaging. Safety concerns are paramount in consumer acceptance of edible packaging. Rigorous adherence to food safety standards, clear labeling, and transparent communication about the manufacturing processes are essential for instilling confidence in consumers. Assurance regarding the suitability of edible packaging for diverse food types and the absence of allergens or harmful substances is pivotal in building positive perceptions. Cultural and regional preferences also play a role in consumer acceptance. The familiarity and acceptance of certain edible packaging materials may vary across different demographics, necessitating a nuanced understanding of consumer preferences [65].

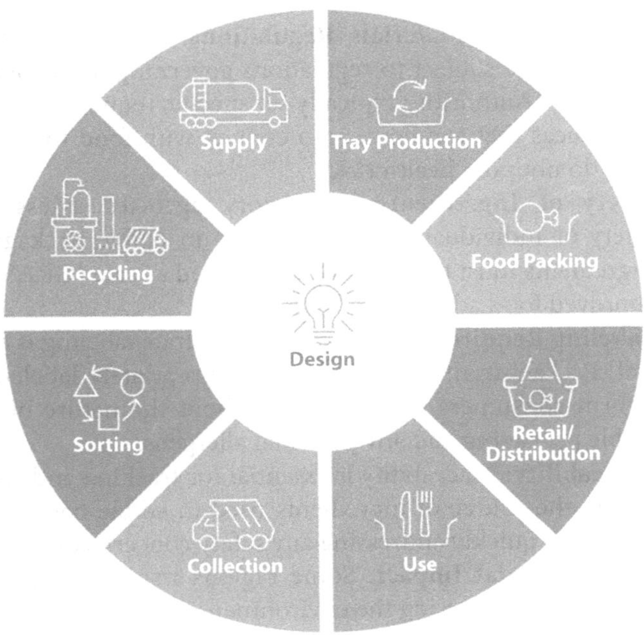

Figure 6.3 Consumer acceptance and perception [66].

6.10 Challenges and Solutions

While edible packaging holds immense promise, its widespread adoption is not without challenges. Regulatory considerations, technical hurdles, cost implications, and the delicate balance between sustainability and functionality are critical aspects that demand attention. Regulatory approval is a crucial step in the adoption of any new packaging material. Edible packaging materials must undergo rigorous testing to ensure their safety for consumption and compliance with food safety regulations [67]. Collaborative efforts between the food industry, regulatory bodies, and researchers are essential to streamline these approval processes and facilitate the integration of edible packaging into the market. The regulatory considerations and approval processes for edible packaging vary across regions and countries [68]. Generally, they involve ensuring the safety of the edible packaging materials, compliance with food regulations, and adherence to specific standards. Here is an overview of the regulatory considerations and approval processes for edible packaging:

6.10.1 Regulatory Considerations

1. **Food Contact Materials Regulations:** Edible packaging materials are subject to regulations governing food contact materials. Authorities typically set specific requirements for substances that can come into contact with food to ensure they do not pose health risks.
2. **Safety of Ingredients:** Regulatory agencies assess the safety of individual ingredients used in edible packaging. Ingredients must be generally recognized as safe (GRAS) or approved food additives.
3. **Labeling Requirements:** Clear and accurate labeling is crucial. Packaging materials must be appropriately labeled to inform consumers and regulators about the nature of the edible packaging and any potential allergens.
4. **Traceability:** Traceability is essential for tracking and identifying the source of ingredients used in edible packaging. It helps in quickly addressing any safety concerns or recalls.
5. **Environmental Impact:** Some regions may have specific regulations addressing the environmental impact of packaging materials. Edible packaging, as a sustainable option, may receive favorable consideration in this regard.

The burgeoning field of edible packaging necessitates close attention to regulatory frameworks and approval processes (Figure 6.4) for safe and responsible market introduction.

1. Regulatory bodies:

- Global: The Codex Alimentarius Commission (CAC) sets international standards for food additives and contaminants, providing a foundational framework [69]. Specific regions have additional regulations:
  - EU: Regulation (EC) No 1333/2008 governs food additives with Annex II listing approved substances [70].
  - US: The Food and Drug Administration (FDA) oversees food contact substances through various regulations including 21 CFR 175-178 [71].
  - China: The National Health Commission (NHC) regulates food additives with GB 2760-2021 outlining permitted substances [72].
  - Canada: Health Canada oversees the safety of food packaging materials through the Food and Drug Regulations.
  - Japan: The Ministry of Health, Labour and Welfare (MHLW) in Japan is responsible for regulating food contact materials.
- Emerging Regulatory Trends:
  - Novelty: Regulatory frameworks may need to adapt to assess the safety of novel materials and production processes associated with edible packaging [73].
  - Sustainability: Sustainability considerations are increasingly integrated into regulations, potentially favoring edible packaging [74].

While offering sustainability and potential food preservation benefits, edible packaging faces various technical challenges in manufacturing and processing.

1. Material Selection and Characterization: Finding edible materials with the desired mechanical, barrier, and functional properties of conventional packaging remains

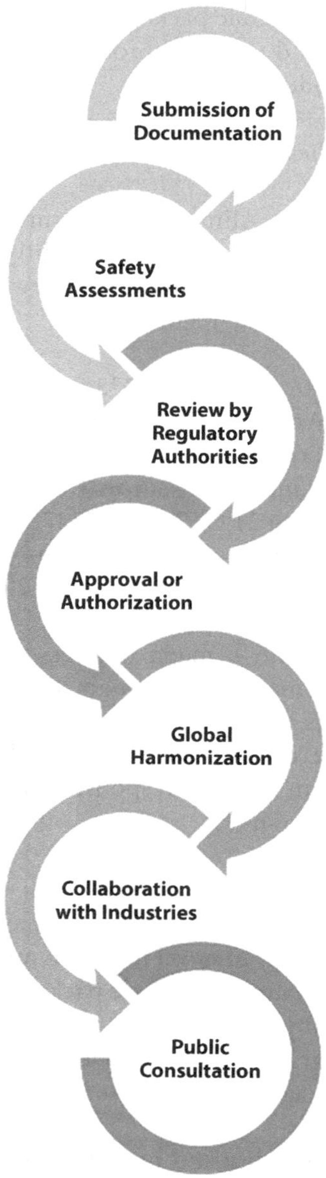

Figure 6.4 Approval process for edible packaging [75].

challenging [76]. Natural materials can exhibit variability in composition and properties, necessitating robust quality control measures [77]. Transitioning from lab-scale to large-scale production of edible packaging materials requires addressing challenges like sourcing consistent raw materials and optimizing processing methods [78].

2. Processing and Forming: Current methods like casting, spraying, and electrospinning can be expensive or have limitations in scalability. Ensuring consistent thickness, strength, and barrier properties across the packaging material is crucial for reliable performance [79]. Integrating edible packaging production into existing food manufacturing lines may require modifications or new machinery [80].
3. Functionality and Performance: Some edible materials show limited water vapor barrier properties compared to conventional plastics, potentially impacting product shelf life [81]. Mechanical strength and handling: Ensuring sufficient mechanical strength to withstand handling and protect the food product throughout the supply chain can be challenging [82]. Balancing oxygen and carbon dioxide permeability to maintain optimal food quality and prevent spoilage requires careful material selection and design [83].
4. Cost and Shelf Life: Compared to conventional packaging, edible materials and processing may be more expensive, impacting product affordability [84]. Edible packaging may have a shorter shelf life than conventional options, requiring careful inventory management and logistics [85]. Building consumer trust and willingness to pay a premium for edible packaging solutions is crucial for market success [86].

6.11 Trends and Application of Edible Packaging

The field of edible packaging is evolving rapidly, driven by the growing demand for sustainable packaging solutions and the desire to reduce environmental impact. Several future prospects and emerging trends are shaping the development and adoption of edible packaging. Table 6.4 shows various trends in edible packaging.

Examining real-world implementations of edible packaging provides valuable insights into its effectiveness and practical applications. Case

Table 6.4 Emerging trends in edible packaging.

Trend	Description	Potential applications	Benefits	Challenges	Reference
Advanced Composite Materials	Blending edible biopolymers, natural additives, and nanomaterials for improved functionality	Coatings for fruits and vegetables, films for meat and cheese, capsules for supplements	Enhanced barrier properties, antimicrobial activity, controlled release of nutrients	Cost-effectiveness, scalability, ensuring food safety	[87]
Engineered Proteins & Polysaccharides	Using precision fermentation and genetic engineering for novel, high-performance materials	Films with tailored barrier properties, edible cutlery and straws, capsules with controlled release profiles	Superior mechanical strength, biodegradability, specific functionalities	Regulatory approval, production costs, consumer acceptance	[88]
Micro & Nanoencapsulation	Encapsulating bioactive compounds within packaging for enhanced preservation and targeted delivery	Antimicrobial coatings, antioxidant films, capsules for nutraceuticals	Extended shelf life, improved food quality, targeted delivery of beneficial compounds	Encapsulation efficiency, stability of actives, potential taste alteration	[89]
3D Printing	Precise customization of packaging shapes and functionalities for diverse food products	Personalized food packaging, edible decorations, customized capsules for specific dosages	Precise control over properties, reduced waste, potential for on-demand production	High initial investment, limited material selection, printing speed and resolution	[90]

(Continued)

Table 6.4 Emerging trends in edible packaging. (*Continued*)

Trend	Description	Potential applications	Benefits	Challenges	Reference
Electrospinning & Electrospraying	Efficient production of thin, uniform films with enhanced barrier properties	Breathable films for fresh produce, oil-resistant coatings, antimicrobial packaging	High surface area, tunable properties, improved barrier functions	Limited scalability, potential for clogging, optimizing processing parameters	[91]
Advanced Drying Methods	Freeze drying and microwave drying for improved quality, shelf life, and energy efficiency	High-value food products, edible coatings, films for sensitive ingredients	Preserves texture and nutrients, reduces energy consumption, improved shelf life	High equipment cost, potential for product shrinkage, optimizing drying parameters	[92]
Food Waste Integration	Utilizing food byproducts like peels or whey protein as feedstock for packaging	Films for bakery products, coatings for snacks, capsules for supplements	Reduced waste, circular economy approach, potential for functional properties	Variability in byproduct composition, ensuring safety and quality, optimizing extraction methods	[93]
Life Cycle Assessment & Optimization	Minimizing environmental impact throughout the entire life cycle	Identifying hotspots, comparing materials and processes, optimizing production and disposal	Sustainable packaging solutions, reduced carbon footprint, increased brand value	Data availability, complex life cycle analysis, stakeholder collaboration	[94]

(*Continued*)

Table 6.4 Emerging trends in edible packaging. (*Continued*)

Trend	Description	Potential applications	Benefits	Challenges	Reference
Sensory Optimization	Addressing taste and texture concerns through flavor masking and modifications	Edible coatings for bitter tasting foods, textural modifications for films and capsules	Improved consumer acceptance, wider application range, reduced waste	Masking effectiveness, maintaining functionality, cost of sensory optimization	[95]
Consumer Education & Awareness	Educating consumers about benefits and functionalities for market adoption	Marketing campaigns, educational initiatives, transparent labeling	Increased demand for sustainable packaging, informed purchasing decisions, overcoming misconceptions	Building trust, addressing concerns about safety and regulation, long-term behavior change	[96]
Collaboration & Partnerships	Collaboration between researchers, manufacturers, food companies, and retailers for scaling up, safety, and trust	Joint research projects, technology transfer, supply chain collaboration	Efficient development, commercialization, and market adoption, ensuring food safety and regulatory compliance	Intellectual property sharing, aligning interests of different stakeholders, building long-term partnerships	[97]

studies and examples from various sectors of the food industry illustrate the diverse ways in which edible packaging can be utilized.

6.11.1 Fresh Produce

Edible coatings and films have been successfully employed to extend the shelf life of fresh fruits and vegetables. For example, a thin layer of chitosan applied to fruits can create a protective barrier against microbial contamination and moisture loss, resulting in prolonged freshness. Table 6.5 shows the application of edible coating on fresh produce.

6.11.2 Bakery Products

Edible films made from starch or protein-based materials have been used in the packaging of bakery products. These films not only contribute to moisture retention, preventing staling, but also act as a barrier against external contaminants. Table 6.6 shows the various bakery product with application of edible packaging.

6.11.3 Dairy Products

Edible packaging materials have found applications in the dairy industry, where they help preserve the texture and flavor of products such as cheese. The controlled release of antimicrobial agents from the packaging material contributes to the inhibition of spoilage microorganisms. Table 6.7 shows the application of edible packaging on various dairy products.

6.11.4 Meat and Poultry

Edible packaging has been explored in the meat and poultry industry to enhance both shelf life and safety. Antimicrobial coatings on meat products can reduce the risk of bacterial contamination during storage. Table 6.8 shows various meat and poultry products with the application of edible packaging.

6.11.5 Convenience Foods

The convenience food sector has embraced edible packaging for its ability to combine sustainability with practicality. Individually wrapped items with edible packaging reduce the environmental impact of single-use plastics

Table 6.5 Edible coatings for fresh produce: applications and limitations.

Produce	Coating material	Beneficial effects	Limitations	References
Avocados	Carboxyl methylcellulose containing moringa plant extracts	Extended shelf life, prevented browning, retained texture	Limited scalability, higher susceptibility to moisture absorption	[98]
Bell peppers	Chitosan and gelatin	Reduced microbial growth, maintained antioxidant activity, extended shelf life	Can mask natural flavor, potential allergic reactions	[99]
Blueberries	Dextran	Reduced weight loss, preserved anthocyanin content, improved visual quality	High cost, limited mechanical strength	[100]
Grapes	Hydroxypropylmethylcellulose edible coating containing propolis extract	Maintained moisture content, reduced shriveling, preserved firmness	High cost of production, limited gas permeability control	[101]
Lettuce	Polysaccharide-based edible coating (alginate, chitosan, and carrageenan)	preserved enzymatic browning, increased antioxidant activity	Potentially unpleasant odor, impact on sensory characteristics	[102]
Mandarin oranges	Pectin	Reduced decay, maintained freshness	Limited scalability, potential interactions with fruit flavor	[103]

(*Continued*)

Table 6.5 Edible coatings for fresh produce: applications and limitations. (*Continued*)

Produce	Coating material	Beneficial effects	Limitations	References
Mangoes	Hydroxypropyl methylcellulose and beeswax	Delayed ripening, maintained firmness, extended shelf life	Potential color changes, difficulty achieving uniform coating	[104]
Mushrooms	Chitosan and guar gum	Reduced microbial growth, maintained sensory quality, extended shelf life	Limited water vapor permeability, potential for film brittleness	[105]
Strawberries	Banana starch-chitosan-Aloe vera gel	Extended shelf life, reduced spoilage, maintained firmness	Potential off-flavor, limited water vapor permeability	[106]
Tomatoes	Chitosan-Pullulan	Reduced fungal decay, controlled ethylene production, delayed ripening	Potential impact on taste, limited effectiveness against all pathogens	[107]

Table 6.6 Edible coatings for bakery products: applications and limitations.

Product	Coating material	Beneficial effects	Limitations	References
Bread	zein-based edible films and coatings	Extended shelf life, improved crumb moisture, delayed staling	Potential impact on taste, limited shelf life extension	[108]
Cakes	Chitosan films incorporating potassium sorbate or vanillin	Reduced mold growth, maintained freshness, enhanced aroma	Potential off-flavor, difficulty achieving uniform coating on complex shapes	[109]
Cookies	Whey, Inulin and Gelatine with *Lactobacillus casei*	Maintained texture, prevented fat migration, extended shelf life	High cost, limited solubility in some baking processes	[110]
Muffins	Cassava Starch and Rosemary essential oil	Reduced microbial growth, preserved flavor, maintained softness	Difficulty achieving consistent adhesion, potential color changes	[111]

Table 6.7 Edible coatings for dairy products: applications and limitations.

Product	Coating material	Beneficial effects	Limitations	References
Cheese	Alginate coating with bacteriocin-producing *Lactococcus* strains	Reduced mold growth, extended shelf life, preserved texture	Potential impact on taste and aroma, limited effectiveness against all pathogens	[112]
Cottage cheese	Chitosan film with pineapple peel extract	Reduced bacterial growth, preserved freshness, maintained nutritional value	Potential for slimy texture, impact on sensory characteristics	[113]
Ice cream	Aloe vera and carrageenan	Reduced ice crystal formation, improved texture, prolonged stability	potential impact on melting point	[114]
Yogurt	Sodium alginate with probiotics	Enhanced probiotic viability, extended shelf life, improved sensory quality	Difficulty achieving uniform coating, potential for syneresis	[115]

Table 6.8 Edible coatings for meat and poultry: applications and limitations.

Product	Coating material	Beneficial effects	Limitations	References
Beefsteak	Pectin coating with guava leaf extract	Reduced lipid oxidation, preserved texture, maintained sensory quality	potential impact on meat aroma	[116]
Chicken breast	Alginate with oregano essential oil/β-cyclodextrin	Reduced microbial growth, extended shelf life, improved color retention	limited effectiveness against all pathogens	[117]
Chicken nuggets	Pomegranate Peel and alginate	Improved crispiness after cooking, reduced oil absorption, maintained texture	High cost of alginate, Limited gas permeability, potential impact on breading adhesion	[118]
Fish fillets	Alginate and chitosan containing sulfated polysaccharide (fucoidan) from *Sargassum angustifolium*	Controlled enzymatic browning, minimized microbial growth, delayed spoilage	Limited water vapor permeability, potential masking of natural fish flavor	[119]
Ground pork	Chitosan/flaxseed soluble dietary fiber/sesamol blend	Improved water retention, reduced drip loss, enhanced product yield	Difficulty achieving uniform coating on complex shapes, potential impact on meat texture	[120]
Ground turkey	*Lallemantia iberica* seed mucilage incorporated with Malva sylvestris leaf bioactive compounds	Delayed oxidative deterioration preserved freshness, maintained nutritional value	Potential for slimy texture, impact on sensory characteristics	[121]
Sausages	Sodium alginate-based film loaded with resveratrol and thymol	Reduced microbial growth, extended shelf life, maintained color stability	High cost of alginate, potential for interfering with traditional smoking processes	[122]

Table 6.9 Edible coatings for convenience foods: applications and limitations.

Product	Coating material	Beneficial effects	Limitations	References
Dried fruits and nuts-Peeled Fresh Almond Kernels	Mastic Gum	Reduced oxidation and rancidity, maintained nutritional value, extended shelf life	Limited scalability, potential impact on product texture	[123]
Pre-cut vegetables	Chitosan with nisin	Extended shelf life, reduced microbial growth, maintained color and texture	Potential off-flavor due to chitosan, limited effectiveness against all pathogens	[124]
Ready-to-eat salads	emulsion-based edible coatings lemongrass essential oil dispersed in different polymers, such as alginate, chia mucilage and chitosan.	Enhanced probiotic viability, extended shelf life, improved gut health benefits	potential masking of salad dressing flavors	[125]

while maintaining the quality and safety of the enclosed products. Table 6.9 shows various convenience products with edible packaging.

Analyzing these case studies provides valuable insights into the versatility of edible packaging across different food categories. Each example highlights the specific challenges addressed, the mechanisms employed for shelf life extension, and the overall impact on food quality.

6.12 Conclusion

Edible packaging emerges as a captivating, yet complex, solution for fostering sustainability in the food industry. While offering exciting opportunities for direct nutrient fortification, potentially slowing down nutrient degradation, and reducing waste, navigating its impact on food texture demands careful consideration. Unwanted nutrient migration and incompatibility with sensitive nutrients necessitate strategic material selection and processing techniques. The challenge of texture retention is equally significant. Edible materials' susceptibility to moisture migration can lead to undesirable changes, impacting palatability. Striking a balance between flexibility and rigidity further complicates design. Interaction with fats and oils, coupled with limitations in current processing techniques, pose additional hurdles. Yet, the future of edible packaging is brimming with potential. Novel materials like protein-based films and nanocomposites with improved textural properties represent promising advancements. Advanced processing techniques like 3D printing and electrospinning offer precise control over structure and functionality, while integration with modified atmosphere packaging holds the potential to further enhance texture retention. Ultimately, overcoming the challenges associated with texture and nutrient compatibility is crucial for realizing the full potential of edible packaging. Through continuous research, innovative material selection, and optimized processing techniques, we can ensure that delicious food experiences and sustainability objectives go hand-in-hand. This pursuit has the potential to create a future where texture and sustainability are not conflicting goals, but rather complementary forces driving a greener and more sustainable food industry. In conclusion, while edible packaging presents a complex, multi-faceted solution, its potential to revolutionize the industry by offering both functionality and environmental benefits is undeniable. Continued research and development efforts focused on overcoming texture and nutrient compatibility challenges are crucial to unlocking this potential and paving the way for a more sustainable future for food packaging.

References

1. Petkoska, A.T., Daniloski, D., D'Cunha, N.M., Naumovski, N., Broach, A.T., Edible packaging: Sustainable solutions and novel trends in food packaging. *Food Res. Intern. 140*, 109981, 2021.
2. Aguirre-Joya, J.A., De Leon-Zapata, M.A., Alvarez-Perez, O.B., Torres-León, C., Nieto-Oropeza, D.E., Ventura-Sobrevilla, J.M., Aguilar, C.N., Basic and applied concepts of edible packaging for foods, in: *Food packaging and preservation*, pp. 1–61, Academic Press, London, 2018.
3. Zheng, H., Cao, L., Huang, S., Neumann, P., Hu, F., Current status of the beekeeping industry in China, in: *Asian beekeeping in the 21st century*, pp. 129–158, 2018.
4. Eom, H., Chang, Y., Lee, E.S., Choi, H.D., Han, J., Development of a starch/gum-based edible coating for rice cakes to retard retrogradation during storage. *Lwt*, *97*, 516–522, 2018.
5. Pirsa, S. and Aghbolagh Sharifi, K., A review of the applications of bioproteins in the preparation of biodegradable films and polymers. *J. Chem. Lett.*, *1*, 2, 47–58, 2020.
6. Yadav, V., Sherly, M.A., Ranjan, P., Tinoco, R.O., Boldrin, A., Damgaard, A., Laurent, A., Framework for quantifying environmental losses of plastics from landfills. *Resour. Conserv. Recycl.*, *161*, 104914, 2020.
7. Khalil, H.A., Saurabh, C.K., Tye, Y.Y., Lai, T.K., Easa, A.M., Rosamah, E., Banerjee, A., Seaweed based sustainable films and composites for food and pharmaceutical applications: A review. *Renew. Sustain. Energy Rev.*, *77*, 353–362, 2017.
8. Ahmed, J., Ravichandran, S., Thakur, R., Sarkar, P., Starch-based edible packaging: Rheological, thermal, mechanical, microstructural, and barrier properties–A review. *Sustain. Food Technol.*, 2, 307–330, 2024.
9. Bairwan, R.D., Yahya, E.B., Gopakumar, D., HPS, A.K., Recent Advances in Poly (3-Hydroxybutyrate-co-3-Hydroxyvalerate) Biocomposites in Sustainable Packaging Applications. *Adv. Mater. Lett.*, *15*, 1, 2401–1739, 2024.
10. Kaur, J., Gunjal, M., Rasane, P., Singh, J., Kaur, S., Poonia, A., Gupta, P., Edible Packaging: An Overview, in: *Edible Food Packaging: Applications, Innovations and Sustainability*, pp. 3–25, 2022.
11. Ramazanidoroh, F., Hosseininezhad, M., Shahrampour, D., Wu, X., Edible Packaging as a Functional Carrier of Prebiotics, Probiotics, and Postbiotics to Boost Food Safety, Quality, and Shelf Life, in: *Probiotics and Antimicrobial Proteins*, pp. 1–21, 2023.
12. Wohner, B., Pauer, E., Heinrich, V., Tacker, M., Packaging-related food losses and waste: an overview of drivers and issues. *Sustainability*, *11*, 1, 264, 2019.
13. Betta, G., Barbanti, D., Massini, R., Food Hygiene in aseptic processing and packaging system: A survey in the Italian food industry. *Trends Food Sci. Technol.*, *22*, 6, 327–334, 2011.

14. Soro, A.B., Noore, S., Hannon, S., Whyte, P., Bolton, D.J., O'Donnell, C., Tiwari, B.K., Current sustainable solutions for extending the shelf life of meat and marine products in the packaging process. *Food Packag. Shelf Life*, *29*, 100722, 2021.
15. Chu, R., Tang, T., Hetherington, M.M., The impact of food packaging on measured food intake: a systematic review of experimental, field and naturalistic studies. *Appetite*, *166*, 105579, 2021.
16. Schifferstein, H.N., de Boer, A., Lemke, M., Conveying information through food packaging: A literature review comparing legislation with consumer perception. *J. Funct. Foods*, *86*, 104734, 2021.
17. Rambabu, L. and Porika, R., Packaging strategies: knowledge outlook on consumer buying behaviour. *J. Ind. Univ. Collab.*, *2*, 2, 67–78, 2020.
18. Ojha, A., Sharma, A., Sihag, M., Ojha, S., Food packaging–materials and sustainability-A review. *Agric. Rev.*, *36*, 3, 241–245, 2015.
19. Brennan, L., Langley, S., Verghese, K., Lockrey, S., Ryder, M., Francis, C., Hill, A., The role of packaging in fighting food waste: A systematised review of consumer perceptions of packaging. *J. Cleaner Prod.*, *281*, 125276, 2021.
20. Eilert, S.J., New packaging technologies for the 21st century. *Meat Sci.*, *71*, 1, 122–127, 2005.
21. Lau, J.H., Recent advances and trends in advanced packaging. *IEEE Trans. Compon. Packag. Manuf. Technol.*, *12*, 2, 228–252, 2022.
22. MacLeod, M., Arp, H.P.H., Tekman, M.B., Jahnke, A., The global threat from plastic pollution. *Science*, *373*, 6550, 61–65, 2021.
23. Li, P., Wang, X., Su, M., Zou, X., Duan, L., Zhang, H., Characteristics of plastic pollution in the environment: a review. *Bull. Environ. Contamination Toxicol.*, *107*, 577–584, 2021.
24. Iroegbu, A.O.C., Ray, S.S., Mbarane, V., Bordado, J.C., Sardinha, J.P., Plastic pollution: a perspective on matters arising: challenges and opportunities. *ACS Omega*, *6*, 30, 19343–19355, 2021.
25. Eissenberger, K., Ballesteros, A., De Bisschop, R., Bugnicourt, E., Cinelli, P., Defoin, M., Schmid, M., Approaches in Sustainable, Biobased Multilayer Packaging Solutions. *Polymers*, *15*, 5, 1184, 2023.
26. Schenker, U., Chardot, J., Missoum, K., Vishtal, A., Bras, J., Short communication on the role of cellulosic fiber-based packaging in reduction of climate change impacts. *Carbohydr. Polym.*, *254*, 117248, 2021.
27. Chandegara, V.K., Cholera, S.P., Nandasana, J.N., Kumpavat, M.T., Patel, K.C., *Plastic packaging waste impact on climate change and its mitigation. Subbaiah R, Prajapati GV, Water management and climate smart agriculture. Adaptation of Climatic Resilient Water Management and Agriculture*, vol. 3, pp. 404–15, Gyan Publishing House, New Delhi, India, 2015.
28. Goel, P., Arora, R., Haleem, R., Shukla, S.K., Advances in Bio-degradable Polymer Composites-Based Packaging Material. *Chem. Afr.*, *6*, 1, 95–115, 2023.

29. Lindh, H., Williams, H., Olsson, A., Wikström, F., Elucidating the indirect contributions of packaging to sustainable development: A terminology of packaging functions and features. *Packag. Technol. Sci.*, *29*, 4-5, 225–246, 2016.
30. Ibrahim, I.D., Sadiku, E.R., Hamam, Y., Kupolati, W.K., Ndambuki, J.M., Jamiru, T., Snyman, J., Recent Recycling Innovations to Facilitate Sustainable Packaging Materials: A Review. *Recycling*, *8*, 6, 88, 2023.
31. Wills, G., Packaging as a Source of Profit. *Int. J. Phys. Distribution*, *5*, 6, 305–334, 1975.
32. Farmer, N., The future: global trends and analysis for the international packaging market in relation to the speed of impact of packaging innovation and likely material changes, in: *Trends in Packaging of Food, Beverages and Other Fast-Moving Consumer Goods (FMCG)*, pp. 288–312, 2013.
33. Meherishi, L., Narayana, S.A., Ranjani, K.S., Sustainable packaging for supply chain management in the circular economy: A review. *J. Cleaner Prod.*, *237*, 117582, 2019.
34. Moustafa, H., Youssef, A.M., Darwish, N.A., Abou-Kandil, A., II, Eco-friendly polymer composites for green packaging: Future vision and challenges. *Composites, Part B*, *172*, 16–25, 2019.
35. Assad, I., Bhat, S.U., Gani, A., Shah, A., Protein based packaging of plant origin: Fabrication, properties, recent advances and future perspectives. *Int. J. Biol. Macromol.*, *164*, 707–716, 2020.
36. Zhu, H., Cheng, J.H., Han, Z., Han, Z., Cold plasma enhanced natural edible materials for future food packaging: Structure and property of polysaccharides and proteins-based films. *Crit. Rev. Food Sci. Nutr.*, *63*, 20, 4450–4466, 2023.
37. Milani, J.M. and Tirgarian, B., An overview of edible protein-based packaging: Main sources, advantages, drawbacks, recent progressions and food applications. *J. Packag. Technol. Res.*, *4*, 103–115, 2020.
38. Arfin, T., Marine polysaccharides: an overview, in: *Marine Polysaccharides*, Ahmed, S., Soundararajan, A. (eds), pp. 1–12, Jenny Standford Publishing, Taylor and Francis group, New York, 1–12, 2018.
39. Yang, L. and Zhang, L.M., Chemical structural and chain conformational characterization of some bioactive polysaccharides isolated from natural sources. *Carbohydr. Polym.*, *76*, 3, 349–361, 2009.
40. Shao, P., Feng, J., Sun, P., Xiang, N., Lu, B., Qiu, D., Recent advances in improving stability of food emulsion by plant polysaccharides. *Food Res. Int.*, *137*, 109376, 2020.
41. Dinel, H., Schnitzer, M., Mehuys, G.R., Soil lipids: origin, nature, content, decomposition, and effect on soil physical properties, in: *Soil Biochemistry*, Bollag, J.M., Stotzky, G. (eds), pp. 397–430, Routledge, Taylor and Francis group, New York, 2017.
42. Zähringer, U., Lindner, B., Rietschel, E.T., Chemical structure of lipid A: recent advances in structural analysis of biologically active molecules, in:

Endotoxin Health and Disease, Helmut Brade (eds), pp. 93–114, CRC Press, Taylor and Francis group, Boca Ranton, 2020.
43. Mohamed, S.A., El-Sakhawy, M., El-Sakhawy, M.A.M., Polysaccharides, protein and lipid-based natural edible films in food packaging: A review. *Carbohydr. Polym.*, *238*, 116178, 2020.
44. Han, J.H., Edible films and coatings: a review, in: *Innovations in food packaging*, pp. 213–255, 2014.
45. Kumar, A., Hasan, M., Mangaraj, S., Pravitha, M., Verma, D.K., Srivastav, P.P., Trends in edible packaging films and its prospective future in food: a review. *Appl. Food Res.*, *2*, 1, 100118, 2022.
46. La Fuente, C., II, do Val Siqueira, L., Augusto, P.E.D., Tadini, C.C., Casting and extrusion processes to produce bio-based plastics using cassava starch modified by the dry heat treatment (DHT). *Innov. Food Sci. Emerg. Technol.*, *75*, 102906, 2022.
47. Ahmed, J., Santhosh, R., Thakur, R., Mulla, M., Sarkar, P., Thermo-mechanical, rheological, microstructural, and barrier properties of gum-based edible packaging: A review. *Food Packag. Shelf Life*, *38*, 101117, 2023.
48. Punia Bangar, S., Chaudhary, V., Thakur, N., Kajla, P., Kumar, M., Trif, M., Natural antimicrobials as additives for edible food packaging applications: A review. *Foods*, *10*, 10, 2282, 2021.
49. Zheng, L., Liu, L., Yu, J., Shao, P., Novel trends and applications of natural pH-responsive indicator film in food packaging for improved quality monitoring. *Food Control*, *134*, 108769, 2022.
50. Beikzadeh, S., Ghorbani, M., Shahbazi, N., Izadi, F., Pilevar, Z., Mortazavian, A.M., The effects of novel thermal and nonthermal technologies on the properties of edible food packaging. *Food Eng. Rev.*, *12*, 333–345, 2020.
51. Ezati, P., Khan, A., Priyadarshi, R., Bhattacharya, T., Tammina, S.K., Rhim, J.W., Biopolymer-based UV protection functional films for food packaging. *Food Hydrocoll.*, 142, 108771, 2023.
52. Petraru, A. and Amariei, S., A Novel Approach about Edible Packaging Materials Based on Oilcakes—A Review. *Polymers*, *15*, 16, 3431, 2023.
53. Kuai, L., Liu, F., Chiou, B.S., Avena-Bustillos, R.J., McHugh, T.H., Zhong, F., Controlled release of antioxidants from active food packaging: A review. *Food Hydrocoll.*, *120*, 106992, 2021.
54. Nair, S.S., Trafiałek, J., Kolanowski, W., Edible Packaging: A Technological Update for the Sustainable Future of the Food Industry. *Appl. Sci.*, *13*, 14, 8234, 2023.
55. Bhat, Z.F., Bhat, H.F., Manzoor, M., Proestos, C., Hassoun, A., Dar, B.N., Bekhit, A.E.D.A., Edible packaging systems for enhancing the sensory quality of animal-derived foods. *Food Chem.*, 428, 136809, 2023.
56. Zhang, W., Jiang, H., Rhim, J.W., Cao, J., Jiang, W., Effective strategies of sustained release and retention enhancement of essential oils in active food packaging films/coatings. *Food Chem.*, *367*, 130671, 2022.

57. Chen, Y.Q., Cheng, J.H., Sun, D.W., Chemical, physical and physiological quality attributes of fruit and vegetables induced by cold plasma treatment: Mechanisms and application advances. *Crit. Rev. Food Sci. Nutr.*, *60*, 16, 2676–2690, 2020.
58. Gürdal, A.A. and Çetinkaya, T., Advancements in edible films for aquatic product preservation and packaging. *Rev. Aquacult.*, 16, 3, 997–1020, 2023.
59. Khalid, S., Hassan, S.A., Javaid, H., Zahid, M., Naeem, M., Bhat, Z.F., Aadil, R.M., Factors responsible for spoilage, drawbacks of conventional packaging, and advanced packaging systems for tomatoes. *J. Agric. Food Res.*, 15, 100962, 2024.
60. Rojas-Graü, M.A., Oms-Oliu, G., Soliva-Fortuny, R., Martín-Belloso, O., The use of packaging techniques to maintain freshness in fresh-cut fruits and vegetables: a review. *Int. J. Food Sci. Technol.*, *44*, 5, 875–889, 2009.
61. Feng, Y., Xu, H., Sun, Y., Xia, R., Hou, Z., Li, Y., Xin, G., Review of packaging for improving storage quality of fresh edible mushrooms. *Packag. Technol. Sci.*, *36*, 8, 629–646, 2023.
62. Salgado, P.R., Ortiz, C.M., Musso, Y.S., Di Giorgio, L., Mauri, A.N., Edible films and coatings containing bioactives. *Curr. Opin. Food Sci.*, *5*, 86–92, 2015.
63. Giles, E.L., Kuznesof, S., Clark, B., Hubbard, C., Frewer, L.J., Consumer acceptance of and willingness to pay for food nanotechnology: a systematic review. *J. Nanopart. Res.*, *17*, 1–26, 2015.
64. Werner, B.G., Koontz, J.L., Goddard, J.M., Hurdles to commercial translation of next generation active food packaging technologies. *Curr. Opin. Food Sci.*, *16*, 40–48, 2017.
65. Lu, P., Parrella, J.A., Xu, Z., Kogut, A., A scoping review of the literature examining consumer acceptance of upcycled foods. *Food Qual. Preference*, 114, 105098, 2024.
66. Schifferstein, H.N., de Boer, A., Lemke, M., Conveying information through food packaging: A literature review comparing legislation with consumer perception. *J. Funct. Foods*, *86*, 104734, 2021.
67. Dey, A., Dhumal, C.V., Sengupta, P., Kumar, A., Pramanik, N.K., Alam, T., Challenges and possible solutions to mitigate the problems of single-use plastics used for packaging food items: A review. *J. Food Sci. Technol.*, *58*, 9, 3251–3269, 2021.
68. Pascall, M.A., DeAngelo, K., Richards, J., Arensberg, M.B., Role and Importance of Functional Food Packaging in Specialized Products for Vulnerable Populations: Implications for Innovation and Policy Development for Sustainability. *Foods*, *11*, 19, 3043, 2022.
69. Laganà, P., Avventuroso, E., Romano, G., Gioffré, M. E., Patanè, P., Parisi, S., ... & Delia, S., The Codex Alimentarius and the European Legislation on Food Additives, in: *Chemistry and Hygiene of Food Additives*, SpringerBriefs in Molecular Science, pp. 23–32, Springer, Cham, 2017.

70. Öztekin, S., Anaya, K., Yurdunuseven-Yıldız, A., Regulation of Natural Food Additives, in: *Natural Additives in Foods,* Valencia, G.A. (eds), pp. 343–372, Springer, Cham, 2023.
71. Tumu, K., Vorst, K., & Curtzwiler, G., Understanding intentionally and non-intentionally added substances and associated threshold of toxicological concern in post-consumer polyolefin for use as food packaging materials. *Heliyon,* 10, 1, 2024.
72. Jie, M., & Clever, J., Chinese Standards for Food Additives-GB2760-2015. Global agricultural information network. Giant report number (FAIRS), 15013, 10-34, 2015.
73. Petraru, A. and Amariei, S., A Novel Approach about Edible Packaging Materials Based on Oilcakes—A Review. *Polymers, 15,* 16, 3431, 2023.
74. Ribeiro, I.S., Maciel, G.M., Bortolini, D.G., Fernandes, I.D.A.A., Volpato, W.M., Pedro, A.C., Haminiuk, C.W., II, Sustainable innovations in edible films and coatings: An overview. *Trends Food Sci. Technol.,* 143, 104272, 2023.
75. Prakash, G. and Mishra, R.K., Edible and bio-based food packaging: A review. *Pharma Innov. J., 7,* 1497–1505, 2023.
76. Putri, T.R., Adhitasari, A., Paramita, V., Yulianto, M.E., Ariyanto, H.D., Effect of different starch on the characteristics of edible film as functional packaging in fresh meat or meat products: A review. *Mater. Today Proc.* 87, 192-199, 2023.
77. Gaspar, M.C. and Braga, M.E., Edible films and coatings based on agrifood residues: a new trend in the food packaging research. *Curr. Opin. Food Sci., 50,* 101006, 2023.
78. Jeevahan, J.J., Chandrasekaran, M., Venkatesan, S.P., Sriram, V., Joseph, G.B., Mageshwaran, G., Durairaj, R.B., Scaling up difficulties and commercial aspects of edible films for food packaging: A review. *Trends Food Sci. Technol., 100,* 210–222, 2020.
79. Cerqueira, M.A.P.R., Pereira, R.N.C., da Silva Ramos, O.L., Teixeira, J.A.C., Vicente, A.A. (Eds.), *Edible food packaging: materials and processing technologies.* CRC Press, Taylor and Francis group, Boca Ranton, 2017.
80. Janjarasskul, T. and Krochta, J.M., Edible packaging materials. *Annu. Rev. Food Sci. Technol., 1,* 415–448, 2010.
81. Avramescu, S.M., Butean, C., Popa, C.V., Ortan, A., Moraru, I., Temocico, G., Edible and functionalized films/coatings—Performances and perspectives. *Coatings, 10,* 7, 687, 2020.
82. Petkoska, A.T., Daniloski, D., D'Cunha, N.M., Naumovski, N., Broach, A.T., Edible packaging: Sustainable solutions and novel trends in food packaging. *Food Res. Int., 140,* 109981, 2021.
83. Chhikara, S. and Kumar, D., Edible coating and edible film as food packaging material: A review. *J. Packag. Technol. Res., 6,* 1, 1–10, 2022.

84. Kumar, N., Prasad, J., Yadav, A., Upadhyay, A., Neeraj, ., Kieliszek, M., Recent Trends in Edible Packaging for Food Applications—Perspective for the Future. *Food Eng. Rev.*, *15*, 4, 718–747, 2023.
85. Wibowo, C., Salsabila, S., Muna, A., Rusliman, D., Wasisto, H.S., Advanced biopolymer-based edible coating technologies for food preservation and packaging. *Compr. Rev. Food Sci. Food Saf.*, *23*, 1, 1–31, 2024.
86. Kusuma, H.S., Yugiani, P., Himana, A., II, Aziz, A., Putra, D.A.W., Reflections on food security and smart packaging. *Polym. Bull.*, *81*, 1, 87–133, 2024.
87. Martins, V.F., Pintado, M.E., Morais, R.M., Morais, A.M., Recent Highlights in Sustainable Bio-Based Edible Films and Coatings for Fruit and Vegetable Applications. *Foods*, *13*, 2, 318, 2024.
88. Singh, A.K., Recent advancements in polysaccharides, proteins and lipids based edible coatings to enhance guava fruit shelf-life: A review. *Int. J. Biol. Macromol.*, 262, 1, 129826, 2024.
89. Han, J., Zhao, Y., Wang, Y., Recent advances in edible films and coatings for fruits and vegetables. *Trends Food Sci. Technol.*, 134, 60–75, 2023.
90. Chen, Y., McClements, D.J., Peng, X., Chen, L., Xu, Z., Meng, M., Jin, Z., Starch as edible ink in 3D printing for food applications: a review. *Crit. Rev. Food Sci. Nutr.*, *64*, 2, 456–471, 2024.
91. Zhang, Y., Min, T., Zhao, Y., Cheng, C., Yin, H., Yue, J., The developments and trends of electrospinning active food packaging: A review and bibliometrics analysis. *Food Control*, 160, 110291, 2024.
92. Suhag, R., Kumar, N., Petkoska, A.T., Upadhyay, A., Film formation and deposition methods of edible coating on food products: A review. *Food Res. Int.*, *136*, 109582, 2020.
93. Khalid, S., Naeem, M., Talha, M., Hassan, S.A., Ali, A., Maan, A.A., Aadil, R.M., Development of biodegradable coatings by the incorporation of essential oils derived from food waste: A new sustainable packaging approach. *Packag. Technol. Sci.*, 37, 3, 167–185, 2023.
94. Notarnicola, B., Sala, S., Anton, A., McLaren, S.J., Saouter, E., Sonesson, U., The role of life cycle assessment in supporting sustainable agri-food systems: A review of the challenges. *J. Cleaner Prod.*, *140*, 399–409, 2017.
95. Matloob, A., Ayub, H., Mohsin, M., Ambreen, S., Khan, F.A., Oranab, S., Ercisli, S., A review on edible coatings and films: Advances, composition, production methods, and safety concerns. *ACS Omega*, *8*, 32, 28932–28944, 2023.
96. Bizymis, A.P. and Tzia, C., Edible films and coatings: Properties for the selection of the components, evolution through composites and nanomaterials, and safety issues. *Crit. Rev. Food Sci. Nutr.*, *62*, 31, 8777–8792, 2022.
97. Akin, M., Eyduran, S.P., Krauter, V., Food packaging related research trends in the academic discipline of food science and technology: A bibliometric analysis. *Cleaner Circ. Bioecon.*, *5*, 100046, 2023.
98. Tesfay, S.Z., Magwaza, L.S., Mbili, N., Mditshwa, A., Carboxyl methylcellulose (CMC) containing moringa plant extracts as new postharvest organic

edible coating for Avocado (Persea americana Mill.) fruit. *Sci. Hortic.*, *226*, 201–207, 2017.

99. Poverenov, E., Zaitsev, Y., Arnon, H., Granit, R., Alkalai-Tuvia, S., Perzelan, Y., Fallik, E., Effects of a composite chitosan–gelatin edible coating on postharvest quality and storability of red bell peppers. *Postharvest Biol. Technol.*, *96*, 106–109, 2014.
100. Davidović, S., Miljković, M., Gordic, M., Cabrera-Barjas, G., Nesic, A., Dimitrijević-Branković, S., Dextran-based edible coatings to prolong the shelf life of blueberries. *Polymers*, *13*, 23, 4252, 2021.
101. Pastor, C., Sánchez-González, L., Marcilla, A., Chiralt, A., Cháfer, M., González-Martínez, C., Quality and safety of table grapes coated with hydroxypropylmethylcellulose edible coatings containing propolis extract. *Postharvest Biol. Technol.*, *60*, 1, 64–70, 2011.
102. Li, L., Yi, P., Li, C., Xin, M., Sun, J., He, X., Tang, Y., Influence of polysaccharide-based edible coatings on enzymatic browning and oxidative senescence of fresh-cut lettuce. *Food Sci. Nutr.*, *9*, 2, 888–899, 2021.
103. Maftoonazad, N. and Ramaswamy, H.S., Application and evaluation of a pectin-based edible coating process for quality change kinetics and shelf-life extension of lime fruit (Citrus aurantifolium). *Coatings*, *9*, 5, 285, 2019.
104. Sousa, F.F., Junior, J.S.P., Oliveira, K.T., Rodrigues, E.C., Andrade, J.P., Mattiuz, B.H., Conservation of 'Palmer'mango with an edible coating of hydroxypropyl methylcellulose and beeswax. *Food Chem.*, *346*, 128925, 2021.
105. Huang, Q., Qian, X., Jiang, T., Zheng, X., Effect of chitosan and guar gum based composite edible coating on quality of mushroom (Lentinus edodes) during postharvest storage. *Sci. Hortic.*, *253*, 382–389, 2019.
106. Pinzon, M., II, Sanchez, L.T., Garcia, O.R., Gutierrez, R., Luna, J.C., Villa, C.C., Increasing shelf life of strawberries (Fragaria ssp) by using a banana starch-chitosan-Aloe vera gel composite edible coating. *Int. J. Food Sci. Technol.*, *55*, 1, 92–98, 2020.
107. Kumar, N., Neeraj, Pratibha, Trajkovska Petkoska, A., Improved shelf life and quality of tomato (Solanum lycopersicum L.) by using chitosan-pullulan composite edible coating enriched with pomegranate peel extract. *ACS Food Sci. Technol.*, *1*, 4, 500–510, 2021.
108. Mouzakitis, C.K., Sereti, V., Matsakidou, A., Kotsiou, K., Biliaderis, C.G., Lazaridou, A., Physicochemical properties of zein-based edible films and coatings for extending wheat bread shelf life. *Food Hydrocolloids*, *132*, 107856, 2022.
109. Sangsuwan, J., Rattanapanone, N., Pongsirikul, I., Development of active chitosan films incorporating potassium sorbate or vanillin to extend the shelf life of butter cake. *Int. J. Food Sci. Technol.*, *50*, 2, 323–330, 2015.
110. García-Argueta, I., Quintero-Salazar, B., Dominguez-Lopez, A., Gómez-Oliván, L.M., Díaz-Bandera, D., Dublán-García, O., Effect of edible coating based on whey, inulin and gelatine with Lactobacillus casei on the textural and sensorial properties of a cracker cookie. *J. Prob. Health*, *4*, 153, 2016.

111. Hassan, S.S., El-Hdide, S., Aboul Anean, H.E.D., Preparation of edible coating from Cassava Starch and Rosemary essential oil and its effect on shelf life and quality of Muffin. *Food Technol. Res. J.*, *2*, 2, 82–97, 2023.
112. Silva, S.P., Ribeiro, S.C., Teixeira, J.A., Silva, C.C., Application of an alginate-based edible coating with bacteriocin-producing Lactococcus strains in fresh cheese preservation. *Lwt*, *153*, 112486, 2022.
113. Jatav, J., Chinchkar, A.V., Bhattacharya, B., Chitosan film with pineapple peel extract in the extension of shelf life of Indian cottage cheese: Release kinetics and bio-accessibility studies. *Food Res. Int.*, *174*, 113580, 2023.
114. Mahajan, K., Kumar, S., Bhat, Z.F., Singh, M., Bhat, H.F., Bhatti, M.A., Bekhit, A.E.D.A., Aloe vera and carrageenan based edible film improves storage stability of ice-cream. *Appl. Food Res.*, *2*, 1, 100128, 2022.
115. Salama, H.H., Trif, M., Rusu, A.V., Bhattacharya, S., Application of Functional and Edible Coatings and Films as Promising Strategies for Developing Dairy Functional Products—A Review on Yoghurt Case. *Coatings*, *12*, 6, 838, 2022.
116. Nascimento, C.P., Lopes, M.V.A., Bessa, J.M., Andrade, A.P., Oliveira, L.S., Eça, K.S., Effect of pectin coating incorporated with guava leaf extract (Psidium guajava L.) on the stability of fresh beef steak. *Int. J. Food Sci. Technol.*, 59, 3, 2068–2078, 2024.
117. Yang, X., Zhao, D., Ge, S., Bian, P., Xue, H., Lang, Y., Alginate-based edible coating with oregano essential oil/β-cyclodextrin inclusion complex for chicken breast preservation. *Int. J. Biol. Macromol.*, *251*, 126126, 2023.
118. Bashir, S., Arshad, M.S., Khalid, W., Nayik, G.A., Al Obaid, S., Ansari, M.J., Karabagias, I.K., Effect of antimicrobial and antioxidant rich pomegranate peel based edible coatings on quality and functional properties of chicken nuggets. *Molecules*, *27*, 14, 4500, 2022.
119. Khorami, F., Babaei, S., Valizadeh, S., Naseri, M., Golmakani, M.T., Bilayer coatings for extension of the shelf life of fish fillets: Incorporating seaweed sulfated polysaccharides in chitosan-alginate LbL structures. *Food Sci. Nutr.*, 12, 4, 2511–2522, 2024.
120. Feng, S., Liu, L., Li, H., Guo, Y., Huo, Y., Wu, X., Xu, J., Characterization, antioxidant and antibacterial activities of chitosan/flaxseed soluble dietary fiber/sesamol blend films and their application in pork preservation. *Food Hydrocolloids*, *147*, 109370, 2024.
121. Mojarradi, F., Bimakr, M., Ganjloo, A., Effect of bio-edible coating based on Lallemantia iberica seed mucilage incorporated with Malva sylvestris leaf bioactive compounds on oxidative stability of turkey meat. *J. Food Meas. Charact.*, 18, 1, 402–412, 2023.
122. Hashemi, M., Aminzare, M., Hassanzadazar, H., Roohinejad, S., Tahergorabi, R., Bekhit, A.E.D.A., Impact of sodium alginate-based film loaded with resveratrol and thymol on the shelf life of cooked sausage and the inoculated Listeria monocytogenes. *Food Sci. Nutr.*, *11*, 12, 7855–7869, 2023.

123. Farooq, M., Azadfar, E., Rusu, A., Trif, M., Poushi, M.K., Wang, Y., Improving the shelf life of peeled fresh almond kernels by edible coating with mastic gum. *Coatings*, *11*, 6, 618, 2021.
124. Yousuf, B., Qadri, O.S., Srivastava, A.K., Recent developments in shelf-life extension of fresh-cut fruits and vegetables by application of different edible coatings: A review. *Lwt*, *89*, 198–209, 2018.
125. Cofelice, M., Iftikhar, A., Lopez, F., De Leonardis, A., Effect of edible coatings on quality parameters and phenol composition of ready-to-eat Salanova lettuce. *Eur. Food Res. Technol.*, 250, 3, 691–700, 2023.

7

Coat the Flavor, Preserve the Quality: Elevating Food Stability with Cutting-Edge Coating Materials

Swetha Priya Gali[1], Reshma R. G.[1], Srinivasan Krishnamoorthy[2] and R. Meenatchi[1]*

[1]Department of Primary Processing & Storage Handling, National Institute of Food Technology Entrepreneurship and Management, Thanjavur (NIFTEM-T), India
[2]Department of Technology Dissemination, National Institute of Food Technology, Entrepreneurship and Management, Thanjavur (NIFTEM-T), India

Abstract
Food safety and stability during processing, storage, transportation, and sales are majorly influenced by microbial contamination and biochemical deterioration. Food coating materials (with a market size of \$3.95 billion in 2021 and a projected CAGR of 6.35% to reach \$6.87 billion by 2030) are eco-friendly biopolymers with unique functional properties and are gaining wide attention for protecting food. Biopolymers, including composite materials, proteins, polysaccharides, and lipids, create edible components through novel techniques such as direct deposition (spraying, panning, fluidized bed, and dipping), casting and extrusion. Recently, nanocoating techniques have been well-recognized to reinforce and improve barrier properties for extended shelf-life. However, certain overarching challenges exist in applying edible coatings to food products. Some of the limitations in creating a sustainable edible coating include (i) poor coating or flavor transmission from organic coating materials to a core food matrix; (ii) lack of efficient adhesion properties to provide biochemical, structural, and mechanical stability complementary to thermoplastic elastomers; (iii) economic aspects in large scale production of biopolymers suitable for safeguarding the food product's quality attributes such as color, flavor, aroma, appearance, taste, nutritional profile, etc. Researchers across multiple disciplines should work together to create sustainable food coating materials ideal for recycling and reuse.

**Corresponding author*: meena@iifpt.edu.in

Mousumi Sen (ed.) Food Coatings and Preservation Technologies, (211–258) © 2025 Scrivener Publishing LLC

Keywords: Edible coating, polysaccharides, proteins, lipids, nanocoating, challenges in edible coating

7.1 Biodegradable Coating Materials

Biodegradable packaging materials have been increasingly recognized for their significance in substituting conventional packaging materials. These environmentally sustainable materials are incorporated into the product or can undergo decomposition by microorganisms, devoid of generating harmful environmental emissions [1]. The primary components essential for creating edible coatings and biodegradable films are film-forming biopolymers derived from plants and animals. Numerous biological substances, including proteins, polysaccharides, and lipids (Figure 7.1) have been documented in scientific literature as viable options for developing biodegradable packaging [2]. The key advantages of using these components are their properties, which include flexibility, colorlessness, antibacterial activity, water solubility, gelling, high mechanical qualities, and moisture resistance. These innovative systems also exhibit similar optical, mechanical, physical, and barrier characteristics to synthetic packaging materials [3].

A thin layer is formed between the food components or the surface using edible substances, and they have varied properties, such as the capacity to control oxygen and carbon dioxide from the surroundings. The biopolymer retains taste and aroma and can hold a variety of food additives. Thus,

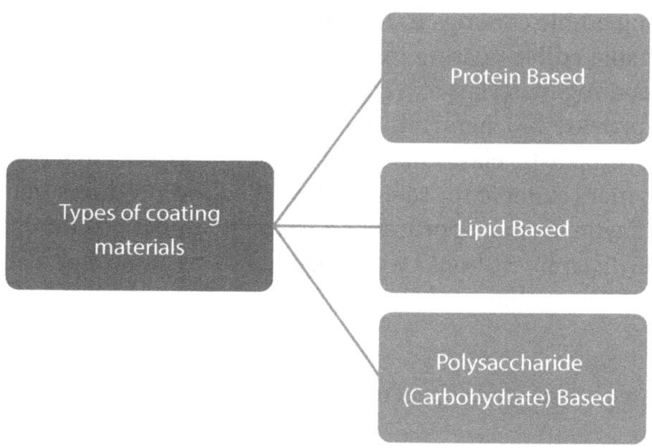

Figure 7.1 Classification of coating materials.

the edible material should be appropriately prepared to achieve all these properties practically [4].

Edible food coatings are commonly made of proteins (like gelatin and whey), polysaccharides (like chitosan and starch), lipids (like vegetable oils and beeswax), and other biodegradable components. By acting as barriers against gasses, moisture, and other impurities, these coatings help to maintain the freshness and calibre of the food underneath [2]. Applications for edible coatings can be found in various culinary products, including meat, confections, fruits, and vegetables. Edible coatings serve a protective purpose but can also support sustainability initiatives by offering a greener substitute for conventional packaging materials. The increasing need for eco-friendly and sustainable packaging options and the goal of improving food product quality and shelf life has led to a surge in research on edible food coatings in recent years. Research in this area is concentrated on creating coatings with certain characteristics, such as controlled release of bioactive chemicals, barrier qualities, and antibacterial activities. The following are some important study topics and fields of inquiry on the characteristics of edible food coatings [5]:

Barrier properties: Scientists have investigated the application of many substances, including lipids, proteins, and polysaccharides, to create edible coatings with superior barrier qualities. By preventing the transfer of gases (such as carbon dioxide and oxygen), moisture, and other impurities, these coatings seek to prolong the shelf life of food items [6].

Biodegradability and Sustainability: Numerous studies address concerns about the environmental effects by concentrating on creating edible coatings from biodegradable materials. Biopolymers made of natural resources, such as proteins (gelatin, whey), polysaccharides (starch, chitosan), and lipids (beeswax, carnauba wax), are used in research [1].

Antimicrobial properties: Using natural antibacterial agents in edible coatings has been interesting. Numerous research works investigate the application of plant extracts, essential oils, and other antimicrobial substances to prevent the growth of germs on the coated surface, thereby improving the safety and quality of food [7].

Controlled release of bioactive compounds: Researchers are developing edible coatings that carry bioactive substances, including vitamins, antimicrobials, and antioxidants. Researchers are working to create controlled release mechanisms to guarantee that these substances release gradually during processing, storage, or ingestion [8].

Nanostructured coatings: Using nanotechnology, edible coatings with nanostructures are being created. Coatings enhance mechanical strength, barrier qualities, and functionality by adding nano-sized materials, including nanoparticles or nanofibers. These nanocoatings might improve the quality and preservation of coated meals [9].
Smart coatings: Intelligent coatings are engineered to react to particular stimuli, including humidity, pH, or temperature variations. To enable real-time food freshness monitoring or to release bioactive substances under specific conditions, research is looking into integrating smart materials into edible coatings [10].
Mechanical properties: Research focuses on comprehending and enhancing the mechanical characteristics of edible coatings, including adhesion, tensile strength, and flexibility. These characteristics are essential for guaranteeing that the coatings offer the intended protective barrier, stick effectively to the food surface, and resist cracking [11].
Sensory and Texture properties. Studies also take into account the sensory characteristics of coated foods. Research aims to create coatings that improve texture, look, taste, and overall sensory experience in addition to their practical benefits [9].
Effect on Nutritional content: Scientists look at how edible coatings affect the meals that are coated in terms of nutrition. The objective is to create coatings without sacrificing the food product's nutritious content [12].
Application methods and Equipment: The research delves into various application techniques and tools for effectively applying edible coatings on a big scale in industrial environments. This covers coating methods such as spraying and dipping [13].

Even if the creation of edible food coatings has advanced significantly, there are still several study gaps that offer chances for additional investigation and creativity. In the area of edible food coatings, some of the major research gaps include long-term stability and performance, standardization and regulation, interaction with different food matrices, scalability and manufacturing processes, Sensory aspects and consumer acceptance, environmental impact and biodegradability, smart coatings, effects on nutrient retention and integration of nanotechnology. Hence, this chapter focuses on (i) various categories of edible food coatings and coating methods, (ii) factors associated with edible food coatings and the challenges encountered, and (iii) the scope for further research and the collaborative work of multidisciplinary fields.

7.1.1 Protein-Based Films and Coating

Proteins can form edible coatings from both animal and plant sources. Proteins include gelatin, collagen, gluten, zein, mung bean protein, casein, and soy protein (Table 7.1). They have potential beyond the food and feed industry for application in packaging. Protein-based polymers are even more advantageous than polysaccharides and lipids because of their potent gas barrier characteristics, but they are not as effective in food waterproofing [14]. Furthermore, protein-based films have better mechanical characteristics than polysaccharide and lipid-based films (Figure 7.2).

Table 7.1 Protein-based food coating materials.

S. no.	Types of protein	Edible coatings	Applications	References
1.	Gelatin	Gelatin derived from animal collagen, is commonly used to create edible coatings	Used for coating fruits, vegetables and confectionery items to reduce moisture loss, extend shelf life, and enhance appearance	[22]
2.	Whey protein	A byproduct of cheese production, are used for their film-forming and emulsifying properties.	Applied to fruits, nuts, and snacks to provide a protective layer, reduce oxidation and improve the overall quality.	[23]
3.	Casein	A milk protein forms a strong and flexible film with good barrier properties.	Used for coating cheese, nuts, and some bakery products to enhance freshness and prevent moisture absorption	[24]

(Continued)

Table 7.1 Protein-based food coating materials. (*Continued*)

S. no.	Types of protein	Edible coatings	Applications	References
4.	Soy Protein	They offer a plant-based alternative with film-forming and adhesion properties.	Used in the coating of various food products, including fruits, vegetables, and meat, to improve texture, appearance and shelf life	[20]
5.	Egg White	Because egg whites are high in protein, they can create a barrier with strong adhesion.	Often used to cover bread and pastries from bakeries to give them a glossy look and prolong their freshness.	[25]
6.	Collagen	The biodegradable and film-forming qualities of collagen, a structural protein present in connective tissues, are the reason for using collagen coatings.	Used to improve texture, stop drying out, and retain moisture in meat products.	[26]

(*Continued*)

Table 7.1 Protein-based food coating materials. (*Continued*)

S. no.	Types of protein	Edible coatings	Applications	References
7.	Plant-based coatings	Plant-based proteins, like those found in rice, wheat, or peas, can be utilized in coating applications.	Good for vegan and plant-based products, providing comparable advantages such as increased shelf life and improved texture.	[27]
8.	Protein nano-particle coatings	Proteins can be digested into nanoparticles and employed as coatings to release functional components under controlled conditions.	Used in the creation of food products that are coated and contain flavors, colors, or bioactive ingredients that are encapsulated.	[28]
9.	Fish gelatin coatings	Fish collagen gelatin serves as a substitute for mammalian gelatin.	Applied to a range of food items, particularly those that need to be certified kosher or halal.	[22]
10.	Functional protein coatings	Proteins can be altered or mixed with other substances to provide particular qualities, including antibacterial or antioxidant activity.	Applied to developing coatings that offer advantages over and beyond the standard protective layer.	[15]

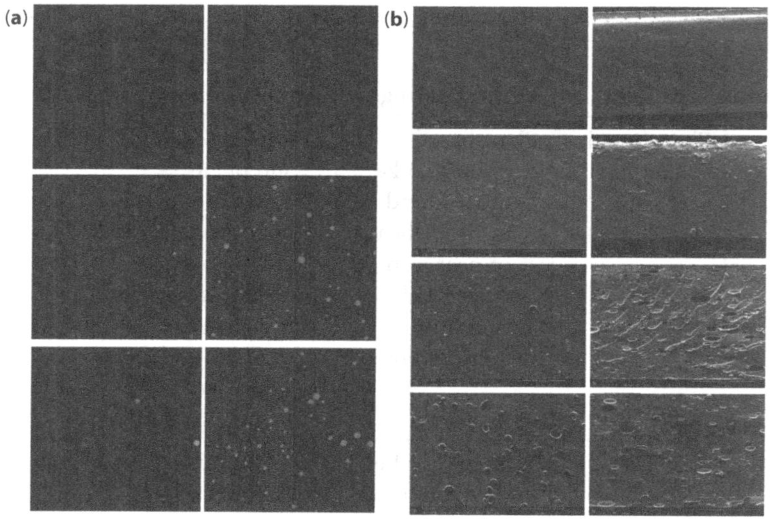

Figure 7.2 Microstructures of protein-based edible food coatings made up of whey protein isolate microbial transglutaminase and sunflower oil at different combinations. (a) Confocal laser scanning microscopy images; (b) Scanning electron microscopic images. Adapted and modified from [29].

7.1.1.1 Gelatin

An incomplete breakdown of collagen can result in the formation of gelatin. Gelatin predominantly comprises proline, 4-hydroxyproline, and glycine. However, the specific composition of these constituents varies depending on the source; in the case of gelatin derived from pigskin, the proportions are 13%, 9%, and 33% for proline, 4-hydroxyproline, and glycine, respectively [15]. Gelatin has emerged as a crucial raw material for the formulation of biodegradable films because of its favorable film-forming characteristics, widespread availability, cost-effectiveness, biocompatibility, and biodegradability. These attributes collectively contribute to the significance of gelatin in developing environmentally friendly and sustainable film materials [16]. Films derived from gelatin exhibit transparency, impermeability to oxygen, and thermos reversible characteristics. Edible films incorporating gelatin tend to have less permeability towards oxygen and can potentially exhibit antioxidant and antimicrobial properties by incorporating agents such as citrus essential oils, carvacrol, and similar additives.

7.1.1.2 Gluten

Gluten proteins are classified into two groups based on their solubility in aqueous alcohols: gliadin (soluble in nature) and glutenin (insoluble in nature). Film-forming properties depend on cohesiveness and elasticity. The gluten-based films possess a glossy surface, high oxygen isolation capacity, poor resistance against water vapor, and reduced mechanical strength.

7.1.1.3 Zein

Zein, a byproduct generated during the corn refining process, contains approximately 45% to 50% of the total proteins found in corn. Zein, a prolamin protein found in corn, can be used commercially in edible packaging [17]. These zein proteins are soluble in alcohol but insoluble in water because nonpolar amino acids are included in their structure. This characteristic contributes significantly to the capacity of zein films to act as barriers against water vapor. Scientific literature indicates that films based on zein exhibit a smooth appearance, high thermal stability, and desirable selective permeability [18]. These films are known for their excellent antibacterial and barrier properties.

7.1.1.4 Casein

Casein, which makes about 80% of milk protein, can create films remarkably. It produces translucent, flexible films excellent as lipid and gas barriers, especially in environments with low relative humidity [19]. The four main subunits that make casein's composition are $\alpha s1$, $\alpha s2$, β, and κ-casein. These subunits contain various amino acids that help caseinates behave favorably during emulsification. Casein is one of the proteins most researchers [20] identified as a viable possibility for creating edible films. Because casein is nanosized and self-assembled, it's a great medium to contain bioactive ingredients.

7.1.1.5 Soy Proteins

Soy protein is derived from soya bean meal, namely soy protein isolate (90% protein) and soy protein concentrate (70% protein). It is an inexpensive, biodegradable source of material that may be used to make edible coatings and films [20]. Soy protein films are neither as mechanically strong nor insulating as synthetic films because of their hydrophilic nature and vulnerability to external factors. It can be improved by combining other biopolymers and developing innovative techniques to create films [21]. Consequently, it is advised that these films be employed in the pharmaceutical and low-moisture food sectors.

7.1.2 Lipid-Based Films and Coating

Traditionally employed as fruit coatings, lipids are recognized for their barrier capabilities. Nevertheless, their utilization has been limited by their insufficient mechanical strength and transparency. To address these limitations, lipids have found effective applications in forming composite edible films. Common edible lipids are paraffin wax, bee wax, carnauba wax, rice bran, candelilla wax, and polyethylene wax [14] (Table 7.2). Waxes are non-polar lipids that are soluble in solvents but insoluble in water. They have a high hydrophobicity. Paraffin wax, derived from petroleum sources, coats raw cheeses, fruits, and vegetables.

Yellow and white are the hues characterizing the cells within the beehive, which produce various varieties of beeswax. Its primary components include diesters, free fatty alcohols, hydrocarbons, monoesters, and free fatty acids. Beeswax exhibits solubility in heated benzene, diethyl ether, and chloroform while remaining insoluble in water. Additionally, it adopts distinct forms at different temperature levels [30]. Brazilian palm tree leaves are used to make carnauba wax. It is soluble in boiling ether and ethanol but insoluble in water. It is white and solid. GRAS wax is the term used for culinary applications. Carnauba is the toughest wax used in glass candies. Like carnauba, candelilla is derived from plants called GRAS wax [30]. During waxing, rice bran wax is separated from rice bran oil. This wax can undergo refinement and bleaching procedures to produce a white variant in both soft and complex forms. Extracting rice oil using hexane, chilling it to 20°C to 25°C, filtering it, and centrifuging it yields rice bran wax.

It may be found in chocolate, gum, fruit and vegetable coatings, and culinary goods as a coating and filler ingredient. It is used in the polymer leather, pharmaceutical, and cosmetics sectors. To extend the shelf life of items, such as crackers, almonds, and snacks, rice bran oil is applied to their surface [31].

7.1.3 Polysaccharides-Based Films and Coatings

Polysaccharides are a potential alternative for edible packaging material in terms of edible food packaging. Since they are natural compounds with

Table 7.2 Lipid-based food coating materials.

S. no.	Lipid-based coatings	Source	Applications	Reference
1	Beeswax	Made by bee hives	For coating cheese, fruits, and vegetables. Beeswax coatings produce a permeable barrier that aids in moisture retention and lessens the risk of dehydration.	[32]
2	Carnauba Wax	Carnauba wax is derived from the leaves of the Copernicia prunifera, a Brazilian palm tree.	Candy and fruits are coated with this substance. Carnauba wax gives coated objects a durable, glossy finish, improving their appearance.	[33]
3	Shellac	Extracted from lac beetle excretions.	Applied to fruits, medications, and confections. Coatings made of shellac offer a glossy appearance and moisture resistance.	[34]

(Continued)

Table 7.2 Lipid-based food coating materials. (*Continued*)

S. no.	Lipid-based coatings	Source	Applications	Reference
4	Chocolate coatings	Cocoa Beans are the source of chocolate coatings.	For coating candies, fruits, and small nibbles. Coatings with chocolate improve the taste and look of coated goods.	[35]
5	Coconut Oil	Removed from the meat or kernel of coconuts.	For coating baked products, popcorn, and munchies. Coatings made with coconut oil have a delicious flavor and scent.	[36]
6	Palm Oil	Derived from the oil palm fruit	For coating snacks, baked foods, and candy products. Coatings containing palm oil enhance flavor and texture.	[37]
7	Olive Oil	Removed from olives.	For coating salads, appetizers, and vegetables. Coatings made with olive oil add a delicious, light layer.	[38]

(*Continued*)

Table 7.2 Lipid-based food coating materials. (*Continued*)

S. no.	Lipid-based coatings	Source	Applications	Reference
8	Sunflower Oil	Extracted from sunflower seeds	For coating a variety of food items, such as baked goods and snacks. The mild flavor of sunflower oil coatings makes them a popular choice.	[39]
9	Soybean Oil	Derived from soybeans	Applied to a range of culinary items to coat them. Coatings made of soybean oil add texture and have a neutral taste	[40]
10	Lecithin Coatings	Usually from eggs, sunflower seeds, or soybeans.	Applied as a stabilizer and emulsifier in coatings for baked products, snacks, and confections.	[41]

minimal toxicity and exhibit selective permeability for O_2 and CO_2, polysaccharides such as chitosan, starch, carrageenan, pullulan, pectin, cellulose, alginate, gellan gum, and xanthan gum were used to manufacture films and coatings that replace conventional plastic packaging (Table 7.3).

7.1.3.1 Chitin and Chitosan

Chitin, a natural biopolymer found in abundance in crustacean exoskeletons and fungal cell walls, possesses a structure of -(1-4)-2-acetamido-D-glucose.

Table 7.3 Polysaccharide-based food coating materials.

S. no.	Polysaccharide-based coatings	Sources	Applications	References
1	Chitosan	Derived from chitin, an organic polymer present in crustaceans' exoskeletons	For coating seafood, fruits, and vegetables. Because of its antibacterial qualities, chitosan coatings can prolong the shelf life of perishable goods.	[66]
2	Aliginate	Derived from brown marine algae.	Can be used to encapsulate liquids as well as coat fruits and vegetables. Alginate coatings produce a layer that resembles gel that aids in moisture retention and shields against outside impurities.	[67]
3	Starch	Starch is mainly taken from potatoes, corn, and other starchy crops.	Fruits, vegetables, and baked goods are frequently coated in starch. They are a barrier against gasses, enhance texture, and reduce moisture loss.	[68]
4	Pectin	The source of pectin is fruits, especially citrus peels.	Used to coat fruits, offer a barrier against drying out, and preserve the firmness of the fruit.	[69]

(Continued)

Table 7.3 Polysaccharide-based food coating materials. (*Continued*)

S. no.	Polysaccharide-based coatings	Sources	Applications	References
5	Gellan Gum	The bacteria Sphingomonas elodea produces it.	For use in coating confections, fruits, and vegetables. Gellan gum coatings offer a flexible and clear layer.	[70]
6	Carrageenan	Derived from red algae.	Dairy products and shellfish are coated with carrageenan coatings. They give coated objects a silky feel and improve their appearance.	[71]
7	Guar Gum	Made from guar beans	Applied on meats and baked goods coating. Coatings with guar gum help frozen goods retain moisture and avoid freezer burn.	[72]
8	Agar	Red algae is the source of agar.	Fruits, vegetables, and confections are coated with agar. They offer a solid, long-lasting gel-like covering.	[73]
9	Arabinoxylan	Several plant cell walls, especially those of cereal grains, contain arabinoxylan	Cereal-based product coatings use this technique to enhance texture and provide moisture resistance.	[74]
10	Konjac Glucomannan	The roots of the konjac plant.	As a carrier for flavor, color additives, and coating fruits and vegetables.	[75]

It has a similar structure to cellulose, but each monomer's hydroxyl group is substituted with an acetamide group [42]. Similarly, chitosan, derived from the deacetylation process of chitin, comprises β-(1 → 4)-2-amino-2-deoxy-D-glucose monomers. Chitosan and chitin, both biodegradable polymers, are recognized for their excellent capacity to form films, exhibit antimicrobial properties [43], and act as barriers against aromas, oxygen, and carbon dioxide. The functional effectiveness of chitosan films and coatings has been evaluated through their application on diverse food items. According to Pavinatto, using chitosan coatings on strawberries exhibited bactericidal effects against bacteria and fungi, effectively maintaining the coated fruits' texture, taste, and visual appeal.

7.1.3.2 Starch

Starch is found abundantly in various food sources like cereals and vegetables. Starch and its derivatives, including amylose and hydroxypropylated high amylose starch, have been identified as cost-effective and biodegradable polysaccharide packaging materials. These materials demonstrate remarkable film-forming capabilities and high water solubility [43]. It protects food from the surrounding atmosphere.

However, a high-water vapor transmission rate alters the internal environmental conditions within the package, creating a favorable environment for the proliferation of microorganisms. Additionally, it leads to significant weight loss in the packaged product, which is a critical concern [44, 45]. Researchers found that films with increased levels of amylose demonstrated greater resistance to water vapor permeability, decreased wettability, and lower water absorption. Despite this, the kind of starch used did not impact the physical properties of the films when exposed to consistent humidity levels (53%). Nonetheless, the decreased wettability traits were linked to a decline in the films' mechanical strength [46]. As a result, starch films are suggested for use in dried and rapidly water-soluble food products.

7.1.3.3 Alginate

Alginate, derived from seaweeds, is a non-toxic, biodegradable, biocompatible linear anionic polysaccharide that dissolves in water but remains indigestible. Several applications of alginate-based film are documented in the literature. For instance, [47] Chen *et al.* introduced thymol into biodegradable sodium alginate films applied to fresh-cut apples to prolong shelf life. The resulting composite films exhibited mechanical solid strength and UV-blocking properties. Despite their limited water vapor permeability, these films effectively minimized weight loss in fresh-cut fruits while preserving nutritional content. Similar applications involving composite coatings and films made from crosslinking of alginates have been observed in various food products such as mushrooms [48], peaches [49], freshly cut cantaloupes, freshly cut mangoes, and lamb meat [50]. These applications extended the products' shelf life, decreased weight loss, and enhanced their functional properties.

7.1.3.4 Carrageenan

Carrageenan, a sulfated water-soluble polysaccharide derived from various red seaweeds within the Rhodophyceae family, originates from marine sources. Carrageenans are used in the dairy, pharmaceutical, and food industries as emulsifying, gelling, and stabilizing agents. Edible films made from carrageenan find applications in diverse areas such as sausage casings, dry solid foods, oily foods, and meat, fish, and poultry to prevent surface dehydration [51].

7.1.3.5 Pullulan

Pullulan is a microbial polysaccharide composed of maltotriose units linked by α (1,6) glycosidic bonds. Aureobasidium pullulans produces it by enzymatic synthesis from starch. Pullulan, well-known for its high water solubility, is used to make edible films. These films have several desired properties, including being odorless, colorless, heat-sealable, tasteless, water-permeable, transparent, and having minimal permeability to oxygen and oil [52].

7.1.3.6 Pectin

Pectin consists of a linear arrangement of poly-α-(1-4)-D-galacturonic acid featuring varying degrees of methyl esterification [53]. Its recognized attributes include biocompatibility, biodegradability, non-toxicity, and, notably, the capacity to create gels [54, 55]. However, pectin films are less desirable than films made from alternative polymers due to their inferior physicochemical characteristics and unstable mechanical properties [54, 56]. Maftoonazad, and Ramaswamy [57] reported that applying a pectin-based coating to lime fruit showed extended shelf life, reduced refrigeration costs, ease of handling, enhanced product safety, and enhanced biodegradability.

7.1.3.7 Cellulose

Cellulose and its derivatives, such as hydroxypropyl (HPC), carboxymethyl cellulose (CMC), and bacterial cellulose (BC), have film-forming solid properties. According to Singh *et al.* [58], it creates transparent, nontoxic, biodegradable films with superior mechanical, thermal, and barrier properties for the oxygen and oil phases. It explains how it might enhance the gas barrier qualities of other biopolymers, such as chitosan and proteins. CMC interacts with the carboxylate group and shows strong gelling ability when the relative humidity increases; nevertheless, these films and coatings lose their mechanical and barrier characteristics due to their hydrophilic nature [59].

7.1.3.8 Gums

Gums, categorized as naturally occurring polysaccharides, can absorb water and create a gel, making them a subject of extensive research. These polysaccharides are cost-effective, safe, environmentally friendly, and renewable for producing edible packaging materials [60, 61]. Furthermore, gums have nutritional and antibacterial properties, contributing positively to human health. Gums are, therefore, often utilized as thickening, binding, emulsifying, suspending, and stabilizing agents in the food and pharmaceutical industries [62]. Tripathi *et al.* [63] tried using methylation to improve the guar gum's mechanical strength, barrier qualities, and film-forming ability. According to their research, methylation promoted guar gum crystallization, which enhanced the films' thermal stability. Furthermore, methylation reduced the water vapour transfer rate by 40 and increased the elongation capacity of guar gum films. This highlights the possibility of altering the chemical structure of gums to explore their uses.

Gellan, a polysaccharide class, is produced by the bacterium Sphingomonas elodea (Pseudomonas elodea). It exhibits distinctive colloidal properties and excellent coatings forming and gelling capabilities [64]. Gellan is increasingly used as a gelling and texturizing agent in the food sector. Gellan also carries various food additives, such as anti-browning and antibacterial agents, colorants, nutraceuticals, and flavors. Gellan gum-based films have a strong and brittle gel structure. Gellan gum-based edible coatings have improved fresh-cut vegetables' quality and shelf life [65].

7.2 Nano-Based Food Coating Materials

Recently, there has been a rise in demand, particularly in the EU, for materials based on nanoparticles for various uses, such as the food sector. The European nanomaterials industry is predicted to produce revenue of over $9 million by 2022, up from more than $2.5 million in 2015. Numerous materials include necessary components; some can withstand high temperatures and pressures, while others are non-toxic [76]. The enhanced characteristics of nanoparticles have facilitated and stimulated the creation of innovations that increase product freshness and shelf life, the creation of instruments and processes for quick on-site examination, and the manufacturing of food packaging that is both biodegradable and safe for the environment. Significant changes in living are possible with the application of nanotechnology.

The creation of better goods, money, health, and quality may all be facilitated by nanomaterials (Figure 7.3). Furthermore, since they are so little, this may lessen the environmental effect. However, food packaging applications of nanotechnology are still in their infancy. The research and implementation of these nanoparticles are also expanding in the food packaging business, boosting packaged food's shelf life and safety [77]. Nevertheless, various functional nanomaterials can enhance packaging materials.

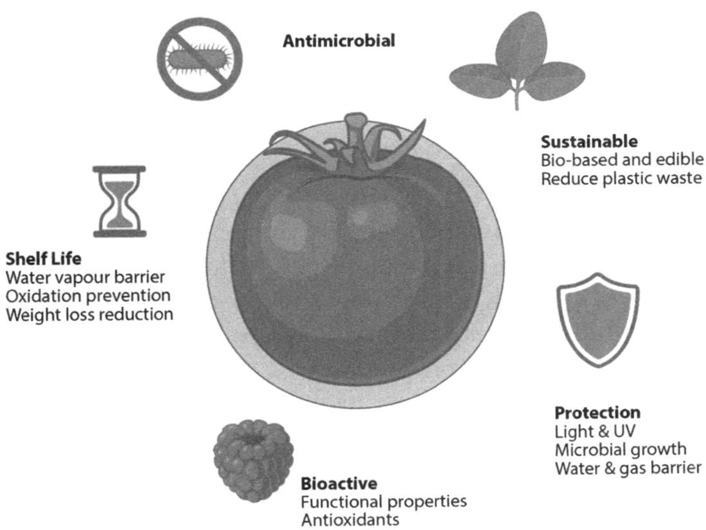

Figure 7.3 Uses of edible food coating materials.

Utilizing nanotechnology methodology, nanoparticles can be mixed with film to create nanofilm that can increase the permeability of some gases to lower the concentration of hazardous gases, like carbon dioxide (CO_2) or oxygen (O_2), which shorten food's shelf life and can also be used as barrier materials to stop microorganisms from spoiling it [78]. Because plastic polymers, which are the most commonly used materials in the food packaging industry and pose a serious threat to both humans and the environment, are not biodegradable, scientific research is also concentrated on creating edible coatings and films from bio nanocomposites or on using them as matrices for incorporating antimicrobial nanoparticles to improve the shelf life and storage quality of fruit and vegetables [79]. Biomaterials have the potential to minimize packaging materials while simultaneously providing a solution to the enormous waste problem. Furthermore, biomaterials have the potential to enhance food safety and quality throughout distribution while also extending the shelf life. The time for consumption and storage may be prolonged [77]. Biomaterials are still not widely used in food packaging.

7.3 Methods of Food Coating

Food coatings can be applied via coagulation, heating, cooling, or drying. The application technique must be carefully selected to get the best possible outcomes, as it can affect the coating's efficiency [80].

There are various methods based on the state of the coating material, which can be liquid, powder, emulsion, or suspension. Common methods are spraying, panning, fluidized bed, dipping, casting, and extrusion.

7.3.1 Spraying

Spraying is a means of dispersing coating droplets over the food's surface. Spraying is classified into numerous types, including air spray atomization, which uses cylindrical spray flow and low-speed, air-assisted airless atomization, which is employed for viscous coatings, and pressure atomization spraying, which uses pressure instead of air to apply coatings [81]. Spraying pressure, viscosity surface temperatures and coating tension are all critical factors that influence ultimate coating performance. One advantage of spraying is the possibility of applying multiple distinct coatings, resulting in a multi-layered coating [82]. This method enables the control of the thickness of the coating layer.

7.3.2 Panning

The pan coating process is placing the food items into a spinning container, where the coating is sprinkled to create an even coating on the food surface. Following that, the coating has dried out. The technique is employed for making thin or thick layers of coatings on rigid surfaces. This method allows for the concurrent coating of numerous food items, including those of diverse sizes, resulting in a transparent, flexible, and sparkling film for coating [80]. The limitation of this method is that it takes time to complete due to the need for water evaporation. Furthermore, the coating must be applied gradually; otherwise, the food products may adhere together inside the revolving container.

7.3.3 Fluidized Bed

The fluidized bed edible coating technique is often employed in food production and research applications. Using a variety of nozzles, the solution of coating and suspension is sprayed onto the fluidized powder surfaces to form a structure resembling a shell in a fluidized coating process a liquid flow that moves upward via a bed of particles fluidizes as it achieves a fast speed to assist the particles without pushing them towards the liquid stream [80].

7.3.4 Dipping (Immersion)

The most common method is immersion, considering it is more affordable and simple to execute. This process has three steps: deposition, immersion and dwell time, and solvent evaporation. The film is in a liquid state, suspension, or emulsion condition [83]. The food has first to be immersed in the coating's solution at a constant rate for the coating to come in contact with its surface and adhere to it. The process of deposition involves obtaining thin coating layers. Following immersion, the food is maintained stationary to enable deposition to remove any extra coating [84]. The last step is solvent evaporation, which comprises evaporation of extra coating, it can be achieved by using drying equipment or maintaining it at room temperature to allow the food to evaporate and dry. The desired outcome of the edible coating and the food product is affected by the immersion period, which can vary from 5 to 30 seconds.

This method has several advantages as it uniformly coats all the food surfaces and may be used on a wide range of surfaces. Despite this, it has a limitation in that the coating becomes too thick, which could prevent respiration and results in the growth of microbes within the coating itself [85].

7.3.5 Casting

In laboratory and pilot scales, the most often used technique for film formation is the casting method, also known as solvent casting. Three steps are involved in this process to create a biopolymer film: i)Biopolymer solubilization in an appropriate solvent, ii) solution casting in a mould, iii) solution drying [80]. The choice of the polymer or polymer combination that forms the foundation of the edible film formulation comes first. This process, known as solubilization, involves dispersing or dissolving the chosen polymer in an appropriate solvent. For instance, soy protein isolate polymer is dissolved in ethanol [86]. Rather than melting, the polymer's solubility determines film formation in solvent casting [87]. During the casting process, the solution is poured into glass plates coated with teflon or into a mould prepared before. The drying step gives the solvent enough time to evaporate. that produces a coating of polymer that sticks to the mould. When casting films, air dryers such hoover driers, microwaves, hot air ovens, and tray dryers are used to remove solvents and peel the film easily [80].

The primary benefit of the casting technique for film creation is its low cost and convenience of manufacture, as it does not need specialized equipment [15]. Due to the wet nature of the casting process, there are fewer and smaller flaws because of improved particle–particle interaction and more uniform particle packing (Yang *et al.*, 2011). Since most materials used in food processing cannot be moulded at higher degrees without causing a permanent structural change in the material, the lower temperature throughout the processing steps may also be advantageous. The main disadvantages of the casting method are as follows: i) forms are restricted (usually only simple sheets and tubes can appear); ii) there is a chance that a toxic solvent will be trapped inside the polymer; and iii) using solvents to denaturize proteins and other molecules added to the polymers [15].

7.3.6 Extrusion

Another approach for creating polymeric films is extrusion, one of the leading commercially utilised polymer processing methods [88]. This process modifies the materials' structures and enhances the extruded material's physiochemical characteristics [89]. Three zones often comprise the extrusion process: the feeding zone, the kneading zone, and the heating zone at the machine's final component or exit [90]. The mixture of the film's components is first moved into the feeding zone and compacted using air. This technique is also known as a dry process, as it functions best when little to no water is present.

7.4 Parameter/Factors in Food Stability

Edible food coatings are essential for preserving the quality and prolonging the shelf life of various food products. These coatings influence respiration and transpiration rates by forming a protective barrier on the food's surface. Fruits and vegetables undergo vital physiological processes, including respiration and transpiration, and applying edible coatings can have various effects on these processes [91]. Transpiration and respiration processes, moisture content, Lipid oxidation, Regulating enzymatic activity, Microbial spoilage.

7.4.1 Transpiration Rate

When discussing fruits and vegetables, transpiration is the process by which water vapor is removed from the surface of the produce. This procedure may cause the product to wilt, dehydrate, and deteriorate. By creating a barrier that prevents water vapor from escaping, edible coatings can aid in the reduction of transpiration. Fruits and vegetables can benefit especially from this in terms of keeping their texture and freshness [92].

Moisture Retention. Edible coatings serve as a barrier, minimizing the food product exposure to the outside world. The coating helps keep the product moist by reducing water loss through transpiration, which keeps it from drying out and preserves its crispness and juiciness [93].
Extended shelf Life. Decreased transpiration can slow the water loss rate, leading to a longer shelf life. This is especially critical for perishable goods like berries, leafy greens, and other fruits, as retaining moisture is essential to preserving quality in storage and during transit [93].

7.4.2 Respiration Rate

The metabolic process of cells breaking down organic substances to produce energy is called respiration. Regarding fruits and vegetables, respiration is linked to the generation of heat, carbon dioxide, and ethylene gas—a plant hormone that affects ripening. The following are some ways in which edible coatings may affect respiration rates:

Gas Permeability. It is possible to modify edible coatings to regulate the permeability of gases, such as carbon dioxide and oxygen. Coatings can reduce the respiration activity of produce by controlling the exchange of these gases. This is especially important for fruits like bananas and apples, where regulating respiration rate helps postpone ripening [10].
Ethylene Absorption. A crucial part of the ripening process, ethylene gas, can be absorbed or blocked by some edible coverings. Coatings can delay ripening by reducing ethylene exposure and keeping the product fresh and high-quality for longer [94].
Temperature control. A thermal barrier that aids in controlling the food product's temperature can be produced via edible coatings. Since metabolic processes frequently depend on temperature, regulating the temperature can affect respiration rates. This is especially crucial to retain the appropriate freshness during storage and transit [95].

7.4.3 Lipid Oxidation

The food industry is very concerned about lipid oxidation since it can cause rancidity, off-flavors, and a reduction in food's nutritional value. Lipid oxidation can be affected by edible coatings in the following ways:

Oxygen barrier. By acting as a barrier, edible coatings lessen the amount of oxygen food products are exposed to. Lipid oxidation is primarily caused by oxygen, which coatings help to slow down by restricting its interaction with lipids. This maintains the product's freshness and quality [6].
Antioxidant incorporation. Plant extracts or tocopherols are antioxidants that can be added to some edible coating recipes. By aiding in the scavenging of free radicals produced during lipid oxidation, these antioxidants prevent the chain reaction that causes lipid degradation [96].
Metal Chelation. Lipid oxidation reactions can be catalyzed by metal ions that certain coatings can chelate. Coatings reduce prooxidative effects and improve the overall stability of fats and oils in food by securing metals [95].

7.4.4 Moisture Content

Maintaining texture, increasing shelf life, and controlling microbiological development depend on moisture control. Moisture content is affected by edible coatings through:

Water vapour permeability. It is possible to engineer edible coatings to regulate the permeability of water vapor. Because of this characteristic, coatings can function as a barrier to prevent excessive moisture gain or loss, assisting in the food product's ability to maintain the appropriate moisture content [10].
Hydrophilic and Hydrophobic properties. Depending on the formulation, coatings can be hydrophilic (attracting water) or hydrophobic (repelling water). This can be adjusted to meet the product's unique requirements, affecting the movement and retention of moisture [97].

7.4.5 Regulation of Enzymatic Activity

Enzymatic activity can negatively impact food quality, particularly in fruits and vegetables. The following are some ways that edible coatings control enzymatic activity:

Temperature control. By acting as a thermal barrier, coatings can assist in controlling the food product's temperature. Since many enzymes are temperature-sensitive, temperature management is crucial for regulating enzymatic processes [95].
Selective Permeability. Certain substances, such as enzymes, can be selectively allowed or prevented from passing through certain coatings. By limiting enzymatic reactions, this regulated permeability can help avoid unfavorable alterations in the food matrix [10].
Enzyme inhibition. Inhibitors that mainly target the enzymes causing unwanted reactions, including browning or softening, may be present in some coatings. This aids in maintaining the food product's appearance and quality [98].

7.4.6 Microbial Spoilage

One of the main causes of food spoiling is microbial development. Edible coatings have several ways to reduce microbial deterioration:

Antimicrobial agents. Certain coatings can be made with bacteriocins, essential oils, or natural antibacterial agents. These compounds prolong the food product's shelf life by preventing the growth of harmful and rotting microbes [7].
pH regulation. The pH of the food surface may be impacted by coatings, which would make the environment less favorable for microbial

development. Because spoiling bacteria prefer certain pH ranges to flourish, this pH adjustment is very effective against them [91].
Physical barrier. Coatings have the potential to alter the pH of the food surface, which would hinder the growth of microorganisms. Since spoilage germs only grow in certain pH ranges, this pH adjustment is very effective against them [11].

7.5 Coating-Specific Quality Attributes

7.5.1 Color

Edible food coatings are essential for maintaining and improving the quality of various food products. They can also significantly alter the color of these goods. One important sensory factor that affects how consumers perceive and accept food is its hue. There are various ways in which edible coatings might affect the color of food products:

Color retention. One of the main purposes of edible coatings is to serve as a barrier against outside elements that can hasten color deterioration, such as oxygen, light, and moisture. . Coatings aid in preserving the food product's original color for a longer time by forming a protective layer [95].
Minimization of oxidation reactions. Color changes can result from oxidation reactions, particularly those involving pigments like the anthocyanins in fruits and the chlorophylls in green vegetables. By preventing oxygen and other prooxidants from entering the coating, edibles can reduce the severity of these oxidation events. This is especially crucial for keeping fruits from browning and for maintaining the vivid green hues of leafy vegetables [99].
Inhibition of enzymatic browning. Fruits and vegetables frequently change color due to enzymatic browning, which is aided by enzymes such as polyphenol oxidase (PPO). Inhibitors that aid in reducing enzymatic browning reactions may be present in edible coatings. For instance, coatings containing ascorbic acid or citric acid can suppress PPO action and maintain the produce's original color [100].
Light protection. Certain food products may lose color due to light exposure, particularly ultraviolet (UV) radiation. Edible coatings can block light, slowing down pigment deterioration and averting color changes. This is especially important for light-sensitive items, like drinks, oils, and some fruits [95].

Uniform coating distribution. Uneven coating distribution can cause food products to change color unevenly. When coatings are applied correctly, they are evenly spread, offering consistent protection and assisting in maintaining a constant color throughout the product.
Addition of natural colorants. Natural colorants, including fruit or vegetable extracts, can be added to edible coatings to change or improve the product's color. This is frequently done to provide aesthetically pleasing coatings that either contrast or enhance the food item's natural hue [80].
Enhancement of gloss and sheen. Certain edible coatings can give the food item a glossy or shiny appearance. This improvement in visual appeal might affect how color is perceived by making the presentation more eye-catching and colorful [101].
Stabilization of color-assisted compounds. The color-associated chemicals carotenoids and anthocyanins, which give many fruits and vegetables red, orange, and purple hues, can be partially stabilised by edible coatings. These hues' richness and brilliance are partly preserved by this stability [11].

7.5.2 Flavor

Edible food coatings can greatly affect food products' flavor and aroma, which can also affect the consumers' overall sensory experience. These coatings accomplish a number of tasks, including as regulating gas exchange, managing moisture content, and forming a protective barrier. This is a thorough examination of how edible food coatings might affect the flavor and scent of food products:

Flavor retention. Edible coatings aid in flavor preservation by forming a barrier that protects the volatile components that give food its distinct flavor. This is especially important for products like fruits and herbs with delicate flavors. Coatings can help keep food tasting as it did when it was first created by reducing its exposure to elements like moisture and oxygen [93].
Moisture control. Many food products' sense of flavor strongly correlates with moisture content. In order to avoid overdrying or moisture absorption, edible coatings help regulate moisture migration. By keeping the product at the ideal moisture content, tastes are released and consumed as intended while also preserving the food's texture and juiciness [6].
Prevention of oxidative changes. Foods can suffer from bad tastes and rancidity due to oxidation reactions. By acting as a barrier against oxygen,

edible coatings can lessen oxidation and the emergence of unwanted tastes. This is especially important for nuts and snacks, including fats and oils [99].

Inhibition of enzymatic reactions. Flavor variations can result from enzymatic reactions, particularly from enzymatic browning and other related processes. Inhibitors that slow down or stop the enzymes causing these reactions can be found in edible coatings. For example, ascorbic acid or citric acid can be added to coatings to prevent fruit enzymatic browning and preserve their original flavors [100].

Masking undesirable tastes. particular edible coatings can be designed to cover up or lessen the flavor or odor of particular meals that people find offensive. This is especially helpful for beneficial substances or components that could taste harsh or unsettling [67].

Retention of aromatic compounds. Aromatic substances play a major role in the whole tasting experience. Volatile scent molecules are kept within the food matrix by edible coatings, which stop them from escaping into the surroundings. This is crucial for commodities where the scent plays a major role in the sensory appeal, such as baked goods and confections [102].

Controlled release of flavors. It is possible to engineer edible coverings to regulate taste releases gradually. This regulated release offers a longer-lasting and more consistent sensory experience by ensuring that the flavors are released gradually during intake. This is especially important for goods like coated candy and chewing gum [8].

Enhancement with flavorings. Certain food products can have their natural or artificial flavorings added to them to improve or balance their natural flavor. This is frequently done to provide distinctive, cutting-edge flavor profiles that appeal to customer tastes [6].

Stabilization of heat-sensitive flavors. Edible coatings that protect against heat and moisture can help maintain the flavors of some foods, including spices and herbs, that are sensitive to heat. This aids in preserving the flavor intensity and authenticity of goods that have undergone thermal processing [93].

Different food products' appearance, flavor, and nutritional profile can all be altered and preserved using edible food coatings. These coatings are used to improve texture, add protection, and regulate elements including oxidation, gas exchange, and moisture content. This is a thorough analysis of how edible food coatings affect the look, flavor, and nutritional value of food products:

Texture improvement. Certain coatings are made to give food items a thin, crisp layer that enhances their feel. This is frequently observed in coated snacks, where the coating adds a pleasing crunch that improves the whole eating experience [12].

Uniformity. The surface of food products can have a consistent and homogenous appearance when edible coatings are applied correctly. This consistency is visually pleasant and helps to create a favorable impression in the eyes of the public [80].

7.5.3 Nutritional Profile

Nutrient retention. By forming a barrier against environmental elements including oxygen, light, and moisture, edible coatings can aid in retaining vital nutrients in food products. Maintaining the nutritional value of vitamins, minerals, and other delicate substances is vital [9].

Fortification. Certain food products can have their edible coatings enhanced with vitamins and minerals to improve the food product's nutritional profile. This is a standard procedure when creating functional foods that meet particular dietary requirements [103].

Reduction of fat absorption. Edible coatings can lessen the amount of fat overall in some fried and coated food by reducing the absorption of cooking oil. This is important when making healthier versions of typically fried items [104].

Stabilization of bioactive compounds. Bioactive substances like polyphenols and antioxidants can be stabilized by edible coatings, maintaining their usefulness and possible health advantages. This is especially crucial in fruits, vegetables, and other plant-based products [97].

7.6 Commercial Applications in Edible Food Coatings

In India and worldwide, edible food coatings have numerous commercial uses in various food-related industries. The multiple advantages of edible coatings, such as increased shelf life, improved sensory qualities, and new functionalities, motivate these uses. The following are a few noteworthy industrial uses for edible food coatings:

Fruits and Vegetables. Worldwide: Edible coatings significantly increase Fruit and vegetable shelf life. They can help keep freshness throughout storage and transportation by reducing oxidation, microbial development, and moisture loss. India: When transporting perishable fruits such as citrus fruits, bananas, and mangoes from farms to markets, edible coatings are used to maintain their freshness [62].

Bakery Products: To improve texture, lessen staling, and give baked goods a glossy look, edible coatings are applied to pastries, cakes, and bread. India: Coatings on baked items prolong their freshness and enhance their aesthetic appeal in India [105].

Nuts and Snacks: The snack industry uses edible coatings to improve crunchiness, lower oil absorption during frying, and form a protective layer. India: The texture and oil absorption of coated foods, such as chips, fried nuts, and extruded goods, are improved [5].

Dairy products: Some cheeses are coated with edible materials to prevent moisture loss and safeguard the product's surface. India: While covering dairy products is not as popular, there is room for creativity in improving the shelf life of specific dairy products [106].

Meat and Poultry: Meat and poultry products can be made more aesthetically pleasing, have less moisture loss, and stop microbial growth using edible coatings. India: Coating meat items there may help them retain their texture and freshness over time [107].

Confectionery: For texture enhancement, protection, and flavor release control, edible coatings, including coated chocolates, candies, and chewing gum, are widely employed in the confectionery business. India: Edible coatings help to ensure product quality, and coated confectionery products are popular in the Indian market [105].

Functional Foods: Bioactive chemicals are encapsulated, their release is regulated, and overall product stability is enhanced by applying edible coatings to functional foods and nutraceuticals. India: In line with the country's rising demand for health-conscious goods, using coatings in functional meals is a developing trend there [108].

Fresh-cut produce: To minimize dehydration and preserve quality, edible coatings are used in the fresh-cut produce sector to increase the shelf life of pre-cut fruits and vegetables. India: To satisfy the desire for quick, ready-to-eat options, coatings applied to fresh-cut vegetables are becoming more popular in India [62].

Fish and Seafood: Using edible coatings can extend fish and seafood products' shelf lives and reduce moisture loss. India: In the Indian market, coatings might help preserve and improve the quality of seafood goods [109].

Pharmaceuticals: Edible coatings are used in the food and pharmaceutical industries, where they are employed for flavor masking and drug delivery in oral treatments. India: Adding edible coatings to pharmaceuticals helps make some medications more palatable and widely accepted [110].

These illustrations show how adaptable edible food coatings are and how widely used they are in various food industries. As research and development efforts concentrate on improving edible coatings' usability, durability, and affordability, commercial applications keep changing.

7.7 Challenges in Edible Coating Application

Although edible food coatings have many advantages, such as better looks, longer shelf lives, and better nutritional profiles, there are drawbacks, such as inadequate coating and taste transmission. These difficulties may impact the general quality of food products. These are a few typical problems:

Uneven coating Application: It can be challenging to apply coatings uniformly, which might result in unequal protection and appearance. Uneven coatings can cause differences in flavor and texture within the same product, impacting the consumer's entire sensory experience. Consequence: An uneven coating distribution could expose some parts of the food product to moisture and air, which could cause problems like inconsistent flavor, uneven browning, and textural variations [11].
Inadequate adhesion: Challenge: Inadequate coating adherence to the food surface may cause partial or total separation. This can be particularly challenging for goods like coated snacks or confectionery when a long-lasting covering is necessary. Consequence: Poor adherence could cause the coating's protective qualities to be lost, exposing the product's quality to outside influences. Furthermore, buyers may have a bad impression of disconnected coatings [13].
Flavor Loss. Challenge: In certain cases, edible coatings can obstruct the taste of food products from reaching the customer's palate. This is especially true if the coating material tastes strongly or creates a barrier restricting the discharge of flavor-producing volatile chemicals. Consequence: A change or loss of flavor can greatly affect how satisfied a customer is with the product overall. Food may taste less appetizing if the coating obscures or modifies the food's inherent flavor [13].
Moisture Incompatability. Challenge: Depending on the food product's moisture content, some coatings could not work well with it. Problems like moisture migration, sogginess, or crispness loss can result from incompatibility, particularly in baked foods or coated snacks. Consequence: Moisture

issues can impact a product's texture and shelf life. It can make eating less enjoyable and hasten the breakdown of the product over time [6].

Limited Barrier properties. Challenge: Depending on the coating material and application technique, the ability of coatings to operate as a barrier against external elements like light or oxygen can vary. Some coatings might not have very strong barriers. The product's shelf life and quality may be jeopardized by inadequate barrier qualities, which can hasten oxidation, microbial development, and other degrading processes [11].

Texture changes. Challenge: Coatings can change a food's texture, particularly if they're meant to give it a certain crunch or mouthfeel. Texture changes that are not desired can be caused by poorly formulated coatings. Consequence: Modifications to texture could cause a decline in sensory appeal, influencing customer approval. The entire eating experience may suffer if the coating doesn't provide the desired texture or gets overly thick [13].

7.7.1 Health and Allergen Concerns

Challenge: For some people, various coating materials may cause allergic responses or give rise to health risks. Coatings containing allergenic chemicals, for instance, can prevent products from being accessible to a broader range of consumers. Consequences: Problems with health and allergies may result in unhappy customers, a decline in market acceptance, and even legal problems [90].

Careful evaluation of coating materials, application techniques, and formulation strategies are necessary to meet these problems. To guarantee consistency, adherence, and compatibility with the intended flavor and texture of the food product, the coating process must be optimized. For edible food coatings to overcome these obstacles and become more effective in various applications, ongoing research and development are necessary [6].

7.7.2 Economic Aspects in the Commercial Scale Production

There are several financial obstacles to large-scale commercial use of edible food coatings. Even though these coatings have many advantages, there may be challenges due to costs, market acceptance, and the realities of mass production. The following are some financial obstacles to the widespread use of edible food coatings on a commercial scale:

Cost of Raw materials. Challenge: The cost of the raw materials necessary to make edible coatings might be high, particularly if those substances are unique or specialized. Purchasing premium, food-grade components may result in higher production costs. Consequence: Higher production costs due to elevated raw material costs may make large-scale deployment less financially feasible. It is critical to weigh the advantages of employing particular materials against the expenses involved [5].

Processing costs. Challenge: Applying edible coatings requires extra processing stages, such as formulating, applying, and curing the coating. These procedures could call for specialized tools and knowledgeable personnel, which would raise the cost of processing. Consequence: Higher processing expenses may impact large-scale production's overall cost-effectiveness. Keeping a competitive edge in the market requires implementing economical and effective coating techniques [5].

Scale up challenges. Challenge: Ensuring quality and consistency while moving from small-scale trials to large-scale production can present difficulties. When scaling up, it could be necessary to make changes to procedures, machinery, and quality control methods. Consequence: Production inefficiencies, waste, and a possible drop in coated product quality can result from scaling up challenges. This may make large-scale implementation less financially feasible [80].

Equipment Investment. Challenge: Large-scale coating applications could call for specialized equipment. For food producers, purchasing this equipment can come with a hefty upfront cost. Consequence: For many businesses, especially smaller ones, the initial equipment investment may be a deterrent to the widespread use of edible food coatings. This may cause the industry's adoption rate to slacken [13].

Market acceptance and Consumer Education. Challenge: It can be difficult and expensive to convince customers of the advantages of edible coatings and to get beyond their reluctance to perceived alterations in flavor, texture, or appearance. Consequence: A low level of market acceptance may impede the adoption of coated products, hence reducing producers' return on investment. It could be required to launch consumer education efforts to emphasize the benefits of coated items [5].

Regulatory Compliance. Challenge: Informing customers about the advantages of edible coatings and removing any reluctance to apparent flavor, texture, or appearance alterations can be difficult and expensive. Consequence: Slow adoption of coated items due to limited market acceptance can reduce producers' return on investment. Consumer education programs can be required to emphasize the benefits of coated items [91].

Supply Chain Complexity. Challenge: Including edible coatings could make the supply chain more complex by requiring more coordination for procuring coating components, inventory management, and on-time delivery.

Consequence: Complex supply chains can raise operating expenses and present logistical difficulties. Cost-effectiveness and efficiency must be maintained by streamlining the supply chain [111].

Competitive pricing. Challenge: It can be difficult to compromise between charging coated items economically in a very competitive market and meeting the expenses related to edible coatings. Consequence: While underpricing could jeopardize profitability, overpricing could discourage customer adoption. Success depends on balancing market developments and consumer expectations [67].

Environmental Concerns. Challenge: Manufacturers may be pressured to adopt more sustainable methods, which could incur additional expendses, if the coating materials or processes impact on the environment. Consequence: Investing in eco-friendly alternatives and practices may be necessary as sustainability receives more attention. It is becoming more and more crucial to strike a balance between environmental responsibility and economic considerations [11].

A thorough strategy that considers every step of the production process—from locating raw materials to shipping the finished product to customers—is needed to navigate these economic hurdles. Before introducing edible food coatings on a broad commercial scale, companies must thoroughly evaluate the market dynamics, scalability, and economic feasibility. Working together with government agencies, business associates, and research institutes can also aid in addressing these issues and promoting innovation in the field.

7.7.3 Safety and Digestibility Issues

Critical factors in the creation and use of edible food coatings are digestibility and safety. Edible coatings are intended to improve the taste and safety of food items. However, ensuring the ingredients are risk-free for ingestion, easily digestible, and safe for consumption is crucial. An outline of the problems with digestibility and safety related to edible food coatings is provided below:

Texture and mouthfeel: Digestion may be impacted by variations in the mouthfeel and texture of coated goods. Overly thick or unappealing coatings might detract from the entire eating experience. Taking into account: It's critical to optimize coating compositions for a pleasing mouthfeel. Improving digestion involves balancing thickness, crispness, and other textural characteristics [112].

Enzymatic compatibility problem: When food is coated, some coating materials might obstruct enzymatic processes, making it more difficult for nutrients to be absorbed and broken down from the meal. Taking into account: Researching the enzymatic compatibility of coating materials and making sure they don't obstruct the digestion and absorption of vital nutrients is crucial for digestibility in general [113].
Effects of Nutrient availability: Nutrient availability and release in the gastrointestinal tract might be affected by coating materials. This may affect how well vitamins, minerals, and other bioactive substances are absorbed. Consideration should be given to understanding how coatings impact the bioavailability of nutrients. It's critical to modify formulations to reduce any adverse effects on nutrient absorption [9].
Potential for gastrointestinal discomfort: The gastrointestinal system may not always tolerate various coating materials, which could cause discomfort or digestive problems. Taking into account conducting clinical research to evaluate the gastrointestinal tolerability of coated products is crucial. This involves watching for any uncomfortable side effects or responses from eating coated foods [91].

Edible food coating safety and digestibility require a multifaceted approach that includes rigorous research, regulatory compliance, and a dedication to label clarity. Cooperation between food scientists, regulatory bodies, and the food sector is crucial to solving these issues and creating edible coatings that improve food safety and retain ideal digestibility.

7.8 Further Research

Edible food coatings have much-untapped promise for increasing food variety, enhancing food safety, developing technological applications, and reducing costs. Future studies could focus on the following areas:

Exploration of Novel coating materials: Researchers can investigate novel and inventive materials like microbial, seaweed, or plant- based polymers for edible coatings. In addition to improving coatings' protective qualities, these materials should also support environmentally responsible and sustainable food sector practices [5].
Natural Antimicrobial agents: Examine whether adding natural antimicrobials to edible coatings can improve food safety. Substances such as plant extracts, essential oils, and other bioactive components can help suppress microbial development to prolong the shelf life of coated products [7].

Biodegradable coatings: Develop biodegradable, edible coverings with the least adverse environmental effects. Studies might look into using naturally decomposing materials to cut waste and meet the growing need for environmentally friendly packaging options [114].

Smart coatings: Examine the creation of intelligent coatings that react to particular environmental circumstances. Coatings that, for example, release antimicrobial compounds as necessary or change color in response to rotting signs may enhance the quality and safety of food [115].

Enhanced flavor delivery system: Investigate ways to enhance the flavor delivery in coated goods. This might entail creating coatings that improve the release of particular flavors, giving users more control over how their food tastes without sacrificing the coating's barrier qualities [116].

Application of robotics and automation: Examine how coating processes might benefit from integrating robotics and automation to increase productivity and save labor expenses. Automated coating systems can improve total production output and guarantee consistent application [117].

Process Optimization for large-scale production: Investigate coating process optimization for large-scale manufacturing. This entails looking at economical techniques for improving adhesion, reducing production times, and applying coatings consistently [80].

Economic Analysis and Cost Reduction Strategies: To find cost-effective manufacturing procedures, sourcing tactics, and cost-effective formulations, do economic evaluations. Comprehending the financial consequences of employing edible coatings extensively might enhance market competitiveness and promote wider acceptance [95].

Collaboration with industry partners: Encourage researchers and industry partners to work together to make implementing research findings easier. Business collaborations may ensure that innovations benefit the whole food business by bridging the gap between lab research and commercial implementation.

Global implementation studies: Research is required to determine how edible coatings are adapted and received in various areas and societies. Factors including supply chain dynamics, legal frameworks, and local dietary preferences should be considered to make worldwide adoption easier.

7.8.1 Need for Interdisciplinary Approach in Novel Edible Film Development

To create sustainable edible food coating materials, scientists from various disciplines, including agriculture, biochemistry, material science, and microbiology, need to collaborate and find packaging solutions. Each discipline's distinct perspectives and proficiencies augment the overall progress towards creating sustainable and efficacious coating solutions. Here's a quick summary of the reasons why multidisciplinary cooperation is crucial.

Agriculture: Finding sustainable and renewable sources for coating materials is a major task for agricultural experts. Investigating natural resources such as plant-based polymers, agricultural byproducts, and other materials that can be utilized to make environmentally friendly coatings is part of this. Importance: Researchers can help design coatings that encourage the use of renewable resources and are environmentally friendly by utilizing their experience in agriculture [118].

Biochemistry: People studying biochemistry thoroughly understand the chemical processes involved in food production and the makeup of edible items. They aid in the formulation and selection of bioactive and safe chemicals for coatings. Importantly, working with biochemists guarantees that coating ingredients are safe to eat while also enhancing the total nutritional value and health advantages of coated food items [90].

Material Science: The characteristics, composition, and usability of materials are the main concerns of material scientists. They aid in creating coatings with the best mechanical, barrier, and sensory qualities for edible coatings. Material scientists play a crucial role in ensuring that coatings perform as intended in terms of protection while preserving desired properties, including texture, appearance, and adhesion to food surfaces [4].

Microbiology: In order to solve the issues surrounding microbiological safety related to edible coatings, microbiologists are essential. They offer their expertise in creating coatings that have inherent antibacterial qualities to increase the coated product's shelf life. Working with microbiologists ensures that coatings efficiently prevent microbial development while maintaining food safety, improving the product's overall quality [119].

7.8.2 Necessity of Collaboration

Holistic approach: Collaborative efforts make a comprehensive approach to coating material development possible. Scholars from various disciplines contribute a variety of viewpoints, taking into account not only the technical aspects of coating formation but also the advantages to society and the environment.

Innovation and creativity: Interdisciplinary cooperation promotes originality and inventiveness. By integrating microbiology, material science, biochemistry, and agriculture knowledge, scientists can create innovative coating materials that tackle several issues simultaneously.

Sustainability considerations: Working together guarantees that sustainability factors are considered at every process stage, from choosing the right material to applying it. This entails locating renewable resources, reducing the influence on the environment, and streamlining the production process.

Efficient problem-solving: Problems with edible food coatings frequently require a diversified strategy. By utilizing the advantages of each profession, collaborative efforts facilitate more effective and expeditious problem-solving.

Translational impact: Working across disciplines makes applying research findings in real-world settings easier. Collaboratively, scientists can create coating materials that are viable for widespread commercial application and are scientifically sound.

Researchers in microbiology, material science, biochemistry, and agriculture should work together to use their combined knowledge to develop sustainable edible food coating materials. This cooperative approach guarantees a thorough comprehension of the opportunities and problems related to edible coatings, resulting in the development of creative, secure, and eco-friendly solutions for the food sector.

7.9 Conclusion

Edible coating matrices composed of lipids, proteins, carbohydrates and bioactive components have tremendous potential to combat post-harvest food losses due to perishability and packaging issues. The type of coating material and the nature of the food product have an influential role in the edible food coatings and impact the food's appearance, taste, and nutritional profile. Despite the multiple benefits offered by edible food coating techniques, poor barrier properties, sensorial quality changes, and contaminating pathogenic growth are a few challenges in industrial adaptation. A multidisciplinary approach combining engineering, materials, and culinary science is a need of the hour to minimize these challenges in commercializing this novel eco-friendly technique. Meanwhile, studies on incorporating nanoparticles in edible films enhance the material strength and increase the shelf life. Food processors, environmental protection agencies, and consumers should promote using biodegradable packaging films to overcome the impacts of modern plastic packaging-based ecological issues.

References

1. Sharma, L., Saini, C.S., Sharma, H.K., Sandhu, K.S., Biocomposite edible coatings based on cross linked-sesame protein and mango puree for the shelf life stability of fresh-cut mango fruit. *J. Food Process. Eng.*, 42, 1, 1–9, 2019, doi: 10.1111/jfpe.12938.
2. Suderman, N., Isa, M.I.N., Sarbon, N.M., The effect of plasticizers on the functional properties of biodegradable gelatin-based film: A review. *Food Biosci.*, 24, June, 111–119, 2018, doi: 10.1016/j.fbio.2018.06.006.
3. Yuan, G., Chen, X., Li, D., *Chitosan films and coatings containing essential oils: The antioxidant and antimicrobial activity, application in food systems*, vol. 89, pp. 117–128, Elsevier B.V, 2016, doi: 10.1016/j.foodres.2016.10.004.
4. Erginkaya, Z., Kalkan, S., Ünal, E., Use of antimicrobial edible films and coatings as packaging materials for food safety, in: *Food Processing: Strategies for Quality Assessment*, pp. 261–95, 2014, doi: 10.1007/978-1-4939-1378-7_10.
5. Galus, S. and Kadzińska, J., Food applications of emulsion-based edible films and coatings. *Trends Food Sci. Technol.*, 45, 2, 273–283, 2015, doi: 10.1016/j.tifs.2015.07.011.
6. Zaritzky, N., Edible Coatings to Improve Food Quality and Safety, in: *Food Engineering Interfaces*, pp. 631–659, Springer, New York, 2011, doi: 10.1007/978-1-4419-7475-4_27.
7. Aloui, H. and Khwaldia, K., Natural Antimicrobial Edible Coatings for Microbial Safety and Food Quality Enhancement. *Compr. Rev. Food Sci. Food Saf.*, 15, 6, 1080–1103, 2016, doi: 10.1111/1541-4337.12226.
8. Saifullah, M., Shishir, M.R.I, Ferdowsi, R., Tanver Rahman, M.R., Van Vuong, Q., Micro and nano encapsulation, retention and controlled release of flavor and aroma compounds: A critical review. *Trends Food Sci. Technol.*, 86, 230–251, 2019, doi: 10.1016/j.tifs.2019.02.030.
9. Fakhouri, F.M. *et al.*, Effect of a gelatin-based edible coating containing cellulose nanocrystals (CNC) on the quality and nutrient retention of fresh strawberries during storage. *IOP Conf. Ser. Mater. Sci. Eng.*, 64, 1, 012024, 2014, doi: 10.1088/1757-899X/64/1/012024.
10. Mishra, B., Khatkar, B.S., Garg, M.K., Wilson, L.A., Permeability of edible coatings. *J. Food Sci. Technol.*, 47, 1, 109–113, 2010, doi: 10.1007/s13197-010-0003-7.
11. Matloob, A. *et al.*, A Review on Edible Coatings and Films: Advances, Composition, Production Methods, Safety Concerns. *ACS Omega*, 8, 32, 28932–28944, 2023, doi: 10.1021/acsomega.3c03459.

12. Dhall, R.K., Advances in Edible Coatings for Fresh Fruits and Vegetables: A Review. *Crit. Rev. Food Sci. Nutr.*, 53, 5, 435–450, 2013, doi: 10.1080/10408398.2010.541568.
13. Jeevahan, J.J. *et al.*, Scaling up difficulties and commercial aspects of edible films for food packaging: A review. *Trends Food Sci. Technol.*, 100, 210–222, 2020, doi: 10.1016/j.tifs.2020.04.014.
14. Vipan, B., Mahajan, C., Tandon, R., Kapoor, S., Sidhu, M.K., Natural Coatings for Shelf-Life Enhancement and Quality Maintenance of Fresh Fruits and Vegetables-A Review. *J. Postharvest Technol.*, 06, 1, 12–26, 2018.
15. Chen, H. *et al.*, Application of protein-based films and coatings for food packaging: A review. *Polymers (Basel)*, 11, 12, 1–32, 2019, doi: 10.3390/polym11122039.
16. Jongjareonrak, A., Benjakul, S., Visessanguan, W., Prodpran, T., Tanaka, M., Characterization of edible films from skin gelatin of brownstripe red snapper and bigeye snapper. *Food Hydrocoll.*, 20, 4, 492–501, 2006, doi: 10.1016/j.foodhyd.2005.04.007.
17. Qiaomei Zhu, L.Y., Lu, H., Zhu, J., Zhang, M., Development and characterization of pickering emulsion stabilized by zein/corn fiber gum (CFG) complex colloidal particles. *Food Hydrocoll.*, 91, 204–13, 2019, doi: https://doi.org/10.1016/j.foodhyd.2019.01.029.
18. Gao, P. *et al.*, Preparation and characterization of zein thermo-modified starch films. *Carbohydr. Polym.*, 157, 1254–1260, 2017, doi: 10.1016/j.carbpol.2016.11.004.
19. Calva-Estrada, S.J., Jiménez-Fernández, M., Lugo-Cervantes, E., Protein-Based Films: Advances in the Development of Biomaterials Applicable to Food Packaging. *Food Eng. Rev.*, 11, 78–92, 2019, doi: 10.1007/s12393-019-09189-w.
20. Tian, H., Guo, G., Fu, X., Yao, Y., Yuan, L., Xiang, A., Fabrication, properties and applications of soy-protein-based materials: A review. *Int. J. Biol. Macromol.*, 120, 475–490, 2018, doi: 10.1016/j.ijbiomac.2018.08.110.
21. Tulamandi, S., Rangarajan, V., Rizvi, S.S.H., Singhal, R.S., Chattopadhyay, S.K., Saha, N.C., A biodegradable and edible packaging film based on papaya puree, gelatin, defatted soy protein. *Food Packag. Shelf Life*, 10, 60–71, 2016, doi: 10.1016/j.fpsl.2016.10.007.
22. Ramos, M., Valdés, A., Beltrán, A., Garrigós, M., Gelatin-Based Films and Coatings for Food Packaging Applications. *Coatings*, 6, 4, 41, 2016, doi: 10.3390/coatings6040041.
23. Kandasamy, S. *et al.*, Application of whey protein-based edible films and coatings in food industries: An updated overview. *Coatings*, 11, 9, 1056, 2021, doi: 10.3390/coatings11091056.
24. Khan, M.R., Volpe, S., Valentino, M., Miele, N.A., Cavella, S., Torrieri, E., Active casein coatings and films for perishable foods: Structural properties and shelf-life extension. *Coatings*, 11, 8, 0899, 2021, doi: 10.3390/coatings11080899.

25. Kavas, N. and Kavas, G., Use of turmeric (Curcuma longa l.) essential oil added to an egg white protein powder-based film in the storage of Çökelek cheese. *J. Food Chem. Nanotechnol.*, 3, 3, 105–110, 2017, doi: 10.17756/jfcn.2017-045.
26. Hashim, P., Mohd Ridzwan, M.S., Bakar, J., Mat Hashim, D., Collagen in food and beverage industries. *Int. Food Res. J.*, 22, 1, 1–8, 2015.
27. Mihalca, V. *et al.*, Protein-based films and coatings for food industry applications. *Polymers (Basel)*, 13, 5, 1–24, 2021, doi: 10.3390/polym13050769.
28. Adeyemi, J.O. and Fawole, O.A., Metal-Based Nanoparticles in Food Packaging and Coating Technologies: A Review. *Biomolecules*, 13, 7, 1092, 2023, doi: 10.3390/biom13071092.
29. Xin, Y., Yang, C., Zhang, J., Xiong, L., Application of Whey Protein-Based Emulsion Coating Treatment in Fresh-Cut Apple Preservation. *Foods*, 12, 6, 1140, 2023, doi: 10.3390/foods12061140.
30. Barman, K., Asrey, R., Pal, R.K., Putrescine and carnauba wax pretreatments alleviate chilling injury, enhance shelf life and preserve pomegranate fruit quality during cold storage. *Sci. Hortic. (Amsterdam)*, 130, 4, 795–800, 2011, doi: 10.1016/j.scienta.2011.09.005.
31. Gunawan, S., Vali, S.R., Ju, Y.H., Purification and identification of rice bran oil fatty acid steryl and wax esters. *JAOCS, J. Am. Oil Chem. Soc.*, 83, 5, 449–456, 2006, doi: 10.1007/s11746-006-1225-8.
32. Chaireh, S., Ngasatool, P., Kaewtatip, K., Novel composite foam made from starch and water hyacinth with beeswax coating for food packaging applications. *Int. J. Biol. Macromol.*, 165, 1382–1391, 2020, doi: 10.1016/j.ijbiomac.2020.10.007.
33. de Oliveira Filho, J.G. *et al.*, Arrowroot starch-based films incorporated with a carnauba wax nanoemulsion, cellulose nanocrystals, essential oils: a new functional material for food packaging applications. *Cellulose*, 28, 10, 6499–6511, 2021, doi: 10.1007/s10570-021-03945-0.
34. Yuan, Y. *et al.*, Shellac: A promising natural polymer in the food industry. *Trends Food Sci. Technol.*, 109, August 2020, 139–153, 2021, doi: 10.1016/j.tifs.2021.01.031.
35. Talbot, G., *Compound coatings*, in: *Science and Technology of Enrobed and Filled Chocolate, Confectionery and Bakery Products*, pp. 80–100, Elsevier Science, Amsterdam, Netherlands, 2009, doi: 10.1533/9781845696436.1.81.
36. Mladenoska, I., The potential application of novel beeswaxedible coatings containing coconut oil in the minimal processing of fruits. *Savrem. Tehnol. / Adv. Technol.*, 1, 2, 26–34, 2012.
37. Min-Hui, P. *et al.*, Preparation of palm oil-based degradable coating materials and the property study. *J. Plant Nutr. Fertil.*, 29, 10, 1966–1976, 2023, doi: 10.11674/zwyf.2023095.
38. Vieira, T.M., Moldao-martins, M., Alves, V., Composite Coatings of Chitosan and Alginate Emulsions with Olive Oil to Enhance Postharvest Quality and Shelf Life of. *Foods*, 10, 4, 718, 2021.

39. Das, B., Konwar, U., Mandal, M., Karak, N., Sunflower oil based biodegradable hyperbranched polyurethane as a thin film material. *Ind. Crops Prod.*, 44, 396–404, 2013, doi: 10.1016/j.indcrop.2012.11.028.
40. Kang, H.J., Kim, S.J., You, Y.S., Lacroix, M., Han, J., Inhibitory effect of soy protein coating formulations on walnut (Juglans regia L.) kernels against lipid oxidation. *LWT*, 51, 1, 393–396, 2013, doi: 10.1016/j.lwt.2012.10.019.
41. Okumuş, E., Bakkalbaşı, E., Javidipour, I., Meral, R., Ceylan, Z., A novel coating material: Ellagitannins-loaded maltodextrin and lecithin-based nanomaterials. *Food Biosci.*, 42, 101158, May, 2021, doi: 10.1016/j.fbio.2021.101158.
42. Hui, C. *et al.*, Chitin degradation and the temporary response of bacterial chitinolytic communities to chitin amendment in soil under different fertilization regimes. *Sci. Total Environ.*, 705, 136003, 2020, doi: 10.1016/j.scitotenv.2019.136003.
43. Amin, U. *et al.*, Potentials of polysaccharides, lipids and proteins in biodegradable food packaging applications. *Int. J. Biol. Macromol.*, 183, February, 2184–2198, 2021, doi: 10.1016/j.ijbiomac.2021.05.182.
44. Basiak, E., Lenart, A., Debeaufort, F., Effect of starch type on the physicochemical properties of edible films. *Int. J. Biol. Macromol.*, 98, 348–356, 2017, doi: 10.1016/j.ijbiomac.2017.01.122.
45. Pluta-Kubica, A., Jamróz, E., Kawecka, A., Juszczak, L., & Krzyściak, P., Active edible furcellaran/whey protein films with yerba mate and white tea extracts: Preparation, characterization and its application to fresh soft rennet-curd cheese. *Int. J. Biol. Macromol.*, 155, 1307–1316, 2020. https://doi.org/10.1016/j.ijbiomac.2019.11.102.
46. Saberi, B., Golding, J.B., Chockchaisawasdee, S., Scarlett, C.J., Stathopoulos, C.E., Effect of Biocomposite Edible Coatings Based on Pea Starch and Guar Gum on Nutritional Quality of 'Valencia' Orange During Storage. *Starch/Staerke*, 70, 5–6, 0–37, 2018, doi: 10.1002/star.201700299.
47. Chen, J. *et al.*, "Characterization of Sodium Alginate-Based Films Incorporated with Thymol for Fresh-Cut Apple Packaging." *Food Control*, 126, 108063, 2021. https://doi.org/10.1016/j.foodcont.2021.108063.
48. Zhu, D., Guo, R., Li, W., Song, J., Cheng, F., Improved Postharvest Preservation Effects of Pholiota nameko Mushroom by Sodium Alginate–Based Edible Composite Coating. *Food Bioprocess Technol.*, 12, 4, 587–598, 2019, doi: 10.1007/s11947-019-2235-5.
49. yu Li, X., long Du, X., Liu, Y., jing Tong, L., Wang, Q., long Li, J., Rhubarb extract incorporated into an alginate-based edible coating for peach preservation. *Sci. Hortic. (Amsterdam)*, 257, July, 108685, 2019, doi: 10.1016/j.scienta.2019.108685.
50. Guerrero, A. *et al.*, Effects of active edible coating based on thyme and garlic essential oils on lamb meat shelf life after long-term frozen storage. *J. Sci. Food Agric.*, 100, 2, 656–664, 2020, doi: 10.1002/jsfa.10061.
51. Ferrara, L., Seaweeds: A food for our future. *J. Food Chem. Nanotechnol.*, 6, 2, 56–64, 2020, doi: 10.17756/jfcn.2020-084.

52. Kumar, N., Pratibha, N., Petkoska, A.T., Al-Hilifi, S.A., Fawole, O.A., Effect of chitosan–pullulan composite edible coating functionalized with pomegranate peel extract on the shelf life of mango (Mangifera indica). *Coatings*, 11, 7, 25–40, 2021, doi: 10.3390/coatings11070764.
53. Priyadarshi, R., Kim S.-M., and Rhim J.-W., Pectin/Pullulan Blend Films for Food Packaging: Effect of Blending Ratio. *Food Chem.*, 347, 129022, 2021.
54. Nisar, T., Wang, Z.C., Yang, X., Tian, Y., Iqbal, M., Guo, Y., Characterization of citrus pectin films integrated with clove bud essential oil: Physical, thermal, barrier, antioxidant and antibacterial properties. *Int. J. Biol. Macromol.*, 106, 670–680, 2018, doi: 10.1016/j.ijbiomac.2017.08.068.
55. Espitia, P.J.P., Du, W.X., de J. Avena-Bustillos, R., de F. F. Soares, N., McHugh, T.H., Edible films from pectin: Physical-mechanical and antimicrobial properties - A review. *Food Hydrocoll.*, 35, 287–296, 2014, doi: 10.1016/j.foodhyd.2013.06.005.
56. Sucheta, Rai, S.K., Chaturvedi, K., Yadav, S.K., Evaluation of structural integrity and functionality of commercial pectin based edible films incorporated with corn flour, beetroot, orange peel, muesli and rice flour. *Food Hydrocoll.*, 91, September 2018, 127–135, 2019, doi: 10.1016/j.foodhyd.2019.01.022.
57. Maftoonazad, N. and Ramaswamy, H.S., Application and evaluation of a pectin-based edible coating process for quality change kinetics and shelf-life extension of lime fruit (Citrus aurantifolium). *Coatings*, 9, 5, 0285, 2019, doi: 10.3390/coatings9050285.
58. Sharma, H.P., Chaudhary, V., Kumar, M., Importance of edible coating on fruits and vegetables: A review. ~ *4104* ~ *J. Pharmacogn. Phytochem.*, 8, 3, 4104–4110, 2019.
59. Achachlouei, B.F. and Zahedi, Y., Fabrication and characterization of CMC-based nanocomposites reinforced with sodium montmorillonite and TiO2 nanomaterials. *Carbohydr. Polym.*, 199, 415–425, 2018, doi: 10.1016/j.carbpol.2018.07.031.
60. Ribeiro, A.J. *et al.*, Gums' based delivery systems: Review on cashew gum and its derivatives. *Carbohydr. Polym.*, 147, 188–200, 2016, doi: 10.1016/j.carbpol.2016.02.042.
61. Viana, R.M., Sá, N.M.S.M., Barros, M.O., de F. Borges, M., Azeredo, H.M.C., Nanofibrillated bacterial cellulose and pectin edible films added with fruit purees. *Carbohydr. Polym.*, 196, May, 27–32, 2018, doi: 10.1016/j.carbpol.2018.05.017.
62. Tahir, H.E., Xiaobo, Z., Mahunu, G.K., Arslan, M., Abdalhai, M., Zhihua, L., Recent developments in gum edible coating applications for fruits and vegetables preservation: A review. *Carbohydr. Polym.*, 224, 115141, August, 2019, doi: 10.1016/j.carbpol.2019.115141.
63. Tripathi, J., Ambolikar, R., Gupta, S., Jain, D., Bahadur, J., Variyar, P.S., Methylation of guar gum for improving mechanical and barrier properties of biodegradable packaging films. *Sci. Rep.*, 9, 1, 1–9, 2019, doi: 10.1038/s41598-019-50991-7.

64. Rodríguez-Gasol, N., Avilla, J., Alegre, S., Alins, G., Food industry by-products valorization and new ingredients: Cases of study. Cases of study, in: *Sustainability of the Food System: Sovereignty, Waste, and Nutrients Bioavailability,* pp. 0–33, 2019, [Online], Available: http://hdl.handle.net/10459.1/65845%0AThe.
65. Almeida, F.S. and Kawazoe Sato, A.C., Structure of gellan gum–hydrolyzed collagen particles: Effect of starch addition and coating layer. *Food Res. Int.*, 121, February, 394–403, 2019, doi: 10.1016/j.foodres.2019.03.057.
66. Wang, H., Qian, J., Ding, F., Emerging Chitosan-Based Films for Food Packaging Applications. *J. Agric. Food. Chem.*, 66, 2, 395–413, 2018, doi: 10.1021/acs.jafc.7b04528.
67. Parreidt, T.S., Müller, K., Schmid, M., Alginate-based edible films and coatings for food packaging applications. *Foods*, 7, 10, 1–38, 2018, doi: 10.3390/foods7100170.
68. Sapper, M. and Chiralt, A., Starch-based coatings for preservation of fruits and vegetables. *Coatings*, 8, 5, 0152, 2018, doi: 10.3390/coatings8050152.
69. Lara-Espinoza, C., Carvajal-Millán, E., Balandrán-Quintana, R., López-Franco, Y., Rascón-Chu, A., Pectin and pectin-based composite materials: Beyond food texture. *Molecules*, 23, 4, 0942, 2018, doi: 10.3390/molecules23040942.
70. Lu, Y., Zhao, X., Fang, S., Characterization, antimicrobial properties and coatings application of gellan gum oxidized with hydrogen peroxide. *Foods*, 8, 1, 1–12, 2019, doi: 10.3390/foods8010031.
71. Sedayu, B.B., Cran, M.J., Bigger, S.W., A Review of Property Enhancement Techniques for Carrageenan-based Films and Coatings. *Carbohydr. Polym.*, 216, 287–302, 2019, doi: 10.1016/j.carbpol.2019.04.021.
72. Kirtil, E., Aydogdu, A., Svitova, T., Radke, C.J., Assessment of the performance of several novel approaches to improve physical properties of guar gum based biopolymer films. *Food Packag. Shelf Life*, 29, May, 100687, 2021, doi: 10.1016/j.fpsl.2021.100687.
73. Martínez-Sanz, M., Martínez-Abad, A., López-Rubio, A., Cost-efficient bio-based food packaging films from unpurified agar-based extracts. *Food Packag. Shelf Life*, 21, June, 100367, 2019, doi: 10.1016/j.fpsl.2019.100367.
74. Salvada, J., Alke, B., Brazinha, C., Alves, V.D., Coelhoso, I.M., Development and Characterisation of Arabinoxylan-Based Composite Films. *Coatings*, 12, 6, 0813, 2022, doi: 10.3390/coatings12060813.
75. Zhou, X. *et al.*, Effect of konjac glucomannan/carrageenan-based edible emulsion coatings with camellia oil on quality and shelf-life of chicken meat. *Int. J. Biol. Macromol.*, 183, 331–339, 2021, doi: 10.1016/j.ijbiomac.2021.04.165.
76. Mohamadian, N., Ghorbani, H., Wood, D.A., Khoshmardan, M.A., A hybrid nanocomposite of poly(styrene-methyl methacrylate- acrylic acid) /clay as a novel rheology-improvement additive for drilling fluids. *J. Polym. Res.*, 26, 2, 33, 2019, doi: 10.1007/s10965-019-1696-6.

77. Kuswandi, B., Environmental friendly food nano-packaging. *Environ. Chem. Lett.*, 15, 2, 205–221, 2017, doi: 10.1007/s10311-017-0613-7.
78. Primožič, M., Knez, Z., Leitgeb, M., (Bio)nanotechnology in food science—food packaging. *Nanomaterials*, 11, 2, 1–31, 2021, doi: 10.3390/nano11020292.
79. Xing, Y. *et al.*, Antimicrobial Nanoparticles Incorporated in Edible. *Molecules*, 24, 9, 1–30, 2019.
80. Suhag, R., Kumar, N., Petkoska, A.T., Upadhyay, A., Film formation and deposition methods of edible coating on food products: A review. *Food Res. Int.*, 136, July, 109582, 2020, doi: 10.1016/j.foodres.2020.109582.
81. Fotovvati, B., Namdari, N., Dehghanghadikolaei, A., On coating techniques for surface protection: A review. *J. Manuf. Mater. Process.*, 3, 1, 0028, 2019, doi: 10.3390/jmmp3010028.
82. Wang, Q., Chen, W., Zhu, W., McClements, D.J., Liu, X., Liu, F., A review of multilayer and composite films and coatings for active biodegradable packaging. *npj Sci. Food*, 6, 1, 18, 2022, doi: 10.1038/s41538-022-00132-8.
83. Tavassoli-Kafrani, E., Shekarchizadeh, H., Masoudpour-Behabadi, M., Development of edible films and coatings from alginates and carrageenans. *Carbohydr. Polym.*, 137, 360–374, 2016, doi: 10.1016/j.carbpol.2015.10.074.
84. Costa, C., Conte, A., Del Nobile, M.A., Effective preservation techniques to prolong the shelf life of ready-to-eat oysters. *J. Sci. Food Agric.*, 94, 13, 2661–2667, 2014, doi: 10.1002/jsfa.6605.
85. Andrade, R.D., Skurtys, O., Osorio, F.A., Atomizing Spray Systems for Application of Edible Coatings. *Compr. Rev. Food Sci. Food Saf.*, 11, 3, 323–337, 2012, doi: 10.1111/j.1541-4337.2012.00186.x.
86. Jensen, A., Lim, L.T., Barbut, S., Marcone, M., Development and characterization of soy protein films incorporated with cellulose fibers using a hot surface casting technique. *LWT*, 60, 1, 162–170, 2015, doi: 10.1016/j.lwt.2014.09.027.
87. Lu, M. *et al.*, Structures of a Na+-coupled, substrate-bound MATE multidrug transporter. *Proc. Natl. Acad. Sci. U.S.A.*, 110, 6, 2099–2104, 2013, doi: 10.1073/pnas.1219901110.
88. Hernandez-Izquierdo, V.M. and Krochta, J.M., Thermoplastic processing of proteins for film formation - A review. *J. Food Sci.*, 73, 2, 30–39, 2008, doi: 10.1111/j.1750-3841.2007.00636.x.
89. Fitch-Vargas, P.R. *et al.*, Physicochemical and Microstructural Characterization of Corn Starch Edible Films Obtained by a Combination of Extrusion Technology and Casting Technique. *J. Food Sci.*, 81, 9, E2224–E2232, 2016, doi: 10.1111/1750-3841.13416.
90. Calderón-Castro, A. *et al.*, Effect of extrusion process on the functional properties of high amylose corn starch edible films and its application in mango (Mangifera indica L.) cv. Tommy Atkins. *J. Food Sci. Technol.*, 55, 3, 905–914, 2018, doi: 10.1007/s13197-017-2997-6.
91. Díaz-Montes, E. and Castro-Muñoz, R., Edible films and coatings as food-quality preservers: An overview. *Foods*, 10, 2, 1–26, 2021, doi: 10.3390/foods10020249.

92. Miele, N.A., Volpe, S., Torrieri, E., Cavella, S., Improving physical properties of sodium caseinate based coating with the optimal formulation: Effect on strawberries' respiration and transpiration rates. *J. Food Eng.*, 331, April, 111123, 2022, doi: 10.1016/j.jfoodeng.2022.111123.
93. Flores-López, M.L., Cerqueira, M.A., de Rodríguez, D.J., Vicente, A.A., Perspectives on Utilization of Edible Coatings and Nano-laminate Coatings for Extension of Postharvest Storage of Fruits and Vegetables. *Food Eng. Rev.*, 8, 3, 292–305, 2016, doi: 10.1007/s12393-015-9135-x.
94. Olivas, G.I. and Barbosa-cánovas, G., *Edible Films and Coatings for Food Applications*, Springer Science + Business Media, LLC, Germany, 2009, doi: 10.1007/978-0-387-92824-1.
95. Valdés, A., Ramos, M., Beltrán, A., Jiménez, A., Garrigós, M.C., State of the art of antimicrobial edible coatings for food packaging applications. *Coatings*, 7, 4, 1–23, 2017, doi: 10.3390/coatings7040056.
96. Eça, K.S., Sartori, T., Menegalli, F.C., Aplicações Green Packaging: Concepts, Materials and. *Brazilian J. Food Technol.*, 17, 2, 98–112, 2014.
97. Tkaczewska, J., Peptides and protein hydrolysates as food preservatives and bioactive components of edible films and coatings - A review. *Trends Food Sci. Technol.*, 106, 298–311, 2020, doi: 10.1016/j.tifs.2020.10.022.
98. Ojeda, G.A., Sgroppo, S.C., Zaritzky, N.E., Application of edible coatings in minimally processed sweet potatoes (Ipomoea batatas L.) to prevent enzymatic browning. *Int. J. Food Sci. Technol.*, 49, 3, 876–883, 2014, doi: 10.1111/ijfs.12381.
99. Sharma, P., Shehin, V.P., Kaur, N., Vyas, P., Application of edible coatings on fresh and minimally processed vegetables: a review. *Int. J. Veg. Sci.*, 25, 3, 295–314, 2019, doi: 10.1080/19315260.2018.1510863.
100. Feng, Z., Wu, G., Liu, C., Li, D., Jiang, B., Zhang, X., Edible coating based on whey protein isolate nanofibrils for antioxidation and inhibition of product browning. *Food Hydrocoll.*, 79, 179–188, 2018, doi: 10.1016/j.foodhyd.2017.12.028.
101. Hassan, B., Chatha, S.A.S., Hussain, A.I., Zia, K.M., Akhtar, N., Recent advances on polysaccharides, lipids and protein based edible films and coatings: A review. *Int. J. Biol. Macromol.*, 109, 1095–1107, 2018, doi: 10.1016/j.ijbiomac.2017.11.097.
102. Rattan, S., Parande, A.K., Ramalakshmi, K., Nagaraju, V.D., Effect of edible coating on the aromatic attributes of roasted coffee beans. *J. Food Sci. Technol.*, 52, 9, 5470–5483, 2015, doi: 10.1007/s13197-014-1707-x.
103. Genevois, C.E., De Escalada Pla, M.F., Flores, S.K., Application of edible coatings to improve global quality of fortified pumpkin. *Innov. Food Sci. Emerg. Technol.*, 33, 506–514, 2016, doi: 10.1016/j.ifset.2015.11.001.
104. Ananey-Obiri, D., Matthews, L., Azahrani, M.H., Ibrahim, S.A., Galanakis, C.M., Tahergorabi, R., Application of protein-based edible coatings for fat uptake reduction in deep-fat fried foods with an emphasis on muscle

food proteins. *Trends Food Sci. Technol.*, 80, 167–174, 2018, doi: 10.1016/j.tifs.2018.08.012.

105. Pellicer, E. *et al.*, Advances in applications of industrial biomaterials. *Adv. Appl. Ind. Biomater.*, 1–214, 2017, doi: 10.1007/978-3-319-62767-0.
106. Di Pierro, P., Mariniello, L., Giosafatto, V.L., Esposito, M., Sabbah, M., Porta, R., Dairy Whey Protein-Based Edible Films and Coatings for Food Preservation [Internet], in: *Food Packaging and Preservation*, pp. 439–456, Elsevier Inc., Amsterdam, Netherlands, 2018, doi: 10.1016/B978-0-12-811516-9/00013-0.
107. Yousefi, M., Azizi, M., Mohammadifar, M.A., Ehsani, A., Antimicrobial coatings and films on meats: A perspective on the application of antimicrobial edible films or coatings on meats from the past to future. *Bali Med. J.*, 7, 1, 87, 2018, doi: 10.15562/bmj.v7i1.759.
108. Peltzer, M.A., Salvay, A.G., Delgado, J.F., Wagner, J.R., Use of edible films and coatings for functional foods developments: A review. *Funct. Foods Sources, Heal. Eff. Futur. Perspect.*, January 1–26, 2016.
109. Dehghani, S., Hosseini, S.V., Regenstein, J.M., Edible films and coatings in seafood preservation: A review. *Food Chem.*, 240, July, 505–513, 2018, doi: 10.1016/j.foodchem.2017.07.034.
110. Maria, C., Freitas, P., Jane, S., Gomes, V., Souza, L., Rita, C., Pharmaceutical Industry : A Review. *Coatings*, 11, 922, 1–22, 2021.
111. Davies, C.R. *et al.*, Evolving challenges and strategies for fungal control in the food supply chain. *Fungal Biol. Rev.*, 36, 15–26, 2021, doi: 10.1016/j.fbr.2021.01.003.
112. Pavlath, A.E. and Orts, W., *Edible Films and Coatings for Food Applications*, Springer Science+Business Media, LLC, Germany, pp. 1–23, 2009, doi: 10.1007/978-0-387-92824-1.
113. Badawy, M.E.I., Rabea, E.I., El-Nouby, M.A.M., Ismail, R.I.A., Taktak, N.E.M., Strawberry Shelf Life, Composition, Enzymes Activity in Response to Edible Chitosan Coatings. *Int. J. Fruit Sci.*, 17, 2, 117–136, 2017, doi: 10.1080/15538362.2016.1219290.
114. Chiralt, A., Menzel, C., Hernandez-García, E., Collazo, S., Gonzalez-Martinez, C., Use of by-products in edible coatings and biodegradable packaging materials for food preservation. *Sustain. Food Syst. Sovereignty Waste, Nutr. Bioavailab.*, 101–127, 2020, doi: 10.1016/B978-0-12-818293-2.00006-9.
115. Valencia, G.A., Luciano, C.G., Fritz, A.R.M., *Polymers for Agri-Food Applications*, pp. 391–416, Springer Nature, Switzerland, 2019, doi: 10.1007/978-3-030-19416-1.
116. Martín-belloso, O., Rojas-graü, M.A., Soliva-fortuny, R., *Edible Films and Coatings for Food Applications*, pp. 295–313, Springer Science+Business Media, LLC, Germany, 2009, doi: 10.1007/978-0-387-92824-1.
117. Mahalik, N.P., Processing and packaging automation systems: A review. *Sens. Instrum. Food Qual. Saf.*, 3, 1, 12–25, 2009, doi: 10.1007/s11694-009-9076-2.

118. Baldwin, E.A., Nisperos-Carriedo, M.O., Baker, R.A., Edible Coatings for Lightly Processed Fruits and Vegetables. *HortScience*, 30, 1, 35–38, 2019, doi: 10.21273/hortsci.30.1.35.
119. Guo, M., Yadav, M.P., Jin, T.Z., Antimicrobial edible coatings and films from micro-emulsions and their food applications. *Int. J. Food Microbiol.*, 263, October, 9–16, 2017, doi: 10.1016/j.ijfoodmicro.2017.10.002.

8

Food Stability Simulation: Accelerated Shelf-Life Mechanism

Somashree Bandyopadhaya, Urmi Sarkar and Debabrata Bera*

Department of Food Technology and Biochemical Engineering, Jadavpur University, Kolkata, India

Abstract

The 2030 agenda stated food security with Sustainable Development Goal (SDG) 3 "good health and well-being" and SDG 2 "zero hunger." Access to safe, nutritious food is essential for food security and imperative to protect public health. In recent years, chemical hazards in food have raised consumer concerns. Additionally, recent food scandals have increased consumer demand for food safety. There is a need to integrate food conservation processes with food processing that aims to extend shelf life, even with scarce technical resources, facilities, and equipment. It is nearly impossible to control every component of an entire process, so the real challenge is controlling different influential dynamics like rancidity, texture changes, moisture fluctuations, staling, browning, and microbial spoilage. It is very important to ensure that food is safe and acceptable for consumption, and the shelf life for which this is maintained is a key component of product design and hygiene. Consumer acceptance will be affected by changes in the sensory and nutritional properties of food products. For health and safety reasons, food must be consumed within a certain period; here a "use by" or "best before" date is required. Generally, a best-before date is applicable to foods whose deterioration affects consumer acceptance but does not affect health and safety. Depending on the type of deterioration the product undergoes, the "use-by" or "best-before" dates are overstated. During real-time testing, the best-before dates mentioned in the products are verified. However, it is not suitable to use real-time testing for improving packed food products with a longer shelf life. The shelf life of those items can be estimated by accelerated stability testing. Analysis of accelerated shelf life test (ASLT) applies to any process of deterioration that can be quantitatively expressed by an Arrhenius relationship. Keeping products at high temperatures,

**Corresponding author*: debabrata.bera@jadavpuruniversity.in

Mousumi Sen (ed.) Food Coatings and Preservation Technologies, (259–300) © 2025 Scrivener Publishing LLC

humidity, or light intensity is typically required to ensure the quality of the product. Various conditions and times are used to inspect products following storage. In some cases, testing beyond routine quality control may be necessary. Statistical and mathematical tools are used in predictive modeling to find relationships between intrinsic, extrinsic, and implicit factors. Also, it examines how it behaves under different combinations of intrinsic and extrinsic factors. Based on these calculations, a predictive shelf life is estimated. In this chapter, methods based on linear kinetic models and recent non-linear ones are reviewed. In conclusion, this significant area of accelerated shelf-life testing is explored in an attempt to suggest future directions. Moreover, it discusses what expectations one should have when developing novel practical and reliable tools that are applicable in the industrial process.

Keywords: Food stability simulation, accelerated shelf life study, food shelf life, accelerated storage conditions

8.1 Introduction

Throughout history, mankind has faced the challenge of preserving food. Harvesting afterward, food begins to break down, causing the major problems. The proper storage of food is therefore crucial for its availability and preventing it from going to waste. The primitive art of food preservation allowed our prehistoric ancestors to settle down in a community and live in one place. Through preserving survivals and foods, every single civilization developed. Also, since people began exploring the earth, food and water were their primary concerns. People understood that fresh harvests or kills no longer had to be consumed right away, but the idea was to keep it preserved for later consumption. Throughout history in different cultures, food has been preserved using some identical basic methods. Food preservation has evolved through the ages since we developed today's methods through advancements in technology and techniques developed by the Egyptians, Mayans, and Greeks. Various methods were found to be used early in the preservation process to remove undesirable or toxic substances, such as cutting, skinning, peeling, and immersion in liquors extracted from plants or salty water. Ancient civilizations brined meat, and fermented fish, barley, milk, and fruit to produce a variety of edible recipes [1]. There is no doubt that the preservation of edibles is all about maintaining a required level of properties while reaping maximum benefits from them. Our food must remain safe for consumption along with maintaining its own natural taste, texture, nutritive value, and appearance so that it can remain intact throughout its entire shelf life, which ultimately represents that the food is still fresh and stable. A lot of food industries are constantly concerned with the stability of their packed products to make sure that they maintain their physical and chemical stability during their production, storage, transport, and retail stores to ensure that the consumer receives all the native tastes and sensations of food. To achieve this, food industries frequently use preservatives and stabilizers to prevent food from deteriorating. Constant research and experimentation gave birth to new efficient and easy-to-implement techniques. Technological superiority has led to food preservation methods improvement over the centuries. By using electronic smoking, the traditional smoking method was replaced, while drying methods were modernized with sophisticated dehydration methods. A new generation of preservation methods has now been developed that improve upon traditional methods by using pulse-electrical power, electro-thermal power, ultrasonic technology, and radiation therapy. The distinct advantage of this process is that it provides both microbial safety and preservation of essential nutrients and bioactives in sensitive foods at the same time. A recent advance in science, nanotechnology, has led to a breakthrough in food preservation, as well as a new dimension of high security in food storage. It is becoming increasingly commonplace for people as technology advances to realize that a combination of preserving methods can provide better chemical and physical stability, and maintain sensory, texture and nutritional properties than using any single preserving method alone [2, 3]. Concerning the new generation of consumers, the food industry has become more aware of the importance of food safety. Store-bought packaged food products in the market now come with instructions on how to handle, store, and consume the products. Consequently, more innovative packaging techniques are being introduced, such as vacuum packaging, which maintains food safety while reducing spoilage. Furthermore, packaging material that contains antioxidants and anti-microbial agents is on the rise [4].

8.2 Fundamentals of Food Shelf Life and Storage Stability

Whenever we mention about the shelf life of foods, we talk about to the period of time in which the food can be consumed without compromising its quality, and safety, and have not deteriorated or been exposed to conditions that would make a consumer uncomfortable while in following the storage instruction given. There are two types of shelf-life items: item with static shelf life that cannot be extended (Type I) and those that can be extended on condition (Type II). It is generally accepted that items of type I would no longer be safe to use for its original purpose and need to be thrown away to ensure health security. Type II items have remarkable extension of shelf-life coding after several scrutinies at regular intervals determining if edible stuffs still be suitable as long as certain criteria have been met. One of the major concerns in the food industry is shelf-life control, and it is a crucial component of any food safety plan developed by a manufacturer. During the course of the shelf life of the product, every food manufacturer expects that the product will deliver the desired experience to the consumer, which may be expressed in terms of taste and texture, odor, color, nutritional content, or characteristics of the product. It was proclaimed in 2017 by the Food Marketing Institute (FMI) and Grocery Manufacturer's Association (GMA) that voluntary guidance would be issued with the direction of the Food and Drug Administration (FDA) in order to simplify product code coding and reduce food waste and consumer confusion by simplifying code naming to two phrases, (a) The best-before date refers to the period of time within which a food can reasonably be expected to maintain its best quality. If the date has been exceeded, the taste, flavor, or texture of the food may not be as expected, but the food should still be considered fit for consumption. (b) A use-by date or freeze-by date specifies how long a food can be safely stored understated storage conditions before it can be eaten. In 2020, Food Safety and Standards (Labelling and Display) Regulations also approved these two phrases symbolizing safe consumption period for labeling packed food items. In conclusion, consuming these foods after the mentioned date may lead to food poisoning. Perishable products must be consumed or discarded before the date indicated on the label. Dairy products, processed meats, and prepared meals and salads have this code on their packaging [4–9].

8.2.1 Factors Affecting Food Shelf Life and Storage Stability

Food spoilage can be brought about by microbial, chemical, or mechanical action. Spoilage caused by one source can stimulate spoilage caused by another. Food spoilage may start with physical damage that leads to the attack of microbes on food, producing enzymes that break down food components and start deterioration. In addition, cross-contamination during the harvest, and post-harvest handling in combination with temperature fluctuations and the relationship between microbial growth and biochemical deterioration parameters under diverse packaging and storage settings. Food spoilage signs such as staling, discoloration, off-flavors, and rancidity can be caused, for example, by enzymatic or microbial activity, oxidation, or chemical spoiling.

The composition of food may be affected differently according to how it is stored, depending on the method used to store it. By performing tests on how the shelf life of the product can be affected by various conditions, it is possible to determine the most suitable condition that can maximize the shelf life of the product. Storage stability and shelf life of food are affected by a number of factors mentioned in Table 8.1. Various storage conditions have been tested as a means of increasing the shelf life of foods in order to determine which storage condition is most suitable to increase their shelf life to maintain the freshness of food for an extended period of time without degrading the food's sensory, nutritional properties, and microbial stability. If a food item is not stored properly, the condition of its physical and chemical stability can deteriorate and cause several undesirable effects on the food item, such as detrimental effects caused by biogenic amines, toxins, and metabolites produced by microorganisms. Organoleptic features such as color, texture change, and flavor changes in the presence of aroma compounds are induced by spoilage organisms or the alteration of the food's biochemical composition over time. These food spoilage problems

Table 8.1 Factors affecting food shelf life and storage stability.

Factors influencing Food shelf life and food stability	Intrinsic Factors	Water Activity, Redox Potential, pH, Chemical Composition, Natural Protectant, Enzymatic Activity	[10, 11]
	Extrinsic Factors	Storage Temperature, Oxygen Concentration, Relative Humidity, Illumination condition	
	Implicit Factors	Level of Contamination, Microbial Growth Character, Microbial-biochemical-environmental Interaction, Type of food processing, Physical damage during handling, Hygiene.	

Table 8.2 Microbial growth influencing factors (all data are provided in the general optimal range).

Microorganism	Water activity	pH	Temperature	Oxygen concentration	Relative humidity	Reference
Bacteria	0.91–1.0	6.5–7.5	5°C–60°C	5–20%	> 60%	[19, 20]
Yeast	0.75-0.85	4.5–6.5	27°C–32°C	0.05–0.5%	80–85%	
Molds	0.8–0.85	2.0–9.0	60°C–80°C	< 0.5%	> 70%	
Fungi	0.6–0.7	4.5–5.0	20°C–62°C	1.5–21%	60–75%	

are caused by a variety of intrinsic factors associated with food and extrinsic factors associated with its storage. Additionally, many implicit factors related to food properties and environmental interactions influence food spoilage. Intrinsic factors are substances that contribute to the decay of food as a result of the decay process itself. Alternatively, any extrinsic factor that causes the food to deteriorate is one that comes from outside the food itself. There are several factors that affect the shelf life of food and alter its rate of deterioration depending on a combination of these factors. There are a lot of methods that can be used to prolong the shelf life of food, including proper storage techniques and preservation processing methods. It is also possible to add certain preservatives to foods to increase the shelf life of them as well. Microorganisms can exhibit specific physiological characteristics such as variable growth rates and interactions between different components. Microbial growth influencing factors such as water activity, temperature, relative humidity, pH varies depending on the type of microbe, such as fungi, molds or bacteria (mentioned in Table 8.2). Level of contamination, growth character, and microbial-biochemical-environmental interaction, type of food processing, physical damage during handling and sustaining proper hygiene are critical in the explanation of how microorganisms' interplay with their environment and each other [10–12].

8.2.1.1 Intrinsic Factors

Water Activity: The existence of life depends on the presence of water. In order to survive, every living organism requires water in order to survive. Water, in the form of entrapped water and intracellular water, can be found inside foods. Foods are categorized into two groups based on their water content, namely perishable foods and non-perishable foods. It is obvious that perishable food are prone to deteriorate within few hours to a few days due to the presence of a large amount of water within them, which gives a chance for different microorganisms to grow in them and make use of the water they contain. In contrast, non-perishable food is technically prone to deteriorating slowly over the period of many months to years. The amount of water that is required by microorganisms to survive and grow varies from one microorganism to another, and that is why foods with a specific level of water activity tend to be more susceptible to infection by bacteria, yeast, mold, and fungi as a result of their water activity. Honey has a low moisture content that prevents water from being available for the growth of bacteria, and a high sugar content which inhibits bacteria from surviving by creating high osmotic pressure, therefore it has long shelf life. Moreover, higher water activity promotes increase rate of chemical spoilage [13, 14].

Redox Potential: Among the most complex indicators of the physiological state of the microorganisms is the redox potential, which can be useful for determining the qualitative and quantitative levels of contamination by microbes in a variety of environments. As the bacterial growth takes place, the redox potential of the medium will decrease, making it ideal for increased rate of other microbial growth. The use of this change in redox potential can enable us to both detect the presence of bacteria and their activity, as well as monitor their growth and metabolism, by monitoring this relative decrease in redox potential. Furthermore, it can also be used to determine the best conditions for the growth of bacteria in a particular food matrix and possibility of lipid oxidation and initiating the signs of rancidity in food. The term "oxidation" refers to the process of losing electrons in a substrate as a result of losing electrons, and the term "reduction" refers to the process of gaining electrons in a substrate. A substrate is generally defined by its redox or oxidation-reduction potential stated in terms of mV, in relation of the efficiency with which electrons can be lost or gained from it. In environments with high oxidation potential, aerobic organisms will grow better than anaerobic organisms, however, in environments with reduced oxidation potential, anaerobic organisms will be the predominant types of organisms. It is suggested that aerobes will grow best at a voltage of +550 mV to +300 mV, facultative anaerobes will grow best at a voltage of +300 to +90 mV, and anaerobes will grow best at a voltage of +100 to −200 mV or lower [11, 15, 16].

pH of the Food: It is also important to consider a food's acidity as another important aspect. There are certain pH ranges necessary for microorganisms to survive and thrive in order for them to thrive and survive. As long as the pH is outside of this range, the microorganism will either not be able to reproduce properly or will die out altogether. If the pH of food is extremely high or extremely low, certain microorganisms can be arrested from growing, while others can grow in such conditions if the pH is in the right range. Several bacterial species multiply under conditions of slight acidity on the pH scale, and most foods fall into this category. Bacteria might require a pH of 5.5 or lower in order to thrive, whereas other bacteria might be able to survive in a pH of 8.0 or higher. There is a high probability that molds that thrive in acidic environments will be the main cause of spoilage in foods with a pH of 5.0 or lower. In general, citrus fruits, berries, tomato, grapes, will not be spoiled by bacteria, but they may be contaminated by mold [10, 11].

Chemical Composition: The food we consume contains carbohydrates, protein, fat, and minerals which are all vital to our survival and provide an energy source for us. Food is the source of nutrients that help all living things survive, including microorganisms that depend on food for their survival. Without food, these microorganisms could not survive. Food also provides other essential nutrients for growth and reproduction. For instance, certain fungi need carbohydrates to grow, bacteria need protein to reproduce. On the other hand, deficiency of specific mineral can hinder growth of bacteria. Stages of food spoilage initiated from carbohydrate degradation then proteins spoilage and lipid hydrolysis. However, amount and availability of carbohydrate, lipid, protein and characteristics of degrading microorganism is significant influence [10, 15].

Presence of Natural Protectant: An outer layer covering of food as well as the presence of antimicrobial components and antioxidants delay the deterioration of the food. For instance, the outer layer of the onion contains sulfur compounds that act as preservatives, preventing bacteria and fungi growth. An egg's bloom is a natural protective layer that reduces the amount of moisture lost through the pores, preventing bacteria from developing and preventing the growth of mold. A number of antioxidants like xanthophylls, betalains, tocopherols, and ascorbic acid are known to prevent oxidative deterioration of food and rancidity. Natural microbial antioxidants nisin, present in cheese is one of the main reasons why cheese does not oxidize over the course of time as the result of exposure to oxygen. A number of natural antimicrobial substances can be found in different spices, herbs, and plants that act as natural antibacterial, anti-fungal agents, such as phenolic compounds, sulfur compounds, terpinenols etc. [16, 17].

Enzymatic Activity: A variety of enzymes are present in food as oxidases, proteases, lipases, and amylolytic enzymes. In the process of ripening, these enzymes help to break down the complex components into simpler compounds that can then be used. Following harvesting, these same enzymes are responsible for the rapid breakdown of the nutrients in the crop to prevent the deterioration of the fresh crop. For example, the polyphenol oxidase enzyme, which is found in some fruits and vegetables, plays a significant role in the undesirable browning of the food as a result of this enzyme. Pectin a complex carbohydrate is responsible for structural rigidity degraded by pectinases, an endogenous enzyme present in fruits and vegetables may get activated upon ripening or mechanical damage may fallouts softening or sogginess [18].

8.2.1.2 Extrinsic Factors

Storage Temperature: To determine the rate at which living organisms grow and multiply, the environmental temperature is, by far, the one of the most significant determinants of how fast their growth occurs. The reason for this is that temperature has an effect on the rate of chemical reactions, which are essential for the growth and survival of any living cell to maintain the uninterrupted metabolism. The rate of chemical reactions increases with increasing temperatures, whereas the rate of chemical reactions decreases with decreasing temperatures. Because of this, it is easy to see why changes in temperature are able to have such a profound influence on microorganisms, including their ability to grow and survive. Extreme temperature disrupts microbial membranes. Microbes also undergo destruction by extreme temperatures as their enzymes, cellular components, and other proteins are got denatured. Very low temperature also inhibits microbial growth as cellular membranes get frozen and enzymes become inactive at that low temperature. There is a significant slowdown in the metabolism of the microbes as a result of this, which prevents them from reproducing and growing. This results in a decrease in the microbial population within the environment, and this in turn reduces the level of contamination within it. The presence of very low temperatures is particularly important in areas where food is manufactured and stored in order to ensure that food is safe to consume. In addition, there is also a possibility that some food borne diseases can also be controlled by using very low temperatures along with controlling the spread of the pathogen and restrict the toxin production [10, 11].

Oxygen Concentration: The amount of oxygen available in the environment can either be a critical growth factor or an inhibitory lethal factor to the microorganisms. Microbes utilize oxygen for their cellular respiration and generate huge amount of highly reactive byproducts such as superoxide and peroxides molecules. These destructive compounds damage the cell by impairing cell membranes via cascade reaction unless they are neutralized. Aerobic bacteria are those grow in O2 rich atmospheres as they produce different superoxide and peroxides neutralizing enzymes such as superoxide dismutase, catalase, peroxidase, that strike unpaired forms of oxygen and break down in to their intermediate derivatives. Anaerobes bacteria are those who cannot grow in O2 rich atmospheres generate ATP via fermentation and anaerobic respiration. Anaerobes cannot synthesize superoxide and peroxides neutralizing enzymes to remove lethal derivatives of O2 hence have no withstands for oxygen rich atmosphere. As a matter of fact reduction in oxygen concentration leads to metabolic modification and halt the typical growth rate. Here comes the technology called modified atmosphere packaging is employed in storage of foods. In this sealed packed system, atmospheric CO2 concentration intentionally increased about 10 - 20% to retard rotting of fruits, vegetables, sweets, fish, meat like food stuffs. CO2 atmospheres packaging lengthen the shelf life of processed foodstuff along with Ozone gas when added as supplementary to the sealed storage environment promote a preservative influence on foods [10, 11, 21, 22].

Relative Humidity: Similar to water activity, relative humidity is a measure of relative moisture content in the storage setting. The percentage of moisture present in an atmosphere is referred to as the percentage of moisture present in the air. In environments with a very high level of relative humidity, foods exhibiting low water activity values will also pick up moisture until equilibrium has been reached as a result of their exposure to higher relative humidity. In addition, foods with a high-water activity value will also lose moisture at elevated temperatures when adverse ambient conditions of the sealed storage packaging system. Depending on the circumstances, the food may become spoiled, or it may lose its flavor and/or a desirable texture and appearance if this occurs. Therefore, it is imperative that when it comes to storage packaging systems, it is important to make sure that they are designed correctly and sealed to prevent moisture loss [10, 11, 21, 22].

8.2.1.3 Implicit Factors

Level of Contamination: Contamination with pathogenic or nonpathogenic microorganisms increases the rate of deterioration. In this case, the product's quality can be affected and it might become unsafe to consume. Furthermore, microbial metabolites can also cause food spoilage, which can result in economic losses. As a consequence of the presence of toxin-producing organisms, it can lead to a fatal outcome if consumed. Thus, monitoring the quality of food products as well as preventing the possibility of contamination is important in order to reduce the risk of these products becoming contaminated. To ensure the safety of food production facilities, regular inspections and cleanings are essential to ensure that they are kept clean. The use of proper packaging materials and the practice of proper hygiene practices are also critical elements in preventing food from spoiling and the development of infections [16].

Microbial-biochemical-environmental Interaction: Interactions between microbes, food biochemicals, and the storage environment play a crucial complex mechanism in microbial growth. A number of interactions between a food product and its environment may influence the growth of spoilage microbes, modify its biochemical properties, influence its stability and shelf life, and affect its ability to withstand contamination. In addition to this, they can also have a significant impact on the taste of the food. Microbes compete with each other to survive. As the name implies, microbial interference occurs when one microorganism is inhibited by another microorganism in the same environment in a general, nonspecific manner by competition for nutrients, adhesion sites, unfavorable environmental changes, or any combination of these [16, 22].

8.2.2 Traditional Approaches to Shelf-Life Determination

A food stability test is a technique for estimating the shelf life of foods based on two criteria: the real-time food stability test or direct shelf-life estimation and the accelerated stability test. As part of the real-time stability testing, the processed food being tested is stored at the appropriate storage conditions until it reaches the requirements. Product quality needs to be monitored over time as well as any changes to it to ensure the product remains safe for consumption by conducting regular testing and analysis. Direct shelf-life determination is a systematic approach since there is no universal protocol. In order to ensure that the shelf-life of a product does not exceed acceptable limits, it is necessary to conduct regular shelf-life tests. The systematic approach of real-time version of shelf-life study follows few phases that can be initiated with identification of the specific deterioration factors to different food stuffs. Several storage conditions and categories of processing are considered when determining the shelf-life estimate, such as frozen foods that can be spoiled by microorganisms, fried or oily foods that can oxidize, staling, vitamin, antioxidant degradation and starch retro-gradation, syneresis that occurs according to the chemical properties of food. Monitoring key attributes of the shelf-life study include physical and chemical analysis, microbial inspection and sensory evaluation throughout the shelf-life study. On the other hand, it should be kept in mind that the utility of any analytical measurement depends on its correlation with changes in sensory perception in order to establish its usefulness. In order to simulate real-life product exposure during storage, supply chain conditions, storage conditions need to be created containing fixed or fluctuating variables such as relative humidity, temperature, and illumination conditions. As part of a systematic study, the selection of testing frequency is also an important phase in evaluating shelf life. A short time frequency shelf-life evaluation entails testing the product for a period of seven to ten days, a moderate time frequency shelf life is evaluated for a period of one to three weeks and a long shelf life involves testing for a period of a year or more. A common way to establish consumer acceptability parameters is to establish key attributes that have a direct correlation with consumer acceptance analytical and sensory attributes that can be correlated with consumer acceptability parameters. Once a reliable analytical indicator has been established, further routine shelf-life studies on the same product can use the analytical indicator to determine the end of the product's shelf life. After performing direct shelf-life study, the estimated shelf-life date is typically assigned as the preceding day of attainment satisfactory sensory score and overall analytical results [23–25].

8.2.2.1 Sensory Evaluation

An organoleptic analysis, which is also known as sensory analysis, explores the quality of an organoleptic property in a food product by using the different senses of the panelists. It has been a standard method of accepting or rejecting food products for centuries through this type of analysis. In terms of the first impression a consumer has of a food product, the characteristics of its texture, flavor, taste, appearance, and smell determine that impression. The method has traditionally been considered to be a complementary measure when assessing the quality of food in conjunction with technological safety indicators. As an independent indicator of shelf life during direct shelf-life estimation, sensory score can be used as part of the analytical process. Additionally, a sensory score can be used for evaluating the quality of food during long-term storage, since it is capable of indicating the degradation in quality and safety of a food item over an extended period. In addition, sensory scores can be used to track changes in the quality of food products over time. To ensure that the stored food meets the desired criterion, a panel of judges evaluates the food at regular intervals of time with in the period of storage life determination study to ensure that it meets the specifications. This criterion score is determined by comparing the sensory score to a reference food. If the sensory score is within the acceptable range, the product's shelf-life is acceptable. There will be a termination of the shelf-life study as soon as the shelf-life criterion score is not met. It should be noted that sensory evaluation is frequently conducted using the following methods: discriminative tests, descriptive tests, duo-trio tests, paired comparison tests, free choice profiles, preference ranking, triangular tests, taste profiles, hedonic rating scales, time intensity analyses [26–28].

8.2.2.2 *Microbial Analysis*

It is also important to point out that microbiological analysis is an important indicator and analytical tool that can be used to determine the estimated shelf life of a food under a specific storage condition. Regularly, it is examined how storage conditions affect the microbial community with the help of several criterion-based reference parameters. There are various kinds of microbial reference parameters reviewed in product analysis laboratory but the most common microbial investigation performed includes are the total viable count estimation, such as standard plate count, viable aerobic colony count, coliform test, dye Reduction tests, and differential investigations to identify spoilage causing microorganisms and potential pathogens including identification of *coliforms, salmonella, clostridium, listeria, staphylococci*, yeasts and molds occurrence. If pathogens associated with foodborne illnesses are found during primary screening, modern techniques such as electrical method, ATP determination, DNA/RNA methodology, and ELISA are applied. The results of microbial analysis can be used in a variety of ways, such as optimizing storage conditions, extending the shelf life of food, grading the quality and promising food safety based on the outcomes [27, 29].

8.2.2.3 *Chemical Analysis*

The analysis of physico-chemical characteristics of food products under commercial storage conditions is essential to make decisions regarding how products should be stored to ensure that either long-term storage or marketing, supply chain storage, or consumer storage will succeed.

During storage, spoilage degrades the proximate principles of the food. Furthermore, microbial spoilage also utilizes proximate constituents, and is accompanied by the presence of microbial metabolites present in the food and the presence of an oxidized component, indicating poor storage conditions. Enzymes produced by microorganisms or enzymes naturally present in food react with components of like proteins, lipids, vitamins, carbohydrates, etc. Several biochemical changes appear after a while in the food product, indicating its degradation. By degrading carbohydrates, oxidizing lipids, and releasing nitrogenous compounds, foods loses its texture, color, and odor over time. Consequently, the food must be discarded immediately and is wasted as it is not suitable for human consumption. Food may become different in color, texture, and flavor as a result of these microbial metabolites and oxidized components. Undesirable metabolites, such as diacetyl compound, cadevarine, putrescine, histamire, trimethylamine (TMA), total volatile base (TVB), total volatile nitrogen (TVN), volatile fatty acid are generated due to proteins undergo putrefaction and lipid oxidation It is known that many flavor changes can be observed, such as nitrogenous flavor or distinct rotten egg smell in protein rich food, souring, alcoholic flavor in spoiled fruit, garlic-like flavor due to the presence of trimethylarsine, potato-like flavor in meat fish egg due to the presence of pyrazine, as well as changes in textural characteristics such as slime, ropiness, perforated holes, fat blooms, softening or rotting, curdling are commonly associated with food spoilage. With the deterioration of food, there can be browning and blackening of color changes. The nutritional value of food can also be diminished by spoilage. In this way, it is possible to identify the quality of the product by analyzing potential risks that may occur by chemical changes such as presence of contamination, toxin, microbial metabolites and others. Moreover, it offers valuable insights into how products should be handled and transported so that they can maximize their shelf life to the maximum extent possible [27, 30].

8.2.3 Limitations of Traditional Approaches

It has been conventional to evaluate shelf-life studies of foods by observing the time taken for undesirable changes in their sensory characteristics or microbiological composition to occur. A simple criterion based on the extreme differences in spoilage rates among the samples is the most logical and simple way to determine shelf life, based on each sample's degradation rate. Moreover, due to consumers' variable sensitivity to smell, taste, and flavor, large panels of consumers are necessary to determine a statistically significant result. This requires a lot of time and effort. Estimated accuracy is established using several different statistical methods available. Statistics of shelf-life testing experiments are based on the assumption that there is a family of probability distributions that define the shelf-life distribution of a food product, which is fundamental to the quantitative analysis of shelf-life testing experiments. Based on experimental data collected from shelf-life testing, it is possible to estimate shelf-life distribution parameters. Using an estimate of the parameters of a shelf-life model, it is possible to predict the probability of various events, such as the failure of the safeguard in the future, after the multifaceted parameters have been estimated. Re-inspection of the product shelf-life is thereafter done using Weibull hazard analysis. In terms of risk analysis methods, the Weibull hazard analysis technology is a graphical model known as the maximum likelihood procedure. It is done by estimating the end of shelf-life in the first place, meaning it becomes an estimate of the point at which the shelf-life reaches its end, which is reasonably close to the actual or desired end of shelf-life. Panelists are asked to rate the product on an acceptable or unacceptable basis at the points at which the total time has been divided into equal time segments. Based on these estimates, it is then possible to predict when the product will reach the end of its shelf life in the future. Once this data is collected, it can be used to determine when to restock a product and when to discontinue it based on the availability of these data [31–33].

8.3 Accelerated Shelf-Life Testing

Producers and retailers of food products are well aware of the significance of precise shelf-life measurement. In the absence of this knowledge, it becomes challenging to ensure that the products are safe to consume and are consumed at the best quality which is needed to meet consumer satisfaction. To overcome the limitations of conventional shelf-life testing procedures, accelerated shelf-life testing has gained much importance. Prediction of shelf-life has thus become an integral part of any new product development process. Food companies currently consider it as a tangible possibility, owing to the enhanced capabilities of controlled storage testing facilities in recent times.

Accelerated shelf-life testing (ASLT) is a method of shelf-life prediction where the food product is stored at elevated environmental conditions (increased humidity, temperature, and modified pH) in order to fasten the deterioration process. The shelf-life of the product is then predicted by using known mathematical relationships between the elevated factor and the rate of degradation of the product. This method is very useful for predicting the shelf-life of food products that remain stable at ambient conditions and have long shelf lives. Examples of such food products include commercially sterile canned foods, dried foods, powdered milk, chocolates, edible oils, baking ingredients, and so on. ASLT offers a significant advantage by expediting the process of a product development and thus it has been adapted diligently by the food manufacturers. Before proceeding with the accelerated tests, there should be a clear understanding of the product that needs to be studied which will result in a precise shelf-life prediction. Nevertheless, it is necessary to conduct real time stability studies in addition to ASLT in order to validate predictive data with real time data. ASLT also serves as a valuable comparative study where the rate of physico-chemical changes in a product can be compared against a similar product whose shelf-life is known [34].

8.3.1 Mechanism of Accelerated Shelf-Life Testing

The first step in the accelerated shelf-testing procedure is the identification of the critical event which has the most significant impact on the quality of food. Different chemical, physical, microbial, and enzymatic processes may occur simultaneously during storage depending on the characteristics of the food and storage environment. Determining the specific physical, chemical, or biological event significantly affecting the shelf-life of food product can be a challenging task. A convenient criterion for selecting the indicator of reduced quality is its early manifestation. In simple terms, the first indicator that undergoes change during storage is often chosen as the monitoring parameter for assessing the changes in food quality during shelf-life prediction. After this first step of selection of the critical event, the influencing factors and their acceptable limits must be established. Generally, three levels of intensity for the accelerating factors are used in accelerated shelf-life studies. The sample size must be selected and analyzed at specific time intervals. The rate of reaction is then analyzed using kinetic approach and Arrhenius model. Shelf-life of the product is estimated by modeling the dependance of degradation reaction on the accelerating factor using suitable mathematical models [35].

8.3.2 Factors Influencing Accelerated Shelf-Life Testing

Several stress factors like temperature, light, oxygen, and relative humidity are used as accelerating factors in ASLT. Of these factors, temperature is the most widely used accelerating factor in the assessment of accelerated shelf-life of a food product. The extensive exploitation of temperature as an accelerating factor can be attributed to the fact that biochemical changes in a food matrix are significantly influenced by changes in temperature. Furthermore, the widespread adoption of Arrhenius model, a globally recognized mathematical relationship, further reinforces the choice of temperature as a key factor in predicting the rate of food degradation during ASLT. Light has also been used as an accelerating factor in ASLT, either as a single factor or in combination with temperature for accelerated shelf-life study of beverages, lemon flavored drinks, etc. In addition to light and temperature, oxygen and relative humidity have also been found in literature where they serve as accelerating factors in the accelerated shelf-life study of different foods like egg white powder, dehydrated cabbage, and powdered infant formula [34, 35].

8.3.3 Mathematical Models in Accelerated Shelf-Life Testing

The fundamental aim of mathematical models is to adequately represent a collection of experimental data. Empirical models prove valuable in simulating complex food systems with intricate reactions and non-uniform structures. Additionally, from a practical standpoint, straightforward mathematical expressions can be conveniently employed for process control purposes. Mathematical models offer several advantages like simplification of processes (improvements in food processing and formulations), reduction in number of experiments for obtaining optimum results, creating prototypes (production of diverse food formulations and packaging films using different materials), and investigation of hypothetical situations.

8.3.3.1 Kinetic Approach

Changes in the quality of foods generally follow either zero order or first order reaction kinetics. Hence, the rate of quality change in foods can be expressed as [36],

$$S = \frac{dP}{dt} = CP^n \tag{8.1}$$

where S denotes the rate of product degradation, P denotes the food property of interest, t denotes time, dP/dt represents the rate of change of P with time, C is the rate constant which depends on water activity and temperature, and n is equal to 0 or 1depending on the order of reaction. A negative sign is used of there is a loss in food quality, and a positive sign is used if there is formation of an undesirable compound during storage. Zero-order reactions are independent of food quality factor. For zero-order food degradation reactions,

$$-\frac{dP}{dt} = C \tag{8.2}$$

Integration of eq (8.2) results in,

$$P = P_0 - Ct \tag{8.3}$$

where, P_0 denotes food property at time zero.

In some cases, P lacks quantification or measurability, and relies solely on human evaluation. In such cases, the value of P at the beginning of shelf-life study is assumed to be 100% quality and the value of P when the shelf-life end is unacceptable. In this case, the rate of degradation is given as,

$$C = \frac{100\%}{t} = constant\%\ per\ day \tag{8.4}$$

Food quality deterioration reactions following zero-order kinetics include enzymatic deterioration (refrigerated foods, fruits and vegetables, refrigerated dough, etc.), non-enzymatic browning (dry dairy products, cereals, and pet-foods), and oxidation of lipids (rancidity in frozen foods, snacks, and dry foods).

Food deterioration reactions following first-order kinetics can be described as,

$$-\frac{dP}{dt} = CP \tag{8.5}$$

which by integration results in,

$$P = P_0 e^{-Ct} \tag{8.6}$$

Changes in food quality following first-order reaction kinetics include growth and inactivation of microorganisms, production of slime and off flavors in fish, meat, and poultry due to microbes, protein degradation in dry foods, and loss of vitamin in dried and canned foods, and different color change in food products. Few data exists in literature describing food quality deterioration by other than zero or first order reaction kinetics. Loss of vitamin C in canned tomato juice has been described using second-order reaction.

8.3.4 Arrhenius Model

Proposed by Svante Arrhenius in 1889, Arrhenius equation is an empirical relationship which has been widely used for modelling of temperature dependant reaction rates. Due to its proven efficacy in estimating the rates of various intricate physical, chemical, and sensory changes in food systems, temperature has been widely used as an accelerating factor in accelerated shelf-life studies. The Arrhenius equation is given as follows [37],

$$C = C_0 e^{\frac{-E_a}{RT}} \tag{8.7}$$

where C represents the rate constant, C_0 represents pre-exponential factor or Arrhenius constant, E_a represents the activation energy in J mol^{-1}, R represents the molar gas constant in J K^{-1} mol^{-1} (8.31 J K^{-1} mol^{-1}), and T represents the absolute temperature in K. The energy of activation is the threshold energy that the molecules need to overcome so that the reaction can take place. The rate constant of a reaction is dependent on temperature. With change in temperature, the rate constant changes and thus the rate of reaction also changes. Thus, by calculating the rate constant from the Arrhenius equation at different temperatures, the rate of food quality degradation can also be evaluated.

Arrhenius equation (8.7) can be re-written as follows,

$$lnC = -\frac{E_a}{RT} + \ln C_0 \tag{8.8}$$

By plotting ln C vs $1/T$, a straight-line curve can be obtained, form where the slope gives the value of $-Ea/R$, and thus the value of activation energy for the degradation reaction can be calculated. The y-intercept will give the value of the Arrhenius constant for that particular degradation reaction. Once the Arrhenius constant and the activation energy are known, a kinetic model predicting the shelf life of the product can be constructed at different conditions.

Accelerated shelf-life studies are conducted to estimate how fast a food deteriorates at elevated temperatures. Knowing the temperature-accelerating factor can be helpful in extrapolating results obtained at elevated temperatures, for prediction of shelf-life at lower temperatures. This temperature-accelerating factor is known as the Q_{10} factor. The Q_{10} factor can be described as [38],

$$Q_{10} = \frac{\theta_s \ at\ T}{\theta_s \ at\ T + 10} \tag{8.9}$$

or,

$$Q_{10} = \frac{C(at\ T + 10)}{C(at\ T)} \tag{8.10}$$

where, T is the temperature in ^{0}C, θ_s is the shelf-life of the food product at indicated temperatures, and C represents the rate constant of the reaction.

If the difference (Δ) in temperature exceeds 10^0c, then Q_{10} factor can be expressed as,

$$Q_{10}{}^{\Delta/10} = \frac{Q_s(T_1)}{Q_s(T_2)} \tag{8.11}$$

8.4 Limitations of Accelerated Shelf-Life Testing

Accelerated shelf-life testing provide several benefits such as requirement of less time and cost effectiveness along with identification of potential product issues, thereby allowing improvements and adjustments in the earlier stages of product and packaging development. Despite these benefits, some limitations of accelerated shelf-life testing method of food products restrict precision and accuracy in the field of shelf-life prediction. Accelerated studies are applicable only for temperature dependent food degradation. Real time conditions are much more complex and nonlinear. This leads to inaccuracy in prediction of shelf life. Accelerated studies based on Arrhenius model do not take into account other factors influencing food degradation like microbial contamination, oxygen levels, and relative humidity. At elevated temperatures some foods may lose their physical integrity such as protein denaturation and coagulation at high temperatures. In such cases, accelerated shelf-life studies at elevated temperatures cannot be employed. For a particular composition of foods, the results obtained from accelerated stability studies cannot be applied to a different composition of the same food type. Inaccurate and inadequate results will be obtained if real world storage conditions are not mimicked properly while conducting accelerated studies. Thus, the predictive model may not match with the real degradation phenomenon. To overcome the challenges of accelerated shelf-life studies, various mathematical and statistical tools have been employed in food technology which can enhance the precision and accuracy of shelf-life prediction studies [39].

8.5 Food Stability Simulation: Predictive Models of Food Deterioration

Predicting the shelf-life of food has emerged as an important tool for consumers and distributors, because prediction enables them to ascertain appropriate storage methods, optimize supply chains, and identify optimal time for food consumption, thereby preventing unforeseen food wastage. To overcome the limitations of accelerated shelf-life testing which can be used to predict only temperature dependent changes in food quality, various simulation models using mathematical and statistical tools have been developed. Mathematical models play a crucial role in predicting the shelf-life of foods. These models use mathematical equations and statistics to simulate complex interactions between factors influencing food spoilage like temperature, humidity, light, oxygen, and microbial activity. These models can also be utilized for optimization of food production and processing methods which in turn affect the quality of the end product. By incorporating mathematics, computation and statistical methods, simulation of real-world scenarios has become possible, which in turn has enhanced the accuracy and precision of shelf-life prediction of foods. Several mathematical models have been used by researchers and industries for prediction and simulation of physical, chemical, and microbial changes in foods. Some of the commonly employed prediction and simulation tools have been described below.

8.5.1 Predictive Microbial Model

Microbial modeling enables the characterization and anticipation of microbial responses within defined environmental parameters. These parameters can be both intrinsic (pH, nutrient content, water activity, redox potential, and antimicrobials) and extrinsic (temperature, relative humidity, oxygen, etc.) factors. Controlled experiments are conducted for assessment of microbial reactions and the findings are translated into mathematical equations, facilitating predictions for untested environmental combinations. Despite various factors influencing microbial growth, emphasis is placed on inclusion of variables with significant impact in order to simplify equations. Microbial predictive models can be categorized into kinetic and probabilistic models, mechanistic and empirical models, and primary, secondary, and tertiary models. Square root and Gompertz models are examples of kinetic models which describe growth, death or survival of micro-organisms under defined conditions. Table 8.3 shows model equations for Gompertz and square root models. Lag time, microbial population size, microbial inactivation, and specific growth rate can be described using kinetic models. Probabilistic models can only describe the likelihood of microbial growth and metabolism rather than rate. Empirical models describe data through mathematical equations and curve fitting techniques. Mechanistic models, also known as deterministic models, describe data based on theory and assess results in relation to known phenomenon. They are preferred over empirical models due to use of fewer parameters. They provide superior data fitting and more logical extrapolation. Primary models describe microbial growth, inactivation or survival as a function of time. Secondary models describe kinetic parameters as a function of affecting factors like temperature, pH, water activity, etc. Tertiary models incorporate primary and secondary models in software tools. Some commonly used software for predictive microbial modeling includes Seafood Spoilage Predictor (SSP), FORECAST, Food Spoilage and Safety Predictor (FSSP), Sym'Previus software and Pathogen Modeling Program (PMP) [40]. Effects of temperature, pH, and sodium chloride concentration on growth of a mixed culture of *Salmonella* sp. in laboratory growth medium have been studied using predictive microbiology model. From the model, generation time, lag time, rate of growth, and population density have been calculated [41].

Table 8.3 Equations for Gompertz and square root model.

Name of model	Equation	Symbols	Reference
Gompertz	$log\, N = \left\{ log\, N_0 - log\left(\frac{N_0}{N_f}\right)\right\}$ $X\, exp\left(-exp\left(\frac{k\, exp(1)}{log\left(\frac{N_0}{N_f}\right)}(L-t)+1\right)\right)$	N = microbial cell density (cfu g^{-1}) at process time t; L = time parameter (s); k = maximum inactivation rate constant; N_0 = initial microbial cell density (cfu g^{-1}); N_f = final microbial cell density (cfu g^{-1}); Application – microbial inactivation model	[42]
Modified Square root	$lnk = 2ln\{b(T-T_{min})\}[1-exp$ $[1-exp\{c(T-T_{max})\}]$	k = growth rate ($time^{-1}$); b = regression coefficient (K^{-1} $time^{-0.5}$); T = temperature (K); T_{min} = notional minimum growth temperature (K); c = regression coefficient (K^{-1}); T_{max} = notional maximum growth temperature (K); Application – bacterial growth rate or lag time as a function of temperature	[43]

8.5.2 Modeling of Low Moisture Packaged Foods

To ensure optimal food product quality during purchase and after delivery, it is essential to ascertain the shelf-life of the packaged food item. Conducting actual storage testing for food products through a prolonged stability study, which entails storing the packaged food product under standard storage conditions and periodically assessing it, proves to be both time consuming and expensive. This is especially true for low moisture food products that often have shelf lives extending over several weeks or even months. Furthermore, protracted shelf-life studies are incongruent with the fast-paced demands of the food industry. Consequently, shelf-life simulations have been devised as an alternative solution. Simulations which are used for predicting shelf-life of packaged food items are fundamentally based on the correlation among moisture absorption by the food item, barrier characteristics of the packaging material and the environment where the food is stored. A moisture sorption isotherm is a graphical or mathematical representation that illustrates the relationship between the equilibrium moisture content of a material and the relative humidity of the surrounding environment at a constant temperature. These moisture sorption isotherms are very important in the determination of shelf-life, texture and stability of foods. Identifying the most suitable equation to match the experimental isotherm data is essential for accurately establishing the correlation between the water activity of foods and equilibrium. Several factors influence the choice of selection of suitable equation like simplicity and easy applicability of the equation along with the characteristics of the studied food item. Different models have been used to describe the moisture sorption isotherm. Table 8.4 shows different model equations for describing the moisture sorption isotherm. These models include empirical models like Oswin and Smith models, semi-empirical models like Henderson, Ferro-Fontan and Halsey models, and kinetic models (multilayer basis) like Guggenheim-Anderson-de Boyer (GAB), and Braunauer-Emmet-Teller models. Among these, GAB stands out as the most flexible model as it has been proven effective in accommodating experimental sorption data of nearly all types of products and across the entire spectrum of water activity. This model also offers knowledge about the monolayer moisture content. GAB model has been employed to determine the shelf-life of packaged rice crackers [44].

Table 8.4 Different model equations describing moisture sorption isotherm.

Name of model	Equation	Symbols	Reference
Oswin	$M_e = (A + Bt)\left(\frac{a_w}{1 - a_w}\right)^C$	M_e = Equilibrium moisture content (%) weight basis (w.b.); A, B, C = model constants; t = temperature (°C); a_w = water activity; Application – whole milk powder	[45]
Smith	$M_{Sw} = M_b - b\ln\left(1 - rh\right)$	M_{Sw} = static equilibrium content w.b. (ratio or %); M_b = bound moisture content w.b. (ratio or %); b = constant; rh = relative humidity (decimal); Application – starch and cellulose	[46]
Modified Henderson	$1 - rh = e^{\left(-aM_{Sd}{}^{b}\right)}$	rh = relative humidity (decimal); a, b = constants; M_{Sd} = static equilibrium moisture content (ratio or %) dry basis (d.b); Application - wheat	
Brunauer-Emmet-Teller (BET)	$\frac{P_V}{V\left(P_{VS} - P_V\right)} = \frac{1}{V_m C} + \frac{\left(C - 1\right)P_V}{(V_m C)P_{VS}}$	P_V = vapor pressure (Pa); P_{VS} = saturated vapor pressure (Pa); V = volume of moisture (m³); V_m = Volume of moisture in monolayer (m³); C = product constant; Application – best suited for 10 to 50% relative humdity	[46]
Guggenheim-Anderson-de-Boyer (GAB)	$M_e = \frac{b_0 b_1 b_2 a_w}{\left(1 - b_2 a_w\right)\left(1 - b_2 a_w + b_1 b_2 a_w\right)}$	M_e = equilibrium moisture content (ratio or %); b_0,b_1,b_2 = constants, a_w = water activity (decimal); Application – brown rice, parboiled rough rice, rough rice, rice bran, half of the fruits, vegetables and meat analyzed	

8.5.3 Computational Fluid Dynamics

Integrating the fields of fluid mechanics, mathematics, and computer science, computational fluid dynamics (CFD) serves as a tool for simulating and modelling of processes frequently used in various industries. CFD is specifically designed for fluids in motion, whose physical properties can be expressed through mathematical equations. These equations are then transformed into computer programs or software packages via programming languages to facilitate their solution. In automotive and aviation industries, CFD plays an important role in the development of product and process. In food industries, it can be employed for simulation and development of several mechanical and thermal processes. It has many advantages over conventional experiments. Traditional experiments enable the examination of data solely from a restricted location, especially where the meters and sensors are placed. On the other hand, these simulations offer researchers the opportunity to investigate any location within a specified region of interest. This allows for the interpretation of performance using a variety of flow and thermal parameters. These simulations can be helpful in designing and modelling various flow and heat transfer processes which cannot be easily carried out in a laboratory setup. Any physical process of interest can be replicated theoretically using CFD, thereby having significant control over the process and isolation of a specific phenomenon for detailed research. By using these simulations, significant adjustments can be made in the initial stages of the method of process designing which can also save time. Some commonly used software for CFD include ANSYS Fluent, ANSYS CFX, OpenFOAM, CFD Module, Autodesk simulation CFD, SimScale, VisualFEA, and Simcenter STAR-CCM+. Optimization of shape, modelling of heat and mass transfer, thermal prototyping, fluid flow motion simulation, and so on have been carried out for different physical processes using these CFD based software [47, 48].

Different unit operations in the food industry like thermal processing (pasteurization and sterilization), cooking, baking, dehydration, smoking, cooling and refrigeration, mixing, extrusion, filling in packaging, etc. have been simulated using CFD simulations. It enables scientists and researchers to simulate proper distribution of temperatures within a food product during thermal operations which is essential for ensuring uniform heat treatment, necessary for inactivation of micro-organisms, thereby maintaining the overall food quality. It has been employed for simulation and determination of temperature profile for bottled beer during pasteurization. It has been concluded that the original thermophysical properties of beer led to close correlation between experimental and predictive data [49].

Mass and heat transfer simulations have been used to predict the extent of acrylamide formation in potato fries [50]. Heat transfer simulations using CFD have been used to design a heat treatment operation for meat and meat products in order to inactivate food borne pathogens like *Escherichia coli* thereby extending the shelf-life of the products [51]. These simulation techniques have been widely employed for simulation of the process of spray drying. Optimization of important parameters like temperature, air flow rate, particle size, and humidity in spray drying are vital because incorrectly executed processes may lead to a reduction in nutritional value due to the breakdown of thermolabile substances, such as degradation of unsaturated fatty acids and vitamins, thereby affecting food stability [52]. CFD has been employed to optimize the construction of smoking chambers (used for smoking of food like fish, meat, vegetables, and fruits) to overcome the challenges of improper smoke utilization, non-uniform smoke distribution, and production of hazardous chemicals like polyaromatic hydrocarbons (PAHs) thereby affecting the quality of the end product [53]. These simulations for air flow calculations are becoming more common in the planning of refrigeration and freezing systems within the food industry. To maintain food stability and quality, it is very critical to choose an appropriate cooling refrigerant and equipment at each stage of a cold chain. CFD simulations have been used in the design of various cooling equipment, and large-scale industrial freezing units, which are utilized for cooling and freezing of various food products like meat and meat products, fish, fruits and vegetables. Optimization of traditional refrigeration processes that operate through forced convection systems, modelling of cooling of beef carcasses, and optimization of precooling conditions through aerodynamic and thermal tests in chamber cooling systems, are among the applications of CFD simulations in food preservation [54–56].

The simulations are also used for designing and modelling of mixers, homogenizers, and extruders. By virtually modelling flow patterns, turbulence, and particle interactions within processing equipment, CFD allows for a detailed analysis of mixing efficiency. This is important in ensuring that ingredients are evenly distributed, preventing non-uniform segregation of components in the final products. In context of food stability, these simulations offer to create virtual models of mixing processes, revealing flow patterns, dead zones (areas with minimal flow), and potential segregation of ingredients. This visualization allows engineers and researchers to pinpoint inefficiencies and identify areas of improvement. By tweaking variables like agitator design, tank geometry, and fluid properties, engineers can test virtually test different mixing scenarios before implementing them in real world production. This helps optimize parameters for efficient and thorough mixing, leading to improved shelf-life, enhanced ingredient functionality, and reduced waste. CFD simulations have also been employed in optimizing yogurt mixing (dairy), dough mixing (bakery), and in creating stable emulsions and suspensions in juices, milk drinks, and other beverages [57]. Prediction of mass transfer models using CFD aids in optimizing the distribution of antioxidants, preservatives, and other stability enhancing agents within the food matrix. By using computational fluid dynamics, food storage chambers or warehouses can be optimally designed to achieve uniform temperature and humidity levels, improved airflow, reduced energy consumption, and enhanced overall storage conditions which will ultimately help in the extension of shelf-life of stored food products. Though the simulations offer many benefits, they are quite complex, time consuming, require powerful machines, and require extensive knowledge of the process, programming, numerical methods, fluid mechanics and, heat and mass transfer.

8.5.4 Finite Element Analysis

Predominantly utilized by scientists, engineers, and mathematicians, finite element analysis (FEA) is a method of computational simulation. It is a mathematical approach which is employed to anticipate the response of a product to various real-world influences such as heat, vibration, fluid flow, and other physical factors. It has the capability to indicate whether a product is prone to failure, deterioration, or if it will perform according to its intended design. It was first used in aeronautical and civil engineering for problems relating to structure analysis and complex elasticity. In recent times, the widespread adoption of computer technology for conducting extensive computations in FEA has facilitated its broad application. It is a valuable modeling approach for food products with irregular geometries, intricate boundary conditions, and diverse compositions. In this simulation technique, a large domain is divided into small elements and equations are formulated for each element. All the equations across the entire domain are combined and solved subsequently. This technique uses an intricate arrangement of points referred to as nodes, forming a grid known as mesh. The mesh is configured to incorporate material properties that dictate the manner in which migration will take place. The three steps in the finite element analysis method are pre-processing, solution or analysis, and post-processing. Some common software that are used for FEA include ABAQUS, COMSOL, MSC MARC, ANSYS, LS-DYNA, and MSC NASTRAN [58].

It has been effectively used for modeling of various food processing methods like freezing and thawing, heating and cooling, drying and mechanical damage, heat and mass transfer, extrusion, baking, microwave heating, aseptic processing, and enhancement of food quality concerning texture, nutrient preservation, microbial inactivation in thermal processing. Freezing process of catfish fillets, bakery products, crab meat and crab claws, beef, and mushroom has been simulated using FEA [59–63]. Drying of potatoes, carrots, and cereal grains has been simulated using FEA. It is now effectively utilized in the designing and improvement of food packaging like paper, paperboards and corrugated paperboard packages [58]. The diffusion process in a packaging material fabricated using nanocomposites, has been successfully simulated and a mathematical model has been established using FEA [64]. In these simulations, incorporation of assumptions is frequently seen as a strategy to decrease the model's complexity and computational time, but this may potentially decrease the model's accuracy. While judiciously constraining assumptions can enhance simulation precision, it might also lead to longer computation times [65].

8.5.5 Artificial Neural Network

Artificial neural network (ANN) is a computational model inspired by the structure and functioning of biological neural networks in the human brain. It is a key component of machine learning and falls under the broader category of artificial intelligence. The processing of data in ANN is analogous to the operation of human brain and its neurons. In an ANN, every neuron is equipped with a tiny quantity of local memory. These neurons or units exclusively operate on their local data and the inputs they receive through unidirectional connections or links. ANN consists of an input layer which directly interfaces with the external world, a hidden layer where computations occur based on provided information, and an output layer which gives the final output. Various applications of food shelf-life prediction may require the utilization of different types of ANNs, and specific types of ANNs may be crucial for specific applications. The selection of the suitable network for prediction primarily relies on the logistical aspects of the problem. Types of ANNs that are mostly used for prediction of food shelf-life include multilayer perceptron neural networks (MLPNN), back propagation neural network (BPNN), and radial basis function neural networks (RBFNN). To establish a precise ANN model for shelf-life prediction, various crucial factors are considered which include choosing input parameters, dividing data, training the model, normalizing data, selecting the architecture and structure of the model, and conducting network training. ANNs are well suited for predicting food shelf-life during storage due to their adaptable and data driven nature. Their effectiveness lie on the fact they do not rely on mathematical formulas or kinetic analysis parameters like activation energy and reaction rate. Moreover, they exhibit tolerance by handling noisy and incomplete data, making them reliable in real world scenarios. Additionally, the non-parametric nature of ANN allows them to model the non-linear and dynamic processes related to food quality changes during storage. It can accurately forecast the shelf-life of foods under various fluctuating conditions by learning from experimental data [66, 67].

ANNs have been successfully employed for predicting quality and shelf-life of fruits and vegetables, meat and meat products, grains and oils, aquatic products (salmon, tilapia, bighead carp), dairy products (processed cheese, butter, milk, kalakand, barfi, etc.), and eggs and egg products [68, 69]. Some limitations of ANN include requirement of substantial number of computational resources, data prerequisites, restricted interpretability, susceptibility to adversarial attacks, and the complexity of hyperparameter tuning.

8.5.6 Molecular Simulation Technology

Over the last few years, molecular simulation technology has found extensive applications in the fields of chemistry, physics, biology, and across various other disciplines. However, it still remains a novel technology in the realm of food sciences. The advancement of computer simulation technology has enabled the acquisition of substantial amounts of crucial information that is challenging to obtain through experimental procedures. Though it does not substitute for experiments, it serves as a valuable reference for researchers and scientists conducting experiments, validating certain hypotheses, and mitigating the uncertainties associated with experimental procedures. Molecular simulation technology is a novel technology which is used for the analysis of interactions between molecules. It has been widely used across various fields to study the intermolecular interactions, changes in the structure of biomolecules, and the mechanisms behind various chemical and physical changes. In food science, molecular simulation technology has been applied to study structural and behavioral changes of carbohydrates, proteins, and lipids, as they undergo various changes and functional modifications during food processing, which eventually affects food stability. Molecular simulation technology provides an opportunity to investigate molecular properties which is otherwise challenging via experimental means. In addition to conventional experiments and theoretical approaches, this technology can also provide additional information for scientific research. The additional information provides a more comprehensive understanding of the function and structure of food materials. First application of molecular dynamics (MD) simulation technology dates back to 1986, marking a significant advancement in comprehending the relationship between structure of carbohydrate and its function. This advancement facilitated a deeper understanding of the interactions among different molecules and carbohydrates. However, unavailability of resources like a suitable computer hardware, and MD simulation software have restricted the simulation of starch to using only 6-8 glucose units for representing single chain starch molecules. Nowadays, some of the commonly used MD simulation software include AMBER, GROMACS, LAMMPS, OPEP, and CHARMM [70].

Carbohydrate solvation influences the behavior of carbohydrates in water, thereby influencing its viscosity, taste, storage, and food gelatinization. Thus, carbohydrate structure is an important area of food research. To have a comprehensive understanding on the behavioral changes of carbohydrate at the atomic scale, molecular dynamics simulation is helpful. Amylose, a major component of starch, is used for the encapsulation of hydrophobic components like flavonoids, polyphenols, and fatty acids. Molecular simulation technique was used to study the interactions and the conformational changes of amylose and linoleic acid complex in water. From the simulation studies, it was found that amylose could make a stable complex with linoleic acid in water. Furthermore, it was also concluded that amylose makes more stable complex with stearic acid than that with linoleic acid. These simulation studies can be helpful in the designing and development of encapsulated drugs and other functional foods as such so that they can remain stable for long duration of time. Though molecular simulation technology has advanced swiftly, the internal structure of biological systems is quite complex. Dynamic factors susceptible to change over time constrain the simulation of biological macromolecules. To overcome these limitations, it is essential to enhance both hardware and computing software to address real world challenges effectively. Consequently, extensive research efforts are necessary to elevate practical significance, and broaden application spectrum [71].

8.5.7 Monte Carlo Simulation

Monte Carlo Simulation (MCS) is a computational mathematical model designed to incorporate risk into quantitative analysis and decision-making processes. Through the inclusion of various risk factors and the execution of numerous simulations, it enables the creation of a realistic representation of risk factors and their associated probabilities. This approach ultimately reduces the chances of occurrences of food-borne illnesses. U.S. Food and Drug Administration (FDA) has introduced a digital tool known as iRISK, for assessing potential risks of food degradation and spoilage. This online interactive tool uses MCS to predict potential risks. The risk assessment is done based on various factors like food type, type of hazard (biological, physical, or chemical), food processing method, consumption patterns, demographic place of interest, dose response, and impact on health. Using this software, food industries can replicate real-world situations by incorporating various types of foods and potential hazards into a single evaluation. Once the relevant input has been given, iRISK promptly produces reports that outline the anticipated risks associated with multiple chemical and microbial food safety hazards. These reports also indicate how risk of contamination of food can increase or decrease based on modifications in the surrounding environment. The Sanghai Food and Drug Administration in China utilized MCS to assess food safety in the 41st World Expo hosted in May 2010 [72].

In recent years, these simulations have been widely employed for the prediction of spoilage in milk and milk products like cheese, and yogurt. This model can predict potential risks by simulating microbial growth, transfer, inactivation, and, or partition of selected microorganisms during storage or processing of food, thereby assessing the concentrations of these microbes in the final product. In a complex system like food, where is there is a high level of uncertainty, MCS has the capability to predict model outcomes with some uncertainty measure as compared to deterministic models providing fixed predictions. Furthermore, within this modelling framework, researchers have the capability to conduct sensitivity analysis and explore what-if scenarios. This enables the assessment of the relative significance of various model variables and the effectiveness of intervention strategies, respectively. These features collectively render that MCS is well-suited for the objectives of risk management [73–75]. Though MCS is a significant tool for risk analysis in food safety, it has certain limitations. The requirement of large sample volume, assumption of variables being independent of each other, lack of prior data, and computational complexity lead to inaccurate results and longer simulation times [76].

8.6 Extension of Food Shelf Life

Food preservation is primarily focused on retaining food quality for an extended period of time by preserving its original physico-chemical and organoleptic characteristics. Traditional food preservation technologies like thermal processing (pasteurization and setrilization), non-thermal processing (cooling, chilling and freezing), dehydration or drying, and usage of chemical preservatives are used globally. However, these conventional processes significantly affect the sensorial and nutritional attributes of foods. Advancements in science and technology have paved the path for modifications of existing technologies and introducing newer ones. Novel food processing technologies have become indispensable to meet consumer demands of environment friendly and minimally processed foods. Some of the novel food processing technologies are irradiation, high pressure processing (HPP), pulsed electric field processing (PEF), ohmic heating (OH), microwave heating, infrared heating, hurdle technology, and cold plasma [77]. The food preservation process has evolved into an interdisciplinary field, involving aspects like cultivations, harvesting, processing, packaging and distribution of food. Thus, employing an integrated method is necessary for preserving food throughout processing and production. Storage system plays an important role in maintaining the life cycle of a crop. Improper storage conditions can lead to deterioration and undesirable food wastage. In this regard, controlled and modified atmosphere storage have significantly helped to increase the shelf-life of harvested produce. Controlled and modified atmosphere storage are widely used for the preservation of freshly harvested fruits and vegetables. In controlled-atmosphere storage (CAS), the levels of oxygen in the storage atmosphere are significantly reduced whereas, the levels of carbon dioxide are increased. Continuous monitoring of storage atmosphere is vital in CAS. On the other hand, in modified atmosphere storage (MAS) provides a modified gaseous atmosphere by either manipulating the levels of oxygen and carbon dioxide, or by inserting inert gases. MAS do not involve continuous monitoring of the gaseous atmosphere. The concept of CAS and MAS have been applied to product CAP (controlled atmosphere packaging) and MAP (modified atmosphere packaging) to preserves different kinds of foods such as meat, fish, fruits, and vegetables. Packaging is a fundamental requirement for ensuring food safety and extending shelf-life, thereby enhancing food security. Active, smart, intelligent, and antimicrobial packaging along with edible films and coatings are some of the technological advancements in food packaging sector enhancing the stability and quality of foods. Active packaging is defined as a type of packaging where interaction between food and active ingredients (moisture absorbers, oxygen scavengers, antimicrobials, antioxidants, ethylene inhibitors, etc.) incorporated in the packaging material can enhance the shelf-life of foods. Smart packaging, also known as intelligent packaging can monitor the quality of a food product over time and make decisions accordingly. In the food sector, nanotechnology emerges as a pivotal force, offering innovative technological solutions to address challenges in food safety, food packaging, and food processing, while also contributing to the development of functional foods. However, application of nanotechnology in food packaging has intensified due to its ability to enhance the safety, quality, and stability of foods. In view of this, nanotechnology can enhance food packaging by ensuring healthier, safer, and higher quality food products with extended shelf-life, without compromising with the physical and sensory attributes of the food items. Several metal or metal oxide nanoparticles like silver, gold, zinc, titanium dioxide, zinc oxide, silicon oxide, etc. have been incorporated into packaging materials for enhancing the shelf-life and quality of food products [78].

8.7 Ethical Considerations in Food Stability Simulation

Any decision that is made for any commodities that are exploited by community should be based on careful consideration of the potential consequences as it is important to ensure ethical considerations. In order to accomplish this, it is important to take into account the impact of all stakeholders, including the environment, society, and the economy. Considering ethical considerations when simulating food stability is an important aspect that should be paid particular attention too. A number of ethical considerations must be taken into account while labeling the estimated shelf life code on packed food items, including ensuring that food safety will not be compromised, that simulations will be fair and accurate, and that the results will not be skewed in favor of any particular interest or group. In order to make a fair and just decision, it must be made on the basis of the principles of fair-mindedness, justice, and respect for the rights of the individual and it must have been made on a basis of individual security. The potential for unintended consequences should also be considered in addition to the potential for intended consequences.

There are a variety of factors that should be considered before making a decision, such as considering the risks and benefits carefully as well as identifying potential solutions to mitigate any negative effects. The implementation of food cautionary policies may also influence manufacturers to reformulate or redesign their products to avoid any threatening. Any minor requisite modification carried out would be aimed at minimizing any potential health risks as well as reducing associated financial loss. A voluntary word of caution on product design can also protect consumers from potential dangers that their products might pose. There are a number of questions that will need to be addressed during this process. Several verifications should be performed, including, but not limited to, (1) a thorough investigation of the mathematical model, (2) a thorough understanding of how the mathematical model interacts with the experimental shelf life study, (3) a thorough investigation of the margins of error, and (4) an in-depth analysis of the risks and benefits of the new technology in sufficient numbers of real-life experiments that will confirm the validity of the technology and reveal any unanticipated outcomes that may emerge. Last but not least, it is essential that this decision is implemented in a way that promotes its effectiveness and success. Any necessary adjustments should be made in a timely manner. Finally, it is important to document the decision-making process and its outcomes. This will ensure that the organization has a record of the decisions taken and the rationale behind them. It will also provide an opportunity to learn from mistakes and successes and to make future decisions more effectively [79, 80].

8.8 Future Scope

Emerging technologies in food stability simulation allow food producers to better understand the behavior of food products under different storage conditions. This allows for more accurate predictions of shelf life, spoilage, and food safety. Additionally, these technologies can also be used to optimize production processes by reducing waste and minimizing losses. Globally, it appears that these technologies are poised to revolutionize the food industry, making it safer and producing better quality food. Food industries are concerned about the efficient manufacturing of food, customer preferences and satisfaction, supply chain management, and so on, to ensure that our food is fresh, and anticipated. The food industry can be made more efficient and productive by implementing AI (artificial intelligence) and machine learning programs to enhance food quality and prevent food waste. The majority of online food service platforms keep track on customer food preferences and buying habits when ordering their food. There is a tremendous opportunity for using AI and machine learning algorithms in food processing industries to improve the efficiency, cost-effectiveness, and flexibility of packaged convenience food production. The use of big data and data science has become increasingly popular over the past few years due to their extensive use in the food business sector to resolve issues regarding product design, production, storage, supply, and marketing activities that result from the usage of these technologies in numerous fields. There is a lot of potential in the area of data science and deep learning technologies to transform food sectors, as it offers food businesses some interesting opportunities to gain market preferences. As a result, there is no doubt that they are exploding in a wide variety of fields. Intelligent optimization techniques, Gaussian mixture models, simulated annealing, genetic algorithm, stochastic simulation, Bayesian network, ANN, tree decision-making in food, and various forecasting models of AI are appeared in the food processing industry is currently going through another period of advancement and progress [81, 82].

8.9 Conclusion

Consumers and distributors have become increasingly aware of the importance of predicting food shelf-lives in order to correct inaccurate and insufficient storage conditions for food, since prediction allows them to determine the best storage methods, optimize supply chains, and identify the best times to consume food, reducing food waste. Several mathematical and statistical tools have been used in the field of food technology to overcome the challenges associated with food spoilage. At this point cutting-edge accelerated shelf-life study techniques help to improve the accuracy and precision of shelf-life prediction studies. Several benefits of accelerated shelf-life testing include requiring less time and cost effectiveness, as well as identifying potential issues in long run, which can lead to improvements and changes in product and packaging development earlier. There have been many advances in early prediction of the shelf life of food together with mathematical models based on complex mathematical equations, curve fitting techniques, mechanistic models, and deterministic models, which are advancing the field of early prediction. In addition to using primary models of shelf life prediction such as microbial and organoleptic prediction, it has also been shown that conducting actual storage tests on food products through prolonged stability studies that require storing and periodically assessing the packaged food product under standard storage conditions is tedious, lengthy, and costly. It is important to recognize that data science and deep learning technologies can have a huge impact on the food industry, as they offer some interesting opportunities for food companies to gain a competitive advantage. Therefore, it is no wonder that they are expanding across a wide range of fields as a result of their popularity. A number of intelligent optimization techniques are now available in the food processing industry. For instance, computational fluid dynamic simulations based on different models that describe the moisture sorption isotherm of packaged food enable us to simulate the proper distribution of temperatures within a food product during thermal operations, to make sure uniform heat treatments has been applied that is necessary to ensure that microorganisms are eradicated, resulting in a high-quality finished product. Other quick techniques include Bayesian networks, artificial neural networks, tree decision-making in food, and different forecasting models based on artificial intelligence used frequently in recent years. Food processing is currently experiencing significant progress and advancement with the introduction of artificial intelligence in the food processing industry, as it has become an efficient tool that is helping to reduce costs and increase efficiency. Furthermore, AI-based expertise is capable of detecting and preventing chance of food-related illnesses, as well as identifying future trends within the food industry. Therefore, it is crucial to use these technologies for the food industry to continue advancing in innovative ways, as well as continue to advance in a sustainable way.

References

1. Knorr, D. and Augustin, M.A., Preserving the food preservation legacy. *Crit. Rev. Food Sci. Nutr.*, 63, 28, 9519–9538, 2023.
2. Gould, G.W., Methods for preservation and extension of shelf life. *Int. J. Food Microbiol.*, 33, 1, 51–64, 1996.
3. Sridhar, A., Ponnuchamy, M., Kumar, P.S., Kapoor, A., Food preservation techniques and nanotechnology for increased shelf life of fruits, vegetables, beverages and spices: a review. *Environ. Chem. Lett.*, 19, 2, 1715–1735, 2021.
4. García-Díez, J., Gonçalves, C., Grispoldi, L., Cenci-Goga, B., Saraiva, C., Determining food stability to achieve food security. *Sustainability*, 13, 13, 7222, 2021.
5. Types of shelf-life items, subpart 101-27.204 of Federal property management regulation, U.S. Department of transportation, 1993.
6. Man, C.M.D. and P., Food storage trials, in: *The stability and shelf life of food*, Subramaniam (Ed.), Part 2, Chapter 6, pp. 171–198, Woodhead Publishing, 2016.

7. Food Safety and Standards (Packaging And Labelling) Regulations, Ministry Of Health And Family Welfare, India, 2011.
8. Torrieri, E., Storage stability: Shelf life testing, in: *Encyclopedia of food and health*, pp. 188–192, Academic Press, 2016.
9. Food Product Shelf-Life Guide for Scaling Businesses, AURI (Agricultural Utilization Research Institute), Minnesota.
10. Awulachew, M.T., Understanding to the shelf-life and product stability of foods. *J. Food Process. Preserv.*, 5, 8, 1–5, 2021.
11. Hamad, S.H., Factors affecting the growth of microorganisms in food, in: *Progress in food preservation*, Bhat, R., *et al.* (Eds.), pp. 405–427, 2012.
12. Odeyemi, O.A., Alegbeleye, O.O., Strateva, M., Stratev, D., Understanding spoilage microbial community and spoilage mechanisms in foods of animal origin. *Compr. Rev. Food Sci. Food Saf.*, 19, 2, 311–331, 2020.
13. Esener, A.A., Bol, G., Kossen, N.W.F., Roels, J.A., Effect of water activity on microbial growth, in: *Scientific and engineering principles*, Moo-Young, M., *et al.* (Eds.), pp. 339–344, 1981.
14. Rahman, M.S. and Labuza, T.P., Water Activity and Food Preservation., in: *Handbook of food preservation*, 2nd edition, Chapter 20, CRC Press, pp. 465–494, 2007.
15. Moral, U., Nagar, P., Kaur, K.A., Growth of Different Types of Microorganism, Intrinsic and Extrinsic Factors of Microorganism and their Affects in Food: A Review. *Int. J. Curr. Microbiol. Appl. Sci.*, 6, 1, 290–298, 2017.
16. Evaluation and Definition of Potentially Hazardous Foods, Food and Drug Administration, in: *Comprehensive Reviews in Food Science and Food Safety*, vol. 2, 2003, https://www.fda.gov/files/food/published/Evaluation-and-Definition-of-Potentially-Hazardous-Foods.pdf.
17. Russell, A.D., Mechanisms of bacterial resistance to antibiotics and biocides, in: *Progress in medicinal chemistry*, vol. 35, Ellis, G.P., *et al.* (Eds.), pp. 133–198, 1998.
18. Singh, B., Suri, K., Shevkani, K., Kaur, A., Kaur, A., Singh, N., M., Enzymatic browning of fruit and vegetables: A review, in: *Enzymes in food technology: Improvements and innovations*, Kuddus, (Ed.), pp. 63–78, 2018.
19. Banwart, G.J., Factors that affect microbial growth in food, in: *Basic Food Microbiology*, 2nd edition, Chapter 4, pp. 101164, 1979.
20. Doyle, M.P., Diez-Gonzalez, F., Hill, C., *Food microbiology: fundamentals and frontiers*, ASM Press, John Wiley & Sons, 5th Edition, 2020.
21. Rolfe, C. and Daryaei, H., In: *Food Safety Engineering. Food Engineering Series*, Demirci, A., Feng, H., Krishnamurthy, K. (eds). Springer, Cham, pp. 3–24, 2020.
22. Gupta, A., Gupta, R., Singh, R.L., Microbes and Environment, in: *Principles and Applications of Environmental Biotechnology for a Sustainable Future*, Singh, R.L. (Ed.), pp. 43–84, 2016.
23. Haouet, M.N., Tommasino, M., Mercuri, M.L., Benedetti, F., Bella, S.D., Framboas, M., Pelli, S., Altissimi, M.S., Experimental accelerated shelf life

determination of a ready-to-eat processed food. *Ital. J. Food Saf.*, 7, 4, 6919, 2019.
24. Taormina, P.J., Purposes and Principles of Shelf Life Determination, in: *Food Safety and Quality-Based Shelf Life of Perishable Foods*, pp. 1–26, 2021.
25. Valero, A., Carrasco, E., García-Gimeno, R.M., Principles and methodologies for the determination of shelf-life in foods, in: *Trends in vital food and control engineering*, Eissa, A. (Ed.), vol. 1, pp. 3–42, 2012.
26. Meilgaard, M.C., Carr, B.T., Civille, G.V., *Sensory evaluation techniques*, CRC Press, 4th edition, 1999.
27. Mihafu, F.D., Issa, J.Y., Kamiyango, M.W., Implication of sensory evaluation and quality assessment in food product development: *Curr. Res. Nutr. Food Sci.*, A review. 8, 3, 690–702, 2020.
28. Amerine, M.A., Pangborn, R.M., Roessler, E.B., *Principles of sensory evaluation of food*, Academic Press, 1965.
29. Jacxsens, L., Kussaga, J., Luning, P.A., Van der Spiegel, M., Devlieghere, F., Uyttendaele, M., A microbial assessment scheme to measure microbial performance of food safety management systems. *Int. J. Food Microbiol.*, 134, 1-2, 113–125, 2009.
30. Zielińska, D., Bilska, B., Marciniak-Łukasiak, K., Łepecka, A., Trząskowska, M., Neffe-Skocińska, K., Tomaszewska, M., Szydłowska, A., Kołożyn-Krajewska, D., Consumer understanding of the date of minimum durability of food in association with quality evaluation of food products after expiration. *Int. J. Environ. Res. Public Health*, 17, 5, 1632, 2020.
31. Fu, B. and Labuza, T.P., Shelf-life prediction: theory and application. *Food Control*, 4, 3, 125–133, 1993.
32. Xiao, M., Liu, S., Jin, H., Xiao, M., Wang, H., Zhang, H., Dai, Q., Evaluating freshness loss of green tea with Q10 method and Weibull hazard analysis under accelerated shelf life testing. *J. Chem.*, 2955839, 1–10, 2022.
33. Sánchez-González, J.A. and Oblitas-Cruz, J.F., Application of Weibull analysis and artificial neural networks to predict the useful life of the vacuum packed soft cheese. *Rev. Fac. de Ing.*, 82, 53–59, 2017.
34. Subramaniam, P.J., Shelf-life prediction and testing, in: *Science and technology of enrobed and filled chocolate, confectionery and bakery products*, pp. 233–254, Woodhead Publishing Series in Food Science, Technology and Nutrition, Chapter 12, pp. 233–254, 2009.
35. Calligaris, S., Manzocco, L., Anese, M., Nicoli, M.C., Accelerated shelf life testing, in: *Food quality and shelf life*, pp. 359–392, Academic Press, Chapter 12, pp. 359–392, 2019.
36. Mizrahi, S., Accelerated shelf-life tests, in: *Understanding and measuring the shelf-life of food*, pp. 318–339, 2004.
37. Gallagher, M.S., Mahajan, P.V., Yan, Z., Modelling chemical and physical deterioration of foods and beverages, in: *Food and beverage stability and shelf life*, pp. 459–481, Woodhead Publishing Ltd., Chapter 14, pp. 459–481, 2011.

38. Perchonok, M., Shelf-Life Considerations and Techniques, in: *Food Product Development: Based on Experience*, vol. 39, pp. 59–73, 2002.
39. Ebrahim, A., DeVore, K., Fischer, T., Limitations of accelerated stability model based on the Arrhenius equation for shelf life estimation of *in vitro* diagnostic products. *Clin. Chem.*, 67, 4, 684–688, 2021.
40. Stavropoulou, E. and Bezirtzoglou, E., Predictive modeling of microbial behavior in food. *Foods*, *8*, 12, 654, 2019.
41. Gibson, A.M., Bratchell, N., Roberts, T.A., Predicting microbial growth: growth responses of salmonellae in a laboratory medium as affected by pH, sodium chloride and storage temperature. *Int. J. Food Microbiol.*, *6*, 2, 155–178, 1988.
42. Gil, M.M., Brandao, T.R., Silva, C.L., A modified Gompertz model to predict microbial inactivation under time-varying temperature conditions. *J. Food Eng.*, *76*, 1, 89–94, 2006.
43. Alber, S.A. and Schaffner, D.W., New modified square root and Schoolfield models for predicting bacterial growth rate as a function of temperature. *J. Ind. Microbiol. Biotechnol.*, *12*, 3-5, 206–210, 1993.
44. Sirpatrawan, U., Shelf-life simulation of packaged rice crackers. *J. Food Qual.*, *32*, 2, 224–239, 2009.
45. Langová, J. and Štencl, J., Moisture sorption isotherms of whole milk powder in the temperature range of 5–35 C and critical values of water activity prediction. *J. Food Qual.*, *83*, 10, 35–40, 2015.
46. Bala, B.K., *Drying and storage of cereal grains,* Wiley-Blackwell, 2nd edition, p. 352, 2016.
47. Szpicer, A., Bińkowska, W., Wojtasik-Kalinowska, I., Salih, S.M., Półtorak, A., Application of computational fluid dynamics simulations in food industry. *Eur. Food Res. Technol.*, *249*, 6, 1411–1430, 2023.
48. Norton, T. and Sun, D.W., Computational fluid dynamics (CFD)–an effective and efficient design and analysis tool for the food industry: a review. *Trends Food Sci. Technol.*, *17*, 11, 600–620, 2006.
49. Xia, B. and Sun, D.W., Applications of computational fluid dynamics (CFD) in the food industry: a review. *Comput. Electron. Agric.*, *34*, 1-3, 5–24, 2002.
50. Chan, D.S., Computer simulation with a temperature-step frying approach to mitigate acrylamide formation in French fries. *Foods*, *9*, 2, 200, 2020.
51. Huang, L., A simplified method for numerical simulation of gas grilling of non-intact beef steaks to eliminate Escherichia coli O157: H7. *J. Food Eng.*, *113*, 3, 380–388, 2012.
52. Kuriakose, R. and Anandharamakrishnan, C., Computational fluid dynamics (CFD) applications in spray drying of food products. *Trends Food Sci. Technol.*, *21*, 8, 383–398, 2010.
53. Kubiak, M.S. and Jakubowski, M., CFD simulations as a supporting tool for process and construction optimization in food industry production practice: A case study of a single truck smoking chamber. *Ital. J. Food Sci.*, 25, 3, 251, 2013.

54. Singh, S. and Chourasia, B.K., Thermal Studies on Effects of Use of Desiccant Cooling in Cold Storage by Using CFD Analysis, in: *Emerging Trends in Energy Conversion and Thermo-Fluid Systems: Select Proceedings of iCONECTS 2021*, Springer Nature Singapore, Singapore, pp. 205–221, 2022.
55. Söylemez, E., Alpman, E., Onat, A., Hartomacıoğlu, S., CFD analysis for predicting cooling time of a domestic refrigerator with thermoelectric cooling system. *Int. J. Refrig.*, *123*, 138–149, 2021.
56. Trujillo, F.J. and Pham, Q.T., A computational fluid dynamic model of the heat and moisture transfer during beef chilling. *Int. J. Refrig.*, *29*, 6, 998–1009, 2006.
57. Connelly, R.K. and Valenti-Jordan, J., Mixing analysis of a Newtonian fluid in a 3D planetary pin mixer. *Chem. Eng. Res. Des.*, *86*, 12, 1434–1440, 2008.
58. Fadiji, T., Coetzee, C.J., Berry, T.M., Ambaw, A., Opara, U.L., The efficacy of finite element analysis (FEA) as a design tool for food packaging: A review. *Biosyst. Eng.*, *174*, 20–40, 2018.
59. Nguyen, T.T.T., Do, Q.K., Vo, C.C., Nguyen, T.V., A numerical approach for fish fillet modeling during freezing process _ Case study from Vietnamese catfish fillets. *J. Phys. Conf. Ser.*, 1457, 1, 012018, 2020.
60. Santos, M.V., Vampa, V., Califano, A., Zaritzky, N., Numerical simulations of chilling and freezing processes applied to bakery products in irregularly 3D geometries. *J. Food Eng.*, *100*, 1, 32–42, 2010.
61. Dima, J.B., Santos, M.V., Baron, P.J., Califano, A., Zaritzky, N.E., Experimental study and numerical modeling of the freezing process of marine products. *Food Bioprod. Process.*, *92*, 1, 54–66, 2014.
62. Huan, Z., He, S., Ma, Y., Numerical simulation and analysis for quick-frozen food processing. *J. Food Eng.*, *60*, 3, 267–273, 2003.
63. Santos, M.V. and Lespinard, A.R., Numerical simulation of mushrooms during freezing using the FEM and an enthalpy: Kirchhoff formulation, in: *Heat and mass transfer*, vol. *47*, pp. 1671–1683, 2011.
64. Cerisuelo, J.P., Rafael, G., Pilar, H.-M., Diffusion modeling in polymer–clay nanocomposites for food packaging applications through finite element analysis of TEM images. *J. Membr. Sci.*, 482, 92–102, 2015.
65. Fadiji, T., Ashtiani, S.-H.M., Daniel, I., Onwude, Z.L., Linus Opara, U., Finite element method for freezing and thawing industrial food processes. *Foods*, 10, 4, 869, 2021.
66. Ahuja, K.K., Goyal, S., Goyal, G.K., Shelf life prediction of paneer tikka by artificial neural networks. *J. Agric. Sci.*, 1, 6, 145–149, 2012.
67. Shi, C., Zhao, Z., Jia, Z., Hou, M., Yang, X., Ying, X., Ji, Z., Artificial neural network-based shelf life prediction approach in the food storage process: A review. *Crit. Rev. Food Sci. Nutr.*, 1–16, 2023.
68. Goyal, S. and Goyal, G.K., Shelflife Prediction of Processed Cheese Using Artificial Intelligence ANN Technique. *Hrvatski* časopis *za prehrambenu tehnologiju, biotehnologiju i nutricionizam*, 7, 3–4184–187, 2012.

69. Goyal, G.K. and Goyal, S., Cascade artificial neural network models for predicting shelf life of processed cheese. *J. Adv. Info. Tech.*, 4, 2, 80–83, 2013.
70. Wang, Y., Liu, T., Xie, J., Cheng, M., Sun, L., Zhang, S., Xin, J., Zhang, N., A review on application of molecular simulation technology in food molecules interaction. *Curr. Res. Food Sci.*, 5, 1873–1881, 2022.
71. López, C.A., de Vries, A.H., Marrink, S.J., Amylose folding under the influence of lipids. *Carbohydr. Res.*, 364, 1–7, 2012.
72. Heffernan, R., Ensuring Food Safety with Monte Carlo Simulation, https://www.riskmanagementmonitor.com/ensuring-food-safety-with-monte-carlo-simulation/, 2014.
73. Griep-Moyer, E.R., Trmčić, A., Qian, C., Moraru, C., II, Monte Carlo simulation model predicts bactofugation can extend shelf-life of pasteurized fluid milk, even when raw milk with low spore counts is used as the incoming ingredient. *J. Dairy Sci.*, 105, 12, 9439–9449, 2022.
74. Rigaux, C., André, S., Albert, I., Carlin, F., Quantitative assessment of the risk of microbial spoilage in foods. Prediction of non-stability at 55 C caused by Geobacillus stearothermophilus in canned green beans. *Int. J. Food Microbiol.*, 171, 119–128, 2014.
75. Qian, C., Martin, N.H., Wiedmann, M., Trmčić, A., Development of a risk assessment model to predict the occurrence of late blowing defect in Gouda cheese and evaluate potential intervention strategies. *J. Dairy Sci.*, 105, 4, 2880–2894, 2022.
76. Limitations of Monte Carlo simulation, https://fastercapital.com/startup-topic/Limitations-of-Monte-Carlo-simulation.html, 2024.
77. Hameed, F., Ayoub, A., Gupta, N., Novel food processing technologies: An overview. *IJCS*, 6, 6, 770–776, 2018.
78. Singh, T., Shukla, S., Kumar, P., Wahla, V., Bajpai, V.K., Rather, I.A., Application of nanotechnology in food science: perception and overview. *Front. Microbiol.*, 8, 1501, 2017.
79. Tongen, A. and Adam, M., Ethics involved in simulation-based medical planning. *Ethics Med.*, 22, 1, 23, 2006.
80. Essex, R., Weldon, S.M., Markowski, M., Gurnett, P., Slee, R., Cleaver, K., Stiell, M., Jagodzinski, L., A systematic mapping literature review of ethics in healthcare simulation and its methodological feasibility, *Clin. Simul. Nurs.*, 76, 48-58, 2022.
81. Kler, R., Elkady, G., Rane, K., Singh, A., Hossain, M.S., Malhotra, D., Ray, S., Bhatia, K.K., Machine learning and artificial intelligence in the food industry: a sustainable approach. *J. Food Qual.*, 8521236, 1–9, 2022.
82. Konfo, T.R.C., Djouhou, F.M.C., Hounhouigan, M.H., Dahouenon-Ahoussi, E., Avlessi, F., Sohounhloue, C.K.D., Recent advances in the use of digital technologies in agri-food processing: A short review. *Appl. Food Res.*, 3, 2, 100329, 2023.

9

Edible Biopolymer Coatings/Films Used for Packaging and Preserving Foods

Latha Sukumaran* **and Negasso Argeta Jano**

Department of Biotechnology, Rathinam College of Arts and Science (Autonomous), Pollachi Road, Eachanari, Coimbatore, Tamil Nadu, India

Abstract

Biopolymers are becoming increasingly popular as a substitute for conventional materials in food preservation and packaging. Their biodegradability, renewable nature, and ability to reduce environmental impact make them an attractive option. This chapter delves into the diverse array of biopolymers utilized in food preservation and packaging applications, including their unique properties, application methods such as dipping, spraying, brushing, challenges encountered during use, and potential solutions. The field of food preservation and packaging has witnessed numerous types of biopolymers gaining significant attention. Polysaccharides, such as chitosan, cellulose, and starch, possess remarkable film-forming properties and barrier properties against moisture and oxygen. Proteins, including gelatin and whey protein, offer exceptional mechanical strength and superior film-forming ability. Lipids like beeswax and shellac provide outstanding water vapor resistance, as well as antimicrobial properties. Moreover, polyhydroxyalkanoates (PHA) obtained from microbial sources exhibit versatile attributes such as biodegradability and biocompatibility. These biopolymers are utilized in diverse food preservation and packaging applications through different application techniques. Dipping entails immersing the food product or substrate into a biopolymer solution to create a protective coating while spraying involves uniformly coating the surface of the food product with a fine mist of biopolymer solution or dispersion. Brushing involves manually applying the biopolymer solution onto the food surface using an applicator or brush. These and other techniques enable the formation of a naturally safe coating that enhances the shelf life of products while reducing packaging waste; they can also enhance nutritional value provided that they are customized for edibility purposes.

**Corresponding author*: lathasukumar586@gmail.com

Mousumi Sen (ed.) Food Coatings and Preservation Technologies, (301–338) © 2025 Scrivener Publishing LLC

The use of biopolymer-based coating or films for food preservation and packaging offer advantages such as simplicity, cost-effectiveness, and versatility. However, despite these benefits, biopolymers face certain challenges due to their mechanical strength, water vapor resistance, and barrier properties that may not match those of synthetic polymers. Biopolymer films are susceptible to moisture and can exhibit poor stability under high humidity conditions. Additionally, sourcing and producing biopolymers on a large scale may pose economic and logistical difficulties. To overcome these challenges, blending biopolymers with other materials can enhance their mechanical properties while incorporating nanoparticles like montmorillonite or silver nanoparticles into the matrix can improve antimicrobial performance. Moreover, optimizing processing conditions alongside exploring novel sources of biopolymers can address economic barriers associated with production logistics. In conclusion, utilizing biopolymers presents promising solutions for food preservation and packaging applications despite the various potential benefits.

***Keywords*:** Biopolymer, packaging, shelf life, preservation, antimicrobial

9.1 Introduction

In recent years, there has been a growing demand for sustainable and environmentally friendly solutions in various industries, including the food industry. Traditional synthetic polymers used in food packaging and preservation have raised concerns due to their non-biodegradable nature and potential adverse effects on human health. As a result, biopolymers have emerged as a promising alternative, offering a range of benefits such as improved food preservation, reduced environmental impact, and enhanced safety [1]. Food packaging serves several important purposes such as protection, preservation, information, and convenience [2]. In the realm of protection, packaging acts as a formidable barrier that shields food from physical harm, contamination, and exposure to undesirable elements such as light, moisture, and air. Its foremost function is to preserve the freshness, flavor profile, and nutritional value of consumables. Preservation aims to prolong the shelf life of perishable items by hindering spoilage processes while curbing chemical reactions that lead to degradation. This process also enables controlled atmosphere packaging (CAP), modified atmosphere packaging (MAP), and vacuum packaging which are all designed to uphold product quality standards. Packaging information provides quintessential details regarding food products including ingredient lists, nutritional facts charts, allergen warnings labels as well as preparation instructions for consumers' convenience. In terms of convenience, biopolymers offer easy handling options during storage or transportation phases for food products through innovative features like resealable closures or portion control systems; these advancements enhance consumer ease of use while simultaneously reducing instances of unnecessary food waste.

Traditional packaging materials, such as polyethylene, polyvinyl chloride, polypropylene, and polystyrene can have significant environmental consequences due to their non-biodegradability and potential for pollution. This has led to an increased focus on sustainable packaging solutions, including biodegradable materials and recycling initiatives [3]. Biopolymers are polymers derived from renewable sources such as plants, animals, and microorganisms, and they can be classified into various categories, including polysaccharides, proteins, and polylactic acid (PLA) [4]. These materials possess unique properties that make them suitable for food preservation and packaging applications. One significant advantage of biopolymers is their biodegradability. Unlike traditional synthetic polymers, biopolymers can be broken down by natural processes, reducing their impact on the environment. When used as packaging materials, biopolymers can help reduce plastic waste and the accumulation of non-recyclable materials in landfills and oceans.

In terms of food preservation, biopolymers offer several advantages. First and foremost, they can create a barrier between the food and the external environment, preventing the ingress of oxygen, moisture, and contaminants. This barrier function helps maintain the freshness, quality of food, and extending its shelf life [5]. Additionally, biopolymers possess antimicrobial properties, inhibiting the growth of microorganisms that cause food spoilage and foodborne illnesses. The preservation ability can reduce the reliance on synthetic chemical preservatives, addressing consumer concerns regarding their potential health risks [6].

Polysaccharides, such as chitosan and cellulose derivatives, are widely used as biopolymer-based food preservatives and packaging materials [7]. Chitosan, derived from chitin, a natural polymer found in crustacean shells, has excellent antimicrobial properties, making it effective in inhibiting the growth of bacteria and fungi. Cellulose derivatives, including methylcellulose and carboxymethylcellulose, have excellent film-forming properties, making them suitable for creating biodegradable food packaging materials with good mechanical strength and barrier properties. Proteins such as casein, whey, and soy proteins, also offer potential as biopolymer-based food packaging materials [8]. They have excellent film-forming capabilities, allowing the production of flexible and biodegradable films that can protect food from environmental factors. These protein-based films can be used for wrapping products, creating pouches, or forming coatings on edible products. Another important biopolymer used in food packaging is polylactic acid (PLA), derived from renewable resources such as cornstarch or sugarcane. PLA films have gained popularity due to their transparency, good barrier properties, and biodegradability. They can be used for various packaging applications, including trays, cups, and wraps [9].

9.2 Types and Composition of Edible Biopolymer Coatings or Films

Various types of biopolymer coatings or films are commonly used in the food industry, each offering unique properties and applications. The most common types of biopolymers coatings include those made from lipids, proteins, and polysaccharides as described in Table 9.1.

9.2.1 Protein-Based Coatings

Proteins can form edible films with excellent mechanical strength and gas barrier properties, making them ideal for use in food products. Gelatin, whey protein, soy protein, and casein are all examples of proteins that can create cohesive and flexible films that offer a protective barrier for food items. These protein coatings are commonly used on fruits, vegetables, and meat products to enhance their shelf life and quality. Gelatin-based coatings have been extensively researched due to their exceptional film-forming properties and their ability to improve various types of food products. By utilizing these proteins to create protective coatings, the industry can ensure that food products remain fresh for longer periods while also retaining their desirable qualities [10].

9.2.2 Polysaccharide-Based Coatings

Polysaccharides are a type of material made from carbohydrates that are commonly used in biopolymer coatings. These materials contribute to the film-forming ability and flexibility of the coatings. Various types of polysaccharides are used, including cellulose derivatives like methylcellulose or carboxymethylcellulose, chitosan (which is derived from chitin), alginate (which comes from seaweed), pectin (which is obtained from fruits), and starch (which is sourced from different plants). Polysaccharides have excellent properties when it comes to water vapor and gas barrier capabilities. They also provide good film-forming ability, flexibility, and gas barrier

Table 9.1 Types of coating, composition properties and measurement technique of the properties of edible films.

Types of coatings	Source used	Properties	Test and measurement	Reference
Protein-based coating or films	Animal-based proteins: gelatin, casein, whey protein, and egg albumin Plant-based proteins: soy protein and zein	Film adhesion	The test involves measuring the force required to separate the protein film from the substrate. The adhesion strength is then quantified in terms of Newtons (N) or pounds-force (IBF) per unit width.	[15]
Starch-based coating or films	Corn Starch (from the endosperm of corn kernels), Potato Starch (obtained from the tubers of potatoes), Cassava Starch (derived from the root of the cassava plant), Wheat Starch (extracted from the endosperm of wheat grains), Rice Starch (Rice starch is obtained from rice grains)	Mechanical strength	Tested and measured using methods such as tensile testing, compression testing, puncture or penetration testing, flexural testing, and adhesion testing. These tests assess properties like tensile strength, compressive strength, puncture resistance, hardness, flexural strength, flexural modulus, and adhesion strength.	[16]
Cellulose-based film or coating	Wood pulp, cotton, bamboo, agricultural residues, algae, microorganisms	Elongation	Typically measured using a tensile testing method. In this test, a sample strip or film of the edible coating is subjected to controlled tension until it breaks or reaches a specified strain. The elongation of the coating is measured by quantifying the change in the length of the sample during the test. The elongation at break or the maximum elongation achieved before failure indicates the coating's flexibility and stretchability.	[17]

(Continued)

Table 9.1 Types of coating, composition properties and measurement technique of the properties of edible films. (*Continued*)

Types of coatings	Source used	Properties	Test and measurement	Reference
Lipid-based coating or film	Vegetable oil, animal fat, dairy fat, cocoa butter, beeswax,	Water vapor permeability	Typically tested using the gravimetric method. A coated sample is placed in a test chamber with controlled humidity gradients. The weight change of the sample over time is monitored, and the water vapor permeability is calculated using equations like Fick's Law of Diffusion. The results are expressed as water vapor transmission rate (WVTR), indicating the amount of water vapor passing through the coating per unit area and time. This measurement helps assess the coating's ability to regulate moisture and preserve food quality.	[18]
Composite	Combination of edible coating	Control gas exchange	Gaseous exchange oxygen, CO_2 permeability testing, respiration rate measurement, Water vapor transmission rate were analyzed by specific testing devices.	[19]

properties. Chitosan, which is derived from chitin, has additional antimicrobial properties that make it useful in inhibiting microbial growth in food products [11]. Alginate-based coatings have been used to preserve fruits and vegetables due to their ability to control respiration rates and reduce water loss [12].

9.2.3 Lipid-Based Coatings

Lipid-based materials have been widely utilized to enhance the water resistance and gloss of biopolymer coatings. These materials consist of various types of lipids, including waxes (such as beeswax, carnauba wax, or shellac), oils (such as vegetable oils or mineral oil), and fatty acids (such as stearic acid or palmitic acid). The hydrophobic properties of lipids play a crucial role in improving the appearance and texture of coated food products. Among all lipid-based materials, beeswax and carnauba wax are commonly used in the food industry due to their exceptional water-repellency and preservation effects. These waxes are known for their ability to resist moisture and prevent microbial growth on food surfaces, thereby extending the shelf life of various food products. In addition to their functional benefits, lipid-based materials can also contribute to the sensory experience of consumers. For instance, these materials can impart a smooth and shiny appearance to coated foods while maintaining their natural flavor and aroma [13].

9.2.4 Composite Coatings

Composite coatings are a novel approach in the field of materials science, which involves combining multiple biopolymers to create a new material with enhanced properties. One such example is the combination of proteins and polysaccharides or proteins and lipids, which can lead to the development of films that exhibit improved mechanical strength, water vapor barrier, and antimicrobial properties. These composite coatings offer an incredibly versatile method for tailoring specific characteristics to suit a wide range of food products. By selecting different combinations of biopolymers, it is possible to obtain a coating that precisely meets the requirements of each product. This technology has great potential for transforming the way to produce and package foods, making them safer, more durable, and longer-lasting while also reducing waste and improving sustainability. With ongoing research and development in this area, it seems likely that composite coatings will continue to play an increasingly important role in food science [14].

9.3 Dual Role of Films/Coating in Food Industry

Edible biopolymers serve a dual purpose as both packaging and preservative materials in the food industry. They can be processed into films,

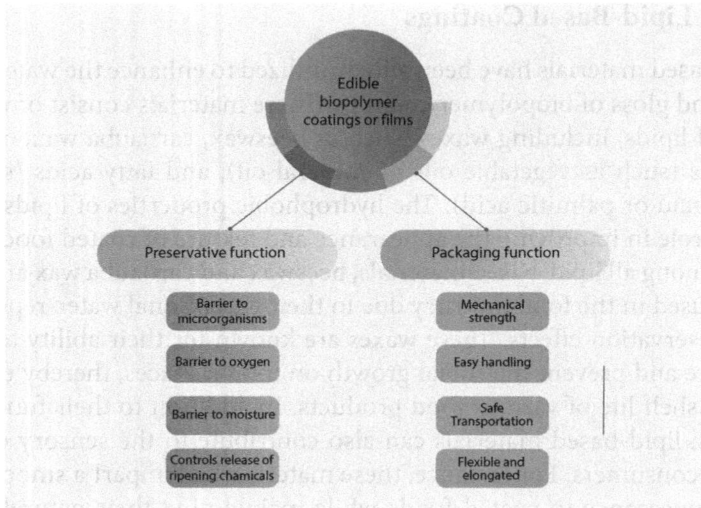

Figure 9.1 Dual role of films/coating in the food industry.

coatings, wraps, pouches, bags, trays, and containers to meet the specific packaging needs of various food products. These materials act as protective barriers, preventing moisture loss, and safeguarding against external contaminants [20]. They regulate moisture levels, maintaining freshness and preserving the desired texture of the food. Edible biopolymers also exhibit gas barrier properties, reducing the permeation of oxygen and carbon dioxide [21]. This slows down oxidative reactions, extends shelf life, and preserves the carbonation of beverages. Additionally, certain edible biopolymers possess inherent antimicrobial properties, inhibiting the growth of spoilage-causing microorganisms. They can also be utilized for the controlled release of additives, such as antioxidants and flavoring agents, further enhancing preservation and sensory qualities. By combining packaging and preservative functionalities, edible biopolymers offer sustainable and efficient solutions that reduce reliance on non-biodegradable packaging materials, contributing to a more environmentally friendly food industry. The scope of the edible films and coatings is elaborated in detail in Figure 9.1.

9.3.1 Shelf Life Extension

Biopolymer coatings are a highly effective solution for preserving perishable foods. These coatings serve as a barrier that shields food products from external factors that can cause spoilage. By minimizing moisture loss and reducing gas exchange, biopolymer coatings create an optimal environment for food preservation. Additionally, they inhibit the growth of harmful microorganisms, which can drastically extend the shelf life of perishable foods. Biopolymer coatings work by forming a protective layer around the food product, preventing it from coming into contact with moisture and oxygen that can cause spoilage. This is particularly useful for products that are highly sensitive to environmental factors, such as fruits and vegetables. By creating an ideal microclimate around the produce, biopolymer coatings can keep them fresh for much longer than traditional storage methods [22].

9.3.2 Quality Retention

Biopolymers are a type of coating that can serve as an effective means to preserve the sensory qualities of food, such as texture, color, flavor, and aroma. By acting as a barrier against dehydration, oxidation, and other chemical reactions that contribute to the decline in food quality over time, biopolymer coatings have the potential to greatly enhance consumer satisfaction while simultaneously decreasing the amount of wasted food. Through their unique properties and composition, these coatings provide an excellent solution for maintaining the integrity and freshness of various types of food products. Whether used in the production of fruits and vegetables or the packaging of meat and dairy items, biopolymer coatings offer a versatile and sustainable approach to preserving food quality for extended periods. With their ability to prevent spoilage and maintain optimal conditions for consumption, biopolymer coatings represent a promising avenue for addressing some of the most pressing challenges facing our global food system today [23].

9.3.3 Natural and Safe

Biopolymer coatings are a type of coating that is commonly made from natural ingredients, such as proteins, polysaccharides, and lipids. These coatings are well-known for being safe for consumption and biodegradable, which makes them an excellent alternative to synthetic packaging materials. Additionally, they offer an eco-friendly solution that is sustainable and environmentally friendly. As consumers become more aware of the impact of their choices on the environment, they may start to prefer edible coatings as a more natural and sustainable option for their packaging needs.

With this in mind, it's clear that biopolymer coatings have a lot to offer in terms of both safety and sustainability [24].

9.3.4 Customization and Versatility

Biopolymer coatings offer a highly adaptable solution to meet the specific needs of diverse food products. These coatings can be customized in terms of their composition and thickness, allowing them to cater to the unique requirements of fruits, vegetables, meats, or bakery items [25]. This versatility enables the development of tailored preservation solutions that are optimized for each product type. The ability to adjust the coating characteristics also enhances product compatibility, leading to improved quality and consumer satisfaction. By harnessing the benefits of biopolymer coatings, food manufacturers can create innovative solutions that maintain product freshness and extend shelf life while meeting consumer demands for high-quality, sustainable products.

9.3.5 Reduction of Packaging Waste

Biopolymer coatings have emerged as a promising solution to reduce the use of additional packaging materials or lessen dependence on non-biodegradable ones. These coatings can be used in the form of edible films, which not only enhance the shelf-life of food products but also minimize the amount of packaging waste generated. By adopting such an approach, we can contribute to promoting sustainability and reducing our carbon footprint [26]. Moreover, biopolymer coatings can also provide an effective barrier against oxygen and moisture, thereby protecting food from spoilage and contamination. This innovation has tremendous potential to revolutionize the way we package our food items and promote a more environmentally conscious future.

9.3.6 Enhanced Nutritional Value

Biopolymer coatings are an innovative solution to preserve the nutritional content of food products. These coatings work by minimizing nutrient losses that may occur during storage and transportation. They act as a protective barrier against harmful factors such as oxygen and light, which can cause the degradation of vitamins and other sensitive compounds in food. In this way, biopolymer coatings help to maintain the nutritional value of food products for longer periods, ensuring that they remain fresh and wholesome when consumed. By reducing the need for preservatives, these coatings offer a more natural alternative to traditional methods of food preservation. Furthermore, biopolymer coatings are environmentally friendly, as they are made from renewable resources and are fully biodegradable. With their ability to enhance the shelf life and quality of food products while preserving their nutritional benefits, biopolymer coatings represent a promising advancement in the field of food science and technology [27].

9.3.7 Marketing and Differentiation

Edible coatings are a versatile and innovative way to differentiate food products in the market. These coatings offer several benefits that can enhance the visual appeal of food items, create unique textures, and add distinctive elements that attract consumer attention. By using edible coatings, manufacturers can provide an opportunity for product differentiation and marketing, which ultimately contributes to a positive product image. The use of such coatings is not limited to a specific type of food product; rather, they can be applied to various foods such as fruits, vegetables, meat, poultry, seafood, confectionery products, and more. With their ability to improve the overall appearance and texture of food products while offering exciting flavor combinations and packaging options, edible coatings are fast becoming one of the most sought-after methods for boosting sales in the highly competitive food industry.

9.4 The Mechanisms of Biopolymer-Based Preservation

Biopolymer coatings and films accomplish all of the aforementioned functions through a variety of mechanisms, which operate in unison or independently to prolong the shelf life of diverse food items while safeguarding their excellence, novelty, and security. In order to preserve the foods the moisture barrier mechanism, gas barrier mechanism, microbial growth inhibition mechanism, and controlling ripening mechanism are the essential factors.

9.4.1 Moisture Barrier Mechanism

Biopolymer coatings are a pioneering solution that can proficiently inhibit moisture transfer from food products, thereby diminishing water loss and minimizing dehydration. By retaining the ideal moisture levels, biopolymer coatings facilitate preserving the quality and texture of food items, which is particularly beneficial for fruits, vegetables, and bakery products. These coatings serve as a potent shield that can impede wilting, softening and staling processes that occur naturally over time. Consequently, foods coated with biopolymers stay fresh for extended periods compared to those without such coatings [28].

9.4.2 Gas Barrier Mechanism

Biopolymer coatings are a type of protective layer that can be used to prevent gases like oxygen and carbon dioxide from penetrating fresh produce. By regulating the respiratory rate of produce, these coatings can slow down the process of respiration, which in turn reduces spoilage due to accelerated respiration. Moreover, biopolymer coatings can limit the amount of oxygen available, thereby preventing oxidative reactions that lead to the degradation of food quality. This includes undesirable changes in color, lipid oxidation, and off-flavor development. With these benefits, biopolymer coatings have become an important tool in extending the shelf life of fresh produce and maintaining its overall quality [29].

9.4.3 Microbial Growth Inhibition Mechanism

Biopolymer coatings are becoming increasingly popular due to their ability to possess antimicrobial properties. This is highly beneficial for perishable food products, such as meats, seafood, and fresh-cut fruits and vegetables. By inhibiting the growth of spoilage microorganisms and foodborne pathogens, these coatings can significantly reduce microbial spoilage and delay the onset of foodborne illnesses. In addition, they can extend the shelf life of these products, ensuring that they remain fresh for longer periods. The antimicrobial activity of edible coatings is especially important in today's world, where food safety is a major concern for consumers and manufacturers. With biopolymer coatings, it is possible to preserve the quality and safety of perishable foods while also reducing waste and improving sustainability. Overall, biopolymer coatings represent an exciting new advancement in the field of food science that has the potential to revolutionize the way we store and consume our favorite [30].

9.4.4 Controlling Ripening Mechanism

Biopolymer coatings have the remarkable ability to regulate and control the release of ethylene, a naturally occurring plant hormone that plays a vital role in fruit ripening. These biodegradable coatings are highly effective in slowing down the ripening process of fruits by controlling the diffusion of ethylene. The regulation of ethylene release is critical as it helps to delay over-ripening, softening, and senescence, thereby extending the shelf life of fruits. This technology is particularly beneficial for fruits that are highly sensitive to ethylene such as bananas, avocados, and tomatoes. By using edible coatings made from biopolymers, farmers can now preserve their fruits for longer periods while maintaining their quality and freshness. This development is a significant breakthrough in the agriculture industry as it not only reduces food waste but also provides consumers with access to fresh produce all year round [31].

9.5 The Role of Biopolymers in Physical Protection During Packaging

Biopolymer coatings are an effective way to protect food products from physical damage, which can occur during handling, transportation, and storage. These coatings serve as a physical barrier that helps minimize abrasions and impacts, thereby reducing the likelihood of spoilage caused by external factors. By preserving the integrity and visual appearance of the product, biopolymer coatings help ensure that it remains fresh for longer periods. In addition, these coatings are edible, meaning that they are safe for consumption and do not pose any health risks to consumers. Overall, biopolymer coatings offer a practical solution for ensuring the quality and safety of food products throughout their lifespan [32].

9.5.1 Properties of Edible Films and Coating as Physical Packaging Material

Biopolymer coatings or films play a crucial role in protection, and sensory enhancement. Several important properties of biopolymer coatings and films affect their performance and suitability for various food applications. The three key properties are mechanical strength, water vapor permeability, and film adhesion. The evaluation of barrier properties, shelf life extension, sensory acceptance, and safety is conducted through a combination of laboratory tests, analytical measurements, and sensory assessments.

Manufacturers and researchers use these performance evaluations to optimize formulation, select suitable coating materials, and ensure the functionality and safety of biopolymer coatings for specific food applications.

9.5.2 Mechanical Strength

The mechanical strength of a biopolymer coating or film refers to its ability to withstand mechanical stresses, such as handling, transportation, or packaging. The mechanical strength of biopolymer coatings can be assessed using techniques like tensile strength testing or puncture resistance testing. These tests measure the coating's resistance to forces like stretching, tearing, or puncturing, which simulates real-world handling conditions. A strong and durable coating protects the food product, preventing physical damage, and maintaining its integrity. The mechanical strength of biopolymer coatings can be influenced by the choice of materials, their concentration, and the application method. It is important to strike a balance between sufficient strength and flexibility to ensure adequate protection without compromising the texture or quality of the food [16].

9.5.3 Water Vapor Permeability

Water vapor permeability is a critical property of biopolymer coatings and films that affects moisture transfer between the food product and its environment. This property can be evaluated using methods such as gravimetric analysis, where the weight change of a coated sample is measured under controlled humidity conditions over time. The permeability of the coating regulates the rate of moisture loss or gain, which is important for maintaining product quality, texture, and shelf life. In some cases, a higher water vapor permeability is desired to allow excess moisture to escape, preventing the formation of condensation and maintaining the crispness of certain food items. Conversely, lower permeability may be preferred to reduce moisture loss and extend the shelf life of products that require moisture retention [33].

9.5.4 Film Adhesion

Film adhesion refers to the ability of the biopolymer coating or film to adhere to the surface of the food product. Strong adhesion ensures that the coating remains intact during handling, transportation, and storage, providing effective protection and retaining the desired attributes. Good adhesion prevents the coating from peeling, cracking, or detaching from the food surface. Factors influencing film adhesion include the compatibility of the coating material with the food substrate, surface preparation techniques, and application methods employed [34]. this can be measured by Barrier properties that determine the coating's ability to block the transfer of gases, such as oxygen or carbon dioxide, and protect the food product. Techniques like gas permeability testing or oxygen transmission rate (OTR) measurements can be used to evaluate the coating's barrier performance.

These properties are interrelated and must be carefully balanced to meet the specific requirements of the food product. For example, a strong coating with excellent mechanical strength may have lower water vapor permeability, which can impact the moisture content and texture of the food. Thus, the selection and formulation of edible coatings should consider a combination of properties that best suit the desired food product, including its sensitivity to moisture, desired shelf life, handling conditions, and sensory attributes. It's worth noting that the desired properties may vary depending on the specific application. For instance, a coating for fresh produce may require high water vapor permeability to maintain freshness, while a coating for a dry snack may need to provide better mechanical strength for protection during transportation and handling. Therefore, understanding the requirements of the food product and tailoring the properties of the edible coatings or films accordingly is crucial for successful application and performance.

9.6 Biopolymer Coatings and Films for Different Classes of Foods

Perishable, semi-perishable and non-perishable foods are classes of foods based on their shelf life and susceptibility to spoilage. It's important to note that the classification of foods into these categories can vary based on factors such as storage conditions, processing techniques, and regional definitions. Therefore, the categorization of specific foods may differ slightly in different contexts. Perishable foods are a category of food items that have a limited shelf life and are prone to microbial growth, enzymatic reactions, and physical degradation. These types of food items require specific storage conditions, such as refrigeration, to preserve their freshness and safety for consumption. Examples of perishable foods include fresh fruits and vegetables, dairy products (such as milk and yogurt), raw meats, seafood, and baked goods without any added preservatives. Hence, the non-perishable foods are an essential part of any food supply chain, as they offer a long-lasting and reliable source of sustenance. These types of foods are known for their ability to resist spoilage and have prolonged shelf lives, making them ideal for emergencies or long-term storage. Non-perishable foods undergo various processing techniques, including canning, drying, or freezing, which help to maintain their freshness and quality. One of the most significant benefits of non-perishable foods is that they do not require refrigeration and can be stored at room temperature without spoiling. The semi-perishable foods are a type of food that falls between perishable and non-perishable foods. They have a moderate shelf life and can spoil if not stored properly, but they can last longer than perishable foods. These foods require specific storage conditions such as refrigeration to maintain their

Table 9.2 Edible coating and films for the preservation of various types of foods.

Food category	Examples	Recommended biofilm or coating	References
Fresh fruits	Apples, berries, and citrus fruit	The thin layer of edible wax or a biodegradable coating made from materials like carnauba wax or shellac, PLA coatings, and Chitosan coatings.	[41, 42]
Leafy greens and vegetables	lettuce, spinach, and herbs	Biodegradable film or spray coating made from materials like chitosan or cellulose, PHA.	[43]
Fresh Meats and Poultry	Beef, chicken, and pork	Protein-based films and PLA coatings.	[44]
Seafood	fish, shrimp, shellfish	Moisture-resistant biodegradable films or wraps made from materials like PLA or chitosan.	[45]
Bakery Products	Bread, cakes, and pastries	The biodegradable film, bag, or paper made from materials like PLA or cellulose (Starch-based films and Cellulose-based films).	[46]
Dairy products	Yogurt, cream, cheese	Protein-based films (whey, collagen) and PLA coatings.	[47]

quality and safety. Preserving perishable and semi-perishable foods products is crucial to maintaining their quality, extending their shelf life, and minimizing waste. The selection of coating or film materials for this purpose depends on various factors such as the type of food, desired preservation method, packaging requirements, and regulatory considerations. Here are some specific biodegradable coating or film materials commonly recommended for preserving these food products (Table 9.2).

9.6.1 Polylactic Acid

Polylactic acid (PLA) is a versatile and eco-friendly biodegradable polymer that can be derived from renewable sources like cornstarch, sugarcane, or other plant-based materials. This innovative material offers impressive barrier properties against gases and moisture, making it an ideal choice for various packaging applications. One of the most popular uses of PLA films is for packaging fresh produce, bakery items, and snacks due to its ability to preserve the quality and freshness of the contents. Additionally, PLA films can be transparent or opaque depending on specific application requirements. The transparency feature allows consumers to easily see the product inside while also providing an excellent marketing opportunity for companies to showcase their products in an aesthetically pleasing way. Overall, PLA films are an excellent sustainable packaging option that not only helps reduce waste but also offers superior performance compared to traditional plastic packaging materials [35].

9.6.2 Polyhydroxyalkanoates

Polyhydroxyalkanoates (PHA), which stands for polyhydroxyalkanoate, is a type of biodegradable polymer that is produced by microorganisms using renewable resources. This remarkable material boasts exceptional barrier properties that make it an ideal choice for use as a coating or film in the food industry. PHA films are not only effective at preserving the freshness and quality of fruits, vegetables, meats, and dairy products but also offer a sustainable alternative to traditional plastic packaging. Thanks to its biodegradability, PHA can help reduce the environmental impact caused by plastic waste and contribute to the development of a more eco-friendly future. With these benefits in mind, it's no wonder that PHA has gained recognition as an innovative solution for food packaging needs [36].

9.6.3 Cellulose-Based Films

Cellulose, a naturally occurring polymer, is commonly found in the cell walls of plants. This remarkable substance can be processed into biodegradable films that possess excellent gas and moisture barrier properties. These cellulose-based coatings or films have found widespread use in wrapping fresh produce, sandwiches, and snacks. They are derived from a variety of sources such as wood pulp, cotton, or agricultural waste. Thanks to their eco-friendly nature and superior performance characteristics, cellulose-based materials are becoming increasingly popular in the packaging industry as an alternative to traditional plastic products. The versatility of this material makes it an ideal candidate for a wide range of applications that require high-performance packaging solutions while also being environmentally sustainable. With the growing demand for eco-friendly alternatives to conventional packaging materials, cellulose-based products are poised to play an increasingly important role in meeting the needs of consumers who care about both quality and sustainability [37].

9.6.4 Chitosan-Based Films

Chitosan, a biopolymer that is naturally derived from chitin, can be found in the shells of crustaceans such as shrimp and crabs. It has been discovered that chitosan-based films possess antimicrobial properties which make them effective in inhibiting the growth of spoilage microorganisms. These films are highly versatile and can be used for packaging seafood, fruits, and vegetables, thereby helping to extend their shelf life. By doing so, they help to reduce waste and minimize food loss in the supply chain. The effectiveness of chitosan-based films has been demonstrated in various studies and they have proven to be a viable alternative to traditional packaging materials. With their ability to preserve the quality of perishable products while also reducing the need for preservatives or other additives, chitosan-based films represent an innovative solution to an age-old problem. As research continues in this area, we will likely see even more exciting developments emerge in the field of sustainable packaging technologies [38].

9.6.5 Starch-Based Films

Starch is an abundant and sustainable biopolymer that can be transformed into films with a range of applications. These films, which are made from starch, possess moderate barrier properties and are suitable for packaging a variety of dry goods, snacks, and certain perishable food items. To optimize their performance, these films are frequently combined with other biodegradable polymers. As such, they constitute an eco-friendly alternative to traditional packaging materials that might otherwise contribute to pollution and waste. By harnessing the versatility and availability of starch-based films, we can take steps toward building a more sustainable future for our planet [39].

9.6.6 Protein-Based Films

Proteins derived from different sources, including soy, whey, and collagen, are highly effective in creating biodegradable films. These films exhibit excellent mechanical properties and can serve as a reliable protective barrier for semi-perishable food items. Protein-based films are commonly used in the packaging of meat and poultry products owing to their superior quality and durability. The use of such eco-friendly materials not only ensures that the food stays fresh for longer periods but also helps reduce plastic waste. Therefore, incorporating protein-based films in food packaging is a sustainable solution that benefits both the environment and consumers alike [40].

9.6.7 Biopolymer Coatings or Films for Nonperishable Foods

To maintain biodegradability and address various functions, a multitude of coatings or films can be employed for non-perishable food packaging. Chitosan-based, Alginate-based, Lipid-based, Protein-based, and Polysaccharide-based coatings are utilized to tackle the following:

i) Protection against external factors: Edible coatings offer an additional layer of defense against physical harm, dust particles, or handling during transportation and storage. Despite being immune to spoilage, non-perishable foods remain vulnerable to damage from external contaminants; however, coatings may help in reducing such risks.
ii) Aesthetic enhancements - Edible coatings can enhance the visual appeal of non-perishable food products by providing an attractive and consistent appearance. This is especially significant when it comes to items displayed on store shelves since appealing packaging has considerable influence over consumer purchasing decisions.
iii) Functional features - Edible coatings can acquire functional properties like improved barrier characteristics

towards light transfer, gas penetration, or odor diffusion. These aspects play a pivotal role in preserving the sensory attributes of the product while preventing any loss in flavor or aroma thereby extending the overall quality of the non-perishable item at hand.

9.7 Application Techniques

Various techniques can be employed to apply biodegradable biopolymer coatings or films onto food products. The selection of the application method hinges on the properties of the coating or film material, the specific food product, and the intended outcome. It is crucial to adhere to appropriate hygiene and safety protocols while applying biodegradable biopolymer coatings or films to guarantee food safety. Moreover, manufacturers ought to take into account the particular requirements and suggestions provided by suppliers of coating materials for optimal application outcomes. Below are some widely used application methods are discussed in Table 9.3.

Table 9.3 Various types of application techniques.

Techniques	Method of application	Type of food	References
Dipping	Immersing the food item in a solution of biodegradable substance and allowing any excess solution to drip off and dry	Fruits, vegetables, and confectioneries	[48]
Spraying	Uniform application of a coating or film material onto the surface of food items using a spray gun or atomizer	Baked goods, Snacks, and meat	[27, 49]
Brushing	Applying a biodegradable coating or film material onto food products using a brush or similar applicator	Pastries, chocolates, fruits	[50]

(*Continued*)

Table 9.3 Various types of application techniques. (*Continued*)

Name of the application	Type of the method	Type of food	References
Vacuum infusion	Placing the food products and a biodegradable coating material in a vacuum-sealed container. By removing air from the Container a pressure difference is created and coating penetrate the food	Fruits, vegetables, meat	[51]
Film forming	Applying flims or sheets formed by the biopolymers onto the food stuffs	Bakery products, pies or pastries,	[52]
Electrostatic coating	Highly effective method of applying biodegradable coatings to food products through the use of an electrostatic charge. This innovative technique involves charging the coating material	Snacks, confectioneries, and dried goods	[53]
3D printing	Innovative technology of 3D printing is poised to revolutionize food industry by allowing the deposition of biodegradable biopolymer coating or film onto food products	Confectioneries, pastries, decorative food	[54]
Electrospray	Electric field to break down a liquid into tiny droplets this process involves the application of electric charge to biodegradable coating material and the specialized nozzle	Delicate food products	[55]
Nano Technology coating	Immense promise for the development of biodegradable biopolymer coatings or films at the nanoscale with the use of nanoparticles or nanostructure	Packed foods	[56, 57]

(*Continued*)

Table 9.3 Various types of application techniques. (*Continued*)

Name of the application	Type of the method	Type of food	References
Spray freeze drying	Merges the benefits of spraying and freeze-drying methods to create coating or film on food items. In the process the food product is sprayed with the biodegradable coating material	Frozen foods	[58]
Magnetic field alignment	Application of a magnetic field to a suspension of biodegradable coating particles	Food production	[59]

9.7.1 Dipping

Dipping is a widely utilized and uncomplicated technique that involves coating or filming food products. This process entails immersing the food item in a solution or suspension of biodegradable coating substance, allowing any excess solution to drip off, and subsequently drying or curing the product to form a coating. Dipping is particularly suitable for small- to medium-sized food items such as fruits, vegetables, and confectioneries. It is an efficient method that ensures uniformity in coating thickness and coverage while maintaining the product's quality and freshness. Additionally, dipping can enhance the food's shelf-life by protecting it from external factors such as moisture loss, oxidation, or microbial spoilage [48].

9.7.2 Spraying

Spraying is a food processing technique that involves the uniform application of a coating or film material onto the surface of food items using a spray gun or atomizer. This process is achieved by spraying a biodegradable solution or suspension evenly over the entire surface of the food product, ensuring complete coverage. Spraying is an ideal method for larger-sized food items such as baked goods, snacks, and meat products because it allows for efficient and uniform application of the coating material. By using this technique, manufacturers can achieve consistent results in terms of both texture and flavor profiles on their products [27, 49].

9.7.3 Brushing

Brushing is a manual process of applying a biodegradable coating or film material onto food products using a brush or similar applicator. This technique is typically used for smaller food items or delicate products that necessitate precise and controlled application. Brushing offers detailed coverage and is frequently employed to apply coatings to pastries, chocolates, or fruits. The method entails dipping the brush into the coating solution and then gently brushing it onto the product surface in a uniform manner, ensuring complete coverage. This approach is particularly useful for creating intricate designs on food items, as it allows for greater control over the application process. Brushing also enables better adhesion of coatings to food surfaces, resulting in improved flavor, texture, and appearance [50].

9.7.4 Vacuum Infusion

Vacuum infusion is a fascinating process that has been used extensively in the food industry to improve the quality and shelf life of fruits, vegetables, and meat products. This technique involves placing the food product and a biodegradable coating material in a vacuum-sealed container. By removing air from the container, a pressure differential is created that helps the coating material penetrate the food product uniformly. The benefits of vacuum infusion are numerous. Firstly, it ensures that the food product's surface is evenly coated with the coating material, resulting in an attractive appearance. Secondly, it enhances the product's flavor by locking in moisture and nutrients while preventing oxidation. Thirdly, it extends its shelf life by inhibiting bacterial growth and spoilage. Vacuum infusion can be used for various types of foods - from fresh produce to processed meats - depending on their desired outcome. For instance, fruits and vegetables can be infused with natural flavors or preservatives to enhance their taste and freshness. Meat products can be infused with marinades or spices to add flavor and tenderness [51].

9.7.5 Film Forming

Biodegradable coating materials offer a sustainable solution for the food industry. One of the ways to apply these coatings is by forming them into films or sheets that can be directly placed onto food products. These preformed films can adhere to the surface of the food item either through natural adhesion or by using food-safe adhesives. This film-forming technique is particularly useful for bakery products, such as pies or pastries, where a protective layer is desired to maintain freshness and extend shelf life. By using biodegradable coatings, we can reduce our reliance on single-use plastics and contribute to a more environmentally friendly approach in the food industry [52].

9.7.6 Electrostatic Coating

Electrostatic coating is a highly effective method of applying biodegradable coatings to food products through the use of an electrostatic charge. This innovative technique involves charging the coating material and then spraying it onto the food item, creating a powerful attraction between the charged particles and the surface of the food. This process ensures that the coating is evenly distributed across the entire surface area of the food product, resulting in a uniform and consistent finish. One of the key benefits of electrostatic coating is its versatility, as it can be used on a wide range of food products including snacks, confectioneries, and dried goods. In addition, this technique offers numerous advantages over traditional coating methods such as dipping or brushing. For example, the electrostatic coating is faster and more efficient than other methods, reducing production time and costs [53].

9.7.7 Potential Techniques

Potential techniques refer to innovative ways of applying biodegradable biopolymer coatings or films onto food products that are not commonly used at present but have the potential to be effective. These techniques are being explored by researchers and scientists to find more sustainable and environmentally-friendly methods of preserving and packaging food. By utilizing these potential techniques, it may be possible to reduce the use of harmful chemicals and materials in the food industry while still maintaining the quality and safety of food products. With further research and development, these techniques could become a valuable tool in promoting sustainability and reducing waste in our food supply chain.

9.7.8 3D Printing

The innovative technology of 3D printing is poised to revolutionize the food industry by allowing the deposition of biodegradable biopolymer coatings or films onto diverse food products. By leveraging food-grade materials in 3D printers, intricate patterns, and designs could be created on the surface of food items, all while providing a protective and visually appealing coating. This technique holds great promise for confectioneries, pastries, and decorative food items that require an extra touch of creativity and elegance. With 3D printing technology, the possibilities for enhancing the presentation and preservation of food are limitless [54].

9.7.9 Electrospray

Electrospray, a fascinating technique, utilizes an electric field to break down a liquid into tiny droplets. This process involves the application of an electric charge to biodegradable coating material and the use of a specialized nozzle. Consequently, the charged droplets can be sprayed onto food products in a fine mist that results in a thin and uniform coating. Electrospray provides exceptional control over droplet size and distribution, making it ideal for delicate food products that require precision. Overall, this innovative technique offers tremendous potential for enhancing food quality and safety by ensuring consistent coatings on products [55].

9.7.10 Nanotechnology Coatings

Nanotechnology is a field that holds immense promise for the development of biodegradable biopolymer coatings or films at the nanoscale. With the use of nanoparticles or nanostructures, it is now possible to create coatings that have improved barrier properties, antimicrobial effects, or enhanced stability. This opens up new avenues for creating innovative food products that are not only safe and healthy but also sustainable. Various techniques such as nanoparticle deposition and self-assembly can be explored to apply these nanotechnology-based coatings to food products. By doing so, we can ensure that our food remains fresh and safe for consumption for longer periods while reducing the environmental impact caused by traditional packaging methods. In addition, the use of nanotechnology in food packaging also has potential benefits in terms of reducing food waste by extending shelf life and improving product quality. It is an exciting time for the field of nanotechnology as we continue to discover new ways to improve our lives and address some of the most pressing challenges facing humanity today [56, 57].

9.7.11 Spray Freeze Drying

Spray freeze-drying is a novel technique that merges the benefits of spraying and freeze-drying methods to create coatings or films on food items.

In this process, the food product is sprayed with the biodegradable coating material, which is then rapidly frozen. Afterward, the water content is removed through freeze-drying, resulting in a thin, dry coating on the surface of the food item. This innovative technique has immense potential for preserving the texture, flavor, and overall quality of delicate food products that are prone to damage during preservation. With spray freeze-drying, food manufacturers can ensure that their products remain fresh and visually appealing for prolonged periods without compromising their nutritional value or taste. This technology offers an exciting new avenue for enhancing the shelf life of perishable foods while reducing waste and boosting sustainability in the food industry [58].

9.7.12 Magnetic Field Alignment

Magnetic field alignment is a fascinating process that involves the application of a magnetic field to a suspension of biodegradable coating particles. This technique allows for the manipulation of the orientation of the magnetic particles, leading to their alignment and deposition onto the food product surface. The resulting coating possesses specific properties that can enhance its barrier properties or target functional effects on the food surface. The use of this innovative technology in food production has been gaining significant attention due to its potential benefits. By aligning and depositing magnetic particles onto food products, it is possible to improve their shelf life by enhancing their barrier properties against external factors such as moisture and oxygen. Additionally, this technique could be explored to create coatings with specific functional effects, such as antimicrobial or antioxidant properties. The process of magnetic field alignment involves several steps, including the preparation of a suspension containing biodegradable coating particles and the application of an external magnetic field to this suspension. The orientation of the magnetic particles is then manipulated using various techniques, such as changing the direction or intensity of the magnetic field [59].

9.8 Current Challenges and Future Perspectives

Biopolymer coatings and films have become increasingly popular in the food industry due to their many benefits, such as enhancing food preservation and improving quality. However, it is important to note that these innovative materials also face certain challenges and limitations that must be taken into consideration. It is crucial to address these challenges to fully maximize the potential of biopolymer coatings and films. Some of the current challenges include:

9.8.1 Limited Barrier Properties

Biopolymer coatings are a popular alternative to traditional synthetic packaging materials due to their eco-friendliness and biodegradability. However, it is important to note that biopolymer coatings may have lower barrier properties compared to their synthetic counterparts. This means that while they can offer some protection against moisture, gases, and microorganisms, their effectiveness as a barrier may not be as high. As a result, highly perishable or sensitive food products may be at risk of reduced protection and shorter shelf life when packaged with biopolymer coatings. It is crucial for manufacturers and consumers alike to carefully consider the intended use and storage conditions of biopolymer-coated packaging to ensure optimal performance and maximum product safety [60].

9.8.2 Sensory Acceptance

The properties of biopolymer coatings, including taste, texture, and appearance, may exhibit differences from traditional packaging materials. These sensory attributes can impact consumer preferences and acceptance levels, ultimately influencing the market adoption of such coatings. The development of coatings that maintain or even improve upon sensory qualities while avoiding any negative impact on the food product itself is a significant challenge to overcome. As such, it requires careful consideration and attention to detail to ensure successful implementation in commercial applications [61].

9.8.3 Stability and Shelf Life

The stability of biopolymer coatings can be a concern during storage, as they may be susceptible to moisture sensitivity, loss of film integrity, and reduced mechanical strength. It is essential for the coating to maintain its effectiveness throughout the desired shelf life of the food product. Furthermore, it is crucial to address the stability of the coating material itself, including potential issues such as polymer degradation or ingredient interactions.

9.9 Possible Solutions and Ongoing Research to Address the Challenges

9.9.1 Solution for Formulation and Optimization

Ongoing research and development endeavors in the field of biopolymer coatings are aimed at optimizing their formulation. This involves carefully selecting suitable coating materials, concentrations, and additives to enhance the barrier properties, mechanical strength, and stability of the coatings. Through a comprehensive understanding of how different components interact with each other, coating formulations can be customized to match specific food products, resulting in enhanced effectiveness. These efforts are geared towards improving the overall quality of food products by ensuring that they remain fresh for longer periods while maintaining their desired taste and texture. By leveraging innovative techniques and technologies, scientists are continuously pushing the boundaries of what is possible when it comes to biopolymer coatings. The result is a growing body of knowledge that is helping to unlock new possibilities for enhancing the shelf life and safety of various food items [62].

9.9.2 Advanced Processing Techniques

When it comes to enhancing the application of biopolymer coatings, delving into advanced processing techniques can prove to be incredibly beneficial. A few examples of such techniques include electrostatic deposition, ultrasonic spraying, and vacuum coating - each of which brings its own set of advantages in terms of coating uniformity, control, and efficiency. By incorporating innovative methods and automation technologies into the process, one can further elevate the coating experience and guarantee consistent application throughout.

9.9.3 Combination Approaches

When it comes to packaging, some limitations can arise with the use of individual coating materials or techniques. However, by combining different materials or methods, these limitations can be overcome. For instance, incorporating multiple layers with varying functionalities or utilizing a combination of biopolymer coatings and traditional packaging materials can lead to improved barrier properties, greater mechanical strength, and extended shelf life for the product in question. By taking advantage of these possibilities, manufacturers can create packaging that is not only more effective at protecting its contents but also more versatile overall [63].

9.9.4 Active and Intelligent Coatings

Biopolymer coatings can be further enhanced by integrating active components like antimicrobial agents or antioxidants that improve their functionality. These active coatings are capable of controlling microbial growth, delaying oxidation, and maintaining the freshness of products. In addition to this, the development of intelligent coatings that respond to environmental stimuli is a promising approach for real-time monitoring and ensuring food safety. These smart coatings can detect changes in temperature, humidity, or pH levels and react accordingly to prevent spoilage or contamination of food products. With these innovative advancements in biopolymer coating technology, we can ensure that our food remains safe and fresh for longer periods while also reducing waste and improving sustainability [64].

9.9.5 Consumer Acceptance and Education

It is crucial to educate consumers about the advantages and safety of biopolymer coatings for them to be widely accepted. To achieve this, it is important to engage in transparent communication and provide extensive information regarding the purpose, safety, and sustainability aspects of edible coatings. By doing so, consumer trust can be built, and the adoption of these coatings can be promoted. It is essential to emphasize that biopolymer coatings are not only safe but also sustainable, as they are made from natural and environmentally friendly materials. Additionally, these coatings offer various benefits such as extending shelf life, improving the quality and appearance of food products, reducing waste, and protecting against contamination. Therefore, it is necessary to raise awareness among consumers about biopolymer coatings to help promote their widespread use in the food industry while ensuring public safety and environmental protection [65].

9.9.6 Collaboration and Knowledge Sharing

Effective collaboration is an essential element in addressing challenges and sharing knowledge among researchers, industry stakeholders, and regulatory bodies. The importance of fostering partnerships cannot be overstated, as it enables researchers to work closely with food manufacturers to develop customized solutions that meet specific product requirements while ensuring compliance with the necessary regulations. This collaboration facilitates the exchange of ideas and expertise, leading to enhanced innovation and improved outcomes for all parties involved. By working together, researchers can leverage each other's strengths and resources, resulting in more robust research findings and better products that benefit consumers and society as a whole. Therefore, creating an environment that supports collaborative efforts is vital for achieving success in today's complex scientific landscape [66].

9.9.7 Sustainability Initiatives

Ensuring sustainability in the development and utilization of biopolymer coatings is imperative for the betterment of our planet. To achieve this, it is crucial to delve deeper into the realm of biodegradable or compostable materials, which can greatly minimize waste during production. Moreover, optimizing end-of-life disposal options can significantly reduce the environmental impact of these coatings. By incorporating circular economy principles, we can further contribute to sustainability goals by utilizing by-products or agricultural waste as coating materials. This approach not only reduces waste but also promotes more efficient use of resources. Therefore, we must prioritize sustainability in every aspect of biopolymer coating development and usage [1].

9.9.8 Ongoing Research

In the realm of biopolymer coatings and films, researchers are constantly striving to overcome obstacles and enhance their performance in a multitude of ways. With this in mind, numerous potential areas of improvement continue to be the focus of ongoing studies. These areas include but are not limited to, durability, flexibility, adhesion, and barrier properties. Additionally, research is being conducted on methods for optimizing the production process and reducing costs associated with these materials. By exploring these different avenues of inquiry, scientists hope to develop more effective biopolymer coatings and films that can meet the demands of a wide range of applications across various industries.

9.9.9 Enhanced Barrier Properties

The improvement of barrier properties in biopolymer coatings is a crucial area of research that has garnered significant attention from scientists.

To achieve this, researchers are actively exploring various approaches to enhance the coatings' resistance to moisture, gases, and UV light. Among these techniques include the integration of nanoparticles, natural antimicrobial agents, or functional additives into the coatings. To create effective barriers, scientists are investigating different strategies such as layer-by-layer deposition and nanostructured coatings. Through these methods, biopolymer coatings can be fortified with additional protective layers that can prevent unwanted substances from penetrating the surface. The incorporation of nanoparticles is especially promising as it offers a unique opportunity to tailor the properties of biopolymer coatings at the nanoscale level. These tiny particles can improve not only barrier properties but also other desirable characteristics such as mechanical strength and durability. Furthermore, natural antimicrobial agents have been identified as another potential additive for enhancing biopolymer coating performance. These agents can help prevent bacterial growth on coated surfaces and thus improve their overall effectiveness in protecting against harmful pathogens [67].

9.9.10 Novel Coating Materials

In the field of biopolymer coatings, researchers are currently investigating the potential use of novel materials that may offer enhanced properties and improved functionality. These innovative materials include proteins, polysaccharides, lipids, and their various combinations. By utilizing these materials, researchers hope to expand the range of possible applications for biopolymer coatings, while also improving their performance in terms of mechanical strength, barrier performance, and overall stability. With ongoing research into these new and exciting materials, we may soon see significant advancements in the field of biopolymer coatings that could have a profound impact on a wide range of industries and applications [68, 69].

9.9.11 Active Packaging

Active packaging is a technique that involves the inclusion of active substances in the coating matrix to provide additional functionality. The primary objective of this technique is to extend the shelf life, enhance food safety, and improve the overall quality of food products. Researchers have been working on incorporating natural antimicrobial agents, antioxidants, and other bioactive compounds into packaging materials to achieve these goals. Active packaging materials are designed to control microbial growth, delay oxidation, and maintain product freshness. This technology has become increasingly popular in the food industry as it provides a more efficient way of preserving food products while ensuring their safety and quality. Moreover, active packaging can also help reduce food waste by preventing spoilage and extending the shelf life of perishable items. The use of active packaging has several benefits for both consumers and manufacturers alike. For instance, it allows manufacturers to create new products with longer shelf lives, which can be transported over long distances without spoiling. Additionally, active packaging helps maintain the nutritional value of foods by preventing nutrient loss due to oxidation or microbial activity [70].

9.9.12 Nanotechnology

Nanotechnology has emerged as a promising avenue for the advancement of biopolymer coatings and films. The use of nanoparticles - such as nano emulsions, nanofibers, or nanocomposites - is being extensively investigated for their potential to enhance barrier properties, antimicrobial activity, and mechanical strength. One of the key advantages of nanotechnology is that it allows for precise control over the coating structure, thus enabling scientists to tailor-make coatings with specific properties. Additionally, nanotechnology offers opportunities for targeted delivery of bioactive compounds, which could have significant implications in various fields such as medicine and agriculture. As research in this area continues to progress, it is becoming increasingly clear that nanotechnology has the potential to revolutionize the way we think about coatings and films [71].

9.9.13 Smart Coatings

The field of intelligent or smart coatings is currently in its nascent stage and has been gaining momentum as a promising area of research. These coatings possess the ability to react to different environmental stimuli, including temperature, humidity, and pH levels, among others. This makes them highly versatile and useful in a wide range of applications. One such application that has garnered significant attention is the use of smart coatings for monitoring food quality and safety. With their real-time information-gathering capabilities, these coatings can detect changes in food freshness, spoilage, or contamination. Such monitoring is crucial for ensuring that food products are safe for human consumption. Smart coatings have the potential to revolutionize the way we approach food safety by providing accurate and timely information about potential hazards.

Moreover, they can help reduce waste by allowing us to identify spoiled or contaminated foods before they reach consumers [72].

9.10 Conclusion

The application of biodegradable coatings and films presents a revolutionary opportunity to revolutionize food packaging practices by providing sustainable alternatives for traditional materials. The progress, investigation, and collaboration between academia, industry, and regulatory entities will facilitate the development of effective, secure, and environmentally friendly coating solutions that contribute to enhancing food preservation techniques while also promoting quality assurance and sustainability in the foreseeable future. The utilization of biodegradable coatings and films is poised to transform the realm of food packaging by offering sustainable substitutes for customary materials. The advancement, exploration, and partnership amongst academic institutions, businesses, and governing bodies will pave the way toward formulating efficacious, secure, and ecologically conscious coating remedies. This will facilitate the enhancement of food preservation standards while promoting quality assurance and sustainability in forthcoming years.

References

1. Rosenboom, J., Langer, R., Traverso, G., Bioplastics for a circular economy. *Nat. Rev. Mater.*, 7, 2, 117–137, 2022, https://doi.org/10.1038/s41578-021-00407-8.
2. Hellström, D. and Olsson, A., *Managing packaging design for sustainable development: a compass for strategic directions*, John Wiley & Sons, Chichester (UK), 2017.
3. Díaz-Montes, E. and Castro-Muñoz, R., Edible films and coatings as food-quality preservers: An overview. *Foods*, 10, 2, 249, 2021.
4. Ibrahim, S., Riahi, O., Said, S.M., Sabri, M.F., Rozali, S., Biopolymers from crop plants, in: *Reference Module in Materials Science and Materials Engineering*, 2019, https://doi.org/10.1016/B978-0-12-803581-8.11573-5.
5. Johansson, C., Functional barriers against migration for food packaging, in: *Multifunctional and Nanoreinforced Polymers for Food Packaging*, pp. 316–344, 2011, https://doi.org/10.1533/9780857092786.1.316.
6. Sivakanthan, S., Rajendran, S., Gamage, A., Madhujith, T., Mani, S., Antioxidant and antimicrobial applications of biopolymers: A review. *Food Res. Int.*, 136, 109327, 2020, https://doi.org/10.1016/j.foodres.2020.109327.

7. Cazón, P., Velazquez, G., Ramírez, J.A., Vázquez, M., Polysaccharide-based films and coatings for food packaging: A review. *Food Hydrocolloids*, 68, 136–148, 2017, https://doi.org/10.1016/j.foodhyd.2016.09.009.
8. Schmid, M. and Müller, K., Whey Protein-Based Packaging Films and Coatings, in: *Whey Proteins*, pp. 407–437, Academic Press, Australia, 2019, https://doi.org/10.1016/B978-0-12-812124-5.00012-6.
9. Shao, L., Xi, Y., Weng, Y., Recent Advances in PLA-Based Antibacterial Food Packaging and Its Applications. *Molecules*, 27, 18, 5953, 2022, doi 10.3390/molecules27185953.
10. Mihalca, V., Kerezsi, A.D., Weber, A., Schmucker, J., Vodnar, D.C., Dulf, F.V., Socaci, S.A., Fărcaş, A., Mureşan, C.I., Suharoschi, R., Pop, O., L. Protein-Based Films and Coatings for Food Industry Applications. *Polymers*, 13, 5, 769, 2021, https://doi.org/10.3390/polym13050769.
11. Rhim J, W., Hong S, I., Park H, M., Ng, P.K., Preparation and characterization of chitosan-based nanocomposite films with antimicrobial activity. *J. Agric. Food Chem.*, 54, 16, 5814–22, 2006, doi 10.1021/jf060658h, PMID: 16881682.
12. Aider, M., Chitosan application for active bio-based film production and potential in the food industry: Review. *LWT Food Sci. Technol.*, 43, 6, 837–842, 2010.
13. Debeaufort, F. and Voilley, A., *Lipid-Based Edible Films and Coatings*, Springer, New York, NY, 2009, https://doi.org/10.1007/978-0-387-92824-1_5.
14. Ghanbarzadeh, B., Almasi, H., Entezami, A.A., Physical properties of edible modified starch/carboxymethyl cellulose films. *Innov. Food Sci. Emerg. Technol.*, 11, 4, 697–702, 2010, https://doi.org/10.1016/j.ifset.2010.06.001.
15. Clardy, C., Han, I., Acton, J., Wardlaw, F., Bridges, W., Dawson, P., Protein-to-film adhesion as examined by amino analysis of protein binding to three different packaging films. *Poult. Sci.*, 77, 5, 745–751, 1998, https://doi.org/10.1093/ps/77.5.745.
16. Lacroix, M., Mechanical and permeability properties of edible films and coatings for food and pharmaceutical applications, in: *Edible films and Coatings for Food Applications*, pp. 347–366, Springer, New York, 2009.
17. Hatmi, R.U., Apriyati, E., Cahyaningrum, N., Edible coating quality with three types of starch and sorbitol plasticizer, in: *E3S Web of Conferences*, vol. 142, EDP Sciences, p. 02003, 2020.
18. Radev, R.S. and Dimitrov, G.A., Water vapor permeability of edible films with different compositions. *Scientific Works of University of Food Technologies*, vol. 64, no. 1, pp. 96–102, 2017.
19. Liyanapathiranage, A., Dassanayake, R.S., Gamage, A., Karri, R.R., Manamperi, A., Evon, P., Merah, O., Recent Developments in Edible Films and Coatings for Fruits and Vegetables. *Coatings*, *13*, 7, 1177, 2023, https://doi.org/10.3390/coatings13071177.
20. Moeini, A., Pedram, P., Fattahi, E., Cerruti, P., Santagata, G., Edible Polymers and Secondary Bioactive Compounds for Food Packaging Applications: Antimicrobial, Mechanical, and Gas Barrier Properties. *Polymers*, 14, 12, 2395, 2022, https://doi.org/10.3390/polym14122395.

21. Hoque, M., Gupta, S., Santhosh, R., Syed, I., Sarkar, P., Biopolymer-based edible films and coatings for food applications, in: *Food, Medical, and Environmental Applications of Polysaccharides*, pp. 81–107, 2021, https://doi.org/10.1016/B978-0-12-819239-9.00013-0.
22. Romanazzi, G., Orçonneau, Y., Moumni, M., Davillerd, Y., Marchand, P.A., Basic substances, a sustainable tool to complement and eventually replace synthetic pesticides in the management of pre and postharvest diseases: reviewed instructions for users. *Molecules*, *27*, 11, 3484, 2022.
23. Lacroix, M. and Vu, K.D., Edible coating and film materials: proteins, in: *Innovations in Food Packaging*, pp. 277–304, Academic Press, CA, USA, 2014.
24. Kumar, L., Ramakanth, D., Akhila, K., Edible films and coatings for food packaging applications: a review. *Environ. Chem. Lett.*, 20, 875–900, 2022, https://doi.org/10.1007/s10311-021-01339-z.
25. Panchal, N., Das, K., Prabhakar, P.K., Ghanghas, N., *Edible Films and Coatings for Fruits and Vegetables: Composition, Functions, and Regulatory Aspects*, Springer, Singapore, 2022, https://doi.org/10.1007/978-981-16-2383-7_10.
26. Fabra, M., López-Rubio, A., Lagaron, J., Biopolymers for food packaging applications, in: *Smart Polymers and their Applications*, pp. 476–509, 2014, https://doi.org/10.1533/9780857097026.2.476.
27. Oduro, K.O.A., Edible coating, in: *Postharvest Technology-Recent Advances, New Perspectives and Applications*, IntechOpen, Ghana, 2021.
28. Dhall, R.K., Advances in edible coatings for fresh fruits and vegetables: a review. *Crit. Rev. Food Sci. Nutr.*, 53, 5, 435–450, 2013.
29. Sahraee, S., Milani, J.M., Regenstein, J.M., Kafil, H.S., Protection of foods against oxidative deterioration using edible films and coatings: A review. *Food Biosci.*, 32, 100451, 2019, https://doi.org/10.1016/j.fbio.2019.100451.
30. Raybaudi-Massilia, R., Mosqueda-Melgar, J., Soliva-Fortuny, R., Martín-Belloso, O., Combinational Edible Antimicrobial Films and Coatings, in: *Antimicrobial Food Packaging*, pp. 633–646, 2016, https://doi.org/10.1016/B978-0-12-800723-5.00052-8.
31. Ebrahimi, A., Zabihzadeh Khajavi, M., Ahmadi, S. *et al.*, Novel strategies to control ethylene in fruit and vegetables for extending their shelf life: A review. *Int. J. Environ. Sci. Technol.*, 19, 4599–4610, 2022, https://doi.org/10.1007/s13762-021-03485-x.
32. Lin, D. and Zhao, Y., Innovations in the Development and Application of Edible Coatings for Fresh and Minimally Processed Fruits and Vegetables. *Compr. Rev. Food Sci. Food Saf.*, 6, 3, 60–75, 2007, doi: 10.1111/j.1541-4337.2007.00018.x.
33. Bertuzzi, M., Castro Vidaurre, E., Armada, M., Gottifredi, J., Water vapor permeability of edible starch-based films. *J. Food Eng.*, 80, 3, 972–978, 2007, https://doi.org/10.1016/j.jfoodeng.2006.07.016.
34. El-Shabasy, M., The perspective of adhesion of thin films. *Periodica Polytechnica Electr. Eng. (Archives)*, 25, 2, 123–1340, 1981, https://pp.bme.hu/ee/article/view/4773.

35. Taib, N.A.A.B., Rahman, M.R., Huda, D., A review on poly lactic acid (PLA) as a biodegradable polymer. *Polym. Bull.*, 80, 1179–1213, 2023, https://doi.org/10.1007/s00289-022-04160-y.
36. Amstutz, V., Hanik, N., Pott, J., Utsunomia, C., Zinn, M., Tailored biosynthesis of polyhydroxyalkanoates in chemostat cultures, in: *Methods in enzymology*, vol. 627, pp. 99–123, 2019, https://doi.org/ doi: 10.1016/bs.mie.2019.08.018.
37. Romão, S., Bettencourt, A., Ribeiro, I.A., Novel Features of Cellulose-Based Films as Sustainable Alternatives for Food Packaging. *Polymers*, 14, 22, 4968, 2022, https://doi.org/10.3390/polym14224968.
38. Xing, Y., Xu, Q., Li, X., Chen, C., Ma, L., Li, S., Che, Z., Lin., H., Chitosan-based coating with antimicrobial agents: preparation, property, mechanism, and application effectiveness on fruits and vegetables. *Int. J. Polym Sci.*, 4, 11–21, 2016, https://doi.org/10.1155/2016/4851730.
39. Onyeaka, H., Obileke, K., Makaka, G., Nwokolo, N., Current Research and Applications of Starch-Based Biodegradable Films for Food Packaging. *Polymers*, 14, 6, 1126, 2022, https://doi.org/10.3390/polym14061126.
40. Thawien, W., Protein-based edible films: characteristics and improvement of properties, in: *Structure and Function of Food Engineering*, IntechOpen, Egypt, 2012, https://doi. org/10.5772/48167.
41. Sharma, H.P., Chaudhary, V., Kumar, M., Importance of edible coating on fruits and vegetables: A review. *J. Pharmacogn. Phytochem.*, 8, 3, 4104–4110, 2019.
42. Bangar, S.P., Purewal, S.S., Trif, M., Maqsood, S., Kumar, M., Manjunatha, V., Rusu, A.V., Functionality and Applicability of Starch-Based Films: An Eco-Friendly Approach. *Foods*, 10, 9, 2181, 2021, https://doi.org/10.3390/foods10092181.
43. Priyadarshi, R. and Rhim, J., Chitosan-based biodegradable functional films for food packaging applications. *Innov. Food Sci. Emerg. Technol.*, 62, 102346, 2020, https://doi.org/10.1016/j.ifset.2020.102346.
44. O'Sullivan, M., The Stability and Shelf Life of Meat and Poultry, in: *The Stability and Shelf Life of Food*, pp. 521–543, 2016, https://doi.org/10.1016/B978-0-08-100435-7.00018-6.
45. Kontominas, M.G., Badeka, A.V., Kosma, I.S., Nathanailides, C.I., Recent Developments in Seafood Packaging Technologies. *Foods*, 10, 5, 940, 2021, https://doi.org/10.3390/foods10050940.
46. Nechita, P., Review on Polysaccharides Used in Coatings for Food Packaging Papers. *Coatings*, 10, 6, 566, 2020, https://doi.org/10.3390/coatings10060566.
47. Nájera, A.I., Nieto, S., R. Barron, L.J., Albisu, M., A Review of the Preservation of Hard and Semi-Hard Cheeses: Quality and Safety. *Int. J. Environ. Res. Public Health*, 18, 18, 323–337, 2021, https://doi.org/10.3390/ijerph18189789.
48. Suhag, R., Kumar, N., Petkoska, A.T., Upadhyay, A., Film formation and deposition methods of edible coating on food products: A review. *Food Res. Int.*, 136, 109582, 2020, https://doi.org/10.1016/j.foodres.2020.109582.

49. Andrade, R.D., Skurtys, O., Osorio, F.A., Atomizing spray systems for application of edible coatings. *Compr. Rev. Food Sci. Food Saf.*, 11, 3, 323–337, 2012.
50. Momin, M.C., Jamir, A.R., Ankalagi, N., Henny, T., Devi, O.B., Edible coatings in fruits and vegetables: A brief review. *Pharma Innov. J.*, 10, 71–78, 2021.
51. Haddad, N.A., Watts, E., Lively, J.A., Evaluation of Post-Harvest Procedures for Quality Enhancement in the Louisiana Commercial Shrimp Industry. *J. Aquat. Food Prod. Technol.*, 31, 1, 96–111, 2022.
52. Xiao, Q., Coating and film-forming properties, in: *Food Hydrocolloids: Functionalities and Applications*, pp. 267–306, Springer, Singapore, 2021.
53. Prasad, L.K., McGinty, J.W., Williams, R.O., Electrostatic powder coating: Principles and pharmaceutical applications. *Int. J. Pharm.*, 505, 1-2, 289–302, 2016. https://doi.org/10.1016/j.ijpharm.2016.04.016.
54. Pulatsu, E. and Lin, M., A review on customizing edible food materials into 3D printable inks: Approaches and strategies. *Trends Food Sci. Technol.*, 107, 68–77, 2021, https://doi.org/10.1016/j.tifs.2020.11.023.
55. Nguyen, T.C. and Choi, W., Electrospray mechanism for quantum dot thin-film formation using an electrohydrodynamic jet and light-emitting device application. *Sci. Rep.*, 10, 1, 1–9, 2020, https://doi.org/10.1038/s41598-020-67867-w.
56. Kondle, R., Sharma, K., Singh, G., Kotiyal, A., Using Nanotechnology for Enhancing the Shelf Life of Fruits, in: *Food Preservation and Packaging - Recent Process And Technological Advancements*, pp. 1–19, Intech Open, India, 2022.
57. Nile, S.H., Baskar, V., Selvaraj, D., Guoyin kai, J.X., Nanotechnologies in Food Science: Applications, Recent Trends, and Future Perspectives. *Nano Micro Lett.*, 12, 45, 2020, https://doi.org/10.1007/s40820-020-0383-9.
58. Ishwarya, S.P., Anandharamakrishnan, C., Stapley, A.G., Spray-freeze-drying: A novel process for the drying of foods and bioproducts. *Trends Food Sci. Technol.*, 41, 2, 161–181, 2015.
59. Akbarzadeh, A., Samiei, M., Davaran, S., Magnetic nanoparticles: Preparation, physical properties, and applications in biomedicine. *Nanoscale Res. Lett.*, 7, 1, 144, 2012, https://doi.org/10.1186/1556-276X-7-144.
60. Gupta, V., Biswas, D., Roy, S.A., Comprehensive Review of Biodegradable Polymer-Based Films and Coatings and Their Food Packaging Applications. *Materials*, 15, 17, 5899, 2022, https://doi.org/10.3390/ma15175899.
61. Zhao, Y. and McDaniel, M., Sensory quality of foods associated with edible film and coating systems and shelf-life extension, in: *Innovations in Food Packaging*, pp. 434–453, 2005, https://doi.org/10.1016/B978-012311632-1/50056-5.
62. Galbraith, S.C., Bhatia, H., Liu, H., Yoon, S., Media formulation optimization: Current and future opportunities. *Curr. Opin. Chem. Eng.*, 22, 42–47, 2018, https://doi.org/10.1016/j.coche.2018.08.004.

63. Ribeiro, A.M., Estevinho, B.N., Rocha, F., Preparation and Incorporation of Functional Ingredients in Edible Films and Coatings. *Food Bioprocess Technol.*, 14, 209–231, 2021, https://doi.org/10.1007/s11947-020-02528-4.
64. Deshmukh, R.K. and Gaikwad, K.K., Natural antimicrobial and antioxidant compounds for active food packaging applications. *Biomass Convers. Biorefin.*, 4, 4419–4440, 2022, https://doi.org/10.1007/s13399-022-02623-w.
65. Siegrist, M. and Hartmann, C., Consumer acceptance of novel food technologies. *Nat. Food*, 1, 6, 343–350, 2020, https://doi.org/10.1038/s43016-020-0094-x.
66. Abbas, Y., Martinetti, A., Rajabalinejad, M., Schuberth, F., van Dongen, L.A.M., Facilitating digital collaboration through knowledge management: a case study. *Knowl. Manage. Res. Pract.*, 20, 6, 797–813, 2022.
67. Sun, X., Wang, J., Dong, M., Zhang, H., Li, L., Wang, L., Food spoilage, bioactive food fresh-keeping films and functional edible coatings: Research status, existing problems, and development trend. *Trends Food Sci. Technol.*, 119, 122–132, 2022, https://doi.org/10.1016/j.tifs.2021.12.004.
68. Bratovcic, A., Physical–Chemical, Mechanical and Antimicrobial Properties of Bio-Nanocomposite Films and Edible Coatings. *Int. J. Res. Appl. Sci. Biotechnol.*, 8, 5, 151–161, 2021.
69. Mkandawire, M. and Aryee, A.N., Resurfacing and modernization of edible packaging material technology. *Curr. Opin. Food Sci.*, 19, 104–112, 2018, https://doi.org/10.1016/j.cofs.2018.03.010.
70. Rodrigues Arruda, T., Campos Bernardes, P., Robledo Fialho e Moraes, A., de Fátima Ferreira Soares, N., Natural bioactive in perspective: The future of active packaging based on essential oils and plant extracts themselves and those complexed by cyclodextrins. *Food Res. Int.*, 156, 111160, 2022, https://doi.org/10.1016/j.foodres.2022.111160.
71. Zambrano-Zaragoza, M.L., González-Reza, R., Mendoza-Muñoz, N., Miranda-Linares, V., Bernal-Couoh, T.F., Mendoza-Elvira, S., Quintanar-Guerrero, D., Nanosystems in edible coatings: A novel strategy for food preservation. *Int. J. Mol. Sci.*, 19, 3, 705, 2018, https://doi.org/10.3390/ijms19030705.
72. Priyadarshi, R., Ezati, P., Rhim, J.W., Recent advances in intelligent food packaging applications using natural food colorants. *ACS Food Sci. Technol.*, 1, 2, 124–138, 2021.

10

Nanoemulsion: A Potential Strategy Toward Edible Coatings

Aditi Negi, Vanmathi Mugasundari A. and Jeyan A. Moses*

Computational Modeling Nanoscale Processing Unit, National Institute of Food Technology, Entrepreneurship and Management-Thanjavur, Tamil Nadu, India

Abstract

Nanoemulsions are colloidal systems that are kinetically stable and include tiny droplets. Compared to traditional emulsions, they offer improved functional qualities. Nanoemulsions are becoming more popular in the food industry due to their physicochemical features for effective encapsulation, solubilization, entrapment of biologically active compounds, controlled distribution, and bioavailability. The food industry utilizes nanoemulsions to deliver antimicrobials, coloring and flavoring ingredients, and nutraceuticals as a natural substitute for synthetic additives, which can also be incorporated into coatings and food packaging films. Studies report that these nano-based biodegradable coatings and packaging films with active component nanoemulsions improve the quality, functionality, nutrient content, and duration of storage of foods. Oil-in-water (o/w) nanoemulsions excel at delivering various compounds like vitamins, flavors, antioxidants, preservatives, and nutraceuticals in edible coatings. These colloidal dispersions consist of two or more immiscible phases held together by a surfactant, with oil droplets typically ranging from 50 to 100 nm in diameter. Notably, they exhibit enhanced resistance to particle aggregation compared to traditional emulsions. To create nanoemulsions for edible coatings, a practical method involves high-energy techniques such as high-pressure homogenization, microfluidization, or sonication. The key components of these nanoemulsions include the continuous phase, disperse phase, and emulsifiers that play a crucial role in stabilizing both phases. Nanoemulsions possess a tiny particle size, a higher surface area, and unique morphological properties, making them essential carriers for the prolonged release of dietary bioactive chemicals. As a way to deliver lipophilic active ingredients, nanoemulsions have a lot of potential for delivering lipophilic active ingredients as a way to deliver them.

**Corresponding author*: moses.ja@iifpt.edu.in

Mousumi Sen (ed.) Food Coatings and Preservation Technologies, (339–366) © 2025 Scrivener Publishing LLC

They have applications in a variety of food products. The encapsulation of natural bioactive compounds within nanoemulsions has the potential to improve their stability and functional performance., both within the food matrix and during *in vivo* conditions such as gastrointestinal conditions. Edible nanoemulsion-based coatings can enhance the shelf life of various foods, such as dairy products, meats, fruits, and vegetables. These coatings also help minimize gas exchange, moisture, and food oxidation by incorporating flavor and coloring components, enzymes, antioxidants, and antimicrobial agents. While liquid food items have been the primary focus of nanoemulsions practical use, the area of food packaging is starting to noticeably benefit from it nowadays. Several recent studies have highlighted the benefits of active edible coatings for the shelf life of various food items, particularly fruits and vegetables. Innovative nanoemulsion-based edible coatings with multiple functions have emerged, combining antimicrobial essential oils and water-soluble fiber materials for low-fat cheese coatings. Notably, these fiber coatings contain a prebiotic component that enhances the nutritional properties of the coated cheese. Despite the numerous advantages, it is essential to acknowledge that certain nanocoatings may possess allergenic qualities and may not be economically viable for widespread use. Due to the possible toxicity effects of nanomaterials, future studies should include the behavior of nanoemulsion-based coatings under gastrointestinal conditions. Also, we need to learn more about the naturally occurring functional compounds that are taken from plants so that we can figure out how they react with different nanoemulsion mixtures that are put on the surfaces of food products. Given these considerations, this chapter focuses on investigating the composition, properties, preparation methods, characterization, challenges associated with nanoemulsion-based edible coatings, potential scalability in the food sector, and beneficial health effects.

Keywords: Nanoemulsion, encapsulation, stability, edible coating, active ingredients, shelf life, antimicrobials

10.1 Introduction

In the modern era, marked by technological advancements, urbanization, and evolving lifestyles, consumers are increasingly seeking food products that not only align with their health-conscious preferences but also offer extended shelf life without compromising the inherent properties of the items. Nanoemulsion-based techniques for edible food coatings have gained considerable interest among these technologies. Nanoemulsions, which are made up of oil-in-water or water-in-oil droplets that are spread out very thinly at the nanoscale level, offer a great way to make food last longer, be safer, and taste better [1, 2].

Recently, the food business has faced difficulties in preserving the quality and condition of perishable food products, such as fruits and vegetables, due to the changing urban environment, technological progress, and shifting customer tastes. Usually, traditional methods of food preservation are not good enough to meet these needs, which is why researchers are looking into new methods like nanoemulsion-based edible coatings. Nanoemulsions are very stable and bioavailable carriers that effectively transport various bioactive substances, such as antimicrobials, antioxidants, and flavor enhancers [3]. Nanoemulsions may safeguard food goods by enclosing these components in tiny droplets, therefore reducing microbial degradation, oxidative processes, and flavor loss on the surface of the food. In addition, nanoemulsion coatings have the benefit of being generated from natural sources, which aligns with the increasing consumer preference for clean-label and sustainable food solutions.

This introduction provides a foundation for comprehending the possibilities of nanoemulsion methods in the context of edible food coatings. This chapter describes composition, properties, approaches for nanoemulsion preparation, and nanoemulsion based edible coating and packaging materials and challenges associated with same.

10.2 Nanoemulsion Structure and Composition

The formulation of nanoemulsions is a complex process that requires solvents, emulsifiers, and functional compounds to achieve stability, bioavailability, and efficient encapsulation of functional compounds. Combining two immiscible liquids with an emulsifier forms nanoemulsion compositions. These compositions exist in two primary forms: oil-in-water (O/W) and water-in-oil (W/O), depending on the spatial arrangement of the aqueous and oil phases [4]. The characteristics, content, and production methods of nanoemulsions, microemulsions, and standard emulsions exhibit variations.

The aqueous phase (generally water) consists of polar solvents and co-solvents, which include proteins, carbohydrates, minerals, alcohols, and polyols such as polyethylene glycols, glycerol, and sorbitol. The oil phase of a nanoemulsion consists of a diverse range of components, including free fatty acids, triacylglycerols, diacylglycerols, and monoacylglycerols. Additionally, as non-polar components, essential oils from plant origin, waxes, oil-soluble vitamins, various lipophilic components and weighting agents can be used as constituents of the oil phase. The former phase determines the polarity, rheology, phase behavior, interfacial tension, and ionic strength of the nanoemulsion, while the latter phase components contribute properties such as refractive index, density, viscosity, phase behavior, and interfacial tension, which play a crucial role in influencing the overall formation, stability, and functional characteristics of the nanoemulsion [5].

Stabilizers such as emulsifiers (lecithin, sodium dodecyl sulfate, and tween), ripening retarders (sodium chloride), weighting agents, surfactants, and texture modifiers (sodium alginate) are incorporated to ensure the kinetic and thermodynamic stability of nanoemulsions [6]. Different proteins (whey proteins, soy proteins), polysaccharides (alginates, carrageenan's, gums, pectin's, starches), and flavonoids are explored as potential stabilizers, offering superior stability as they form dense layers at the O/W interface. They lower interfacial tension by attaching to the point where droplets meet the dispersion medium. This stops droplets from sticking together and makes the mixture more stable. Ionic surfactants introduce electrical charges into droplets, while non-ionic surfactants create steric barriers.

Nanoemulsions are made up of functional compounds that make it possible to package a wide range of lipophilic substances, such as drugs, flavors, antimicrobials, antioxidants, and bioactive lipids. This encapsulation enhances bioactivity, desirability, and palatability. Notable categories of lipophilic functional compounds encompass fatty acids, carotenoids, antioxidants, and phytosterol [7, 8].

Nanoemulsions are kinetically stable systems that show physicochemical stability. However, their long-term stability is compromised by destabilizing mechanisms that are chemical (oxidation, hydrolysis) and physical (coalescence, flocculation, and gravitational separation). The gravitational separation leads to creaming or sedimentation, influenced by relative densities and particle size. Employ techniques like high-pressure homogenization to adjust relative densities and control particle size for mitigating gravitational separation. Maintaining nanoemulsion particle diameters below 70 nm optimizes Brownian motion's effectiveness in preventing creaming. Choosing stabilizing agents that minimize attractive interactions between droplets can address the challenge of droplet aggregation, such as flocculation or coalescence. Managing colloidal interactions, including the use of suitable surfactants, helps maintain stability. To counter Ostwald ripening, strategies include controlling droplet size by manipulating factors such as bulk phase solubility and interfacial elasticity. Incorporating antioxidants or selecting oil phases with lower susceptibility to oxidation and hydrolysis enhances chemical stability. Protective packaging or additives that mitigate the impact of UV or visible light on clear nanoemulsions can further contribute to their long-term stability. Employing a comprehensive approach that considers both physical and chemical factors is essential in designing nanoemulsions with improved stability [9].

10.3 Approaches for the Preparation of Nanoemulsions

Nanoemulsions, while thermodynamically unstable, exhibit kinetic stability owing to small dimension of the droplets and stability for a longer time. The methods for nanoemulsion preparation are categorized into high-energy and low-energy approaches based on the energy input involved. These approaches, by changing the composition of the emulsion system and operating conditions, have a significant impact on droplet size and instability mechanisms. Overall, the preparation of nanoemulsions requires significantly less surfactant (5–10%) than that of microemulsions (20% or more) [10]. The selection of specific methods and materials plays a pivotal role in achieving optimal particle sizes and ensuring the efficiency of the nanoemulsion preparation process. All the approaches are summarized in Figure 10.1.

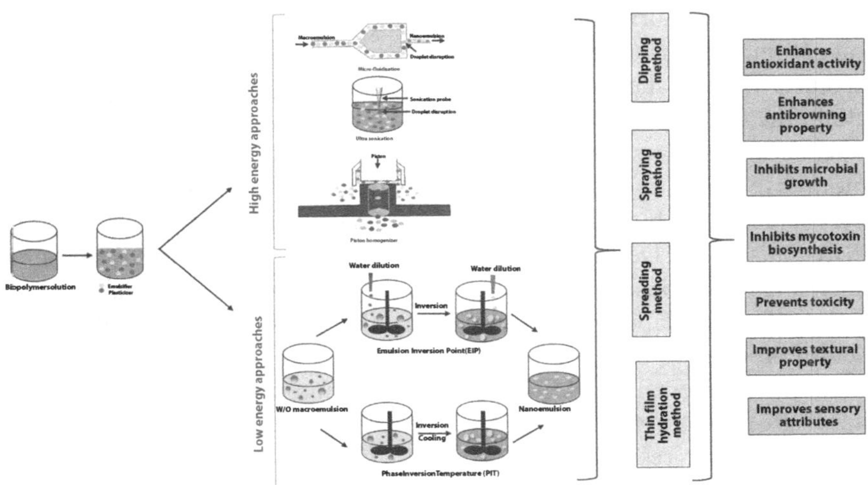

Figure 10.1 Nanoemulsion preparation and application in food packaging.

10.3.1 High-Energy Approaches

High-energy methods are very important for making nanoemulsions because they use complex mechanical tools that can create strong forces that can mix and break up the oil and water phases in a very complex way. These methods, including high-pressure valve homogenizers (HPH), microfluidizers, and sonication, are characterized by their ability to produce fine oil droplets efficiently [11]. The success of these high-energy methods depends on a variety of factors, including the intensity and duration of energy input, the type and concentration of emulsifiers, and the interfacial tension between the dispersed and continuous phases.

The rotor stator emulsification, commonly used in the food industry for nanoemulsion preparation before subsequent stages like high-pressure homogenization [12]. However, achieving droplet sizes below 1 μm is challenging with this method. Common approaches involve modifying rotor design, adjusting gap size, increasing rotor speed, and extending droplet residence time. With advanced gear-rim mechanism with a repeated cycles allows variation in mean droplet size (150-270 nm) by fine-tuning speed of rotor and time, with high-pressure homogenization used for achieving smaller droplet sizes [13].

High-pressure homogenization (HPH), the predominant method for nanoemulsion production, involves subjecting the emulsion to extremely high pressure between 500 to 5000 psi and multiple run cycles to achieve the desired nanoemulsion droplet size. The macroemulsions are forced through a small orifice, where process combines hydraulic shear stress, cavitation, and turbulent flow conditions, effectively breaking down large droplets into smaller, more stable ones [14]. The process is repeated to attain the desired droplet size and polydispersity index (PDI), which indicates droplet size uniformity. A lower PDI reflects higher uniformity, with monodisperse samples having PDI below 0.08, 0.08 to 0.3 indicating a narrow size distribution, and PDI greater than 0.3 suggesting a broad size distribution [15]. Despite its efficacy, the HPH process requires substantial energy and may pose challenges, particularly in terms of potential deterioration of thermolabile compounds such as proteins, enzymes, and nucleic acids [14, 16].

Microfluidizers, resembling high-pressure homogenizers in design, utilize high-pressure displacement pumps to propel the emulsion through microchannels, resulting in the production of stable nanoemulsions. Repeating the process and adjusting the operating pressure ensures uniform nanoemulsion production. The interaction chamber, where crude emulsion jets collide, is powered by a pneumatically driven pump compressing air at pressures between 150 to 650 MPa, generating significant shearing forces and producing fine emulsions [17].

Ultrasonication is another approach to employing high energy and high intensity ultrasonic waves to induce disruptive strong forces for transforming immiscible solutions into fine and stable droplets [18]. Ultrasonic emulsification involves two mechanisms: the acoustic field creating interfacial waves to disperse the oil phase into droplets, and ultrasound inducing acoustic cavitation, leading to microbubble formation and collapse, generating turbulence and disrupting large droplets into sub-micron size above 20 kHz or higher frequencies. High-power ultrasonic devices, such as focusing horns, cause extreme shear and cavitation, breaking up droplets. However, concerns about sonication methods include the potential for inducing protein denaturation, polysaccharide depolymerisation, and lipid oxidation [17].

Combining high-speed homogenization with other above mentioned methods such as HPH or ultrasonication, often enhances its effectiveness in achieving stable nanoemulsions. The widespread adoption of high-energy approaches in industrial food operations is attributed to their capability for large-scale production and their ability to confer superior stability to nanoemulsions with finely tuned properties for various applications [19].

10.3.2 Low-Energy Approaches

Low-energy approaches exploit the inherent changes in composition or chemical energy within a system for nanoemulsion synthesis, providing an alternative to high-energy approaches. These techniques, relying on spontaneous formation and phase transitions, offer alternative and cost-effective approaches for tailored nanoemulsion synthesis with diverse applications in various fields [20, 21]. This category includes physiochemical factors which are affecting nanoemulsion formation; spontaneous or solvent diffusion emulsification, membrane emulsification, and phase inversion (PI) with variations such as PI temperature (PIT), PI composition or PI point (PIC), and emulsion PI (EPI) [17]. Researchers acknowledge membrane emulsification for its simplicity and scalability in producing stable nanoemulsions with controlled droplet size distributions [22, 23].

Spontaneous emulsification or self-emulsification, works through diffusion mechanisms, driven by a chemical potential gradient when immiscible liquids are mixed under non-equilibrium conditions. This method, adaptable through temperature changes or interfacial property adjustments and does not necessitate specialized equipment and allows for nanoemulsion development at ambient conditions [24]. The process involves combining an organic phase (oil and surfactant) with an aqueous phase (co-surfactant and water) to transfer water-miscible components. The breakdown of the bi-continuous microemulsions phase leads to the spontaneous formation of fine oil drops, with solvents expediting this process (ouzo effect) [25]. While the utilization of this method at an industrial level is still in its infancy, it shows promising economic viability [11, 26]. This method offers a promising alternative for nanoemulsion formation, despite typically resulting in larger droplet sizes compared to other approaches [21].

Phase inversion, a dynamic process during emulsification, involves transforming O/W emulsions to W/O emulsions or vice versa. This phenomenon, driven by surfactant-induced phase transitions, is influenced by altering the surfactant's spontaneous curvature through temperature changes and system composition adjustments. Widely used in the food industry, techniques like phase inversion composition, phase inversion temperature, and emulsion phase inversion (EPI) are employed. Parameters such as temperature, salt concentration, oil fraction, water content, and energy input, as well as formulation variations, play key roles in inducing emulsion phase inversion [17, 27].

The emulsion PI point, or PIC, implicates manipulating surfactants hydrophilic-lipophilic behavior through ingredient variations, such as adding salt to nanoemulsions for a transition to (W/O) emulsions [5]. Similarly, water-in-oil emulsions can shift to oil-in-water by gradually adding water to the oil phase, creating bicontinuous phases and forming O/W nanoemulsions [28]. This approach, although requiring longer preparation time but can create fortified nanoemulsion with vitamin E acetate [29]. The PIC method involves changing composition at a constant temperature, suitable for large-scale production by consistently adding water or oil to oil-surfactant or water-surfactant mixtures. It achieves a balance in surfactant properties, resulting in small-sized metastable oil-in-water droplets [30, 31].

The PIT method induces the transition from (O/W) to (W/O) emulsions by leveraging the temperature-dependent solubility of non-ionic surfactants, such as polyethoxylated surfactants. This technique, driven by changes in hydrophilic-lipophilic balance (HLB) as a function of temperature at a constant composition, results in nanoemulsions with low polydispersity and high emulsification efficiency [32]. The stability of nanoemulsion near the PIT is temperature-sensitive, requiring co-surfactants for stabilization. Gradual heating followed by rapid cooling results in stable O/W nanoemulsions, especially with non-ionic surfactants. However, this method may not be suitable for nonionic and ionic surfactants less sensitive to temperature fluctuations [31, 33].

10.3.3 Nanoemulsion Coating Approaches

Various approaches methods have been developed for coating application to the surfaces of food products, with spraying, dipping, spreading, panning, and edible coating techniques being widely adopted. Spraying involves the application of a low-viscosity solution at high pressure (60–80 psi), ensuring small droplet sizes of around 20 μm for even coverage on the product's surface. The uniformity of the coating depends on factors such as the application technique, drying time, and temperature [34]. Spreading, suitable for high-viscosity coating solutions, may lead to uneven coating distribution. Dipping entails immersing food products in a solution for 10 to 30 seconds, resulting in a uniform coating layer after drying at ambient temperature. Careful monitoring is necessary to avoid the formation of thick coatings, which can create anaerobic conditions [35]. In the panning technique, a coating mixture is continuously applied in a vessel or pot for uniform coverage. This method is used for both pharmaceutical and food items, particularly in confectionery for a finishing touch. Another emerging technique is fluidized bed coating, accepted not only in the food industry but also in the pharmaceutical sector for polishing tablets. This method involves applying a coating uniformly by suspending particles in an upward-moving air stream, demonstrating versatility in various industries [36]. In the realm of active packaging, where materials interact with the packaged food to extend shelf life or monitor freshness, coatings play a vital role. Edible coatings, a subset of active packaging, provide an additional layer of protection, enhancing the preservation of food quality and extending shelf life. These coatings may contain antimicrobial agents, antioxidants, or other bioactive compounds that contribute to maintaining the freshness and safety of the packaged products. This integration of edible coatings into the coating application arsenal further exemplifies the industry's commitment to innovative and sustainable packaging solutions.

10.4 Application of Nanoemulsion-Based Food Packaging Materials

Integration of ingredients into food formulations has gained significant attention in the creation of important edible nanoemulsion packaging. This innovation has become crucial for developing novel delivery systems that effectively protect, release, and encapsulate active compounds [37]. With the rising prevalence of health concerns and the accelerated pace of consumer lifestyles, there is a growing demand for fresh and nutritious food products. Additives like antibrowning/antioxidants, antimicrobial agents, volatile precursors, firming agents, preservatives, color, nutraceuticals, and flavors play a pivotal role in enhancing the nutritional, visual, rheological, and organoleptic properties of these nanoemulsion-based coatings. Some of these advancement in the food industry, emphasizing the importance of edible nanoemulsion packaging are summarized here along with Table 10.1.

Table 10.1 Application of nanoemulsions in food packaging.

Food material	Nanoemulsion composition and preparation method	Types of packaging	Coating functionality	Key findings	References
Pork loin	Oregano EO, Resveratrol EO, Pectin, and Tween 80 using Homogenization	Coating	Texture and shelf life extension	Extended shelf life and slower physiochemical shifts. The coating has improved stability and preservation properties.	[66]
Ground beef	Thyme oil, calcium alginate, and propylene glycol using Ultrasonication	Film	Antimicrobial	Superior antimicrobial activity against coliforms, *S. aureus*, yeast, mold, and lactic acid bacteria. Increased water-vapor permeability, opacity, wettability, and thickness	[67]
Chicken Fillets	Curcumin-Cinnamon EO, Curcumin-Garlic EO, and Curcumin-Sunflower Oil using Inversion point method	Coating	Texture & shelf life extension	Extension of shelf life up to 12 days, improved microbiological quality parameters, and improved sensory, physical, and chemical properties	[68]
Chicken	Cumin EO, sunflower oil, and oxygen absorber using Homogenization	Packaging (O2 Absorber)	Antimicrobial & shelf life extension	Total mesophilic and psychotropic bacteria growth have retarted. Increased shelf-life.	[69]

(*Continued*)

Table 10.1 Application of nanoemulsions in food packaging. (*Continued*)

Food material	Nanoemulsion composition and preparation method	Types of packaging	Coating functionality	Key findings	References
Chicken breast	Ginger EO, sodium caseinate using Ultrasonication	Coating	Antimicrobial & shelf life extension	In refrigerated condition, shelf life increased to 12 days. Reduced activity of psychrophilic bacteria	[70]
Shrimp	Aloe vera and eugenol using Ultrasonication	Coating	Shelf life extension	Shrimp quality is maintained for up to 7 days in cold storage conditions.	[71]
Tomato	Cardamom essential oil (CMC) using Ultrasonication	Coating	Antibacterial & shelf life extension	Increased antibacterial and antibiofilm action against *E. coli* and *L. monocytogens*. Improved quality, reduced oxidative damage, and increased shelf life of tomatoes	[72]
Tomato	Sweet Orange EO using Ultrasonication	Edible coating	Texture & antimicrobial	33% increase in firmness, antibacterial, reduced *Salmonella*, and *Listeria*. Weight loss was tripled compared to uncoated tomatoes. Increased sensory acceptance scores	[73]
Banana	Tea-tree oil, TiO_2 nanoparticles, and sodium alginate using Ultrasonication	Double coating	Texture	Improved the water vapor, oxygen barrier, and retention properties. Improved the film's barrier and UV-blocking qualities. Improved quality and decreased anthracnose in preserved bananas.	[74]

(*Continued*)

Table 10.1 Application of nanoemulsions in food packaging. (*Continued*)

Food material	Nanoemulsion composition and preparation method	Types of packaging	Coating functionality	Key findings	References
Fresh-cut cucumber slices	Carvacrol and chitosan with pulsed light treatment using High-pressure homogenization	Coating	Antimicrobial	Antimicrobial effect and 4-log contamination reduction against *E. coli* and *Listeria innocua,*	[56]
Plum	Lemongrass oil using Dynamic high pressure (DHP)	Coating	Antimicrobial	Inhibition of *Salmonella* and *E. coli* O157:H7 contamination reduced weight loss and ethylene production. Coated plums showed higher firmness than the uncoated ones at 4 °C.	[75]
Grape Berry	Lemongrass oil using Dynamic high pressure (DHP)	Coating	Antimicrobial	Antimicrobial activity against *S. typhimurium and E. coli* O157:H7 Increased anthocyanin and phenolic content, firmness, and extended shelf life at 4 and 25 °C for 28 days.	[53]
Green Beans	Mandarin Essential Oil using High hydrostatic pressure	Coating	Antimicrobial & shelf life extension	Achieving a 4 log reduction in the initial microbial load results in increased shelf life. Preservation, microbial analysis, antimicrobial effect, firmness, and color	[64]
Cabbage	Carvacrol using Ultrasonication	Coating	Antimicrobial	Antimicrobial activity against *E. coli* and *P. pastoris*	[76]

(*Continued*)

Table 10.1 Application of nanoemulsions in food packaging. (*Continued*)

Food material	Nanoemulsion composition and preparation method	Types of packaging	Coating functionality	Key findings	References
Okra	Basil oil, alginate, and tween 20 using Ultrasonication	Coating	Antimicrobial & shelf life extension	It is effective against spoilage fungi. *Penicillium chrysogenum*, *Aspergillus flavus*, and shelf life extension	[77]
Aqueous and non-aqueous foods stimulate	Curcumin, banana starch, and glycerol (Tween 80) using Emulsion phase inversion	Film	Texture	Enhanced elongation at break and decreased permeability of the film to water vapor	[78]
Active food packaging plays a crucial role in extending the shelf life of food products.	Cinnamon and Rutin, Chitosan/Gelatin Polymer Matrix using solution casting	Film	Shelf life extension	Greater antioxidant and antibacterial activities, as well as enhanced thermal and UV-blocking capabilities	[79]
Fresh-cut pineapple	Citral, sesame oil, and sodium alginate using Ultrasonication	Coating	Antimicrobial & shelf life extension	Retention of color, reduced respiration rate, and microbial growth—around a 2-log reduction in *Salmonella enterica* and *Listeria monocytogenes contamination*—extended shelf life up to 12 days when stored at 4 °C.	[80]
Kasar Cheese	Myrtle (*Myrtus communis L.*) EO, alginate, glycerol using Ultrasonication	Coating	Antimicrobial	*L. monocytogenes* reduced population in cheese stored under refrigerated conditions by 0.42 and 0.88 log cfu/g, with 1 and 2% EO coating, respectively.	[81]

10.4.1 Antimicrobial Coating

The integration of nanoemulsions into edible films and coatings shows significant prospects in packaging of food and its applications. These nano based packaging and coating material are made up of a biopolymer matrix and act as the continuous phase, keeping the nanoemulsion droplets stable and preventing them from spreading out. The elevation of viscosity in the continuous phase plays a role in minimizing droplet coalescence, as explained by Artiga-Artigas *et al.* [38]. Freshly cut fruits and vegetables, vulnerable to microbial contamination, encounter safety concerns arising from the cutting processes. Promoting healthier lifestyles has contributed to a significant surge in demand for these products, resulting in increased consumption in developed countries. To address safety concerns, innovative techniques, such as spraying or dipping with antimicrobial agents, are employed to extend the shelf life of the products. The application of food antimicrobials, chemical compounds that inhibit the growth of microorganisms, has been a longstanding practice in the food industry. For example, the use of preservatives like sodium benzoate or sorbic acid effectively inhibits microbial activity, thereby extending the shelf life and enhancing the safety of various food products [39]. Responding to consumer preferences, there is a shift towards additive-free, fresh-cut produce. Researchers explore the antimicrobial properties of essential oils rich in non-volatile and volatile compounds [40]. Challenges, including the hydrophobic nature of these antimicrobial compounds, are addressed through nanotechnology, particularly nanoemulsions. This method improves the effectiveness of antimicrobials by enclosing, protecting, and controlling their release, which could help make coatings work better [41, 42].

The researchers prepared edible films using the biopolymer pectin that incorporated antimicrobial nanoemulsions of clove essential oil and cinnamaldehyde. These pectin films exhibited antimicrobial characteristics. The films also demonstrated poor water and vapor permeability, attributed to a reduced ratio of hydrophilic to hydrophobic components, as described by Otoni *et al.* and Sasaki *et al.* [43, 44]. Alexandre *et al.* (2016) reported that the films with antioxidant activity were created by incorporating ginger essential oil with canola oil nanoemulsion with porcine gelatin [45]. Further, nanoemulsion-based coatings and films containing essential oils like carvacrol, mandarin oil, bergamot oil, and lemon oil that have been prepared using chitosan have also demonstrated good antimicrobial properties [46]. Artiga-Artigas *et al.* (2017) applied nanoemulsion coatings with oregano essential oil, sodium alginate, and mandarin fiber on low-fat cut cheese to extend its shelf life. These coatings demonstrated antimicrobial activity against *Staphylococcus aureus*, inhibited native microbiota growth during refrigerated storage for more than 6 days, and preserved the cheese's appearance. The research also highlights the feasibility of using mandarin fiber and sodium alginate as effective texturizing agents [47].

Researchers immersed foods in nanoemulsions to evaluate their efficacy as independent coating agents, separate from a polymer matrix. Applied the chitosan solution containing mandarin essential oil nanoemulsions to the green beans. They exposed the samples to γ-irradiation, UV-C radiation, and ozonized water treatment. As shown by Severino *et al.* (2014), the synergistic activity of the nanoemulsion coating in conjunction with UV-C irradiation was efficient in reducing the development of *Listeria innocua* in the green beans [48]. As reported by Shadman *et al.* (2017), the rainbow trout from fresh water was submerged for 15 minutes in a nanoemulsion formulation that included essential oils from *Zataria multiflora* Boiss. It was then packaged in polyethylene bags and stored under refrigerated conditions at 4°C. This treatment extended the shelf life of the trout and maintained its sensory qualities throughout the storage period [49]. Freshly cut Fuji apples were immersed in an edible coating formulation of alginate-based nanoemulsion infused with lemongrass essential oil for a duration of 2 minutes. Afterward, the apples that had been treated were stored under refrigerated conditions at 4°C for two weeks. These coated samples exhibited inhibitory effects on *E. coli,* as reported by Salvia-Trujillo *et al.* [42]. An edible coating made from quinoa protein and chitosan, incorporating a thymol nanoemulsion, effectively suppressed fungal growth on cherry tomatoes over a period of 7 days at storage temperatures of 5°C [50].

Research has investigated the efficacy of natural antimicrobial agents encapsulated in nanoemulsion systems, such as essential oils from a variety of sources, including clove, geranium, lemongrass, marjoram, palmarosa, rosewood tea tree, thyme, mint, oregano oil, and plant by-product extracts [41, 42, 51]. Nanoemulsion based edible coating with lemongrass oil exhibited a bactericidal effect against *Salmonella typhimurium*, *Staphylococcus aureus*, *Listeria monocytogenes*, and *Escherichia coli*. These coatings have shown promise in inhibiting microbial growth and preserving the quality of fresh-cut fruits and vegetables during storage [51–57].

10.4.2 Antioxidant and Antibrowning Coating

Polyphenol oxidase enzymes frequently contribute to the oxidation process, affecting the color and appearance of fresh-cut fruits and vegetables, which are critical quality aspects. When these enzymes interact with oxygen, phenolic chemicals undergo transformation into dark pigments, causing undesirable surface alterations on cut produce [58]. A common method to regulate enzymatic browning involves using ascorbic acid, wherein chopped fruits and vegetables are submerged in an aqueous solution containing antioxidant and antibrowning chemicals [59]. However, drawbacks such as the loss of valuable compounds, low solubility, and susceptibility to oxidation in the environment limit the effectiveness of such treatments. Hence, there is a current challenge in developing innovative technologies that can reduce the dosage of these compounds required for fresh-cut produce.

Edible coatings based on nanoemulsions provide a practical approach to encapsulating natural antioxidants or antibrowning substances like α-tocopherol and carotenoids in the lipophilic component [7]. This method has demonstrated a reduction in the browning index of freshly cut vegetables and fruits when compared to the use of antioxidants alone [40]. It was also show by Zambrano-Zaragoza *et al.* that larger particles caused system aggregation and uneven coatings, which shows how important particle size is for getting a consistent and uniform coating. This study further revealed that nanoemulsion-based edible coatings with the addition of α-tocopherol and chitosan served as a robust oxygen barrier, preventing polyphenol oxidase from oxidizing apples and significantly reducing the browning index during storage compared to coatings using conventional methods [57]. Over a 5-month storage period, Jo *et al.* (2014) observed that a carnauba-shellac wax-based nanoemulsion with lemongrass oil coatings effectively minimized color changes in "Fuji" apples [54]. Similarly, in case of fresh-cut Fuji apples coated with a nanoemulsion of lemongrass essential oil exhibited a lower browning index than untreated ones [41, 42]. Researchers like Martínez-Hernandez *et al.*, and Radi *et al.* found that putting nanoemulsion on orange and carrot slices stops the colors from changing while they are being stored. This shows that these materials are good at blocking oxygen [52, 55].

10.4.3 Texture Enhancer

The lower acceptability of fresh-cut fruits often stems from integrity loss of texture and cell wall during storage, primarily as a result of enzymatic degradation [60]. Water loss further contributes to changes in texture, reducing cell turgidity. Nanoemulsion-based edible coverings emerge as promising solutions to enhance the texture of fresh produce. Xanthan gum, a texture enhancer based on carbohydrates, enhances nanoemulsions. Research indicates that nanoemulsions containing xanthan gum effectively delay the firmness loss in freshly cut apples [57]. Nanoemulsions comprising xanthan gum and carnauba wax preserved the guava texture for up to 30 days under specific storage conditions [61].

Nanoemulsion coatings with α-tocopherol, xanthan gum, and nopal mucilage extracts were shown in several studies to lower the activity of pectin methylesterase and polyphenol oxidase in fresh-cut apples and guava [57, 62]. Using a nanoemulsion with xanthan gum changed the metabolism of cell wall disintegration, which slowed down enzyme activity and increased firmness retention. These are important enzymes that help fruits soften. Nanoemulsions with nanosized droplets exhibited superior functionality on the cell wall of fruits at the molecular level compared to traditional coatings [41, 42, 61]. Additional studies reported positive outcomes when adding nanoemulsion loaded with texture enhancers to fresh-cut apples [41, 42, 54].

Chitosan-based nanoemulsion coatings have been shown to effectively prevent water loss and reduce texture loss in strawberries, zucchini, and green beans [48, 63, 64]. In order to change the rheological properties of nanoemulsions, stabilizers like xanthan, pectin, carrageenan, alginate, gelatine, whey protein isolate, and esters gum must be added. These stabilizers play a crucial role in reducing water loss and establishing a cross-linked network that reinforces the mechanical strength of the products, ensuring texture preservation.

Nanoemulsions were formulated using clove and cinnamon, varying the amount of soy lecithin surfactant. These nanoemulsions displayed robust antioxidant activity, and when applied to muffins, they effectively preserved the texture despite undergoing weight and moisture loss. The introduction of these nanoemulsions extended the components' shelf life to six days, ensuring longevity without compromising the overall quality of the product [65].

10.4.4 Sensory, Toxicity and Safety Considerations

Coating materials are commonly used in the preparation of cut fruits and vegetables, which necessitates the addition of antimicrobials, antioxidants, and texture enhancers to ensure consumer acceptability. Using essential oils as antimicrobials in conventional edible coatings presents difficulties due to their strong flavors and the risk of toxicity when used in higher concentrations. The sensory quality of fresh-cut apples coated with lemongrass oil and oregano essential oil [82], and it was observed that overall preference for oregano oil-coated samples was less due to its lingering herbal aroma. On the other hand, Radi *et al.* (2018) observed that nanoemulsions containing 1% orange peel essential oil when coated on orange slices exhibited superior sensory attributes compared to untreated and microemulsion-coated samples [52]. According to Kim *et al.*, the amount of lemongrass oil in the nanoemulsion had no effect on the grape berry taste, highlighting the minimal amounts of functional chemicals required in nanoemulsion-based coatings [53].

Researchers have not yet established standardized testing methods for the potential toxicity of food-grade nanoemulsions. Various factors, such as droplet size, emulsifier type, charge, and concentrations of bioactive chemicals, can influence nanoemulsion toxicity. Initial research indicates that nanoemulsions containing essential oils at concentrations effective for antimicrobial purposes are biocompatible with various human epithelial cells [83] and do not exhibit toxicity effects on mononuclear cells in the peripheral circulation [84]. However, additional research is required to obtain a more comprehensive understanding of the safety and toxicity profiles of nanoemulsions, as these results are preliminary.

10.5 Challenges Associated with Food Grade Nanoemulsions

The application of edible coatings to fresh vegetables and fruits presents both promising advancements and challenges as presented in Figure 10.2. While these coatings can enhance the shelf life and quality of produce, concerns arise regarding the alteration of internal atmospheres, particularly with the potential disorders resulting from changes in oxygen and carbon dioxide levels. The use of nano-systems in coatings introduces regulatory complexities, demanding clear guidelines to ensure safety and compliance [85]. Allergenic components in some coatings pose a risk, necessitating transparent labeling and exploration of hypoallergenic alternatives. Cost considerations also loom large, urging the exploration of economical alternatives and efficient application methods. Additionally, the lack of expertise, information on compositions, and suitable machinery hinders widespread adoption. Overcoming these challenges requires collaborative efforts, including robust research, regulatory frameworks, and technological investments, to establish edible coatings as a viable and safe solution for preserving fresh produce.

Similarly, the integration of nanoemulsion-based coatings in the food industry presents a promising yet complex frontier. Safety concerns arise due to the nanometric size of droplets, particularly when surfactants are involved, demanding a thorough understanding of potential long-term

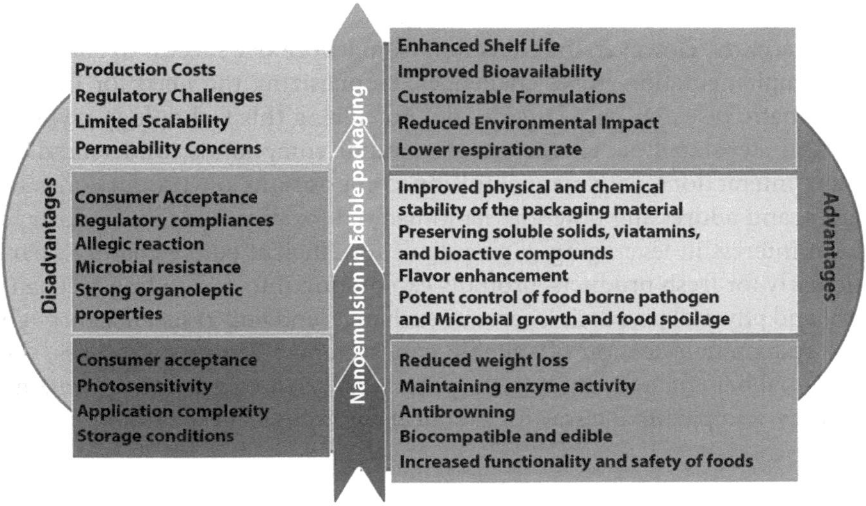

Figure 10.2 Advantages and disadvantages of nanoemulsion based food packaging.

health effects. Interactions with the gastrointestinal tract and their impact on digestive system functionality necessitate a thorough investigation. Limited *in vivo* studies leave gaps in our knowledge regarding the broader physiological impact of nanoemulsions, and those containing antimicrobial agents may disrupt the balance of the intestinal microbiota, raising concerns for overall health [86, 87]. Regulatory challenges persist, with a lack of international standards tailored to nanoemulsions requiring the development of guidelines by regulatory bodies. Scaling up production while maintaining droplet size consistency poses challenges, and ensuring long-term stability and cost-effectiveness is crucial. Collaborative efforts are needed to ensure the safe and sustainable integration of both edible coatings and nanoemulsion-based coatings into the food market, accompanied by consumer education on their benefits and safety [2].

10.6 Conclusion

The development of food-grade nanoemulsions offers a promising opportunity for enhancing edible coatings with natural active ingredients like antimicrobials, antioxidants, and texture enhancers. Recent studies, aligning with consumer preferences for natural and healthier options, showcase improved quality, safety, and shelf life for fresh products utilizing nanoemulsion-based coatings compared to their conventional counterparts. However, the transition from laboratory success to commercial implementation poses challenges, emphasizing the need for further systematic research on a larger scale. Advancing this technology involves critical steps such as characterizing natural compounds, understanding their interactions with nanoemulsion compositions on production surfaces, and addressing potential nanomaterial toxicity. The food industry's keen interest in leveraging nanosystems for efficient edible coatings, particularly for fresh products, prompts exploration into controlling mechanical and physical properties, gaseous exchange, and heat resistance through the systematic selection of suitable nanosystems. Despite challenges, the potential benefits warrant systematic research, ensuring safety, optimizing efficacy, and paving the way for practical applications in the future.

References

1. Al-Tayyar, N.A., Youssef, A.M., Al-Hindi, R.R., Edible coatings and antimicrobial nanoemulsions for enhancing shelf life and reducing foodborne pathogens of fruits and vegetables: A review. *Sustain. Mater. Technol.*, 26, e00215, 2020, doi.org/10.1016/j.susmat.2020.e00215.
2. Tripathi, A.D., Sharma, R., Agarwal, A., Haleem, D.R., Nanoemulsions based edible coatings with potential food applications. *Int. J. Biobased Plast.*, 3, 1, 112–125, 2021, doi.org/10.1080/24759651.2021.1875615.
3. Salvia-Trujillo, L., Soliva-Fortuny, R., Rojas-Graü, M.A., McClements, D.J., Martín-Belloso, O., Edible nanoemulsions as carriers of active ingredients: A review. *Annu. Rev. Food Sci. Technol.*, 8, 439–466, 2017, doi: 10.1146/annurev-food-030216-025908.
4. McClements, D.J. and Rao, J., Food-grade nanoemulsions: formulation, fabrication, properties, performance, biological fate, and potential toxicity. *Crit. Rev. Food Sci. Nutr.*, 51, 285–330, 2011, doi: 10.1080/10408398.2011.559558.
5. Aswathanarayan, J.B. and Vittal, R.R., Nanoemulsions and their potential applications in food industry. *Front. Sustain. Food Syst.*, 3, 95, 2019, doi: 10.3389/fsufs.2019.00095.
6. Shakeel, F., Baboota, S., Ahuja, A., Ali, J., Shafiq, S., Celecoxib nanoemulsion for transdermal drug delivery: Characterization and *in vitro* evaluation. *J. Dispersion Sci. Technol.*, 30, 6, 834–842, 2009, doi: 10.1080/01932690802644012.
7. Mcclements, D.J., Critical review of techniques and methodologies for characterization of emulsion stability. *Crit. Rev. Food Sci. Nutr.*, 47, 611–649, 2007, doi: 10.1080/10408390701289292.

8. Chen, L., Remondetto, G.E., Subirade, M., Food protein-based materials as nutraceutical delivery systems. *Trends Food Sci. Technol.*, 17, 272–283, 2006, doi: 10.1016/j.tifs.2005.12.011.
9. Jin, W., Xu, W., Liang, H., Li, Y., Liu, S., Li, B., Nanoemulsions for food: properties, production, characterization, and applications, in: *Emulsions*, pp. 1–36, Academic Press, Cambridge, MA, USA, 2016, doi: doi.org/10.1016/B978-0-12-804306-6.00001-5.
10. Bhosale, R.R., Osmani, R.A., Ghodake, P.P., Shaikh, S.M., Chavan, S.R., Nanoemulsion: a review on novel profusion in advanced drug delivery. *Indian J. Pharm. Biol. Res.*, 2, 122–127, 2014, doi: 10.30750/ijpbr.2.1.19.
11. Espitia, P.J., Fuenmayor, C.A., Otoni, C.G., Nanoemulsions: Synthesis, characterization, and application in bio-based active food packaging. *Compr. Rev. Food Sci. Food Saf.*, 18, 1, 264–285, 2019, doi: doi.org/10.1111/1541-4337.12405.
12. Van der Schaaf, U.S. and Karbstein, H.P., Fabrication of nanoemulsions by rotor-stator emulsification, in: *Nanoemulsions*, pp. 141–174, Academic Press, Cambridge, MA, USA 2018.
13. Scholz, P. and Keck, C.M., Nanoemulsions produced by rotor–stator high speed stirring. *Int. J. Pharm.*, 482, 1-2, 110–117, 2015, doi.org/10.1016/j.ijpharm.2014.12.040.
14. Setya, S., Talegaonkar, S., Razdan, B.K., Nanoemulsions: formulation methods and stability aspects. *World J. Pharm. Sci.*, 3, 2, 2214–2228, 2014.
15. Zhou, Y., Yue, W., Luo, Y., Luo, Q., Liu, S., Chen, H., Zhang, Q., Preparation and stability characterization of soybean protein isolate/sodium alginate complexes-based nanoemulsions using high-pressure homogenization. *LWT*, 154, 112607, 2022, doi.org/10.1016/j.lwt.2021.112607.
16. Chime, S.A., Kenechukwu, F.C., Attama, A.A., Nanoemulsions— advances in formulation, characterization and applications in drug delivery, in: *Application of Nanotechnology in Drug Delivery*, vol. 3, pp. 77–126, 2014, doi: 10.5772/58673.
17. Mushtaq, A., Wani, S.M., Malik, A.R., Gull, A., Ramniwas, S., Nayik, G.A., Bari, A., Recent insights into Nanoemulsions: Their preparation, properties and applications. *Food Chem.:* 17, 100684, 2023, doi.org/10.1016/j.fochx.2023.100684.
18. Wang, Z., Neves, M.A., Isoda, H., Nakajima, M., Preparation and characterization of micro/nano-emulsions containing functional food components. *Jpn. J. Food Eng.*, 16, 4, 263–276, 2015, doi.org/10.11301/jsfe.16.263.
19. Karthik, P., Ezhilarasi, P.N., Anandharamakrishnan, C., Challenges associated in stability of food grade nanoemulsions. *Crit. Rev. Food Sci. Nutr.*, 57, 7, 1435–1450, 2017, doi.org/10.1080/10408398.2015.1006767.
20. Qian, C., Decker, E.A., Xiao, H., McClements, D.J., Physical and chemical stability of β-carotene-enriched nanoemulsions: Influence of pH, ionic strength, temperature, and emulsifier type. *Food Chem.*, 132, 3, 1221–1229, 2012, doi.org/10.1016/j.foodchem.2011.11.091.

21. Saberi, A.H., Fang, Y., McClements, D.J., Fabrication of vitamin E-enriched nanoemulsions: Factors affecting particle size using spontaneous emulsification. *J. Colloid Interface Sci.*, 391, 95–102, 2013, doi.org/10.1016/j.jcis.2012.08.069.
22. Ribeiro, H.S., Rico, L.G., Badolato, G.G., Schubert, H., Production of O/W emulsions containing astaxanthin by repeated premix membrane emulsification. *J. Food Sci.*, 70, 2, E117–E123, 2005, doi.org/10.1111/j.1365-2621.2005.tb07083.x.
23. Laouini, A., Fessi, H., Charcosset, C., Membrane emulsification: A promising alternative for vitamin E encapsulation within nano-emulsion. *J. Membr. Sci.*, 423, 85–96, 2012, doi.org/10.1016/j.memsci.2012.07.031.
24. Maali, A. and Mosavian, M.H., Preparation and application of nanoemulsions in the last decade (2000–2010). *J. Dispersion Sci. Technol.*, 34, 1, 92–105, 2013, doi.org/10.1080/01932691.2011.648498.
25. Komaiko, J.S. and McClements, D.J., Formation of food-grade nanoemulsions using low-energy preparation methods: A review of available methods. *Compr. Rev. Food Sci. Food Saf.*, 15, 2, 331–352, 2016, doi.org/10.1111/1541-4337.12189.
26. Solans, C., Morales, D., Homs, M., Spontaneous emulsification. *Curr. Opin. Colloid Interface Sci.*, *22*, 88–93, 2016, doi.org/10.1016/j.cocis.2016.03.002.
27. Jasmina, H., Džana, O., Alisa, E., Edina, V., Ognjenka, R., Preparation of nanoemulsions by high-energy and low energy emulsification methods, in: *CMBEBIH 2017: Proceedings of the International Conference on Medical and Biological Engineering*, Springer Singapore, pp. 317–322, 2017.
28. Perazzo, A., Preziosi, V., Guido, S., Phase inversion emulsification: Current understanding and applications. *Adv. Colloid Interface Sci.*, *222*, 581–599, 2015, doi.org/10.1016/j.cis.2015.01.001.
29. Mayer, S., Weiss, J., McClements, D.J., Vitamin E-enriched nanoemulsions formed by emulsion phase inversion: factors influencing droplet size and stability. *J. Colloid Interface Sci.*, *402*, 122–130, 2013, doi.org/10.1016/j.jcis.2013.04.016.
30. Solans, C. and Solé, I., Nano-emulsions: Formation by low-energy methods. *Curr. Opin. Colloid Interface Sci.*, 17, 5, 246–254, 2012, doi.org/10.1016/j.cocis.2012.07.003.
31. Anandharamakrishnan, C., *Techniques for nanoencapsulation of food ingredients*, vol. 8, pp. 65–67, Springer, New York, 2014.
32. Jintapattanakit, A., Preparation of nanoemulsions by phase inversion temperature (PIT) method. *Pharm. Sci. Asia*, 45, 1, 1–12, 2018.
33. Koroleva, M.Y. and Yurtov, E.V., Nanoemulsions: the properties, methods of preparation and promising applications. *Russ. Chem. Rev.*, 81, 1, 21, 2012, doi.org/10.1070/RC2012v081n01ABEH004219.
34. Ju, J., Xie, Y., Guo, Y., Cheng, Y., Qian, H., Yao, W., Application of edible coating with essential oil in food preservation. *Crit. Rev. Food Sci. Nutr.*, 59, 15, 2467–2480, 2019, doi.org/10.1080/10408398.2018.1456402.

35. Andrade, R.D., Skurtys, O., Osorio, F.A., Atomizing spray systems for application of edible coatings. *Compr. Rev. Food Sci. Food Saf.*, 11, 3, 323–337, 2012, doi.org/10.1111/j.1541-4337.2012. 00186.x.
36. Chaudhary, P., Fatima, F., Kumar, A., Relevance of nanomaterials in food packaging and its advanced future prospects. *J. Inorg. Organomet. Polym. Mater.*, 30, 5180–5192, 2020, doi.org/10.1007/s10904-020-01674-8.
37. Abinash, V., Rahul, T., Antoniraj, M.G., Moses, J.A., Anandharamakrishnan, C., Nanotechnology approaches for food fortification, in: *Food, Medical, and Environmental Applications of Polysaccharides*, pp. 161–186, Elsevier, Amsterdam, The Netherlands, 2021, doi.org/10.1016/B978-0-12-819239-9.00007-5.
38. Artiga-Artigas, M., Lanjari-Pérez, Y., Martín-Belloso, O., Curcumin-loaded nanoemulsions stability as affected by the nature and concentration of surfactant. *Food Chem.*, 266, 466–474, 2018, doi.org/10.1016/j.foodchem.2018.06.043.
39. Davidson, P.M., Critzer, F.J., Taylor, T.M., Naturally occurring antimicrobials for minimally processed foods. *Annu. Rev. Food Sci. Technol.*, 4, 163–190, 2013, doi.org/ 10.1146/annurev-food-030212-182535.
40. Rojas-Graü, M.A., Soliva-Fortuny, R., Martín-Belloso, O., Edible coatings to incorporate active ingredients to fresh-cut fruits: a review. *Trends Food Sci. Technol.*, 20, 10, 438–447, 2009, doi.org/10.1016/j.tifs.2009.05.002.
41. Salvia-Trujillo, L., Rojas-Graü, A., Soliva-Fortuny, R., Martín-Belloso, O., Physicochemical characterization and antimicrobial activity of food-grade emulsions and nanoemulsions incorporating essential oils. *Food Hydrocolloids*, 43, 547–556, 2015a, doi.org/10.1016/j.foodhyd.2014.07.012.
42. Salvia-Trujillo, L., Rojas-Graü, M.A., Soliva-Fortuny, R., Martín-Belloso, O., Use of antimicrobial nanoemulsions as edible coatings: Impact on safety and quality attributes of fresh-cut Fuji apples. *Postharvest Biol. Technol.*, 105, 8–16, 2015b, doi.org/10.1016/j.postharvbio.2015.03.009.
43. Otoni, C.G., deMoura, M.R., Aouada, F.A., Camilloto, G.P., Cruz, R.S., Lorevice, M.V., Antimicrobial and physicalmechanical properties of pectin/papaya puree/cinnamaldehyde nanoemulsion edible composite films. *Food Hydrocolloids*, 41, 188–194, 2014, doi: 10.1016/j.foodhyd.2014.04.013.
44. Sasaki, R.S., Mattoso, L.H., De Moura, M.R., New edible bionanocomposite prepared by pectin and clove essential oil nanoemulsions. *J. Nanosci. Nanotechnol.*, 16, 6, 6540–6544, 2016, doi.org/10.1166/jnn.2016.11702.
45. Alexandre, E.M.C., Lourenço, R.V., Bittante, A.M.Q.B., Moraes, I.C.F., Sobral, P.J.A., Gelatine based films reinforced with montmorillonite and activated with nanoemulsion of ginger essential oil for food packaging applications. *Food Packag. Shelf Life*, 10, 87–96, 2016, doi: 10.1016/j.fpsl.2016.10.004.
46. Severino, R., Ferrari, G., Vu, K.D., Donsì, F., Salmieri, S., Lacroix, M., Antimicrobial effects of modified chitosan based coating containing nanoemulsion of essential oils, modified atmosphere packaging and gamma irradiation against *Escherichia coli* O157: H7 and *Salmonella Typhimurium*

on green beans. *Food Control*, 50, 215–222, 2015, doi: 10.1016/j.foodcont.2014.08.029.

47. Artiga-Artigas, M., Acevedo-Fani, A., Martín-Belloso, O., Improving the shelf life of low-fat cut cheese using nanoemulsion-based edible coatings containing oregano essential oil and mandarin fiber. *Food Control*, 76, 1–12, 2017, doi.org/10.1016/j.foodcont.2017.01.001.
48. Severino, R., Vu, K.D., Donsì, F., Salmieri, S., Ferrari, G., Lacroix, M., Antibacterial and physical effects of modified chitosan based-coating containing nanoemulsion of mandarin essential oil and three non-thermal treatments against *Listeria innocua* in green beans. *Int. J. Food Microbiol.*, 191, 82–88, 2014, doi.org/10.1016/j.ijfoodmicro.2014.09.007.
49. Shadman, S., Hosseini, S.E., Langroudi, H.E., Shabani, S., Evaluation of the effect of a sunflower oil-based nanoemulsion with *Zataria multiflora* Boiss. essential oil on the physicochemical properties of rainbow trout (*Oncorhynchus mykiss*) fillets during cold storage. *LWT Food Sci. Technol.*, 79, 511–517, 2017, doi: 10.1016/j.lwt.2016.01.073.
50. Robledo, N., Vera, P., López, L., Yazdani-Pedram, M., Tapia, C., Abugoch,L. Thymol nanoemulsions incorporated in quinoa protein/chitosan edible films; antifungal effect in cherry tomatoes. *Food Chem.*, 246, 211–219, 2018, doi: 10.1016/j.foodchem.2017.11.032.
51. Bhargava, K., Conti, D.S., da Rocha, S.R.P., Zhang, Y., Application of an oregano oil nanoemulsion to the control of foodborne bacteria on fresh lettuce. *Food Microbiol.*, 47, 69–73, 2015, doi: 10.1016/j.fm.2014.11.007.
52. Radi, M., Akhavan-Darabi, S., Akhavan, H.R., Amiri, S., The use of orange peel essential oil microemulsion and nanoemulsion in pectin-based coating to extend the shelf life of fresh-cut orange. *J. Food Process. Preserv.*, 42, 2, e13441, 2018, doi.org/10.1111/jfpp.13441.
53. Kim, I.H., Oh, Y.A., Lee, H., Song, K.B., Min, S.C., Grape berry coatings of lemongrass oil-incorporating nanoemulsion. *LWT Food Sci. Technol.*, 58, 1, 1–10, 2014, doi.org/10.1016/j.lwt.2014.03.018.
54. Jo, W.S., Song, H.Y., Song, N.B., Lee, J.H., Min, S.C., Song, K.B., Quality and microbial safety of 'Fuji'apples coated with carnauba-shellac wax containing lemongrass oil. *LWT Food Sci. Technol.*, 55, 2, 490–497, 2014, doi.org/10.1016/j.lwt.2013.10.034.
55. Martínez-Hernández, G.B., Amodio, M.L., Colelli, G., Carvacrol-loaded chitosan nanoparticles maintain quality of fresh-cut carrots. *Innov. Food Sci. Emerg. Technol.*, 41, 56–63, 2017, doi.org/10.1016/j.ifset.2017.02.005.
56. Taştan, O., Pataro, G., Donsì, F., Ferrari, G., Baysal, T., Decontamination of fresh-cut cucumber slices by a combination of a modified chitosan coating containing carvacrol nanoemulsions and pulsed light. *Int. J. Food Microbiol.*, 260, 75–80, 2017, doi.org/10.1016/j.ijfoodmicro.2017.08.011.
57. Zambrano-Zaragoza, M.L., Mercado-Silva, E., Gutiérrez-Cortez, E., Cornejo-Villegas, M.A., Quintanar-Guerrero, D., The effect of nano-coatings with α-tocopherol and xanthan gum on shelf-life and browning index of fresh-cut

"Red Delicious" apples. *Innov. Food Sci. Emerg. Technol.*, 22, 188–196, 2014, doi.org/10.1016/j.ifset.2013.09.008.
58. Zawistowski, J., Biliaderis, C.G., Eskin, N.A.M., *'Polyphenol oxidase', Oxidative enzymes in foods*, pp. 217–273, Elsevier, London, 1991.
59. McEvily, A.J., Iyengar, R., Otwell, W.S., Inhibition of enzymatic browning in foods and beverages. *Crit. Rev. Food Sci. Nutr.*, 32, 3, 253–273, 1992, doi.org/10.1080/10408399209527599.
60. Alandes, L., Hernando, I., Quiles, A., Pérez-Munuera, I., Lluch, M.A., Cell wall stability of fresh-cut Fuji apples treated with calcium lactate. *J. Food Sci.*, 71, 9, S615–S620, 2006, doi.org/10.1111/j.1750-3841.2006.00180.x.
61. Zambrano-Zaragoza, M.L., Mercado-Silva, E., Ramirez-Zamorano, P., Cornejo-Villegas, M.A., Gutiérrez-Cortez, E., Quintanar-Guerrero, D., Use of solid lipid nanoparticles (SLNs) in edible coatings to increase guava (*Psidium guajava* L.) shelf-life. *Food Res. Int.*, 51, 2, 946–953, 2013, doi.org/10.1016/j.foodres.2013.02.012.
62. García-Betanzos, C.I., Hernández-Sánchez, H., Bernal-Couoh, T.F., Quintanar-Guerrero, D., de la Luz Zambrano-Zaragoza, M., Physicochemical, total phenols and pectin methylesterase changes on quality maintenance on guava fruit (*Psidium guajava* L.) coated with candeuba wax solid lipid nanoparticles-xanthan gum. *Food Res. Int.*, 101, 218–227, 2017, doi.org/10.1016/j.foodres.2017.08.065.
63. Donsì, F., Cuomo, A., Marchese, E., Ferrari, G., Infusion of essential oils for food stabilization: Unraveling the role of nanoemulsion-based delivery systems on mass transfer and antimicrobial activity. *Innov. Food Sci. Emerg. Technol.*, 22, 212–220, 2014, doi.org/10.1016/j.ifset.2014.01.008.
64. Donsì, F., Marchese, E., Maresca, P., Pataro, G., Vu, K.D., Salmieri, S., Ferrari, G., Green beans preservation by combination of a modified chitosan based-coating containing nanoemulsion of mandarin essential oil with high pressure or pulsed light processing. *Postharvest Biol. Technol.*, 106, 21–32, 2015, doi.org/10.1016/j.postharvbio.2015.02.006.
65. Prastuty, Kaur, G., Singh, A., Shelf life extension of muffins coated with cinnamon and clove oil nanoemulsions. *J. Food Sci. Technol.*, 59, 1878–1888, 2021, doi.org/10.1007/s13197-021-05202-5.
66. Xiong, Y., Li, S., Warner, R.D., Fang, Z., Effect of oregano essential oil and resveratrol nanoemulsion loaded pectin edible coating on the preservation of pork loin in modified atmosphere packaging. *Food Control*, 114, 107226, 2020, doi.org/10.1016/j.foodcont.2020.107226.
67. Almasi, L., Radi, M., Amiri, S., McClements, D.J., Fabrication and characterization of antimicrobial biopolymer films containing essential oil-loaded microemulsions or nanoemulsions. *Food Hydrocolloids*, 117, 106733, 2021, doi.org/10.1016/j.foodhyd.2021.106733.
68. Abdou, E.S., Galhoum, G.F., Mohamed, E.N., Curcumin loaded nanoemulsions/pectin coatings for refrigerated chicken fillets. *Food Hydrocolloids*, 83, 445–453, 2018, doi.org/10.1016/j.foodhyd.2018.05.026.

69. Hasani-Javanmardi, M., Fallah, A.A., Abbasvali, M., Effect of safflower oil nanoemulsion and cumin essential oil combined with oxygen absorber packaging on the quality and shelf-life of refrigerated lamb loins. *LWT*, 147, 111557, 2021, doi.org/10.1016/j.lwt.2021.111557.
70. Noori, S., Zeynali, F., Almasi, H., Antimicrobial and antioxidant efficiency of nanoemulsion-based edible coating containing ginger (*Zingiber officinale*) essential oil and its effect on safety and quality attributes of chicken breast fillets. *Food Control*, 84, 312–320, 2018, doi.org/10.1016/j.foodcont.2017.08.015.
71. Sharifimehr, S., Soltanizadeh, N., Hossein Goli, S.A., Effects of edible coating containing nano-emulsion of Aloe vera and eugenol on the physicochemical properties of shrimp during cold storage. *J. Sci. Food Agric.*, 99, 7, 3604–3615, 2019, doi.org/10.1002/jsfa.9581.
72. Das, S.K., Vishakha, K., Das, S., Chakraborty, D., Ganguli, A., Carboxymethyl cellulose and cardamom oil in a nanoemulsion edible coating inhibit the growth of foodborne pathogens and extend the shelf life of tomatoes. *Biocatal. Agric. Biotechnol.*, 42, 102369, 2022, doi.org/10.1016/j.bcab.2022.102369.
73. Das, S., Vishakha, K., Banerjee, S., Mondal, S., Ganguli, A., Sodium alginate-based edible coating containing nanoemulsion of Citrus sinensis essential oil eradicates planktonic and sessile cells of food-borne pathogens and increased quality attributes of tomatoes. *Int. J. Biol. Macromol.*, 162, 1770–1779, 2020, doi.org/10.1016/j.bcab.2022.102369.
74. Yang, Z., Li, M., Zhai, X., Zhao, L., Tahir, H.E., Shi, J., Xiao, J., Development and characterization of sodium alginate/tea tree essential oil nanoemulsion active film containing TiO_2 nanoparticles for banana packaging. *Int. J. Biol. Macromol.*, 213, 145–154, 2022, doi.org/10.1016/j.ijbiomac.2022.05.164.
75. Kim, I.H., Lee, H., Kim, J.E., Song, K.B., Lee, Y.S., Chung, D.S., Min, S.C., Plum coatings of lemongrass oil-incorporating carnauba wax-based nanoemulsion. *J. Food Sci.*, 78, 10, E1551–E1559, 2013, doi.org/10.1111/1750-3841.12244.
76. Sow, L.C., Tirtawinata, F., Yang, H., Shao, Q., Wang, S., Carvacrol nanoemulsion combined with acid electrolysed water to inactivate bacteria, yeast *in vitro* and native microflora on shredded cabbages. *Food Control*, 76, 88–95, 2017, doi.org/10.1016/j.foodcont.2017.01.007.
77. Gundewadi, G., Rudra, S.G., Sarkar, D.J., Singh, D., Nanoemulsion based alginate organic coating for shelf life extension of okra. *Food Packag. Shelf Life*, 18, 1–12, 2018, doi.org/10.1016/j.fpsl.2018.08.002.
78. Sanchez, L.T., Pinzon, M.I., Villa, C.C., Development of active edible films made from banana starch and curcumin-loaded nanoemulsions. *Food Chem.*, 371, 131121, 2022, doi.org/10.1016/j.foodchem.2021.131121.
79. Roy, S. and Rhim, J.W., Fabrication of bioactive binary composite film based on gelatin/chitosan incorporated with cinnamon essential oil and rutin. *Colloids Surf., B*, 204, 111830, 2021, doi.org/10.1016/j.colsurfb.2021.111830.

80. Prakash, A., Baskaran, R., Vadivel, V., Citral nanoemulsion incorporated edible coating to extend the shelf life of fresh cut pineapples. *LWT*, 118, 108851, 2020, doi.org/10.1016/j.lwt.2019.108851.
81. Polat Yemiş, G., Sezer, E., Sıçramaz, H., Inhibitory effect of sodium alginate nanoemulsion coating containing myrtle essential oil (*Myrtus communis* L.) on Listeria monocytogenes in Kasar cheese. *Molecules*, 27, 21, 7298, 2022, doi: 10.3390/molecules27217298.
82. Rojas-Graü, M.A., Raybaudi-Massilia, R.M., Soliva-Fortuny, R.C., Avena-Bustillos, R.J., McHugh, T.H., Martín-Belloso, O., Apple puree-alginate edible coating as carrier of antimicrobial agents to prolong shelf-life of fresh-cut apples. *Postharvest Biol. Technol.*, 45, 2, 254–264, 2007, doi.org/10.1016/j.postharvbio.2007.01.017.
83. Bonferoni, M.C., Sandri, G., Rossi, S., Usai, D., Liakos, I., Garzoni, A., Ferrari, F., A novel ionic amphiphilic chitosan derivative as a stabilizer of nanoemulsions: Improvement of antimicrobial activity of *Cymbopogon citratus* essential oil. *Colloids Surf., B*, 152, 385–392, 2017, doi.org/10.1016/j.colsurfb.2017.01.043.
84. de Godoi, S.N., Quatrin, P.M., Sagrillo, M.R., Nascimento, K., Wagner, R., Klein, B., Ourique, A.F., Evaluation of stability and *in vitro* security of nanoemulsions containing *Eucalyptus globulus* oil. *Biomed Res. Int.*, 10, 2723418, 2017, doi.org/ 10.1155/2017/2723418.
85. Yaashikaa, P.R., Kamalesh, R., Kumar, P.S., Saravanan, A., Vijayasri, K., Rangasamy, G., Recent advances in edible coatings and their application in food packaging. *Food Res. Int.*, 173, 113366, 2023, doi.org/10.1016/j.foodres.2023.113366.
86. Gaspar, M.C. and Braga, M.E., Edible films and coatings based on agrifood residues: A new trend in the food packaging research. *Curr. Opin. Food Sci.*, 50, 101006, 2023, doi.org/10.1016/j.cofs.2023.101006.
87. Ribeiro, A.M., Estevinho, B.N., Rocha, F., Preparation and incorporation of functional ingredients in edible films and coatings. *Food Bioprocess Technol.*, 14, 209–231, 2021, doi.org/10.1007/s11947-020-02528-4.

11

Long-Term Enhancement of Food Stability Using Encapsulation-Based Coating

Snigdha Homroy[1], Monika Chand[1], Binanshu Talwar[1], Aishwarya Dhiman[1], Priyanka Kumari Singh[2], Abdul Wahid[1] and Rajni Chopra[1]*

[1]*Department of Food Science and Technology, National Institute of Food Technology Entrepreneurship and Management, Kundli, Haryana, India*
[2]*Department of Food and Nutrition & Food Technology, Institute of Home Economics, University of Delhi, Delhi, India*

Abstract

"Long-term enhancement of food stability using encapsulation-based coating" is a comprehensive exploration of encapsulation techniques' vital role in enhancing food processing and preservation. This book discusses the mechanisms and applications of encapsulation for preventing food deterioration, stabilizing food products, and ensuring their safety and quality. After briefly introducing edible coatings and food safety, this chapter explains encapsulation techniques and their significance in the food industry. Readers will gain insights into appropriate encapsulation materials for various food applications, enabling the formulation of efficient and targeted encapsulation strategies. The second section focuses on the mechanisms of encapsulation for preventing food deterioration. By addressing the underlying causes of food spoilage, including oxidation, moisture uptake, and enzymatic reactions, this section clarifies how encapsulation-based coatings act as protective barriers, thereby effectively limiting these deterioration mechanisms. Research-based case studies are examples of the successful implementation of encapsulation in food preservation. The third section explores the encapsulation methods and processes that are currently in use. Readers will gain a comprehensive understanding of microencapsulation methods, such as spray drying, coacervation, and extrusion, as well as nanoencapsulation techniques, such as emulsion-based methods and liposomes. Emerging novel technologies, such as electrospinning and fluidized bed coating, have also been described for prospective food stability enhancement. The fourth section explores specific encapsulation

**Corresponding author*: rajnichopra.niftem@gmail.com

Mousumi Sen (ed.) Food Coatings and Preservation Technologies, (367–402) © 2025 Scrivener Publishing LLC

matrices for various food applications and preservation. This chapter analyzes different encapsulation matrices, such as lipids, proteins, and polysaccharides, and highlights their compatibility with various food products. Techniques for modifying encapsulation materials to suit specific storage conditions and food types are also discussed. Furthermore, readers will learn about the importance of encapsulation efficiency, evaluation of the release and control of bioactive compounds in the gastrointestinal tract and impact of encapsulation on sensory attributes and nutritional quality. Chapter also discusses regulatory and safety considerations, covering food safety regulations and guidelines for edible coatings and ingredients permitted in food products. The health and environmental implications of encapsulation materials were critically evaluated, and consumer perceptions of encapsulated food products were discussed. The final section discusses future possibilities and challenges for enhancing food stability through encapsulation. It explores potential advancements and innovations in food encapsulation technologies, focusing on cost-effectiveness and scaling up encapsulation processes for industry-level operations. "Long-term enhancement of food stability using encapsulation-based coating" provides a valuable resource for researchers, food scientists, and professionals seeking to learn the potential of encapsulation-based coatings to preserve and enhance the stability of food products.

Keywords: Edible coating, encapsulation, preservation, antioxidant, antimicrobial

11.1 Introduction

Edible coating is a nonconventional method used for the protection of food products against physiological disorders and microbial contamination. The edible coating is an eco-friendly packaging that replaces synthetic or plastic packaging [1] since single-use plastic is untenable and unsustainable [2, 3].

Various techniques are employed for food preservation, edible coating is one of them. In multilayered containers that contain nonedible films, the edible films can be employed as an inner layer that interacts directly with the foods. They can be used to retain the interior moisture and solute flow of foods like pizza, pies, and candies [4]. This preservation technique can be employed either by direct application or by spraying. Edible coating is a material used to cover the surface of food products, which is safe to consume and protects the food from mechanical damage, light, and other contaminants. This can also be defined as a thin layer of an edible substance that is adhered to the surface of the food to avoid the transportation of different molecules. These coatings can not only extend the shelf life but also reduce the risk of pathogen growth on the surface of food products [5, 6]. Some of the target properties required from the edible coatings are transparency, inert nature, low gas permeability and prevention of moisture migration in the food products [7–9]. Encapsulation in edible films is the entrapment of a substance (often bioactive) within a substance, where the former substance is the core material and the latter substance is the wall material [10]. On the basis of their dimension size, the film-forming materials created by encapsulation can be divided into two groups, microencapsulation and nanoencapsulation [11]. The encapsulation method works on the edible coating's embedding effect to establish a microenvironment in the structural matrix that can regulate interactions between the internal and external components [12]. Some of the encapsulation systems are spray drying, spray chilling, co-extrusion, spray drying being the most common technique. It is defined as liquid droplets enclosed in edible coating. For this method, the substance must be emulsified with an encapsulating agent [6]. The main advantages of edible coatings includes low nutrient loss, improved sensory qualities and enhanced consumer acceptability with lower susceptibility to microbial contamination.

Edible coatings are made using a variety of biopolymers, including lipids, proteins, polysaccharides, and composite materials. Nowadays, the combination of these molecules is also used to bring in the synergistic effect with respect to preservation. Therefore, the covering material must still be entirely edible [13]. Applying techniques also varies depending on food products, surfaces, and coating targets [6].

Edible coating has emerged as a sustainable option due to change in market dynamics caused by consumer awareness of safe and chemical-free meals nowadays. The storage life of highly perishable foods can be enhanced further via creative and novel types of edible coatings. Selecting the ideal composition of several polymers for a certain agricultural product significantly lengthens its shelf life. This chapter is about encapsulation based edible coating and its role in food preservation with shelf life enhancement. For specialized uses in food and stability enhancement, encapsulating technologies and product shapes have also been investigated. The chapter concludes with the future perspective and challenges in food stability and enhancement.

11.2 Role of Encapsulation in Shelf Life Enhancement

Encapsulation in food is the process of trapping food ingredients within a shell or coating as shown in Figure 11.1. This can be done for a variety of reasons, including

- To protect the ingredient from degradation: Encapsulation can protect ingredients from light, heat, moisture, and other

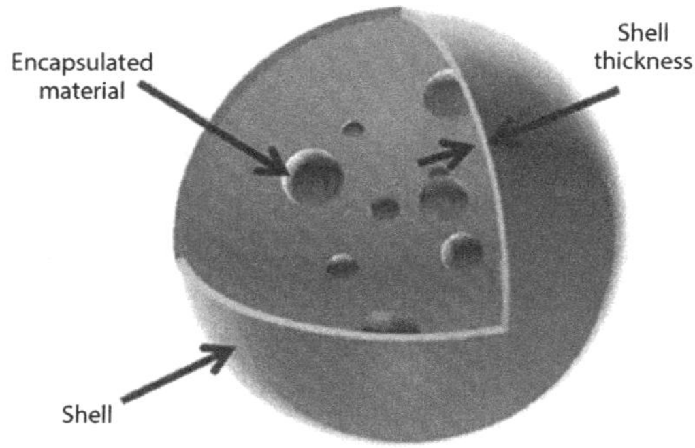

Figure 11.1 Encapsulation (source: Taouzinet *et al.*, 2023).

environmental factors that can cause them to degrade. This can help to extend the shelf life of food products.

- To improve the delivery of the ingredient: Encapsulation can improve the delivery of ingredients to specific parts of the body or to specific cell types. This can be useful for delivering nutrients, probiotics, or other beneficial compounds.
- To mask the taste or odour of the ingredient: Encapsulation can be used to mask the taste or odour of unpleasant ingredients, such as bitter vitamins or minerals. This can make food products more palatable.

Encapsulation is a versatile technique that can be used to improve the quality, shelf life, and delivery of a variety of food ingredients. It is used in a wide range of food products, including beverages, dairy products, bakery products, and confectionery products. Figure 11.2 shows a graphical representation of the possible application of encapsulation technology in food industry.

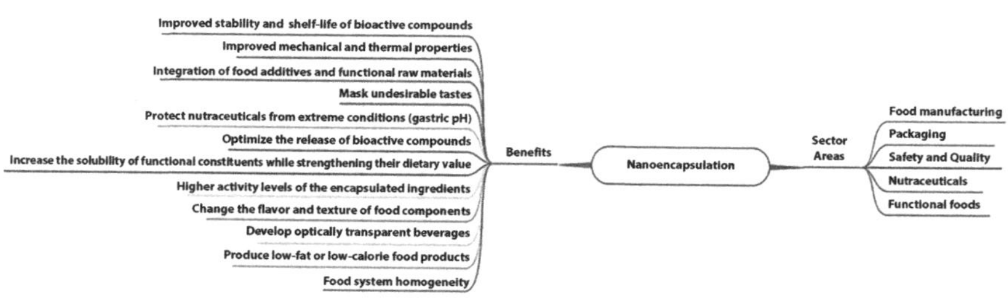

Figure 11.2 General areas covered by nanoencapsulation in the food sector and their benefits (source: Taouzinet *et al.*, 2023).

11.2.1 Encapsulation Technology

The utilization of encapsulation technology has shown to be a multifaceted and pioneering methodology that has made substantial contributions to the advancement of functional and fortified food products. The method described here pertains to the encapsulation of bioactive chemicals, flavours, colours, and other delicate elements within protective matrices. This process serves to improve the stability, bioavailability, and controlled release of these substances. Encapsulation finds application in a range of food-related contexts, encompassing the preservation of food quality during processing and storage, the generation of innovative textures, and the precise delivery of nutrients and functional components. The initial iterations of microcapsules exhibited impermeability and were often disrupted using mechanical methods in order to activate the encapsulated contents. The examples encompassed the controlled release of flavours and smells, perfumes, pharmaceuticals, detoxifying agents, fertilizers, and textile and printing precursors.

11.2.2 Appropriate Encapsulation Materials for Food Applications

Encapsulation materials have a significant impact on the food business as they contribute to the preservation, enhancement of flavour, and controlled release of bioactive substances in diverse food products. Encapsulation refers to the process of enclosing active components or compounds within a protective matrix or shell, which can be composed of diverse materials. In recent years, there has been a growing recognition of the significance of this process, as it holds the potential to enhance food quality, prolong the duration of shelf life, and augment the entire customer experience. This document aims to examine the importance of encapsulating materials in food-related contexts, the various categories of materials employed, their respective characteristics, and their wide-ranging uses.

11.3 Role of Encapsulation in Prevention of Food Deterioration

The food sector is continuously expanding and in such a scenario, the preservation of the quality and shelf life of the food products is of utmost importance. It is essential to maintain the nutritional integrity and freshness of foods ranging from raw fruits and vegetables to processed foods to dietary supplements, both for customer satisfaction and sustainability. Encapsulation is one such novel technology which had aimed at assuring the same. Food preservation is crucial for minimizing food waste along with ensuring food safety, national security and stability of world food supply. It plays a vital role in contemporary society by tackling various social, economic and environmental issues. It is now a widely used technology in the food sector which can be used to protect the delicate and volatile food components from external deteriorants causing degradation or spoilage. Encapsulation creates an oxygen barrier around the delicate food components like vitamins, bioactive substances, and flavors, which serves as a shield for them and prevents them from oxidative damage. The exposure to oxygen is the main factor responsible for food deterioration and degradation, further leading to rancidity, color change and nutrient loss. The secondary matrix occurs as a shell which serves as a physical barrier against air, light, moisture, and other elements causing damage. For instance, vitamin C is shielded from photochemical oxidation upon encapsulation. It also helps in controlling the nutrient release of the bioactive compounds. Furthermore, the amount of moisture absorbed in foods is also reduced (mainly in powdered food components), thus keeping such products from clumping or getting soggy and losing their texture. The additional upsides of encapsulation include the features such as, regulated nutrient release, improved solubility, flavor retention, oxidation prevention, prolonged shelf life, and defense against photochemical reactions. Depending on the particular needs of the food product and the required level of encapsulation, a variety of techniques, such as spray drying, coacervation, extrusion, and emulsification, can be used to accomplish encapsulation. It has potential application in the food industry to protect delicate volatile ingredients, enhance the products' overall quality and extend its shelf life. The delicate food components (i.e., bioactive compound/volatile compound/antimicrobial agent) are isolated from the external environment by the process of encapsulation, resulting in formation of a capsule [14]. The primary function of the secondary shell, matrix or encapsulant is to safeguard this ingredient by encasing it and restricting its interaction with degradants such as moisture, air, acids and enzymes. Encapsulation is largely employed in the food industry to safeguard delicate food ingredients (such as flavors, vitamins, omega-3 fatty acids, and probiotics) and to facilitate release of the ingredient at the desired time and location. Unsaturated essential fatty acids, for instance omega-3 fatty acids, are highly prone to oxidation, which lowers their nutritional value and lead to formation of various off odors and off flavors due to rancidity. These can also be prevented by encapsulation. During encapsulation, the PUFA rich oil is mixed with an aqueous dispersion of the encapsulating matrix, containing an emulsifier, which is subsequently subjected to heating along with addition of preservatives (antioxidants or antimicrobials) to improve the stability. The encapsulated fat's rate of lipid oxidation is influenced by a number of variables, including the nature of the lipid interface, the concentration of encapsulates in the lipid phase, oxygen availability, water activity, capsule storage conditions, and the method of incorporation [15]. To turn these emulsions into a free-flowing powder, further drying methods are used, such as spray drying, vacuum drying, fluid bed drying, or freeze drying. Another important element that contributes to food spoilage is microbial contamination. Different food products can become contaminated by bacteria and fungi, which can have a serious detrimental impact on the sensory (flavor, color, and texture) qualities of food. More importantly, foodborne illnesses, which are brought on by microbial contamination in food commodities and are influenced by both internal and external factors (such as pH, oxygen content, and water content of a food product), pose a serious threat to food safety. To increase food safety, a number of foodborne pathogens must be taken into account, including *Staphylococcus aureus, Listeria monocytogenes, Escherichia coli O157, Salmonella typhimurium, Bacillus cereus, Campylobacter jejuni, Clostridium perfringens,* and *Aspergillus niger* [16]. By integrating natural antimicrobials into particles using the encapsulation technique, target compounds are protected from hostile environments and their release is also controlled, which increases the antimicrobial properties of the compounds [17]. Lavender essential oil's antibacterial activity was around three times greater against *S. aureus*, *Candida albicans*, and *E. coli* when it was encapsulated in hydroxypropyl-cyclodextrin (HPCD) [18]. The preservation of fruit and vegetable juices, meat and dairy-based products cereal-based goods, and edible coatings (for fruits and vegetables) have all benefited from the use of such encapsulated antimicrobials. Table 11.1 lists a few of these instances.

11.4 Encapsulation Methods and Product Forms

Over the years, significant advancements have been made in encapsulation methods to improve the efficiency and effectiveness of food products. Originally, techniques such as spray drying and freeze drying were commonly employed for ingredient encapsulation. However, more recent approaches including extrusion, fluidized bed technology, and nanoencapsulation have emerged as significant alternatives. These advancements

Table 11.1 Enhancement in shelf life through encapsulation-based products.

S. no.	Food product	Antimicrobial compound	Encapsulation system	Function/Result	Reference
1	Tomato Juice	Rosemary essential oil	β- cyclodextrin	Against *L. monocytogenes*, *S. typhimurium*, *Candida tropicalis*, and *Saccharomyces pastorianus*	[19]
2.	Orange juice	Nisin	Chitosan monomethyl fumaric acid nanoparticles	Against several food borne pathogens	[20]
3.	Milk	Thyme Oil	Nanoemulsion	Against *S. aureus and E.coli*	[21]
4.	Milk	Garlic Extract	Nanoliposome	Against *Listeria*	[22]
5.	Beef	Zataria multiflora Boiss essential oil	Liposomes	Against *E.coli*	[23]
6.	Bread	Carvacrol	Electrospan zein and PLA fibers	Against mold and yeast	[24]
7.	Essential Oil/ application in orange juice	Lemon grass	Emulsification/ sonication	Antibacterial effect in orange juice similar to sodium benzoate	[25]
8.	Apple slices	Cinnamon essential oil	Emulsification	Shelf life increased for 6 days with 1500-2000 ppm essential oil zein-protein nanoemulsion without any change in flavor	[26]

have had transformative effects on the food industry by improving shelf life through innovative solutions like coatings based on encapsulation for fruits and vegetables.

11.4.1 Coacervation

Coacervation is an encapsulation technique with a long history and wide usage. It relies on the electric attraction between two biopolymers carrying opposing charges. The formation of coacervates takes place within a specific pH range. In this method, the polymer rich coacervate phase splits up from the liquid phase. The process can be categorized as simple or complex. In simple coacervation, the polymer is either desolvated through the incorporation of a lyophilic nonsolvent (such as ethanol), or precipitated through addition of electrolytes (like sodium sulfate), or regulated changes in temperature. These conditions facilitate interactions between macromolecules and provide favorable conditions for producing microcapsules containing water-repelling substances like essential oils. Simple coacervation presents significant benefits in terms of cost-effectiveness and operational flexibility compared to complex coacervation. In the case of simple coacervation, phase separation is induced utilizing affordable inorganic salts, whereas complex coacervation proves to be highly susceptible even to minor alterations in pH levels. Additionally, the utilization of relatively expensive hydrocolloids is required for complex coacervation processes [27].

Complex coacervation involves the interlinkage of multiple polymers with opposite charges. The initial step in microencapsulation through complex coacervation is the preparation of a solution containing an aqueous mixture of bioactive compounds and different types of polymers, typically including a protein and a polysaccharide. This solution is maintained at a pH and temperature above the isoelectric an gelling point of the protein. Subsequently, by reducing the pH below the isoelectric point,

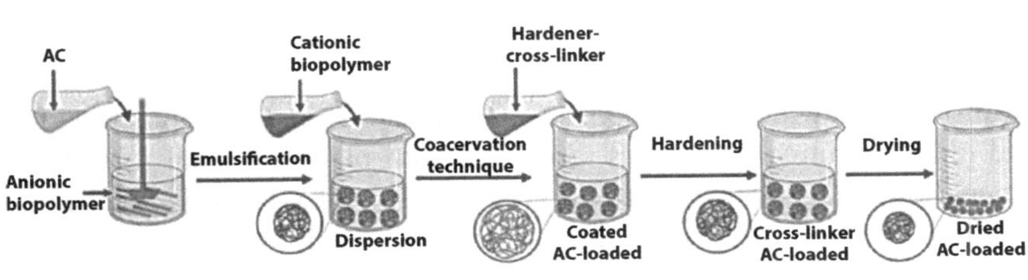

Figure 11.3 Coacervation and phase separation (source: Taouzinet *et al.*, 2023).

electrostatic attraction occurs between oppositely charged polymers which leads to separation between the liquid and insoluble polymer-rich phase. The formation of wall around hydrophobic droplets then occurs as these polymer-rich phases are deposited onto them, followed by controlled cooling below gelling temperature for further solidification. To enhance wall hardening in microcapsules, cross-linking agents can be added into this process [28]. Figure 11.3 illustrates the process of complex coacervation.

Polyelectrolytes, which are large macromolecules, contain a many of functional groups that can either acquire a charge under appropriate conditions or are charged already. Polyelectrolyte complexes (PECs) form when oppositely charged polyelectrolytes are mixed together in a solution. PECs exhibit properties of being generally nontoxic, well-tolerated by the body, and biocompatible. The primary interactions between two polyelectrolytes involve dipole-dipole and strong electrostatic interactions as well as hydrogen bonding. Several neutral polymers such as hydroxypropyl methylcellulose, chitosan (cationic polymer), κ-carrageenan (anionic polymer), and sodium alginate have been employed in creating various PEC combinations like sodium alginate-chitosan, polyacrylic acid-chitosan, and chitosan-carrageenan for the purpose of designing controlled release formulations [27].

Gelatin is widely employed as a microencapsulating agent for both simple as well as complex coacervation. However, its high viscosity even in low concentrations necessitates the exploration of substitution systems, such as glucose, caseinate or plant proteins and resistant starches.

In contrast to spray dried emulsions, the complex coacervation technique offers significant benefits in the production of microcapsules. Microcapsules created through coacervation exhibit superior controlled release properties and heat resistance. This method stands out for its high loading capacity, simplicity, scalability, cost-effectiveness, lack of solvent usage, and ability to consistently obtain encapsulated extracts. Consequently, complex coacervation holds great potential for large-scale industrial applications involving encapsulation-based coating fabrication.

11.4.2 Emulsification

Emulsions find widespread application in various fields, including the food and pharmaceutical industries. Emulsification plays a crucial role in the process of encapsulating compounds [29]. This technique is commonly employed for encapsulating bioactive substances in aqueous solutions, which can be used as is or converted into powders through drying methods such as spray-drying or freeze-drying after emulsification. Thus, it forms an integral part of the overall encapsulation procedure. In instances like spray-drying and freeze-drying encapsulation, emulsion techniques are utilized to prepare both the core and wall materials before final drying takes place. Moreover, during extrusion processes or coacervation processing, emulsion droplets can be incorporated into matrices or serve as templates respectively.

In essence, an emulsion comprises minimally two immiscible liquids (typically oil and water), with one liquid dispersed as small spherical droplets within the other liquid. The size of the individual droplets found within food systems span a range from nano to micrometre. An oil-in-water emulsion is a system comprising of oil droplets dispersed in an aqueous phase, while a water-in-oil emulsion refers to a system where water droplets are dispersed in an oil phase. These two types of emulsions display distinct composition and dispersion characteristics, making them significant in various fields of study. Moreover, the development of oil-in-water-in-oil or water-in-oil-in-water emulsions have also been reported.

In order to achieve a kinetically stable solution, it is common practice to incorporate emulsifiers or texture modifiers into the emulsion system. To create these emulsions, a mechanical device known as a homogenizer is employed to thoroughly mix together oil, water, and an appropriate amount of emulsifier molecules. Particularly for oil-in-water type of emulsion structures, small oil droplets are dispersed evenly throughout an aqueous medium while being enveloped by a thin interfacial layer consisting of said relevantly added surfactant agents. These systems offer several advantages including ease of preparation and cost-effectiveness; however, they do come with some limitations: they may not remain physically stable when subjected to conditions like heating/cooling cycles, pH fluctuations, drying processes, and exposure high salt concentrations. Furthermore, the control over release also gets compromised due to the broken emulsion.

11.4.3 Spray Drying

Spray-drying is a cost-effective method widely used in large-scale microencapsulation. It offers several advantages, including its simplicity and affordability compared to other encapsulation techniques. For decades, the food and pharmaceutical industries have successfully utilized spray-drying to enclose various bioactive substances. This technique benefits from readily available equipment and lower production costs compared to alternative methods. Spray-drying has emerged as an effective solution for drying challenges due to its efficiency, rapidity and consistency making it suitable for easy scaling-up, distinguishing it as a preferred choice within the industrial sector. Additionally, this process allows flexibility in choosing different matrices for encapsulation while maintaining high-quality particle output [28].

Spray-drying has been employed by the food industry to achieve microbiological stability and enhance product functionality. This process involves the rapid evaporation of water from emulsions, achieved by atomizing them into a high-temperature drying chamber (Figure 11.4). The resulting fast formation of crusts and instantaneous entrapment of compounds aids in preserving products by minimizing chemical and biological degradation risks. Additionally, spray-drying offers benefits such as cost-effective storage and transportation options for food manufacturers while obtaining desired functional properties for their products.

In the process, the initial droplets in the emulsion range from nanometers to micrometers. It is important for the emulsion to remain stable during a specific timeframe before undergoing spray-drying. Additionally, it is necessary for the viscosity of the emulsion to be low enough so as not to introduce air into particles and keep core droplets small. The viscosity of the emulsion and its particle size distribution play key roles in microencapsulation through spray-drying. Higher viscosities can lead to elongated and larger droplet formation, which adversely affects drying rate by impeding atomization processes. Therefore, effective operating conditions and proper composition of solution containing active components are

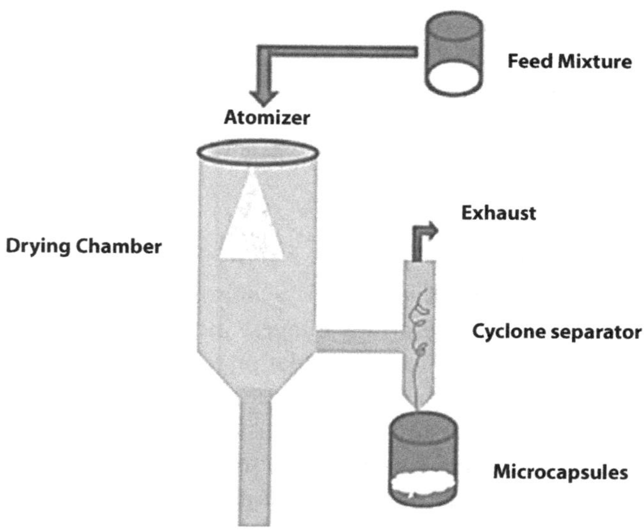

Figure 11.4 Spray drying techniques of encapsulation (source: Taouzinet *et al.*, 2023).

crucial requirements for successful spray-drying procedures in this context. The temperature of the feed has an impact on the viscosity and fluidity of the emulsion, which in turn affects its ability to be evenly sprayed. When the feed temperature is raised, it leads to a decrease in viscosity and droplet size. However, it is important to consider that higher temperatures can potentially cause volatility or degradation of certain heat-sensitive components. volatility or degradation of certain heat-sensitive components. To make sure that each droplet that is sprayed, achieves optimal drying levels, adjustments are made to the feed rate before entering the drying chamber. Increasing the levels of wall materials in the solution can enhance encapsulation efficiency by promoting shell formation around core material particles. A wide range of natural (such as starch, dextrins, protein isolates) and synthetic polymers can be chosen as wall materials based on their compatibility with core materials and desired characteristics for final product outcomes. As most spray-drying processes used in the food industry utilize an aqueous feed formulation, it becomes essential for the wall material to be readily soluble in water. The selection of a suitable wall material primarily depends on its physicochemical characteristics including solubility, molecular weight, glass/melting transition temperature, crystallinity, diffusibility, film forming ability and emulsifying properties. Additionally, cost considerations should also factor into the decision-making process.

One of the primary constraints of the spray-drying method in encapsulation is its reliance on a limited range of wall materials that possess favorable water solubility. It is essential to acknowledge an additional drawback associated with spray-drying, which pertains to the production of a finely powdered form of microcapsules requiring subsequent agglomeration processing. Moreover, it should be noted that elevated temperatures utilized during atomization contribute to diminished stability and bioactivity of the compound.

11.4.4 Lyophilization

Lyophilization, also known as freeze drying, is a widely utilized process for dehydration that is particularly suitable for heat-sensitive materials and aromas such as essential oils and water-soluble vitamins [30]. Prior to the drying stage, bioactive compounds or carrier core material are dissolved in water and frozen at temperatures ranging from -80°C to -40°C. Subsequently, through reducing the surrounding pressure while applying sufficient heat, the frozen water within the substance undergoes direct sublimation from solid to gas phase. In addition to its ability to safeguard heat-sensitive core materials during processing, lyophilization offers simplicity and ease of operation. Compared with spray-drying methods, freeze-dried materials generally exhibit optimal retention of volatile compounds. Moreover, research indicates that freeze-dried samples display enhanced resistance against oxidation; although it should be noted that this potentially results in lower encapsulation efficiency.

The major drawbacks of freeze-drying methods are the significant consumption of energy, extended processing durations, and elevated production expenses in comparison to alternative drying techniques. Additionally, freeze-dried ingredients may exhibit increased porosity which allows for greater exposure of the core material to its surroundings. However, this porous nature also presents an advantage as it facilitates a higher release rate for bioactive compounds within freeze-dried products.

11.4.5 Extrusion

Extrusion techniques have been widely utilized in the food industry to encapsulate various plant extracts and essential oils, such as olive, clove, thyme, and cinnamon oils [27]. Specifically, these techniques have predominantly focused on employing a carbohydrate matrix for encapsulation purposes. The process of extrusion microencapsulation typically has three distinct processes: melt injection, melt-extrusion, and coextrusion. These sequential processes allow for efficient encapsulation of the desired substances within a protective matrix [31].

The melt injection process involves dispersing core component into a carbohydrate based matrix, and then forcing it through one or more openings into a cold bath, for instance of liquid nitrogen and isopropanol. As the wall material cools and solidifies in the liquid, it forms an encapsulating matrix that traps the core material within. The resulting granules can be collected either through filtration or centrifugation. To remove any remaining solvent, air-drying or vacuum-drying methods are employed. Another method for encapsulation into a solid matrix is the continuous screw-extrusion process. This process resembles melt-injection but differs in two major aspects; firstly, it is a vertical screwless procedure involving surface-washed particles instead of horizontally-oriented particles without surface washing. Secondly, this technique bears similarity to the manufacturing approach utilized for producing expanded cereal snack products.

Melt-extrusion utilizes extruders, which are cylindrical devices housing thermomechanical mixers comprising one or more screws. In this process, commonly preferred extruders used for encapsulation are twin-screw extruders with self-wipe screws. Studies have proven that introducing the core material into the shell matrix towards the ending stages of the

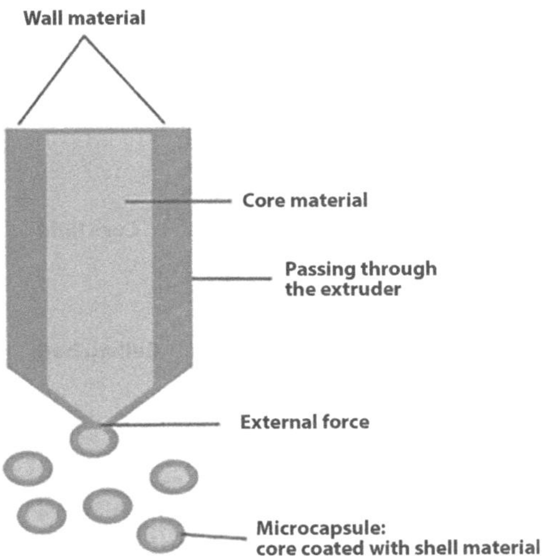

Figure 11.5 Extrusion-based encapsulation (source: Taouzinet *et al.*, 2023).

screw-extrusion procedure effectively shields sensitive compounds from extrusion's high heat and pressure conditions. Centrifugal extrusion, also known as coextrusion, is another specialized form of this technique. It involves employing a concentric feed tube to pump wall and core materials separately towards different nozzles openings located on the equipment's outer. The wall matrix is directed via the outer tube while the core through a centre tube as shown in Figure 11.5.

One significant benefit of utilizing extrusion-based encapsulation is the enhanced stability of antioxidant compounds, compared to spray-dried essential oils. Additionally, this process effectively reduces the rate at which essential oils evaporate. However, it should be noted that extrusion-based encapsulation comes with a higher cost when compared to spray drying. Moreover, the presence of relatively larger particles, measuring in upto millimeters, poses limitations on the applications where extruded essential oils can be used effectively.

11.4.6 Electrospray System

The coaxial electrospray system has emerged as an innovative technology in the food industry [32]. Over the past few years, researchers have explored the utilization of single-fluid electro spraying for developing drug-loaded systems with varying release profiles, be it sustained or rapid. However,

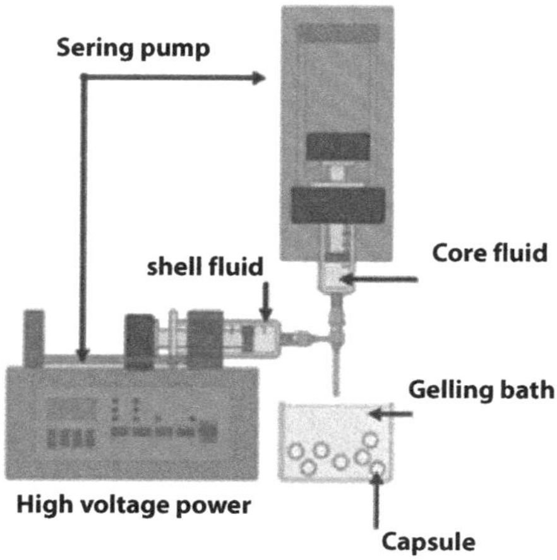

Figure 11.6 Coaxial electrospray system (source: Taouzinet *et al.*, 2023).

a more advanced approach known as two-phase coaxial electrospray has been developed where two separate solutions - wall and core - are simultaneously sprayed through different feeding channels into a single nozzle. During the process, a core-wall structure is made at the tip of the spray nozzle, where the core liquid is encapsulated by the outer polymeric wall matrix. Figure 11.6 shows the coaxial electrospray equipment. The notable advantages of this method include its simplicity, speediness, and efficiency in retaining maximum amounts of core material. It should be noted that both encapsulation efficiency and microcapsule stability can be influenced by the choice of wall materials used in this technique's application.

Firstly, in the process of electrospraying, the liquid wall material is carefully prepared using a syringe pump while the core liquid is injected using a syringe pump. The nozzle needle used has a larger outer diameter and a smaller inner diameter. Each syringe is connected to a programmable syringe pump for precise control. A high voltage along with a limiting current generated by a generator, is applied to the nozzle of the coaxial setup. Coaxial electrospray offers various benefits including homogeneous size distribution, high encapsulation efficiency, and efficacious preservation of bioactivity.

11.4.7 *In Situ* Polymerization

In situ polymerization has emerged as a widely utilized approach for fabricating microcapsules and functional fibers. Through this method, a wall is formed by introducing a reactant into the internal or external of the core

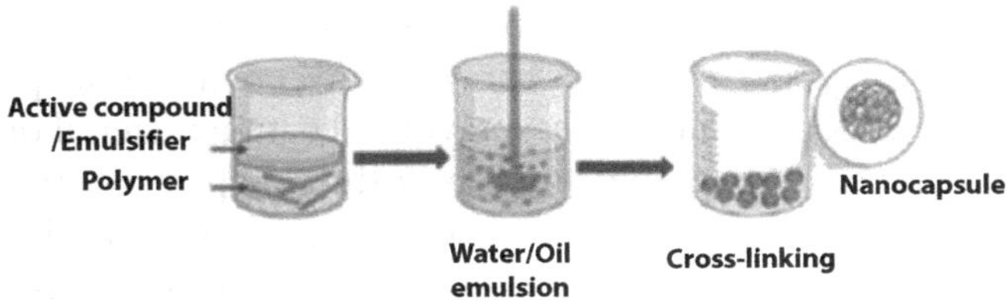

Figure 11.7 Principe of emulsion polymerizations (source: Taouzinet *et al.*, 2023).

material. Unlike other encapsulation techniques, what distinguishes *in situ* polymerization is its exclusion of reactants within the core material itself. This characteristic enables the preservation of the original chemical composition of the core material—a key factor in applications like food coating that aim to extend fruit and vegetable shelf life [33].

The polymerization process takes place primarily within the continuous phase rather than just at the junction between continuous phase and the core. To form microcapsules, a two-step process is followed. In the first step, a solution of bioactive compound in a melamine-formaldehyde resin is sonicated to create an emulsion. The addition of resin is done while stirring, and the pH of the emulsion is adjusted to acidic conditions. This leads to shell formation by promoting cross-linking reactions between melamine and formaldehyde at oil droplet interfaces, resulting in microcapsules with a cross-linked polymer layer of melamine-formaldehyde as their shells. Figure 11.7 explains the principal process of emulsion polymerizations.

The *in situ* polymerization method has proven to be successful in the creation of microcapsules containing various essential oils such as peppermint and tea tree oil, which are commonly used for fragrances and food packaging. One of the key advantages of using a polymer-based microcapsule shell is its notable strength and stability. To form these capsules, monomers and copolymer of melamine-formaldehyde are initially precross-linked in water. The core material is then added to this mixture, followed by an increase in emulsion temperature that triggers polymerization. As a result, the monomers cross link to form polymers and hence create a protective shell layer over the core droplets. Research shows that it is possible to significantly reduce formaldehyde levels which would slightly affect mechanical properties, which maintains optimal encapsulation efficiency.

11.4.8 Fluidized Bed Coating

Fluidized bed coating, also known as "air suspension coating" or "spray coating," is a highly efficient method of applying coatings. It has gained popularity and is now widely utilized in various applications within the food industry. This process involves spraying the core material's particles in a stream of air while simultaneously spraying a coating material onto them. Over time, a thin layer of coating builds over the core particle. The viscosity and film-forming capabilities of the coating material are crucial factors for successful implementation. Additionally, the duration that particles spend in the chamber determines how much they will be coated. By utilizing this method, materials can be efficiently mixed, granulated, and dried in a single vessel compared to alternative wet granulation methods [27, 34].

Fluid-bed coaters utilize different techniques namely top, bottom and tangential spray systems. In the top spray method, an air-assisted coating solution is sprayed in a downward direction onto the fluidized bed. This causes solid or porous particles to become encapsulated as they move towards the coating area. The opposite direction of flows between the shell and core solution enhances encapsulation efficiency while also preventing cluster formation. Compared to bottom or tangential sprays, top-spray systems yield higher amounts of microencapsulated particles. The bottom spray method is a common technique employed for encapsulating microcapsules. Figure 11.8 shows the bottom spray method of fluidized bed coating. It involves the use of a chamber equipped with a cylindrical nozzle and a plate at the bottom. During the process, the cylindrical stainless-steel nozzle sprays the coating material onto the particles. As these particles traverse upwards through the bottom plate, they come into contact with the zone where the nozzle operates. At this point, they become enveloped by the shell solution, which accumulates at

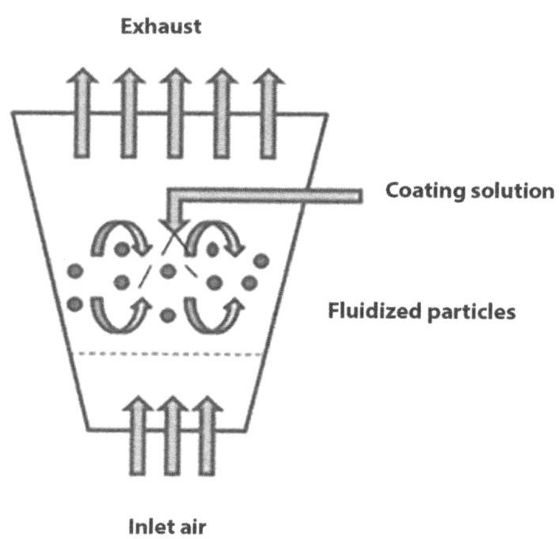

Figure 11.8 Fluidized-bed coating (source: Taouzinet *et al.*, 2023).

the surface either through cooling of the encapsulated particle or solvent evaporation. This sequence is repeated to achieve the desired thickness on each coated particle.

The multilayer coating method effectively reduces the occurrence of particle defects, despite being a time-intensive procedure. The tangential spray technique involves utilizing a coating chamber that features a rotatable base with equivalent diameter to the chamber itself. The drum is lifted to provide space between drum and the chamber's edge. The coating substance is sprayed by a tangential nozzle that is positioned above the rotating drum. Particles then move through this opening and into a space where they are encapsulated.

During the operation of the tangential spray system, particle movement, mixing, and granulation are influenced by three mechanical forces. These include the lifting force produced by process air passing through the adjustable gap in the drum, the centrifugal force produced by the rotating drum, and gravity. The combined action of these helical-like forces facilitates effective mixing and leads to symmetrical particles with desirable content characteristics. It is ideal for particles that need to be coated in a fluid bed to possess spherical shape with high density as well as a narrow size distribution and excellent flowability. Nonspherical particles have larger surface areas compared to spherical ones which results in greater coating material requirement for achieving similar shell thicknesses. Additionally, sharp edges on nonspherical particles can cause damage during handling due to their abrasive nature.

Fluidized-bed coating is a highly effective method for applying coatings, and it has seen increasing usage in the food and pharmaceutical sectors. This process allows for mixing, granulation, and drying to occur simultaneously in one vessel, resulting in reduced material handling and processing durations when compared to alternative wet-granulation techniques.

11.4.9 Spray Chilling

Spray chilling, also referred to as spray congealing or spray cooling, is a newly developed technique for the synthesis of microcapsules. In contrast to spray drying, which applies energy to the droplet, resulting solvent evaporation, spray chilling encapsulation removes energy from the droplet, causing the shell substance to solidify. This method offers several advantages including its cost-effectiveness and ability to encapsulate thermosensitive constituents without exposing them to high temperatures. Nonetheless, some technological challenges arise due to lower efficiency

Table 11.2 Microencapsulation process, principle, and nature of core material.

Microencapsulation process	Nature of core material	Principle
Spray drying	Solids and liquids	The process of emulsion or dispersion formation and subsequent atomization of the mixture within the drying chamber.
Spray cooling	Solids and liquids	The process involves the dispersion of the core material within a liquefied shell substance, which is subsequently sprayed through a heated nozzle into a carefully regulated and chilly environment.
Fluidized bed coating	Solids	The process involves the suspension of solid or powdered particles of core material within an unobstructed airflow, followed by the application of a molten polymer coating.
Coacervation	Solids and liquids	The formation of three immiscible chemicals. The process of applying a coating. The process of rigidizing a coating
Coextrusion	Solids	The process entails the passage of a high-pressure emulsion consisting of core and wall materials through a die. The mixture is subjected to extrusion within a solvent bath at low temperatures, resulting in the formation of pellets.
Emulsification	Solids and liquids	The emulsion is formed through the combination of the core and the wall material, and its stability is enhanced by the introduction of an emulsion stabilizer.

in terms of encapsulation and the hydrophobic properties exhibited by the particles during this process [35].

Spray chilling is a lipid-based encapsulating technology that employs hydrophobic carriers such as fatty acids and hydrogenated oils. The process involves atomizing a solution or emulsion that contains active ingredients and a molten lipid carrier into a cold chamber, where solid lipid microparticles are formed. Cold air or liquid nitrogen is introduced to facilitate the formation of solid lipid microparticles. These microparticles serve as carriers for phytosterols, aiding in their preservation and safeguarding against degradation [36]. Table 11.2 summarises the different methods of encapsulation and their principles.

11.5 Encapsulation Matrix for Specific Food Applications and Stability Enhancement

Different types of materials can be utilised as encapsulation matrices in food-related contexts, each possessing distinct characteristics and benefits. In the food sector, several often employed materials for encapsulating purposes include as follows:

- Lipids, including phospholipids and triglycerides, are frequently employed in the process of encapsulation owing to their biocompatible nature and their capacity to encapsulate substances with both hydrophilic and hydrophobic properties. Liposomes, characterised as spherical vesicles consisting of lipid bilayers, and lipid nanoparticles are frequently employed for this particular objective.
- Polymers, such as polysaccharides like alginate and chitosan, as well as proteins like gelatin and whey protein, are commonly utilised for the purpose of encapsulation. A diverse array of features is exhibited by these substances, including biodegradability, film-forming capacity, and pH sensitivity, rendering them well-suited for a wide range of applications.
- Emulsifiers, also known as emulsifying agents, are utilised to form emulsions in which active components are uniformly disseminated within a continuous phase. Examples of emulsifiers include lecithin and mono- and diglycerides. The aforementioned components are frequently employed

in the manufacturing process of emulsified food items, such as salad dressings and mayonnaise.

- In the realm of inorganic materials, nanoparticles composed of substances such as silica and calcium phosphate have garnered significant interest due to their capacity to encapsulate and then release bioactive chemicals. They are frequently employed in applications that necessitate regulated release and stability.
- Coatings, which can be derived from natural waxes such as beeswax or synthetic substances like shellac, are employed for the purpose of encapsulating specific confectionery items, fruits, and nuts. The application of these coatings has the potential to improve the visual aesthetics, tactile qualities, and longevity of the encased objects.

11.5.1 Wall Matrix

The selection of wall material for encapsulation must possess certain characteristics. Firstly, it should have the ability to form a cohesive covering over the core material, thereby stabilizing it and enhancing the strength of the encapsulated droplet. Additionally, the chosen material should be inert, meaning it should not react with the core material. Furthermore, it should not impart any specific taste to the product. Moreover, the wall matrix should be impermeable; however, it should also possess the capability to release the bioactives at the core in response to specific treatment. There are many polymer based wall material utilized for encapsulation and the common ones have been listed in Table 11.3. Each of the wall material hold varied techno-functional property and suitable for different techniques. Table 11.4 gives a glimpse of the different shell material and how they differ based on properties and suitability.

11.5.2 Properties of Encapsulation Material

The selection of encapsulation materials is contingent upon the specific demands of the food application. Different materials possess unique characteristics that render them suitable for specific purposes. Several essential properties of encapsulation materials can be delineated:

- Biocompatibility: Encapsulation materials must be safe for consumption, devoid of any health risks. Ensuring

Table 11.3 Commonly used wall materials for encapsulation of food components.

Polymers	Carbohydrates	Fat and waxes	Proteins
Polyacrylamide	Gum Arabic	Lecithin	Zein
Polyacrylic acid	Modified starches	Hydrogenated vegetable oils	Caseins
Polyanhydrides	Cyclodextrins	Bee wax	Soy proteins
Polycaprolactone	Cellulose derivatives	Medium chain triglycerides	Whey protein
Polyethylene glycol	Maltodextrin		Sodium caseinate
Polyglycolides	Chitosan		Gelatin
Polylactides	Pectin		
Polyorthoesters	Alginates		
Polyvinyl alcohol	Carrageenan		

biocompatible materials is imperative to safeguard the well-being of consumers in relation to food products.

- Stability: The stability of encapsulation materials is crucial under the processing and storage conditions of the food product. These materials need to shield the encapsulated compounds from deterioration caused by factors, such as temperature, light, and oxygen.
- Solubility: The encapsulation material should exhibit compatibility with the food matrix, allowing for proper dissolution or dispersion upon consumption of the food product. Inadequate solubility can lead to undesirable issues like grittiness or an unwelcome texture.
- Encapsulation Efficiency: The encapsulation process should yield a high level of encapsulation efficiency, ensuring that a substantial portion of the active ingredient is securely enclosed within the matrix.
- Release Profile: It is imperative that the encapsulation material permits the controlled release of the active ingredient at the desired rate. This property assumes paramount importance in applications where a gradual release is necessitated.

Table 11.4 Coating materials, their sources, properties, and the techniques they are suitable for.

Coating material	**Source**	**Techniques used**	**Properties**
Gums	Gum Arabic, Carrageenan, Sodium Alginate.	Extrusion, Spray drying, Micro-coacervates, Phase separation Emulsification.	Form soft elastic gels, low tensile strength, Hydrophilic colloids.
Carbohydrates	Sucrose, Starch, Dextran.	Fluidized bed coating, Spray drying, Freeze drying, Extrusion.	Hydrophilic colloids, Higher tensile strength compared to gums.
Proteins	Gelatin, Albumin	Spray drying, Micro coacervates, Freeze drying Emulsification, Extrusion.	Emulsification, Water binding capacity, Gelation, Foaming capacity.
Lipids	Phospholipids Bee's wax (Cera alba), Stearic acid.	Fluidized bed coating method, Spray chilling, Extrusion.	Plasticizing properties, Good barrier to water vapor and gases.
Celluloses and their derivatives	Plant cells	Emulsification and precipitation, Spray drying, Fluidized bed coating, , Micro coacervates, Extrusion.	Good film forming ability, Good surface activity, Hydrophilic.
Chitosan	Shells of Crustaceans (Arthropod)	Emulsification, Spray drying, Micro coacervate.	Good barrier to water vapor and gases

- Compatibility: The chosen material should harmonize with the sensory attributes of the food product, including taste, texture, and appearance. It should refrain from exerting any adverse effects on the overall quality of the food.

11.6 Safety Evaluation

Food safety regulations and guidelines for encapsulation-based coatings vary by region and country, but they generally aim to ensure the safety, quality, and integrity of food products. Compliance with these regulations is essential when using encapsulation technologies in the food industry. Coating or encapsulating solids, liquids, or gases of various types and qualities can be achieved using a wide variety of substances. However, the laws governing food additives are stricter than those governing pharmaceuticals. The number of compounds that are broadly accepted for drug encapsulation has not been approved for use in the food sector because many of these substances have not been certified as "generally recognized as safe" (GRAS) materials. The entire food process should be designed to meet the safety requirements of governmental agencies such as the European Food Safety Authority (EFSA) or the Food Safety and Standards Authority of India (FSSAI).

Food additives, including microcapsules, have proven to be useful and require meticulous scientific safety evaluation before approval [37]. The Joint Expert Committee on Food Additives (JECFA) was formed by the World Health Organization (WHO) and Food and Agriculture Organization (FAO) at the international level. JECFA has been working since 1955 and plays an important scientific advisory role to the WHO, FAO, and the Codex Alimentarius Commission. It plays two roles: to recommend the Acceptable Daily Intake (ADI) and to assess the risks associated with consumption of additives to food with human health and to recommend Acceptable Daily Intake (ADI) levels. In 1961, the ADI approach to toxicological evaluation was launched by the WHO Expert Committee on Food Additives and the Joint FAO. It entails gathering the correct information, verifying the accuracy of the data that is already available, analyzing the no-effect level using the sensitive indicator of toxicity, also known as the no-observed-adverse-effect level (NOAEL), and then recommending an acceptable safety factor to create an ADI for humans. The ADI is an extensive safety margin that represents the daily intake of a food additive over a lifetime without any negative health effects. The

Codex Alimentarius Commission, a joint FAO/WHO initiative that develops global food safety guidelines and is currently developing new "General Standards for Food Additives" (GSFA), aims to establish a practical, unified, and unchallengeable standard for international trade. Additives that have been tested and approved by JECFA can only be used as food additives. Food additives are now considered safe components of the human diet owing to the strict regulations and extensive tests that have helped the global food supply evolve. According to the Joint Expert Committee on Food Additives' Compendium of Food Additive and Flavoring Agent Specifications, each officially recognized and listed food additive has its own specification card [38]. The specification cards in the compendium offer detailed details and definitions related to the chemical composition, physical composition, and functional characteristics of the additives.

Furthermore, each card provides comprehensive information, including standardized identification methods, purity requirements for factors such as dry content and heavy metals, and specific microbiological criteria. Chemical Technical Assessment (CTA) underscores the significance of chemical characterization in the evaluation of food additive safety. The reports provide fundamental details concerning the identity, purity, and usage of food additives in the context of risk assessment. EU regulations mandate research that examines the spectrum of consumption levels within a population and considers shifts in consumption habits. Infrequent intake exceeding the ADI is unlikely to result in harm owing to the substantial 100-fold safety margin.

Information pertaining to the regulation of materials already approved or potentially applicable for encapsulation in the food industry is consistently refreshed and made available through standard reference sources such as the General Standard for Food Additives 1995 and the Codex Alimentarius 2008. The ADI was estimated using JECFA/OFAS based on either human studies or animal toxicology (Table 11.5). It is typically denoted as a numerical value (mg/kg body weight per day) or, in some cases, categorized as 'A' for acceptable 'or 'ND' for not determined, NS' for not specified (indicating very low toxicity), and 'NL' for not limited.

11.6.1 Health and Environmental Implications of Encapsulation Materials

The selection of encapsulation materials carries notable health and environmental implications across various industries. In the realm of health, these materials can influence the bioavailability and allergenicity of encapsulated substances, impacting the efficacy and safety of pharmaceuticals,

Table 11.5 Compilation of JECFA-reviewed food additives with INS numbers and applied or potential Acceptable Daily Intake (ADI) values within the scope of encapsulation technologies, as per the 1995 General Standard for Food Additives.

Sr. no.	Name	ADJ	INS/E'number
1	Agar	NL	406
2	Alginic acid	NS	400
3	Gum tragacanth	NS	4 13
4	Locust/carob bean gum (LBG)	NS	410
5	Ammonium alginate	NS	403
6	Calcium alginate	NS	404
7	Propylene glycol alginate	70 mg/kg	405
9	Sodium alginate	NS	401
10	Carrageenan	NS	407
11	Xanthan gum	NS	4 15
12	Guar gum	NS	412
13	Tara gum	NS	4 17
14	Gum Arabic	NS	4 14
15	Potassium alginate	NS	402
16	Karaya gum	NS	4 16
17	Curdlan	NS	424
18	Gellan gum	NS	4 18
19	Konjac flour glucomannan	NS	425
20	Pectins (nonamidated and amidated)	NS	440
21	Cyclodextrin, gamma	NS	458
22	Cyclodextrin, alpha	NS	45
23	Methyl cellulose	NS	461
24	Ethyl cellulose	NS	462

(Continued)

Table 11.5 Compilation of JECFA-reviewed food additives with INS numbers and applied or potential Acceptable Daily Intake (ADI) values within the scope of encapsulation technologies, as per the 1995 General Standard for Food Additives. (*Continued*)

Sr. no.	Name	ADJ	INS/E'number
25	Hydroxypropyl methyl cellulose	NS	464
26	Sodium carboxymethy l cellulose	NS	466
27	Methyl ethyl cellulose	NS	465
28	Ethyl hydroxyethyl cellulose	NS	467
29	Cross-Jinked sodium carboxymethy l	NS	468
30	Beeswax	90 1	A
31	Carnauba wax	903	7 mg/kg
32	Candelilla wax	902	A
34	Shellac	904	A
35	Polydextrose A and N	1,200	NS
36	Gum benzoic	906	ND
37	Dextrins, white and yellow roasted	1,400	NS
38	Insoluble polyvinylpyrrolidone	1,202	NS
39	Beeswax, white and yellow	90 1	A
42	Microcrystalline wax	905ci	20 mg/kg

**a"INS/E International Numbering and E-number systems are employed to categorize food additives. This INS system was established by Codex Committee on Contaminants and Food Additives to establish a globally recognized numerical system to identify these food additives among the ingredient lists, serving as an alternative to specifying their specific names. Meanwhile, an E-number signifies that additive has received approval from the European Union. In order to obtain the E-number, an additive must undergo a comprehensive safety evaluation conducted bv either the Europoean Food Safetv Authoritv or the SCF. In this context, 'E' denotes 'edible.'*

dietary supplements, and cosmetics. Moreover, the potential toxicity and degradation of encapsulation materials are critical concerns, as certain materials can release harmful compounds over time, posing risks to consumers. On the environmental front, the choice of encapsulation materials plays a pivotal role in sustainability efforts. Biodegradable and eco-friendly materials are increasingly favored to reduce waste and limit pollution. Additionally, resource use and the eco-toxicity of nonbiodegradable materials must be considered, as they can have far-reaching consequences on ecosystems and wildlife. To address these implications, rigorous material selection, life cycle assessments, adherence to regulations, and ongoing innovation in encapsulation technologies are essential to ensure the health and environmental well-being of both consumers and the planet.

11.6.2 Consumer Perception of Encapsulated Food Products

Consumer perception of encapsulated food products is shaped by a combination of factors that impact their acceptance and preferences. Encapsulation is often associated with benefits such as flavor retention, extended shelf life, and controlled release of nutrients or active ingredients. As a result, consumers often view encapsulated food products positively in terms of quality and convenience. The perception of these products is also influenced by the degree of transparency and labeling clarity, with consumers valuing information about the encapsulation process and the materials used. However, there can be concerns regarding the use of artificial or synthetic encapsulation materials, leading to preferences for natural or organic alternatives. Sustainability considerations have gained prominence, with consumers increasingly seeking eco-friendly packaging options. In essence, consumer perception of encapsulated food products is closely tied to a balance between perceived benefits, ingredient transparency, and sustainability, highlighting the importance of clear communication and responsible product development in meeting consumer expectations.

11.7 Future Perspectives and Challenges in Food Stability Enhancement

Despite encapsulation technology having a history of over six decades across various industries, there have been notable advancements in both the scientific understanding and practical applications of this technique since its initial commercial use. It has gained increasing popularity in the pharmaceutical, nutraceutical, and functional food sectors as an exceptionally effective method that serves a range of purposes. These include extending the shelf life of active ingredients, concealing undesirable flavors, colors, and tastes, and regulating the release of bioactive compounds. Continual research is actively exploring new encapsulation methods for emerging bioactive substances, aiming to enhance the process and improve product characteristics. Scientist are also working to develop new food-grade products, focused on reducing production costs while meeting stringent regulation and meeting with consumer needs.

11.7.1 Potential Advancements and Innovations in Food Encapsulation Technologies

Various technologies used in encapsulation holds a promise in the development of food products. Bioavailability and addressing health issues make it effective in combating many deficiencies. Encapsulated bioactive compounds should play an important role in improving the effectiveness of energy and energy products.Advances in microencapsulation technology to improve bioavailability, address health concerns, and combat deficiencies ultimately help restore greater health and wellness [39].

Various technology used in encapsulation holds great promise in enhancing bioavailability and addressing specific health issues, making it a valuable tool for combatting various deficiency diseases. Encapsulated bioactive compounds are poised to play a significant role in augmenting the advantages of functional and fortified foods.

This technology's potential to improve the delivery of bioactive compounds in a safe and effective manner is a promising development for the food industry. Researchers are developing encapsulation systems that respond to specific triggers such as pH, temperature, or enzymatic activity. These "smart" capsules release their contents at precise moments, enhancing the targeted delivery of bioactives [40, 41]. Nanoencapsulation techniques in encapsulation also gaining momentum. By reducing particle size to the nanoscale, these systems offer improved stability, solubility, as well as bioavailability for encapsulated compounds. Nanotechnology is a promising avenue for encapsulation, with its potential tied closely to the effectiveness of the drying process. While further research is required to understand how encapsulates behave in the human gastrointestinal tract and to address potential nanoparticle toxicity concerns, it's evident that encapsulation technology will continue to establish itself as an efficient means of delivering bioactive compounds. Researchers are working on encapsulation systems that can simultaneously deliver multiple bioactive compounds, offering synergistic health benefits. This approach is particularly relevant in functional foods and supplements [42].

11.7.2 Industry Level Operations and Cost Effectiveness

Encapsulation technologies hold great promise, but their industrial-scale practicality presents challenges. To achieve cost-effectiveness and scalability, various strategies are under exploration. Scientist are optimizing processes to minimize waste, reduce energy consumption, and streamline production, crucial for large-scale operations. Affordable and sustainable encapsulation materials, often sourced from natural and renewable origins, are being investigated. Automation and high-throughput screening accelerate technique evaluation, expediting the identification of economically viable approaches. Industry-academia collaborations facilitate innovation and reduce development costs. When assessing novel microencapsulation solutions for food industry challenges, factors like reproducibility, cost, and scalability play critical roles in predicting their adoption. The future will reveal which techniques and materials gain wider acceptance in the food industry.

11.8 Conclusion

There is strong potential for improving the longevity of fresh fruits and vegetables using encapsulation-based food coatings. This approach involves encapsulating plant-based bioactive or antibiotic compounds and incorporating them into a protective food coating, which can synergistically prolong the freshness and quality of perishable produce. The regulated release of bioactive compounds within the coating provides additional protection against microbial degradation and other deteriorative processes. Furthermore, this approach is in line with the requirements of food safety and regulations, making it a feasible solution to preserve fresh produce. Embracing this innovative technique can help producers and industries to reduce food waste and ensure that nutritious fresh produce remains accessible to the population for extended periods. Therefore, encapsulation-based food coatings hold great promise in revolutionizing how we store and consume fresh fruits and vegetables, fostering a sustainable future.

References

1. Moghadam, M., Salami, M., Mohammadian, M., Khodadadi, M., Emam-Djomeh, Z., Development of antioxidant edible films based on mung bean

protein enriched with pomegranate peel. *Food Hydrocoll.*, 104, 105735, Jul. 2020, doi: 10.1016/j.foodhyd.2020.105735.

2. Ju, J., Xie, Y., Guo, Y., Cheng, Y., Qian, H., Yao, W., Application of edible coating with essential oil in food preservation. *Crit. Rev. Food Sci. Nutr.*, 59, 15, 2467–2480, Aug. 2019, doi: 10.1080/10408398.2018.1456402.
3. Marsh, K. and Bugusu, B., Food Packaging—Roles, Materials, and Environmental Issues. *J. Food Sci.*, 72, 3, 39–55, Apr. 2007, doi: 10.1111/j.1750-3841.2007.00301.x.
4. Hassan, B., Chatha, S.A.S., Hussain, A.I., Zia, K.M., Akhtar, N., Recent advances on polysaccharides, lipids and protein based edible films and coatings: A review. *Int. J. Biol. Macromol.*, 109, 1095–1107, Apr. 2018, doi: 10.1016/j.ijbiomac.2017.11.097.
5. Dehghani, S., Hosseini, S.V., Regenstein, J.M., Edible films and coatings in seafood preservation: A review. *Food Chem.*, 240, 505–513, Feb. 2018, doi: 10.1016/j.foodchem.2017.07.034.
6. Quirós-Sauceda, A.E., Ayala-Zavala, J.F., Olivas, G.I., González-Aguilar, G.A., Edible coatings as encapsulating matrices for bioactive compounds: a review. *J. Food Sci. Technol.*, 51, 9, 1674–1685, Sep. 2014, doi: 10.1007/s13197-013-1246-x.
7. Bhagath, Y. and Manjula, K., Influence of composite edible coating systems on preservation of fresh meat cuts and products: a brief review on their trends and applications. *Int. Food Res. J.*, 26, 2, 377–392, 2019.
8. Lacroix, M. and Vu, K.D., Edible Coating and Film Materials. in: *Innovations in Food Packaging*, pp. 277–304, Elsevier, 2014, doi: 10.1016/B978-0-12-394601-0.00011-4.
9. Umaraw, P. and Verma, A.K., Comprehensive review on application of edible film on meat and meat products: An eco-friendly approach. *Crit. Rev. Food Sci. Nutr.*, 57, 6, 1270–1279, Apr. 2017, doi: 10.1080/10408398.2014.986563.
10. Fang, Y., Zhang, H., Nishinari, K. (Eds.), *Food Hydrocolloids: Functionalities and Applications*, Springer Singapore, Singapore, 2021, doi: 10.1007/978-981-16-0320-4.
11. Ye, C. and Chi, H., A review of recent progress in drug and protein encapsulation: Approaches, applications and challenges. *Mater. Sci. Eng., C*, 83, 233–246, Feb. 2018, doi: 10.1016/j.msec.2017.10.003.
12. Ravichandran, K. *et al.*, Effects of different encapsulation agents and drying process on stability of betalains extract. *J. Food Sci. Technol.*, 51, 9, 2216–2221, Sep. 2014, doi: 10.1007/s13197-012-0728-6.
13. Vital, A.C.P. *et al.*, Effect of Edible and Active Coating (with Rosemary and Oregano Essential Oils) on Beef Characteristics and Consumer Acceptability. *PLoS One*, 11, 8, e0160535, Aug. 2016, doi: 10.1371/journal.pone.0160535.
14. Augustin, M.A. and Sanguansri, L., Use of encapsulation to inhibit oxidation of lipid ingredients in foods, in: *Oxidation in Foods and Beverages and Antioxidant Applications*, pp. 479–495, Elsevier, 2010, doi: 10.1533/9780857090331.3.479.

15. Drusch, S. *et al.*, Role of glycated caseinate in stabilisation of microencapsulated lipophilic functional ingredients. *Food Hydrocoll.*, 23, 3, 942–948, 2009.
16. Boukhatem, M.N., Kameli, A., Saidi, F., Essential oil of Algerian rose-scented geranium (Pelargonium graveolens): Chemical composition and antimicrobial activity against food spoilage pathogens. *Food Control*, 34, 1, 208–213, Nov. 2013, doi: 10.1016/j.foodcont.2013.03.045.
17. Dima, C., Assadpour, E., Dima, S., Jafari, S.M., Bioactive-loaded nanocarriers for functional foods: from designing to bioavailability. *Curr. Opin. Food Sci.*, 33, 21–29, Jun. 2020, doi: 10.1016/j.cofs.2019.11.006.
18. Yuan, C., Wang, Y., Liu, Y., Cui, B., Physicochemical characterization and antibacterial activity assessment of lavender essential oil encapsulated in hydroxypropyl-beta-cyclodextrin. *Ind. Crops Prod.*, 130, 104–110, Apr. 2019, doi: 10.1016/j.indcrop.2018.12.067.
19. Garcia-Sotelo, D. *et al.*, Antimicrobial activity and thermal stability of rosemary essential oil:β–cyclodextrin capsules applied in tomato juice. *LWT*, 111, 837–845, Aug. 2019, doi: 10.1016/j.lwt.2019.05.061.
20. Khan, I., Tango, C.N., Miskeen, S., Oh, D.-H., Evaluation of nisin-loaded chitosan-monomethyl fumaric acid nanoparticles as a direct food additive. *Carbohydr. Polym.*, 184, 100–107, Mar. 2018, doi: 10.1016/j.carbpol.2017.11.034.
21. Zhang, Y. and Zhong, Q., Physical and antimicrobial properties of neutral nanoemulsions self-assembled from alkaline thyme oil and sodium caseinate mixtures. *Int. J. Biol. Macromol.*, 148, 1046–1052, Apr. 2020, doi: 10.1016/j.ijbiomac.2020.01.233.
22. Pinilla, C.M.B., Noreña, C.P.Z., Brandelli, A., Development and characterization of phosphatidylcholine nanovesicles, containing garlic extract, with antilisterial activity in milk. *Food Chem.*, 220, 470–476, Apr. 2017, doi: 10.1016/j.foodchem.2016.10.027.
23. Khosravi-Darani, K., Khoosfi, M.E., Hosseini, H., Encapsulation of Zataria multiflora Boiss. Essential oil in liposome: antibacterial activity against E. coli O157: H7 in broth media and minced beef. *J. Food Saf.*, 36, 4, 515–523, 2016.
24. Altan, A., Aytac, Z., Uyar, T., Carvacrol loaded electrospun fibrous films from zein and poly (lactic acid) for active food packaging. *Food Hydrocoll.*, 81, 48–59, 2018.
25. Tyagi, A.K., Gottardi, D., Malik, A., Guerzoni, M.E., Chemical composition, *in vitro* anti-yeast activity and fruit juice preservation potential of lemon grass oil. *LWT Food Sci. Technol.*, 57, 2, 731–737, Jul. 2014, doi: 10.1016/j.lwt.2014.02.004.
26. Gautam, G., Rani, R., Badwaik, L.S., Mahanta, C.L., Chitosan-Based Films and Coatings, in: *Biopolymer-Based Food Packaging*, 1st ed, S. Kumar, A. Mukherjee, J. Dutta (Eds.), pp. 110–146, Wiley, 2022, doi: 10.1002/9781119702313.ch4.

27. Bakry, A.M. *et al.*, Microencapsulation of Oils: A Comprehensive Review of Benefits, Techniques, and Applications: Encapsulation of marine, vegetable, essential oils. *Compr. Rev. Food Sci. Food Saf.*, 15, 1, 143–182, Jan. 2016, doi: 10.1111/1541-4337.12179.
28. Mukurumbira, A.R., Shellie, R.A., Keast, R., Palombo, E.A., Jadhav, S.R., Encapsulation of essential oils and their application in antimicrobial active packaging. *Food Control*, 136, 108883, Jun. 2022, doi: 10.1016/j.foodcont.2022.108883.
29. Kakran, M. and Antipina, M.N., Emulsion-based techniques for encapsulation in biomedicine, food and personal care. *Curr. Opin. Pharmacol.*, 18, 47–55, Oct. 2014, doi: 10.1016/j.coph.2014.09.003.
30. Mudalip, S.K.A., Khatiman, M.N., Hashim, N.A., Man, R.C., Arshad, Z.I.M., A short review on encapsulation of bioactive compounds using different drying techniques. *Mater. Today Proc.*, 42, 288–296, 2021, doi: https://doi.org/10.1016/j.matpr.2021.01.543.
31. Bamidele, O.P. and Emmambux, M.N., Encapsulation of bioactive compounds by 'extrusion' technologies: a review. *Crit. Rev. Food Sci. Nutr.*, 61, 18, 3100–3118, Oct. 2021, doi: 10.1080/10408398.2020.1793724.
32. Zhang, L., Huang, J., Si, T., Xu, R.X., Coaxial electrospray of microparticles and nanoparticles for biomedical applications. *Expert Rev. Med. Devices*, 9, 6, 595–612, Nov. 2012, doi: 10.1586/erd.12.58.
33. Nguon, O., Lagugné-Labarthet, F., Brandys, F.A., Li, J., Gillies, E.R., Microencapsulation by *in situ* Polymerization of Amino Resins. *Polym. Rev.*, 58, 2, 326–375, Apr. 2018, doi: 10.1080/15583724.2017.1364765.
34. Bratovcic, A. and Suljagic, J., Micro- and nano-encapsulation in food industry. *Croat. J. Food Sci. Technol.*, 11, 1, 113–121, May 2019, doi: 10.17508/CJFST.2019.11.1.17.
35. Pavani, M., Singha, P., Dash, D.R., Asaithambi, N., Singh, S.K., Novel encapsulation approaches for phytosterols and their importance in food products: A review. *J. Food Process. Eng.*, 45, 8, e14041, Aug. 2022, doi: 10.1111/jfpe.14041.
36. Zabot, G.L. *et al.*, Encapsulation of Bioactive Compounds for Food and Agricultural Applications. *Polymers*, 14, 19, 4194, Oct. 2022, doi: 10.3390/polym14194194.
37. Saltmarsh, M., *Essential guide to food additives*, Royal Society of Chemistry, 2013.
38. Wandrey, C., Bartkowiak, A., Harding, S.E., Materials for encapsulation, in: *Encapsulation Technologies for Active Food Ingredients and Food Processing*, pp. 31–100, 2010.
39. Timilsena, Y.P., Haque, M.A., Adhikari, B., Encapsulation in the Food Industry: A Brief Historical Overview to Recent Developments. *Food Nutr. Sci.*, 11, 06, 481–508, 2020, doi: 10.4236/fns.2020.116035.

40. Taouzinet, L. *et al.*, Trends of Nanoencapsulation Strategy for Natural Compounds in the Food Industry. *Processes*, 11, 5, 1459, May 2023, doi: 10.3390/pr11051459.
41. Yang, M., Liang, Z., Wang, L., Qi, M., Luo, Z., Li, L., Microencapsulation Delivery System in Food Industry—Challenge and the Way Forward. *Adv. Polym. Technol.*, 2020, 1–14, Jul. 2020, doi: 10.1155/2020/7531810.
42. Đorđević, V. *et al.*, Trends in Encapsulation Technologies for Delivery of Food Bioactive Compounds. *Food Eng. Rev.*, 7, 4, 452–490, Dec. 2015, doi: 10.1007/s12393-014-9106-7.

12

Role of Bioactive Carrier in Edible Films: Encapsulation Theory

Sudarshan Ramanathan, Leya B., Freeda Blessie R.* and M. M. Pragalyaashree

Division of Food Processing Technology, School of Agricultural Sciences, Karunya Institute of Technology and Sciences, Coimbatore, India

Abstract

Packaging is regarded as a protective mechanism that postpones degradation and stops the introduction of foreign particles. Tremendous growth is seen in packaging technology, especially in food and food-related commodities. Without packaging, commercially produced food cannot be handled or delivered safely, which is why packaging is seen as a crucial component of today's technology. One of the fastest-growing technologies in packaging is edible packaging and edible films. The benefits of employing edible films and coatings include maintaining the food items' quality, reducing microbial contamination and oxidation processes, and extending shelf life while keeping them in their fresh and processed forms. Therefore, the role of bioactive in edible films is crucial for the development of novel food packaging materials with enhanced functionality and sustainability. Being a flexible material, edible film may be enhanced or improved by adding different pigments, coatings, and infusions of different bioactive ingredients, such as phenolic compounds, vitamins, nutraceuticals, and probiotics, which function as carriers for antimicrobial agents. Bioactive carriers are used to encapsulate bioactive compounds in order to protect them from degradation and increase their bioavailability. These carriers, which can be either natural or synthetic, are frequently utilized in the manufacturing of edible films. The process of encasing a bioactive substance inside a carrier material is referred to as encapsulation. It is a method for enhancing the distribution of active ingredients in foods and pharmaceuticals. Encapsulation's major goal is to contain or release the bioactive substances under regulated conditions, shielding them from the environment and enhancing their effectiveness. To these inventions, encapsulation theory has given wings

**Corresponding author*: freedablessier@karunya.edu

Mousumi Sen (ed.) Food Coatings and Preservation Technologies, (403–446) © 2025 Scrivener Publishing LLC

to make the process quite efficient as well as reliable. Active ingredients that are encapsulated improve the finished goods' stability, which is tracked throughout the process. One of the most promising applications of edible films is the encapsulation of bioactive ingredients, such as vitamins, antioxidants, and antimicrobials. It also has the benefit of reducing evaporation and the loss of volatile active ingredients like aroma. Encapsulation is also used to cover up the bitter flavours and astringents that food's polyphenols create. The food sector uses a variety of encapsulation processes. By encapsulating these compounds into edible films, they are protected from degradation and can be released slowly throughout the digestive system, ensuring maximum absorption in the body. This has been considered significant for both the food industry and consumers, as it facilitates the development of healthier and more nourishing food products that can enhance health and well-being. With the combination of both technologies, the production of good antimicrobial edible film can be achieved with the most advanced benefits of delayed deterioration. This chapter deals with the overall concept of encapsulation, the mechanism, followed by the types of bioactive carriers included in the edible films along with their uses and limitations. It also focuses on technologies and methods of nutrient retention, their level of retention and their ability to improve shelf-life.

Keywords: Encapsulation, bioactive carriers, packaging

12.1 Introduction

The addition of bioactive compounds, such as probiotics, essential oils, and plant extracts, to edible films can enhance their functionality by providing antimicrobial, antioxidant, and other beneficial properties [1]. The historical evolution of bioactive carriers in edible films has also been influenced by the continuous exploration of various raw materials and their properties for the production of viable packaging materials. This has led to the utilization of protein-based biopolymers, plant proteins, and hydrocolloids for the production of edible films and coatings, which have gained recognition for their ability to produce films and potential as carriers for bioactive compounds [2–4].

Additionally, incorporating compounds with beneficial functions into edible films has been a focus of research to enhance their properties and enrich their characteristics with bioactive functionality, leading to the development of bioactive edible films with enhanced properties [5, 6]. The historical background of bioactive carriers in edible films reflects a continuous effort to improve the functionality and efficacy of edible films as carriers of bioactive compounds for various food applications, thereby contributing to the advancement of food packaging and preservation technologies [7–9].

Bioactive carriers in edible films serve as a vehicle for encapsulating and delivering bioactive compounds, such as peptides, antioxidants, and antimicrobial agents. These carriers are made from edible and biodegradable materials, meeting the requirements of the food industry for safety and sustainability [2]. The use of bioactive carriers in edible films is essential for the encapsulation theory. By incorporating bioactive compounds into edible films, the functional properties of these films are enhanced [10].

By integrating bioactive substances into edible films, it becomes possible to regulate the release of these compounds, both in packaged foods and in the consumer's gastrointestinal system [11]. This controlled release ensures that the bioactive compounds are delivered at a targeted rate and location, maximizing their effectiveness. Furthermore, bioactive carriers in edible films contribute to the improvement of food quality by enhancing factors such as stability, flavor, color, and texture [12]. Moreover, bioactive carriers in edible films also contribute to the overall performance of the food product by enhancing its nutritive value and functionality.

The use of bioactive carriers in edible films has gained significant attention in the food industry since they have the ability to enhance the food's quality and functionality. products while also enhancing their shelf life, reducing lipid peroxidation, inhibiting bacterial growth, and improving flavor profiles. In conclusion, bioactive carriers in edible films are integral to the encapsulation theory [13]. They serve as a vehicle for encapsulating bioactive compounds, allowing for their controlled release and improving the functional properties of edible films.

The importance of accurate weather forecasts in today's dynamic environment cannot be overstated. Multiple studies have shown that it is possible to create edible films using various structural ingredients, including lipids, polysaccharides, and proteins, either individually or in combination [14]. Protein-based edible films have garnered significant interest because of their notable mechanical qualities and diverse functional attributes. They are particularly advantageous because proteins have a distinctive structure that confers better mechanical properties, making them suitable for encapsulating and delivering bioactive compounds [6]. The potential benefits of using bioactive carriers in edible films to helps improve food quality, safety, and shelf life in the rapidly developing field of biodegradable materials.

By addressing the issues with encapsulation theory and enhancing the bioavailability and functioning of the encapsulated compounds, the use of bioactive carriers in edible films provides enhanced protection and regulated release of bioactive substances.

12.2 Definition and Types of Bioactive Carriers

Bioactive carriers in the context of edible films refer to substances or components that are capable of encapsulating bioactive compounds and releasing them gradually over time. These carriers come from a variety of origins, such as lipids, polysaccharides, and proteins [7]. In edible films, the function of bioactive carriers is to transport bioactive substances, allowing for their encapsulation and controlled release [15]. Bioactive carriers play a crucial role in the delivery of bioactive agents, such as drugs, proteins, and essential oils, to target sites in the body. These carriers are designed to protect the bioactive agents, enhance their stability, and control their release kinetics [16]. There are various types of bioactive carriers, each with its own unique characteristics and advantages. Some of the majors are listed below:

12.2.1 Polysaccharide-based bioactive carriers
12.2.2 Protein-based bioactive carriers
12.2.3 Lipid-based bioactive carriers
12.2.4 Other minor bioactive carriers

12.2.1 Polysaccharide-Based Bioactive Carriers

Polysaccharide-based carriers are another type of bioactive carrier that has been widely investigated. Polysaccharide hydrogel carriers, polysaccharide-protein carriers, and polysaccharide-based micelles have shown promise in the delivery of bioactive compounds [17]. These carriers can enhance the therapeutic efficacy of associated drugs and improve the targeting ability of the carrier system [18].

Polysaccharide-based bioactive carriers have gained significant attention in the development of edible food packaging technology due to their renewable, biodegradable, and biocompatible properties. These materials also exhibit antioxidant and antimicrobial activities, making them suitable for enhancing food preservation and safety [19]. It has been noted that the use of polysaccharide components in edible film and coating formulations is a sustainable method that provides biodegradable and environmentally friendly packaging materials that can be consumed with food directly, meeting the growing demand for eco-friendly packaging solutions [20].

Moreover, polysaccharide-based materials have demonstrated promise in the newest food packaging developments, such as intelligent films, edible coatings, and thermo-insulated aerogel packaging. These materials include hemicelluloses, marine polysaccharides, and bacterial exopolysaccharides, indicating their versatility and adaptability in addressing diverse packaging needs [21]. Incorporating bioactive ingredients into polysaccharide-based carriers has been explored for the delivery of both hydrophobic and hydrophilic bioactive compounds, demonstrating the potential of these materials in encapsulation, protection, and regulated release of food bioactive substances [22].

Additionally, the development of edible films and coatings from algal polysaccharides has evolved as a promising alternative to traditional plastic-based packaging materials, offering nontoxic, biodegradable, biocompatible, and bioactive characteristics that align with the growing consumer preference for sustainable packaging options [23]. The utilization of polysaccharide-based films in the food packaging industries has been found to improve product stability, safety and shelf-life, while also contributing to environmental sustainability through the use of biodegradable materials [24].

12.2.2 Protein-Based Bioactive Carriers

Protein-based carriers have attracted substantial attention in recent years due to their simple architectures and convenient production techniques. Protein–ligand nanocomplexes, oil-in- water (O/W) emulsions, and micro/nanoparticles are among the protein-based carriers that have been extensively researched. These carriers have great stability, matrix compatibility, and dose flexibility, making them appropriate for the co-encapsulation of various bioactive substances [16].

The potential for improving food safety and preservation has drawn a lot of interest to the development of protein-based bioactive carriers in edible food packaging technologies. Natural biopolymers like polysaccharides, proteins and lipids are frequently used to develop these carriers because they have the benefits of being palatable, biodegradable, and environmentally benign [25]. These protein-based food packaging materials can be used for a wide range of food products because their mechanical qualities can be improved by adding different ingredients like plasticizers, surfactants, crosslinkers, and bioactive compounds, such as antioxidant and antimicrobial compounds [3].

Additionally, the use of proteins in the development of bioactive food packaging has been highlighted, with protein isolates being utilized to generate protein hydrolysates with technical and bioactive qualities, acting as matrices for the fabrication of biodegradable, and edible as well active films for food [26].

12.2.3 Lipid-Based Bioactive Carriers

Lipid-based carriers, such as liposomes, have also been extensively studied as efficient delivery systems for bioactive compounds. Lipid nanostructured carriers offer controlled release capabilities and can be administered intra-articularly for the treatment of conditions like osteoarthritis [27]. These carriers provide a protective environment for the bioactive compounds and can improve their bioavailability.

Lipid-based bioactive carriers have gained significant attention in the development of edible films for various applications, particularly in the food industry. These carriers have been shown to effectively retard lipid oxidation and microbial changes in meat products, thereby improving their stability during storage. The incorporation of bioactive ingredients, such as *Tinospora cordifolia* and *Terminalia arjuna*, into edible films has demonstrated strong antimicrobial and antioxidant properties, contributing to the preservation of meat products [28].

Additionally, lipid-based nanoparticles, such as solid lipid nanoparticles and nanostructured lipid carriers, have been identified as promising carriers for pharmaceutical drug delivery, offering advantages such as increased bioavailability and controlled release of encapsulated active ingredients [29, 30]. Furthermore, the use of lipid-based carriers has extended to the development of edible films incorporated with probiotics, demonstrating the potential for enhancing the viability of beneficial microorganisms in food packaging materials [7]. The versatility of lipid-based carriers is further exemplified in their potential applications beyond food and pharmaceuticals. These carriers have been explored for the delivery of microRNAs, gene carriers, and plant extracts, showcasing their broad utility in diverse fields such as intercellular communication, gene therapy, and functional packaging.

12.2.4 Other Minor Bioactive Carriers

In addition to protein, polysaccharide, and lipid carriers, other types of bioactive carriers have been explored. For example, nanoparticles have been investigated as carriers for therapeutic protein delivery. These carriers have advantages such as high stability, controlled release, and the ability to protect proteins from degradation [31]. Natural macromolecules, such as chitosan and sugar beet pectin, have also been used as carriers for hydrophilic bioactive compounds [32]. These carriers offer promising potential for the delivery of bioactive agents.

When selecting a bioactive carrier, several factors must be considered, including the type of bioactive compound, its solubility and surface properties, the target site, and the desired release profile [4]. The carrier should be compatible with the bioactive compound and provide optimal protection and release characteristics. Additionally, the carrier's handling properties, yield, and bioactive properties should be taken into account [33].

12.3 Influence of Bioactive Carriers on Food Preservation

The influence of bioactive carriers on food preservation is a subject of great notice in the food industry. Bioactive carriers, such as nano emulsions, edible films, and composite films, have been developed to encapsulate and protect lipophilic functional components, probiotic cells, essential oils, and other bioactive compounds [34]. These carriers offer numerous advantages in terms of preserving fresh and treated meats, fruits, vegetables, fish, seafood, and other foods [35]. They act as suitable carriers for probiotic cells, providing a beneficial strategy to address restrictions connected with the usage of bioactive compounds in or on food products [36]. Additionally, they possess bacteriostatic activities and can be developed as bioactive packaging for preserving food [37].

The use of bioactive carriers in food preservation also aligns with the increasing demand for environmentally friendly packaging that offers food preservation. Edible films, for example, have gained attention in the food industry due to their potential to improve food preservation while reducing environmental impact. These films can be optimized by incorporating natural dyes, which not only enhance their visual appeal but also contribute to their preservation properties [38]. Furthermore, the pH of foods can influence the effectiveness of bioactive carriers, as it can favour the solubilization of hydrophobic essential oils [39].

Nanotechnology plays a significant role in the development of bioactive carriers for food preservation. Nano emulsions, nanoliposomes, and tocosomes are examples of nanocarriers that can encapsulate and deliver bioactive compounds [34]. These nanocarriers offer advantages such as improved stability and controlled release of bioactive compounds [40]. Moreover, nanotechnology has the potential to enhance microbial food safety by providing high-potential tools for food packaging, safety, and preservation [41].

12.3.1 Antimicrobial Activity of Bioactive Carriers in Food Preservation by Food Packaging

Antimicrobial activity of bioactive carriers in food preservation by food packaging is an important topic of investigation. Incorporating antimicrobial agents into packing materials can successfully limit microbial contamination and improve the shelf life of food products [42]. Nanoparticles and biodegradable polymer composites have been widely studied for their antimicrobial properties in active food packaging applications [43]. These materials can reduce the growth rate and maximum population of microorganisms, extend the lag phase, and inactivate the microorganisms upon contact [42]. Fruit wastes have also been utilized as additives in food products and as encapsulating agents in edible packaging materials, contributing to the preservation and enhancement of food quality [44]. Biobased and biodegradable polymer nanocomposites, such as starch, cellulose, and chitosan, have shown promise in active food packaging due to their significant antimicrobial effects and low environmental impact [45]. Additionally, antimicrobial packaging with biodegradable polymers and bacteriocins has been explored as a means to prevent the growth of infective microorganisms and improve food safety [46]. Silver-based nanocomposites have also been employed as antimicrobial films for food packaging, thanks to their significant toxicity to a varied range of microorganisms [47].

12.3.2 Antioxidant Activity of Bioactive Carriers in Food Preservation by Food Packaging

Antioxidant activity plays a crucial role in food preservation, and bioactive carriers in food packaging have been explored as a means to enhance this activity. The addition of antioxidants to packaging materials can help encompass the shelf life of food products by delaying oxidation. However, it is important to rigorously evaluate the retained antioxidant activity and stability of these bioactive carriers after being incorporated into packaging materials. The selection of appropriate dynamic agents and application approaches should also take into consideration the characteristics of the food products [48].

Chitosan-based edible films and coatings have been studied for their antimicrobial activity and physical properties. These films and coatings, which can contain antioxidants, have shown potential in food preservation by prolonging the shelf life of products [49]. Similarly, the addition of caffeic acid, rosemary extract, and ascorbic acid into beef flesh prior to high-pressure treatment did not offer safety against pressure-induced production of radicals [50]. This highlights the need for further research to optimize the antioxidant activity of bioactive carriers in food packaging. The use of natural antioxidants in meat and meat products has also gained attention. Natural antioxidants, such as those derived from rosemary, have been shown to postpone oxidation induced by high-pressure processing [50]. Additionally, the incorporation of essential oils, such as clove, thyme, or cinnamon, into bioactive edible coatings has been studied for their antioxidant activity. These essential oils have shown potential in preserving the value of food products by inhibiting oxidation [51].

The influence of natural antioxidants incorporated into biodegradable materials, such as chitosan films, has become a focus of research in food packaging. These materials have the potential to enhance the antioxidant activity of the packaging, thereby improving food preservation [52]. Furthermore, the use of oxygen absorbers in food preservation has been effective in improving the quality of packaged foods by reducing oxygen levels [53]. This technology, combined with the antioxidant activity of bioactive carriers, can contribute to the preservation of food quality.

12.3.3 Other Mechanisms of Action of Bioactive Carriers in Food Preservation

Other mechanisms of action of bioactive carriers in food preservation can be achieved through food packaging. Active packaging technologies, such as antimicrobial packaging, have been developed to enhance the preservation of food products. These technologies involve the usage of materials that have antimicrobial qualities, which can help limit the growth of pathogens and lengthen the shelf life of food [54].

Another approach is the use of food-grade nanoemulsions, which are formulated to advance the stability and delivery of bioactive components in food. Nano emulsions can alter the biological fate of these components within the gastrointestinal tract, potentially enhancing their bioavailability and efficacy. However, there may also be hazards connected with the consumption of nano emulsions, such as the possible toxicity of certain of the components utilized in their manufacturing [34].

Edible films and coatings can also serve as carriers for bioactive components in food preservation. These films can provide physical protection, amplify and protect food properties, and prolong shelf life [36]. They can be tailored to specific food products and can be categorized based on the materials used. Similarly, composite films loaded with probiotics have been developed as bioactive packaging for preserving food, with notable bacteriostatic activities [37].

Smart delivery systems using stimuli-responsive carriers have also been explored for the controlled release of bioactive compounds in food. These carriers can respond to physical, chemical, and biological stimuli in the food environment, leading to the release of encapsulated bioactive molecules. This on-demand release mechanism allows for targeted and efficient delivery of bioactive compounds [55]. Hydrogels have also been investigated as alternative structures for food packaging systems. These materials can help regulate moisture levels and prevent physicochemical changes in packaged foods. They can remove excess moisture or exudates generated by vegetables or prevent water vapor permeation [56].

12.4 Encapsulation Theory

Encapsulation theory refers to the use of encapsulation technology in the development of edible films [15]. It is a fundamental idea in the arena of food packaging, as it involves the process of enclosing food ingredients or active compounds within a protective barrier. This technology involves incorporating bioactive compounds into edible films, which act as carriers for these compounds. The integration of bioactive compounds into edible films allows for the precise release of these compounds, delivering them at targeted areas and rates [57]. This controlled release is crucial for improving the functional properties of edible films and increasing the overall performance of food products.

By encapsulating bioactive compounds, edible films provide a protective barrier against external factors such as oxidation, microbial growth, and moisture, thereby extending the shelf life of packaged foods [58]. Various methods of encapsulation have been developed and utilized in the food industry, including fluidized bed-coating, spray-drying, extrusion, freeze-drying, molecular inclusion, co-crystallization, and coacervation [59]. These techniques allow for the controlled release of bioactive substances, the improvement of new packaging products, the development of functional products, and the detection of pathogens [60].

Encapsulation has been widely applied in active food packaging, where active compounds such as antimicrobial agents, enzymes, and antioxidant agents are loaded into encapsulation systems. These encapsulated active compounds can provide enhanced functionality and preservation of food products [61]. Microcapsules, nano capsules, solid-lipid nanoparticles, metal nanoparticles, and nanostructured lipid carriers are examples of encapsulation techniques for antimicrobial food packaging components. Vitamins, antioxidants, food coloring, tastes, and antimicrobials have all been encapsulated for use in culinary applications using these encapsulation techniques [62].

In addition to encapsulating active ingredients, edible films can also be used for the microencapsulation of bioactive compounds. Microencapsulation involves enclosing bioactive compounds, such as vitamins, probiotics, or nutraceuticals, within small particles or capsules that are then incorporated into the edible film matrix. The encapsulation process helps protect these compounds from degradation and enhances their stability. Microencapsulation techniques, such as spray drying or coacervation, are commonly employed to achieve this. For example, arrowroot starch films have been developed with microparticles of bioactive compounds for active packaging applications. This technique enables the controlled release of these compounds, thus enhancing their bioavailability and functionality in the food product.

Various techniques are used for encapsulating active ingredients in edible films, including emulsion-based methods, nano emulsions, and incorporation of microparticles. These techniques enable the uniform distribution of the active ingredients within the film matrix, ensuring their effective encapsulation. The properties of edible films, such as mechanical strength, barrier properties, and release kinetics, can be influenced by the encapsulation process and the type of encapsulating agent used. The choice of encapsulating agent, such as alginate, soy protein concentrate, chitosan, or arrowroot starch, can impact the physical and functional properties of the edible film.

12.5 Mechanism of Encapsulation Theory

Bioactive encapsulation is a technique that involves the incorporation of active complexes, such as antioxidants, antimicrobials, vitamins, and nutraceuticals, into edible films. These films serve as a carrier for these bioactive compounds, protecting them from degradation and ensuring their controlled release. Encapsulation in edible films refers to the process of entrapping one ingredient, typically a bioactive component, within another ingredient, often a polymer matrix or film. The mechanism of encapsulation in edible films involves creating a protective barrier around the encapsulated ingredient, providing various benefits such as controlled release, improved stability, and enhanced functionality [63–65].

Encapsulation is the process of entrapping or enclosing one substance within another substance. In the case of edible films, the goal is to encapsulate active compounds or ingredients within a film matrix, which can be consumed along with the food product. This encapsulation process offers several advantages such as improved stability, controlled release, and enhanced bioavailability of the encapsulated material [66].

12.5.1 Steps Involved in Development of Encapsulated Edible Films

The mechanism of encapsulation in edible films involves various steps that are crucial for the successful formation and functionality of the films. These steps can be broadly categorized as:

12.5.1.1 Material Selection
12.5.1.2 Film Formation
12.5.1.3 Encapsulation Process

12.5.1.1 Material Selection

Material selection is a critical step in the encapsulation method. The choice of film-forming materials determines the physical and biochemical properties of the edible film. Common materials used for edible film formation include proteins (such as gelatin or whey protein), polysaccharides (like starch or alginate), lipids (such as waxes or liposomes), and composite materials. The assortment of materials be subject to on the desired properties of the film, the encapsulated material, and the intended application.

12.5.1.2 Film Formation

Film development involves the preparation of a film-forming solution or dispersion, which is then cast or coated onto a substrate to form a thin film. The film-forming solution typically contains the film-forming material, plasticizers (to improve flexibility), and other additives (such as antimicrobial agents or antioxidants). The film can be formed by various procedures, including hot-melt extrusion, solvent casting, or electrostatic deposition.

12.5.1.3 Encapsulation Process

The encapsulation process is the key step in creating an edible film with encapsulated materials. This process can be achieved through different techniques as follows:

i. **Emulsification**
   Emulsification involves the dispersion of one liquid phase within another immiscible liquid phase to form an emulsion. In this mechanism, bioactive compounds are dispersed in a continuous phase (such as a polymer solution) to form small droplets. These droplets are then entrapped within the edible film matrix, providing protection and controlled release of the bioactive compounds. Research has shown that emulsification-based encapsulation can be achieved using techniques such as solvent evaporation, solvent extraction, or coacervation [67].
ii. **Coacervation**
   Coacervation is a process where a polymer solution undergoes phase separation, leading to the formation of a coacervate phase rich in bioactive compounds. This coacervate phase is then encapsulated within the edible film matrix. Coacervation-based encapsulation offers advantages such as high encapsulation efficiency and the ability to encapsulate hydrophilic and hydrophobic compounds simultaneously [68].

The encapsulated materials can vary widely, depending on the desired functionality of the edible film. These materials can include flavors, colors, antioxidants, antimicrobial agents, vitamins, enzymes, probiotics, or even active pharmaceutical ingredients. The encapsulation process protects these materials from degradation, evaporation, or reaction with other food components, thus enhancing their stability and functionality.

12.5.2 Different Mechanisms of Encapsulation in Edible Films

12.5.2.1 Core-Shell Encapsulation

This mechanism involves encapsulating the active ingredient within a protective shell or coating. The active ingredient is surrounded by a matrix or film material, forming a core-shell structure. This encapsulation method provides a barrier against exterior factors, such as moisture, oxygen, and light, thereby preserving the encapsulated ingredient and controlling its release.

The core-shell encapsulation mechanism in edible film technology involves the entrapment of active ingredients within a protective matrix, leading to enhanced functionality of the film [59]. This encapsulation strategy is widely utilized for controlling the stability, delivery, and release of bioactive complexes, such as essential oils and oxygen, in edible films [69, 70]. The core- shell structure, where the core material is encapsulated within a shell, provides advantages such as controlled release rates and prevention of leaching of toxic core elements into the surrounding medium [71, 72]. The encapsulation process can be achieved through various methods, including electrospinning, electro spraying, and emulsion-based techniques, using materials such as calcium alginate, zein, and milk proteins [73–75]. Additionally, the barrier and mechanical properties of edible films can be enhanced by incorporating bioactive molecules and aroma compounds through microencapsulation [76–78]. The resulting edible films exhibit improved stability, controlled release, and enhanced barrier properties, making them suitable for various food, pharmaceutical, and packaging applications.

The process flowchart for the core-shell encapsulation mechanism in edible film technology involves several key steps. Firstly, the selection of suitable wall materials and core ingredients is crucial for achieving the desired functionality and stability of the encapsulated bioactive compounds [64, 70, 75]. Subsequently, the encapsulation process, which can be carried out using techniques such as electrospinning, electro spraying, or emulsion-based methods, involves the formation of core-shell structures with controlled surface chemistry and regulated release properties [69, 72, 74]. The encapsulated core materials, such as proteins, essential oils, or antioxidants, are entrapped within the protective shell, leading to improved stability and controlled release rates [71, 76, 78]. Finally, the incorporation of these encapsulated bioactive substances into edible films results in enhanced mechanical, barrier, and functional properties, creating them appropriate for diverse applications in the food, pharmaceutical, and packaging industries [65, 79, 102].

12.5.2.2 Emulsion-Based Encapsulation

It involves dispersing the active ingredient within an emulsion system. The active ingredient is typically discrete in an oil phase, which is then emulsified with a water phase containing emulsifiers. The resulting emulsion is then incorporated into the edible film matrix, providing a means of encapsulation. Emulsion-based encapsulation offers advantages such as improved solubility, uniform distribution, and controlled release of the encapsulated ingredient.

Emulsion-based encapsulation in edible film technology involves the preparation of edible films by encapsulating bioactive compounds within emulsions. These emulsions serve as the continuous phase for the encapsulation process [80]. The qualities of emulsion-based films rely on the structure of the dried emulsion, the chain length of the fatty acids, and the composition of the lipids, which creates the film [81]. The encapsulation process is crucial for controlling the release of flavor compounds, as demonstrated in the encapsulation of aroma substances in biopolymeric emulsion-based edible films allows for flavor release [82]. Additionally, the use of emulsions for encapsulation in edible films has been shown to enhance the stability and release behaviour of active aroma compounds, protecting them against moisture and oxygen [77]. Furthermore, the use of emulsions in the production of active edible films offers a novel technique to adjust transport qualities and release active chemicals while enhancing transparency, mechanical resistance, and antioxidant and antibacterial activity [83].

The process of emulsion-based encapsulation involves the stabilization of emulsions using various emulsifiers and the subsequent development of edible films and coatings with desired water permeability characteristics [84]. It has been demonstrated that the various mechanisms of action of bioactive compounds encapsulated in emulsion-based edible films vary depending on their composition, concentration, type of microorganism targeted, and method of encapsulation or incorporation in the biopolymer film-networks [85]. Moreover, the application of emulsion-based technologies as edible delivery methods by the food and other industries has been researched, including multiple emulsions, standard emulsions, multilayer emulsions, filled hydrogel particles, and solid lipid particles [86]. These technologies offer advantages in controlling the release of sensitive ingredients, protecting them from chemical degradation, and masking undesirable sensory attributes of bioactive ingredients [87].

12.6 Factors Affecting Encapsulation Efficiency

The proportion of the core substance retained within an encapsulate to the original core content injected into an encapsulation process is referred to as encapsulating efficiency. This was computed following Equation (12.1)

$$\text{Encapsulation Efficiency}(\%) = \frac{W_t}{W_i} \times 100\% \rightarrow \qquad (12.1)$$

while Wt denotes the quantity of integrated fundamental material and Wi denotes the total quantity of core material included throughout preparation [88]. The factors that influence the effectiveness of encapsulation in edible films differ based on the mixture and encapsulation methodology employed.

12.6.1 Composition of the Film

The elements employed in the film composition can have an impact on encapsulation efficiency. In one instance, tiny particles of cellulose have been utilized in the film manufacturing process, although their dimension has no direct effect on encapsulation efficiency [89]. Encapsulation efficiency is occasionally affected by the pace that the film absorbs water throughout the encapsulation process. In liposome production, for instance the rate of wetting of the thin layer might influence the encapsulation effectiveness of hydrophilic medicines. Increased encapsulation efficiency has been linked to decreased hydration rates.

The proportion of particles of the encapsulated material is potentially influenced by the film structure, therefore can affect encapsulation efficiency. Shorter sizes of particles could boost material dispersion and integration throughout the film matrix, resulting in better encapsulation efficiency [90].

The chemical structure of the lipids utilized in the film may impact encapsulation efficiency for lipid-based encapsulation systems. The firmness, dimensions, and encapsulating capability of liposomes or lipid nanoparticles might vary depending on the lipid composition. The chemical composition that forms the buffer applied throughout encapsulation might also impact encapsulation efficiency. Various studies have demonstrated that various buffer formulations may possess variable impacts on the encapsulation effectiveness of proteins as well as enzymes in liposomes [91]. The presence of additives in the film composition, which might include plasticizers or stabilizers, may have an effect on encapsulation efficiency. These chemical compounds can change the qualities of the film, especially elasticity and rigidity, which affects encapsulation efficiency [92].

12.6.2 Encapsulation Technique

The encapsulating technology used can have an effect on encapsulation efficiency. The effectiveness of encapsulation can be influenced by factors such as the technique of trapping and the long-term viability of the encapsulated substances [93]. Microencapsulation, liposomal encapsulation, spray drying, emulsification, and gelation are the various encapsulation techniques that are capable of having an impact on encapsulation efficiency. Each process has benefits and disadvantages, and the method used is determined by the chemical composition of the substance being contained and the desired attributes of the outcome [94].

Encapsulation efficiency is frequently affected by the degree of concentration of the polymers employed in the manufacturing procedure. Larger polymer levels may result in greater encapsulation efficiency owing to better material encapsulation as well as retention inside the polymer matrix's structure. The type of solvent utilized throughout the encapsulation process is capable of having an effect on encapsulation efficiency. The polymer's dissolution within the solvent, along with the solubility of the item getting encapsulated, can both affect encapsulation efficiency. The rate that the solvent that was used is withdrawn around the encapsulation process may possess an impact on encapsulation efficiency. Delayed and regulated elimination of solvent may improve encapsulation efficiency by permitting adequate production and stability of the encapsulated substance [95].

12.6.3 Properties of the Encapsulated Film

The features of the molecules being encapsulated, especially their hydrophilicity or lipophilicity, might influence encapsulation efficiency. The durability of the encapsulated chemicals may additionally impact encapsulation efficiency. The qualities of the film as a whole, such as stiffness, area of coverage, and hydrodynamic amount, can have an effect on encapsulation efficiency.

Encapsulation efficiency can be influenced by the machine-driven properties of the film, like flexibility, tensile strength, and elasticity. Films having suitable mechanical qualities offer greater safety along with durability to the encapsulated substance, resulting in increased encapsulation efficiency [64]. Chemical affinity among the film and the encapsulated substance is critical for encapsulation efficiency. When a film seems in equilibrium with the encapsulated substance, this may avert contacts or responses that could result in encapsulated substance breakdown or deterioration. The qualities of the film might affect the encapsulation production, that signifies the quantity of material effectively enclosed. Encapsulation productivity and effectiveness may be increased by using films with suitable characteristics including high attachment, rigidity, as well as compatibility with the enclosed substance [96]. The barrier qualities of the film, especially its susceptibility to water, oxygen, as well as other gases, could have an influence on encapsulation efficiency. Films that have excellent barrier qualities can limit the intrusion of external elements that may deteriorate or impact the integrity of the encapsulated substance, resulting in better encapsulation efficiency [97].

12.6.4 Environmental Factors

Environmental conditions involving temperature, humidity, light exposure, oxygen exposure and air quality might have an impact on the long-term viability and effectiveness of the encapsulated substances. The abovementioned environmental factors that affects encapsulation efficiency are

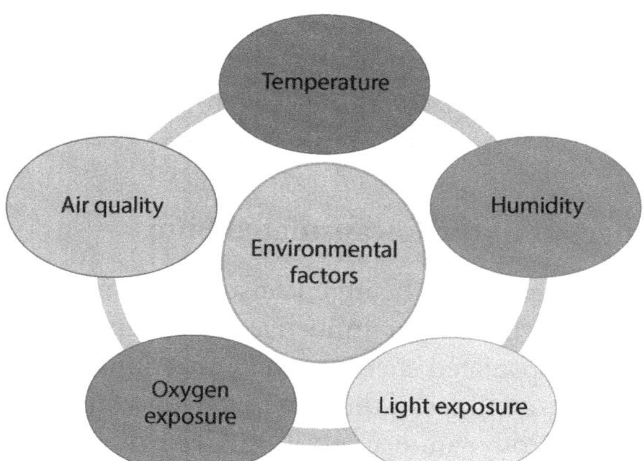

Figure 12.1 Environmental factors affecting encapsulation efficiency [98].

represented in the Figure 12.1. The atmosphere of films also influences the pace of active material emission. The dielectric connections between the medium and the solvent remain unchanged with rising temperatures, but the splitting process is impulsive, leading to a drop in molecular energy via exothermic processes [98].

12.6.4.1 Temperature

The temperature of the encapsulated chemicals may determine their durability and potency. Increasing the temperature may allow the encapsulated chemicals to degrade or evaporate, whilst freezing temperatures may modify the physical characteristics of the film and the method of encapsulation [99].

Changes in temperature can also have an effect on the chemical stability for encapsulated substances. High temperatures might accelerate chemical processes, resulting in deterioration or loss of effectiveness of the encapsulated chemicals. Similarly, low temperatures might have an impact on the stability of highly volatile chemicals. The resulting shifts in chemical stability might lead to decreased encapsulation efficiency [100].

Temperature fluctuations throughout the encapsulation procedure might reduce encapsulation efficiency. In particular, managing the temperature is critical in some encapsulation treatments, such as microencapsulation, to guarantee optimal creation and persistence of the encapsulating material. Variation in temperature can alter the shape and characteristics of the encapsulating substance, affecting encapsulation effectiveness [101].

12.6.4.2 Humidity

The stability of encapsulated chemicals may become affected by humidity conditions. Moisture absorption is possible due to high humidity, resulting in alterations to film characteristics and probable deterioration of the encapsulated chemicals [102]. Increased humidity can induce variances in wetness pressure, which can compromise the homogeneity of some films, especially casein films [103].

Moisture-triggered discharge devices, like microcapsules, are able to be influenced by conditions that promote solidification. This process thus reduces encapsulation efficiency [104]. The barrier qualities of packing films might be affected by moisture content and external variables such as humidity. Temperature and humidity have an impact on the water vapor transmission rate (WVTR) of films, having values that are lower which indicates greater humidity prevention [105].

12.6.4.3 Light Exposure

Radiation, particularly UV radiation, can trigger the breakdown of sensitive substances. To ensure the stability of the encapsulated substances, they must be protected from direct light exposure [106].

Exposure to ultraviolet (UV) light might trigger the active ingredients in edible films to deteriorate. Specific saccharide matrices, such as cyclodextrin, are capable of reducing degradation and retain the viability of active chemicals even when exposed to high levels of UV radiation [107].

12.6.4.4 Oxygen Exposure

Oxygen may react with some chemicals, causing them to oxidize and degrade. A high level of oxygen can shorten the duration of storage and decrease the stability of encapsulated substances. As a result, it is critical to limit oxygen exposure throughout the encapsulation and storage processes [99].

Encapsulation methods, especially microencapsulation, are capable of preventing encapsulated components from being exposed to oxygen while in storage [99]. Oxygen scavenging films have the purpose to eliminate oxygen out of the packing surroundings, keeping the substance from coming into touch with food and preserving the quality of the item [108]. Exposure to oxygen may result in the breakdown or removal of active chemicals in films that are edible. Encapsulation safeguards delicate molecules against oxygen exposure throughout manufacture, storage spaces. and gastrointestinal transit, including probiotics and bioactive chemicals. The adoption of proper encapsulating methods and supplies can improve active chemical stability while safeguarding against oxygen exposure [109].

12.6.4.5 Air Quality

Air pollution has a tendency to alter the look of packaging materials for food. The use of plastics that are not biodegradable in packaging use, notably food packaging, has prompted concerns about environmental contamination. The encapsulation efficiency can also be affected by the condition of the air around it. Pollutants or contaminants in the air may come into contact with the encapsulated substances, reducing their stability [89].

12.7 Edible Film Technology

Edible film technology is a forefront method to food packaging that makes use of films created from organic compounds such as proteins, polysaccharides, and lipids. The aforementioned biopolymers are easily biodegradable and abundant, offering them an environmentally acceptable alternative to standard packaging materials [110].

Filmogenic properties are found in a number of food manufacturing by-products, such as chitin and related compounds from crustacean shells, polysaccharides (alginates, carrageenan, and agar) from seaweeds, whey protein from cheese production, corn zein from the production of ethanol, collagen and Gelatin from animal skins, and potato starch from recyclable potato chips. These films are dried, prefabricated thin sheets normally between 50 m and 250 m thick which are utilized for wrapping food products or that may be transformed into pouches and bags, or simply set among the layers of meals [111].

These protective films have a variety of functional features, including moisture, oxygen, and light barrier protection, which can greatly improve the shelf life of food materials [112]. Edible film technology is aimed at the development and implementation of films that are safe for ingestion and may be utilized in a variety of dietary applications. Edible packaging can be applied in two ways: 1. Direct applying of film forming solution on the food product or 2. Premade film enfolded around the food product [113].

When in conjunction with bioactive components like vitamins, minerals, and polyphenols, they might carry out other functions. Another advantage of edible packaging is that it is ecologically outgoing because it is made from renewable resources. Food may be consumed with no the need to remove it from and discard the package. Renewable resources are not always edible. Thus, edible packaging is created by means of only safe for consumption components, i.e., plasticizers and additional additives to create the film-forming matrix, as well as the solvent, which must be widely recognized as safe, according to a formal method recognized by the US FDA [111].

The following two steps are commonly used for the formation of films:

1. Wet formation
2. Dry formation

Wet formation is usually done by mixing the edible material in a liquid medium, such as water or a solvent, and then casting or spreading the mixture onto a surface to produce a thin layer [114]. Probably the wet process, that primarily consists of the casting technique (or solvent casting), is the least often utilized approach to laboratory or pilot-scale film making. The above procedure is divided into three phases. Typically, the initial step in producing films is to dissolve or disperse the biopolymers in solvents [115].

To improve the stability and flexibility of the matrix substance, additives, such as plasticizers, antibacterial agents, and coloring agents or flavoring are added. A conceivable requirement is to adjust the pH and heat the solutions over their melting point to improve their solubility. In contrast, dry formation entails the direct deposition of the edible material in a dry condition, as it depends upon the intrinsic thermoplastic such as using extrusion or deposition processes (compression, molding) [116].

There are four methods used for the film formation:

12.7.1 Casting method
12.7.2 Extrusion method
12.7.3 Compression molding, and
12.7.4 Injection molding.

12.7.1 Casting Method

Preparing a film-forming solution by liquifying the biopolymer in a solvent, followed by casting the solution onto a flat surface and allowing it to dry, is the casting process as shown in the Figure 12.2 [113]. This strategy is commonly utilized in laboratory and pilot settings since it is easy and does not require any specific equipment. The film production process consists of three primary stages:

(1) the dissolution of the biopolymer in the solvent; (2) casting of the solution in the mold; and (3) drying of the solution. Different compounds with low adhesion are utilized as forms for drying film-forming solutions. Teflon glass coating (which can be coated with a polyester film), Petri dishes, Plexiglass, and polycarbonate plates are the most frequent. Films are dried out in a variety of means, including climatic circumstances, hot air, infrared radiation, and microwave energy, all of which effect the development of physicochemical and mechanical properties.

The persistent casting process allows the films to be created by continuously transferring the biopolymer solution on a conveyor belt and then drying them. The benefit of this process is the option to select modes such as conveyor rapidity and solution casting cycles, allowing you to modify the film viscosity, temperature, and drying length. This approach allows for

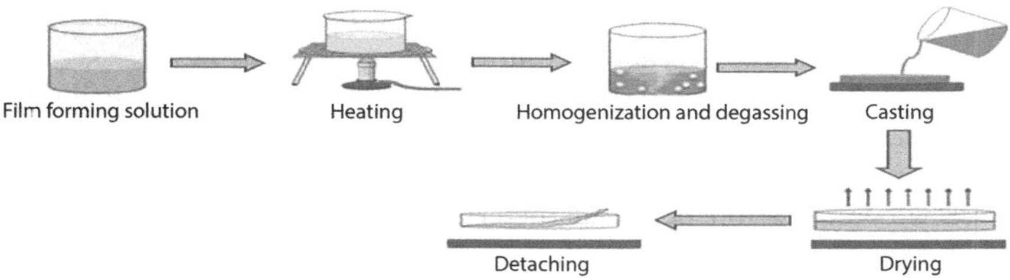

Figure 12.2 Casting method diagram [113].

the creation of multilayer films [117]. 300 mL of the extruded composition was obtained and heated for 10 minutes at 80°C on a plate for the production of edible film. The second stage is where the casting technique was applied to yield the edible films. To make edible film, 25 mL of the solidified fluid was poured into acrylic and baked in a 60°C oven for 2 hours. A digital micrometre was used to determine the thickness of the film [118].

12.7.2 Extrusion Method

To make a continuous film, the biopolymer is melted and extruded via a die. It is appropriate for creating films with uniform thickness and high output [119]. The basic components for the edible film are normally blended simultaneously and fed into an extruder during extrusion. The extruder is made out of a barrel that houses a spinning screw. The screw turns and pushes the raw material combination forward as it enters the barrel. The spinning of the screw produces heat and shear forces, which aid in the melting and homogenization of the liquid. This method depends on polymer thermoplastic characteristics during the plasticization and heating over their glass changeover temperature under low water content surroundings. As a result, the thermoplastic properties of film-forming materials must be thoroughly investigated before producing the films. Extrusion is simply one of the ways utilized to make edible films. Reliant on the precise needs and desired qualities of the film, other processes like casting and compression molding are also applied [117].

Extrusion is a extra active process as compared to solvent casting and compression. The technique is separated into three chief steps: feeding, kneading, and heating. The method is divided into three major steps: feeding, kneading, and heating. Initially, the film-forming material and NAMAs are introduced into the feeding zone, where they are degassed using compressive force. Furthermore, the supplies are further crushed by raising the pressure and temperature to achieve certain physical properties.

Finally, heating happens in the last part of the extruder, when factors such as temperature, shear rate, and pressure are maximum [120].

As a result, it is correct to state that the physical and chemical attributes of the produced film are a result of the occurrences of the heating phase. The ultimate application of the edible film may adapt parameters such as screw configuration, screw diameter and length ratio, rate of feeding, moisture percentage, screw speed, and so on [115].

12.7.3 Compression Molding

Compression molding is the process of depositing biopolymer powder or granules in a mold and using heat and pressure to produce a film as depicted in the Figure 12.3. Extrusion is frequently used for shaping the material before the primary thermoforming process. The thermoplastic procedure used in compression molding allows proteins to break down and cross-link throughout thermal processing, leading to the creation of films with desirable characteristics. The films might be completed in a single step [121].

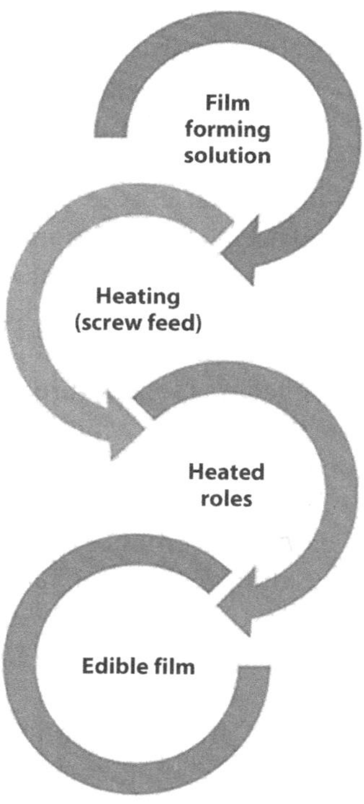

Figure 12.3 Flowchart for extrusion process [121].

The film-forming ingredients are bonded into the desired shape and thickness using in two ways ultrasonic compression or thermo-compression. The ultrasonic welder is the tool used for welding formerly developed film materials. Following compression, the welded components are chopped and treated to create viable edible packaging solutions [122].

Compression-mold films with yerba mate extract may extend food shelf life, making them an appropriate active food packaging technology. This approach has not so far achieved fame for producing edible films, although it is a quick and cost-effective process that should be adopted to the edible film packaging companies [120].

Extrusion and compression molding were used to create an edible film using yerba mate extract and starch. The films made with yerba mate extract using the compression-mold technology can extend the shelf life of foods, making them a viable active food packaging material [115].

12.7.4 Injection Molding

Injection molding is a popular method for bulk producing plastic objects, but it may likewise be used to produce edible films. It is a technique in which a biopolymer is melted and injected under high pressure into a mold. This process involves three major stages: filling, packing, and chilling. The most significant factors are injection pressure, pre-injection pressure temperature and molding temperature [115].

Twin-screw extrusion technique is used for producing particles from plasticized wheat starch with or without polyethylene oxide. During the following phase, injection molding was applied to make films [123]. Instead of treating with compression molding method to wheat gluten and glycerin pellets on a three-phase screw injection molding machine, the Injection Molding System was used to make pea isolate films with glycerol addition [117].

12.7.5 Application of Edible Films

Major decreases in fresh fruit quality and quantity exist throughout harvesting, production and food intake. In the food sector, edible films offer a broad spectrum of applications. Here are some examples of common uses [124].

12.7.5.1 Food preservation
12.7.5.2 Packaging
12.7.5.3 Enhanced appearance

12.7.5.1 Food Preservation

Edible films are often applied to rise the storage lifetime of fresh fruits and vegetables, meat, fish, and other perishable food products by acting as a block against oxygen, moisture, and other external elements that trigger decomposition [125].

12.7.5.2 Packaging

Edible films are suitable for helping package a variety of food products, including snacks, dry ingredients, and frozen foods. They can assist inhibit oxidation, moisture migration, and taste or color absorption while also acting as a barrier [126].

12.7.5.3 Enhanced Appearance

Edible films with a shine or matt texture can enhance the visual attractiveness of food items. They may boost the overall look and appeal of food that is packaged [125].

12.8 Recent Advances in Edible Film Encapsulation

Encapsulation is becoming more popular for usage not just in food items, but also in edible films, which offer an environmentally friendly choice to traditional food packaging techniques. Attempts are now being undertaken to strengthen the sustainable packaging system by integrating bioactive compounds and probiotics, which raise the nutritional content, functionality, and overall performance of edible films. Bioactives in encapsulated films not only promote edible film stability but also alter food product properties such as rate of lipid peroxidation, bactericidal/bacteriostatic effects, browning index, shelf life, and taste profile [126].

12.8.1 Development of Encapsulation Strategies

Various encapsulation strategies have been investigated in direction to upsurge the stability and controlled release of bioactive substances in edible films. Microcapsules, nano emulsions, or Pickering emulsions are used to encapsulate functional chemicals in these procedures [127]. Enclosing or encapsulating microscopic particles/droplets in an outer covering or shell is a method used in microencapsulation. It has gotten a lot of interest recently because of its broad range of uses in sectors including medicines, food, in addition cosmetics. For medication delivery methods, microencapsulation was one that was intensively researched. It has benefits including increased medication stability, controlled release, and targeted distribution. Microencapsulation might enhance medication bioavailability and absorption, as well as offer continuous and regulated distribution of active components [128]. Microencapsulation is currently utilized in the food business to preserve delicate components, improve flavor, and extend product shelf life. Probiotics, which are helpful bacteria, can be for instance, microencapsulated to make sure their continued existence and regulated discharge in the intestines [129].

12.8.2 Application in Nutraceuticals and Functional Foods

Nutraceuticals, functional meals, and immune boosters have all benefited from edible film encapsulation. Bioactive chemicals are better protected from degradation when they are encapsulated in edible films.

In the advancement of innovative food packaging materials, functional edible films have established interest. These films, which can be based on fruits and vegetables, have several benefits such as nutritional preservation, microbial contamination reduction, and shelf-life extension. They can be employed as transporters for bioactive chemicals encapsulated inside the film matrix, such as antioxidants, vitamins, and antimicrobials [130].

12.8.3 Extension of Shelf Life

To improve the shelf life of new and minimally treated foods, edible film encapsulation has been employed. Encapsulated edible films help protect the quality and freshness of food items by acting as a blockade against oxygen, moisture, and other external elements [131].

Biopolymer-based edible films, such as polysaccharides and proteins, have received interest for their potential to maintain and increase the shelf life of fruits and vegetables. These coatings operate as a barrier, protecting the food from external forces such as moisture loss, microbial contamination, and oxidation [132].

12.8.4 Functional Edible Films

New developments have concentrated on the creation of functional-edible films which not just give physical defence but also offer extra health advantages. Combining curcumin into edible films, for example, has demonstrated possibilities as an antioxidant and anti-inflammatory functional film [133]. Edible films can operate as carriers for bioactive chemicals, enabling regulated discharge and focused administration of functional components. Biopolymeric edible films, for instance, have been investigated as transporters for bioactive substances such as curcumin, that exhibit antioxidant and anti-inflammatory effects [2].

Engineered functional edible films may boost barrier qualities like oxygen and moisture resistance. By reducing oxidation and moisture loss, technology can assist to retain the nutritional value and quality of food goods. To improve the mechanical, physical, and barrier qualities of edible films, composite edible packaging combining various active ingredients has been created. To improve the functioning of the film matrix, active substances such as plant extracts, cross-linking agents, essential oils, and nanomaterials are included [134].

12.8.5 Antimicrobial Properties

Antimicrobial edible films have gained popularity in food packaging. Applying antimicrobial compounds into biopolymer-based edible films aids in the inhibition of spoiling bacteria development and promotes food safety [111].

Nanotechnology-improved food packaging solutions are more environmentally friendly than traditional alternatives that usage of plastic barriers, whereas functional elements such as antibacterial representatives enable for lengthier shelf life for the product. It may also identify other rotting indications, such as the creation of off-flavors, color, and harmful food toxins. Films containing nisin, a naturally occurring antimicrobial peptide, have been shown to effectively prevent Listeria growth in food items [135].

12.8.6 Use of Plant Mucilages

Plant mucilages, which include seed mucilage, are being investigated as possible ingredients for the creation of edible films. These mucilages thicken, emulsify, and create films, improving the functionality of edible films [131].

Concerns regarding food safety have grown in prominence in recent decades, heightening public knowledge and concern. The food sector is taking steps to accommodate the increasing need for secure food sources. There are new technologies for identifying harmful viruses, as well as rules for food cultivation, transportation, and packaging. Food containers made from recyclable materials are growing more and more common.

Biodegradable materials are becoming more popular in packaging and food service [136].

12.9 Advantages and Limitations of Bioactive Carriers in Edible Films

12.9.1 Advantages of Bioactive Carriers in Edible Films

Bioactive carriers in edible films offer several advantages in food applications. One advantage is the controlled dispersal of active compounds at the superficial of the food, which helps to preserve food quality. By incorporating bioactive compounds into the edible films, the diffusion of these compounds can be regulated, allowing for a gradual release of antimicrobial or antioxidant agents onto the food surface [85]. This controlled release can aid to extend the shelf life of food products and inhibit the growth of spoilage microorganisms.

Another advantage of bioactive carriers in edible films is the reduction of preservatives added to the food. Edible films can act as a barrier between the food and the external environment, reducing the need for additional preservatives [85]. By incorporating bioactive compounds into the films, the need for synthetic preservatives can be minimized, leading to cleaner label products.

Furthermore, the incorporation of bioactive components can also affect the mechanical properties of the edible films. The connections among biopolymer molecules in the films can be altered, resulting in changes in tensile strength, elongation at break, and Young's modulus [64]. This allows for the customization of the film properties to suit specific food applications, such as films with enhanced flexibility for wrapping delicate food items or films with increased strength for packaging heavy products.

In addition to these advantages, bioactive carriers in edible films have the potential to serve as probiotic carriers. Probiotic-loaded edible films can deliver beneficial bacteria to the consumer, promoting gut health and overall well-being [1]. This opens up opportunities for the development of functional foods that provide both nutritional and health benefits.

12.9.2 Limitations of Bioactive Carriers in Edible Films

Bioactive carriers in edible films have limitations that need to be considered. One limitation is the control of dispersion of active compounds at the surface of the food. Edible films offer the advantage of controlling the proclamation of bioactive compounds, such as antimicrobial or antioxidant agents, at the surface of the food [85]. This controlled release can benefit to extend the shelf life of food products and enhance their quality. However, it is important to note that the diffusion of these compounds may not be uniform, which can affect their effectiveness in inhibiting microbial growth or oxidative reactions.

Another limitation is the number of preservatives added to the food. Edible films can reduce the need for adding preservatives directly to the bulk food by incorporating them into the film matrix [85]. This can be beneficial in terms of reducing the overall number of preservatives in the food product. However, it is important to ensure that the amount of the bioactive compounds in the film is sufficient to provide the desired antimicrobial or antioxidant effect without negatively impacting the sensory properties of the food.

Additionally, the viability of bioactive compounds, such as probiotics, in edible films is a crucial factor to consider. The feasibility of probiotics in edible films must be sufficiently high to ensure their effectiveness as carriers of these beneficial microorganisms [7]. The stability of probiotics in edible films can be prejudiced by aspects such as the kind of biopolymer used, the presence of other ingredients, and the processing conditions [137]. Therefore, it is important to optimize the formulation and processing parameters to maintain the practicability of probiotics in the edible film.

Furthermore, the incorporation of bioactive compounds into edible films can also affect the structure and properties of the film itself. The type and concentration of the bioactive compounds can influence the film's mechanical, barrier, and sensory properties [138]. It is significant to sensibly consider the compatibility between the bioactive compounds and the film matrix to ensure that the anticipated functional properties of the film are maintained.

12.10 Future Perspectives

The utilization of edible films and coatings as transporters for bioactive compounds has a lot of promise in the food business. One viewpoint is the extending of product shelf life while maintaining quality. Edible biopolymer films and coatings have been intensively researched for their capacity to improve the stability and preservation of food items [139]. These films may protect perishable goods from environmental elements such moisture, oxygen, and microbial contamination, increasing their shelf life [130].

Furthermore, the incorporation of bioactive compounds within these films might improve their functioning and help to food quality preservation [2].

Another perspective is the possibility of improving food safety and nutritional characteristics. Edible films and coatings have the added benefit of integrating several active substances into the polymer matrix, which may then be taken with the food [130]. This brings up options to improve food safety by introducing antimicrobial drugs that may suppress pathogen and spoilage microbe development [140]. Furthermore, the encapsulation of bioactive compounds in edible films, such as antioxidants and probiotics, might contribute to the nutritional value of food items [1, 141]. These compounds may offer customers health advantages as well as enhance the sensory properties of food.

In terms of future developments, research into the creation of edible films and coatings with better characteristics and functions is continuing. This involves investigating various biopolymers and their combinations in order to improve film qualities such as mechanical strength, barrier properties, and biodegradability [64, 76]. The incorporation of new bioactive compounds and the creation of controlled release mechanisms inside films are further topics of research [1, 2]. These developments are intended to progress the performance and adaptability of edible films and coatings for a variety of food applications.

12.11 Summary and Conclusion

Edible packaging and edible films are establishing themselves as one of the most rapidly increasing packaging solutions. These films have various advantages, including protecting food quality, decreasing microbial contamination and oxidation processes, and increasing the duration of storage while preserving foodstuffs intact after being processed. The incorporation of bioactive substances into edible films is critical for the creation of innovative food packaging supplies with enhanced efficiency along with durability. Edible films are able to be improved by integrating various colors, coatings, and bioactive ingredient treatments which include phenolic compounds, vitamins, nutraceuticals, and probiotics. Such bioactive transporters preserve antimicrobial agents alongside other active substances by encapsulating them, inhibiting deterioration and increasing bioavailability. Encapsulation, or the practice of encapsulating bioactive chemicals within carrier supplies, is a critical approach for improving the transportation and efficacy of active components in foods and medications. Encapsulation aids in the regulation of bioactive material discharge, protection against surroundings, and enhancement of storage. This is very promising for encapsulating bioactive substances such as vitamins, antioxidants, and antimicrobials. Encapsulation may additionally disguise harsh odors and astringency induced by polyphenols in food through minimizing evaporation in addition to elimination of volatile active components such as fragrance compounds. The blend of edible films and encapsulating technologies enables the creation of antimicrobial edible films that have greater deterioration-delay advantages. It has significant consequences for the food business and consumers, since it allows for the production of superior and more nutritious food items, which could enhance the quality of life. Nutrient preservation and extension are critical aspects in assuring food product quality and safety. In the end, edible films and encapsulation technologies provide novel approaches to maintaining food and packaging. The aforementioned methods aid in the creation of functional and environmental food packaging materials by adding bioactive substances and applying encapsulation processes. The implementation of edible films and encapsulation may increase food item stability, quality, and shelf life, benefiting either the food sector and customers.

References

1. Pavli, F., Tassou, C., Nychas, G., Chorianopoulos, N., Probiotic incorporation in edible films and coatings: bioactive solution for functional foods. *Int. J. Mol. Sci.*, 19, 1, 150, 2018, https://doi.org/10.3390/ijms19010150.
2. Oliveira Filho, J.G. and Egea, M.B., Edible Bioactive Film with Curcumin: A Potential "Functional" Packaging? *Int. J. Mol. Sci.*, 23, 10, 5638, 2022, https://doi.org/10.3390/ijms23105638.
3. Shah, Y., Bhatia, S., Al-Harrasi, A., Afzaal, M., Saeed, F., Anwer, M., Faisal, Z., Mechanical properties of protein-based food packaging materials. *Polymers*, 15, 7, 1724, 2023, https://doi.org/10.3390/polym15071724.
4. Gomes, A. and Sobral, P., Plant protein-based delivery systems: an emerging approach for increasing the efficacy of lipophilic bioactive compounds. *Molecules*, 27, 1, 60, 2021, https://doi.org/10.3390/molecules27010060.
5. Praseptiangga, D., Ferichani, I., Mufida, N., Development and characterization of bioactive edible films based on semi-refined kappa carrageenan incorporated with honey and kaempferia galanga l. essential oil. *Trends Sci.*, 19, 17, 5761, 2022, https://doi.org/10.48048/tis.2022.5761.
6. Bourbon, A., Costa, M., Maciel, L., Pastrana, L., Vicente, A., Cerqueira, M., Active carboxymethylcellulose-based edible films: influence of free and encapsulated curcumin on films' properties. *Foods*, 10, 7, 1512, 2021, https://doi.org/10.3390/foods10071512.

7. Lee, Y., Yusof, Y., Pui, L., Development of Milk Protein Edible Films Incorporated with Lactobacillus rhamnosus GG. *BioResources*, 15, 3, 6960–6973, 2020, https://jtatm.textiles.ncsu.edu/index.php/BioRes/article/view/BioRes_15_3_6960_Lee_De velopment_Milk_Protein_Edible_Films.
8. Nurdiani, R., Prihanto, A., Firdaus, M., Aini, F., Nabilah, F., Talib, R., Huda, N., Physicochemical characteristics of pangasius sp. skin-gelatin-based-edible film enriched with silver nanoparticles. *F1000research*, 12, 160, 2023, https://doi.org/10.12688/f1000research.129024.1.
9. Fadilla, F., Rochima, E., Pratama, R., Rostini, I., Physical characteristics of biocomposite edible film based on starch and nanochitosan with the addition of oil and acids: a review. *Asian J. Fish. Aquat. Res.*, 22, 1, 1–9, 2023, https://doi.org/10.9734/ajfar/2023/v22i1560.
10. Yanat, M., Colijn, I., Schroën, K., Chitin nanocrystals provide antioxidant activity to polylactic acid films. *Polymers*, 14, 14, 2965, 2022, https://doi.org/10.3390/polym14142965.
11. Rico, D., Albertos, I., Martinez-Alvarez, O., Lopez-Caballero, M.E., Martin-Diana, A.B., Use of sea fennel as a natural ingredient of edible films for extending the shelf life of fresh fish burgers. *Molecules*, 25, 22, 5260, 2020, https://doi.org/10.3390/molecules25225260.
12. Otálora González, C.M., Schelegueda, L.I., Pizones Ruiz-Henestrosa, V.M., Campos, C.A., Basanta, M.F., Gerschenson, L.N., Cassava Starch Films with Anthocyanins and Betalains from Agroindustrial by-Products: Their Use for Intelligent Label Development. *Foods*, 11, 21, 3361, 2022, https://doi.org/10.3390/foods11213361.
13. Siddiqui, S.A., Singh, S., Bahmid, N.A., Mehany, T., Shyu, D.J., Assadpour, E., Jafari, S.M., Release of Encapsulated Bioactive Compounds from Active Packaging/Coating Materials and Its Modeling: *A Systematic Review. Colloids Interfaces*, 7, 2, 25, 2023, https://doi.org/10.3390/colloids7020025.
14. Pereira, J.O., Soares, J., Sousa, S., Madureira, A.R., Gomes, A., Pintado, M., Edible films as carrier for lactic acid bacteria. *LWT*, 73, 543–550, 2016, https://doi.org/10.1016/j.lwt.2016.06.060.
15. Manzanares, P., Gandía, M., Garrigues, S., Marcos, J.F., Improving health-promoting effects of food-derived bioactive peptides through rational design and oral delivery strategies. *Nutrients*, 11, 10, 2545, 2019, https://doi.org/10.3390/nu11102545.
16. Cheng, H., Chen, W., Jiang, J., Khan, M., Liang, L., A comprehensive review of protein-based carriers with simple structures for the co-encapsulation of bioactive agents. *Compr. Rev. Food Sci. Food Saf.*, 22, 3, 2017–2042, 2023, https://doi.org/10.1111/1541-4337.13139.
17. Stevanovic, Z., Sieniawska, E., Glowniak, K., Obradović, N., Pajic-Lijakovic, I., Natural macromolecules as carriers for essential oils: from extraction to biomedical application. *Front. Bioeng. Biotechnol.*, 8, 563, 2020, https://doi.org/10.3389/fbioe.2020.00563.

18. Zhang, N., Wardwell, P., Bader, R., Polysaccharide-based micelles for drug delivery. *Pharmaceutics*, 5, 4, 329–352, 2013, https://doi.org/10.3390/pharmaceutics5020329.
19. Zhao, Y., Li, B., Li, C., Xu, Y., Luo, Y., Huang, C., Comprehensive review of polysaccharide-based materials in edible packaging: a sustainable approach. *Foods*, 10, 8, 1845, 2021, https://doi.org/10.3390/foods10081845.
20. Kumar, N. and Neeraj, Polysaccharide-based component and their relevance in edible film/coating: a review. *Nutr. Food Sci.*, 49, 5, 793–823, 2019, https://doi.org/10.1108/nfs-10-2018-0294.
21. Nešić, A., Cabrera-Barjas, G., Dimitrijević-Branković, S., Davidovic, Z., Radovanović, N., Delattre, C., Prospect of polysaccharide-based materials as advanced food packaging. *Molecules*, 25, 1, 135, 2019, https://doi.org/10.3390/molecules25010135.
22. Wei, Z. and Huang, Q., Assembly of protein–polysaccharide complexes for delivery of bioactive ingredients: a perspective paper. *J. Agric. Food. Chem.*, 67, 5, 1344–1352, 2019, https://doi.org/10.1021/acs.jafc.8b06063.
23. Wang, H., Cao, Z., Yao, L., Feng, T., Song, S., Sun, M., Insights into the edible and biodegradable ulvan-based films and coatings for food packaging. *Foods*, 12, 8, 1622, 2023, https://doi.org/10.3390/foods12081622.
24. Kong, I., Degraeve, P., Pui, L., Polysaccharide-based edible films incorporated with essential oil nanoemulsions: physico-chemical, mechanical properties and its application in food preservation—a review. *Foods*, 11, 4, 555, 2022, https://doi.org/10.3390/foods11040555.
25. Mohamed, S.A., El-Sakhawy, M., El-Sakhawy, M.A.M., Polysaccharides, protein and lipid-based natural edible films in food packaging: A review. *Carbohydr. Polym.*, 238, 116178, 2020, https://doi.org/10.1016/j.carbpol.2020.116178.
26. de Oliveira Filho, J.G., Bertolo, M.R.V., Rodrigues, M.A.V., Marangon, C.A., da Cruz Silva, G., Odoni, F.C.A., Egea, M.B., Curcumin: A multifunctional molecule for the development of smart and active biodegradable polymer-based films. *Trends Food Sci. Technol.*, 118, 840–849, 2021, https://doi.org/10.1016/j.tifs.2021.11.005.
27. Craciunescu, O., Icriverzi, M., Florian, P., Roseanu, A., Trif, M., Mechanisms and pharmaceutical action of lipid nanoformulation of natural bioactive compounds as efficient delivery systems in the therapy of osteoarthritis. *Pharmaceutics*, 13, 8, 1108, 2021, https://doi.org/10.3390/pharmaceutics13081108.
28. Kalem, I., Bhat, Z., Kumar, S., Noor, S., Desai, A., The effects of bioactive edible film containing terminalia arjuna on the stability of some quality attributes of chevon sausages. *Meat Sci.*, 140, 38–43, 2018, https://doi.org/10.1016/j.meatsci.2018.02.011.
29. Puri, A., Loomis, K., Smith, B., Lee, J., Yavlovich, A., Heldman, E., Blumenthal, R., Lipid- based nanoparticles as pharmaceutical drug carriers:

from concepts to clinic. *Crit. Rev. Ther. Drug Carrier Syst.*, 26, 6, 523–580, 2009, https://doi.org/10.1615/critrevtherdrugcarriersyst.v26.i6.10.
30. Aldosari, B., Alfagih, I., Almurshedi, A., Lipid nanoparticles as delivery systems for rna-based vaccines. *Pharmaceutics*, 13, 2, 206, 2021, https://doi.org/10.3390/pharmaceutics13020206.
31. Marante, T., Viegas, C., Duarte, I., Macedo, A., Fonte, P., An overview on spray-drying of protein-loaded polymeric nanoparticles for dry powder inhalation. *Pharmaceutics*, 12, 11, 1032, 2020, https://doi.org/10.3390/pharmaceutics12111032.
32. Tang, X., Wang, Z., Meng, H., Lin, J., Guo, X., Zhang, T., Yu, S., Robust w/o/w emulsion stabilized by genipin-cross-linked sugar beet pectin-bovine serum albumin nanoparticles: co-encapsulation of betanin and curcumin. *J. Agric. Food. Chem.*, 69, 4, 1318–1328, 2021, https://doi.org/10.1021/acs.jafc.0c05212.
33. Gawałek, J. and Domian, E., Tapioca dextrin as an alternative carrier in the spray drying of fruit juices—a case study of chokeberry powder. *Foods*, 9, 8, 1125, 2020, https://doi.org/10.3390/foods9081125.
34. McClements, D. and Rao, J., Food-grade nanoemulsions: formulation, fabrication, properties, performance, biological fate, and potential toxicity. *Crit. Rev. Food Sci. Nutr.*, 51, 4, 285–330, 2011, https://doi.org/10.1080/10408398.2011.559558.
35. Kong, F., Kang, S., Zhang, J., Jiang, L., Liu, Y., Yang, M., Yue, X., The non-covalent interactions between whey protein and various food functional ingredients. *Food Chem.*, 394, 133455, 2022, https://doi.org/10.1016/j.foodchem.2022.133455.
36. Pop, O., Pop, C., Dufrechou, M., Vodnar, D., Socaci, S., Dulf, F., Suharoschi, R., Edible films and coatings functionalization by probiotic incorporation: a review. *Polymers*, 12, 1, 12, 2019, https://doi.org/10.3390/polym12010012.
37. Dai, L., Yuan, Y., Song, Z., Qiu, Y., Yue, T., Preparation and characterization of lactobacilli-loaded composite films with sustaining antipathogenic activity and preservation effect. *J. Food Sci.*, 83, 10, 2511–2519, 2018, https://doi.org/10.1111/1750-3841.14348.
38. Andriani, D., Karina, M., Priatni, S., Indriyati, I., Apriyana, A., Srikandace, Y., Optimization of edible film production: natural dyes addition. *IOP Conf. Ser. Earth Environ. Sci.*, 1201, 1, 012093, 2023, https://doi.org/10.1088/1755- 1315/1201/1/012093.
39. Nieto, G., A review on applications and uses of thymus in the food industry. *Plants*, 9, 8, 961, 2020, https://doi.org/10.3390/plants9080961.
40. Zarrabi, A., Abadi, M., Khorasani, S., Mohammadabadi, M., Jamshidi, A., Torkaman, S., Rasti, B., Nanoliposomes and tocosomes as multifunctional nanocarriers for the encapsulation of nutraceutical and dietary molecules. *Molecules*, 25, 3, 638, 2020, https://doi.org/10.3390/molecules25030638.

41. Rahmati, F., Hosseini, S., Safai, S., Lajayer, B., Hatami, M., New insights into the role of nanotechnology in microbial food safety. *Biotech*, 10, 10, 425, 2020, https://doi.org/10.1007/s13205-020-02409-9.
42. Kenawy, E., Worley, S., Broughton, R., The chemistry and applications of antimicrobial polymers: a state-of-the-art review. *Biomacromolecules*, 8, 5, 1359–1384, 2007, https://doi.org/10.1021/bm061150q.
43. Omerović, N., Djisalov, M., Živojević, K., Mladenović, M., Vunduk, J., Milenkovic, I., Vidić, J., Antimicrobial nanoparticles and biodegradable polymer composites for active food packaging applications. *Compr. Rev. Food Sci. Food Saf.*, 20, 3, 2428–2454, 2021, https://doi.org/10.1111/1541-4337.12727.
44. Sroy, S., Miller, F., Fundo, J., Silva, C., Brandão, T., Freeze-drying processes applied to melon peel: assessment of physicochemical attributes and intrinsic microflora survival during storage. *Foods*, 11, 10, 1499, 2022, https://doi.org/10.3390/foods11101499.
45. Vieira, I., Carvalho, A., Conte-Junior, C., Recent advances in biobased and biodegradable polymer nanocomposites, nanoparticles, and natural antioxidants for antibacterial and antioxidant food packaging applications. *Compr. Rev. Food Sci. Food Saf.*, 21, 4, 3673–3716, 2022, https://doi.org/10.1111/1541-4337.12990.
46. Gumienna, M. and Górna, B., Antimicrobial food packaging with biodegradable polymers and bacteriocins. *Molecules*, 26, 12, 3735, 2021, https://doi.org/10.3390/molecules26123735.
47. Azeredo, H., Nanocomposites for food packaging application. *Food Res. Int.*, 42, 9, 1240–1253, 2009, https://doi.org/10.1016/j.foodres.2009.03.019.
48. Realini, C. and Marcos, B., Active and intelligent packaging systems for a modern society. *Meat Sci.*, 98, 3, 404–419, 2014, https://doi.org/10.1016/j.meatsci.2014.06.031.
49. Vásconez, M., Flores, S., Campos, C., Alvarado, J., Gerschenson, L., Antimicrobial activity and physical properties of chitosan–tapioca starch based edible films and coatings. *Food Res. Int.*, 42, 7, 762–769, 2009, https://doi.org/10.1016/j.foodres.2009.02.026.
50. Kumar, Y., Yadav, D., Ahmad, T., Narsaiah, K., Recent trends in the use of natural antioxidants for meat and meat products. *Compr. Rev. Food Sci. Food Saf.*, 14, 6, 796–812, 2015, https://doi.org/10.1111/1541-4337.12156.
51. Sánchez-González, L., Vargas, M., González-Martínez, C., Cháfer, M., Use of essential oils in bioactive edible coatings: a review. *Food Eng. Rev.*, 3, 1, 1–16, 2011, https://doi.org/10.1007/s12393-010-9031-3.
52. Rivero, S., García, M., Pinotti, A., Microstructural characterization of chitosan films used as support for ferulic acid release. *Adv. Mater. Lett.*, 5, 10, 578–586, 2014, https://doi.org/10.5185/amlett.2014.5582.
53. Cichello, S., Oxygen absorbers in food preservation: a review. *J. Food Sci. Technol.*, 52, 4, 1889–1895, 2014, https://doi.org/10.1007/s13197-014-1265-2.
54. Suppakul, P., Miltz, J., Sonneveld, K., Bigger, S., Active packaging technologies with an emphasis on antimicrobial packaging and its applications.

J. Food Sci., 68, 2, 408–420, 2003, https://doi.org/10.1111/j.1365-2621.2003.tb05687.x.

55. Shishir, M., Gowd, V., Suo, H., Wang, M., Wang, Q., Chen, F., Cheng, K., Advances in smart delivery of food bioactive compounds using stimuli-responsive carriers: responsive mechanism, contemporary challenges, and prospects. *Compr. Rev. Food Sci. Food Saf.*, 20, 6, 5449–5488, 2021, https://doi.org/10.1111/1541-4337.12851.
56. Batista, R., Espitia, P., Quintans, J., Freitas, M., Cerqueira, M., Teixeira, J., Cardoso, J., Hydrogel as an alternative structure for food packaging systems. *Carbohydr. Polym.*, 205, 106–116, 2019, https://doi.org/10.1016/j.carbpol.2018.10.006.
57. Rihayat, T., Hadi, A.E., Aidy, N., Safitri, A., Siregar, J.P., Cionita, T., Fitriyana, D.F., Biodegradation of Polylactic Acid-Based Bio Composites Reinforced with Chitosan and Essential Oils as Anti-Microbial Material for Food Packaging. *Polymers*, 13, 22, 4019, 2021, https://doi.org/10.3390/polym13224019.
58. Bilecen, D. and Kilic, B., Determining the effects of encapsulated polyphosphates on quality parameters and oxidative stability of cooked ground beef during storage. *Food Sci. Technol.*, 39, 341–347, 2018, https://doi.org/10.1590/fst.03518.
59. Shahidi, F. and Han, X., Encapsulation of food ingredients. *Crit. Rev. Food Sci. Nutr.*, 33, 6, 501–547, 1993, https://doi.org/10.1080/10408399309527645.
60. Sekhon, B., Nanotechnology in agri-food production: an overview. *Nanotechnol. Sci. Appl.*, 7, 31–53, 2014, https://doi.org/10.2147/nsa.s39406.
61. Zhang, C., Li, Y., Wang, P., Zhang, H., Electrospinning of nanofibers: potentials and perspectives for active food packaging. *Compr. Rev. Food Sci. Food Saf.*, 19, 2, 479–502, 2020, https://doi.org/10.1111/1541-4337.12536.
62. Becerril, R., Nerín, C., Silva, F., Encapsulation systems for antimicrobial food packaging components: an update. *Molecules*, 25, 5, 1134, 2020, https://doi.org/10.3390/molecules25051134.
63. Medina-Pérez, G., Estefes-Duarte, J.A., Afanador-Barajas, L.N., Fernández-Luqueño, F., Zepeda-Velázquez, A.P., Franco-Fernández, M.J., Campos-Montiel, R.G., Encapsulation preserves antioxidant and antidiabetic activities of cactus acid fruit bioactive compounds under simulated digestion conditions. *Molecules*, 25, 23, 5736, 2020, https://doi.org/10.3390/molecules25235736.
64. Chaudhary, V., Thakur, N., Kajla, P., Thakur, S., Punia, S., Application of encapsulation technology in edible films: carrier of bioactive compounds. *Front. Sustain. Food Syst.*, 5, 734921, 2021, https://doi.org/10.3389/fsufs.2021.734921.
65. Bulkaini, B., Sumirtha, M., Ariana, T., Sumadi, K., Wrasiati, W., Sutama, S., Chemical-physical characteristics of edible film incorporated bali cattle hide gelatin and encapsulated gaharu leaf extract. *AJARCDE Asian J. Appl. Res.*

Community Dev. Empowerment, 7, 1, 69–73, 2022, https://doi.org/10.29165/ajarcde.v7i1.228.
66. Malekjani, N. and Jafari, S.M., Release modeling of nanoencapsulated food ingredients by mechanistic models, in: *Release and Bioavailability of Nanoencapsulated Food Ingredients*, pp. 247–271, Academic Press, Cambridge, Massachusetts, 2020, https://doi.org/10.1016/B978-0-12-815665-0.00007-2.
67. Bruni, G.P., de Oliveira, J.P., Gómez-Mascaraque, L.G., Fabra, M.J., Martins, V.G., da Rosa Zavareze, E., López-Rubio, A., Electrospun β-carotene–loaded SPI: PVA fiber mats produced by emulsion-electrospinning as bioactive coatings for food packaging. *Food Packag. Shelf Life*, 23, 100426, 2020, https://doi.org/10.1016/j.fpsl.2019.100426.
68. Jafari, S.M. (Ed.), *Nanoencapsulation of food bioactive ingredients: Principles and applications*, Academic Press, Cambridge, Massachusetts, 2017.
69. Warner, N., Črnivec, I., Rana, V., Cruz, M., Scherman, O., A platform approach to protein encapsulates with controllable surface chemistry. *Molecules*, 27, 7, 2197, 2022, https://doi.org/10.3390/molecules27072197.
70. Filippidi, E., Patel, A., Bouwens, E., Voudouris, P., Velikov, K., All-natural oil-filled microcapsules from water-insoluble proteins. *Adv. Funct. Mater.*, 24, 38, 5962–5968, 2014, https://doi.org/10.1002/adfm.201400359.
71. Samanta, A., Deng, Z., Liu, Y., Aqueous synthesis of glutathione-capped cdte/cds/zns and cdte/cdse/zns core/shell/shell nanocrystal heterostructures. *Langmuir*, 28, 21, 8205–8215, 2012, https://doi.org/10.1021/la300515a.
72. Mistry, P., Aied, A., Alexander, M., Shakesheff, K., Bennett, A., Yang, J., Bioprinting using mechanically robust core–shell cell-laden hydrogel strands. *Macromol. Biosci.*, 17, 6, 1600472, 2017, https://doi.org/10.1002/mabi.201600472.
73. Lam, L., Schilling, K., Romas, S., Misra, R., Zhou, Z., Caton, J., Electrospun core-shell nanofibers with encapsulated enamel matrix derivative for guided periodontal tissue regeneration. *Dent. Mater. J.*, 40, 5, 1208–1216, 2021, https://doi.org/10.4012/dmj.2020-412.
74. Zhang, X., Hu, B., Zhao, Y., Yang, Y., Gao, Z., Nishinari, K., Fang, Y., Electrostatic interaction-based fabrication of calcium alginate–zein core–shell microcapsules of regulable shapes and sizes. *Langmuir*, 37, 35, 10424–10432, 2021, https://doi.org/10.1021/acs.langmuir.1c01098.
75. Am, S., Gopikrishna, G., Ac, P., Aj, Y., Milk protein based edible films and coatings–preparation, properties and food applications. *J. Nutr. Health Food Eng.*, 8, 2, 219–226, 2018, https://doi.org/10.15406/jnhfe.2018.08.00273.
76. Garcia-Henao, C., Valderrama-Sanchez, V., Arboleda-Murillo, J., Pinzon, M., Sanchez, L., Villa, C., Bioactive food coating: a review. *Packag. Technol. Sci.*, 36, 1, 3–13, 2022, https://doi.org/10.1002/pts.2689.
77. Hambleton, A., Debeaufort, F., Beney, L., Karbowiak, T., Voilley, A., Protection of active aroma compound against moisture and oxygen by encapsulation in biopolymeric emulsion-based edible films. *Biomacromolecules*, 9, 3, 1058–1063, 2008, https://doi.org/10.1021/bm701230a.

78. Marcuzzo, E., Debeaufort, F., Sensidoni, A., Tat, L., Beney, L., Hambleton, A., Voilley, A., Release behavior and stability of encapsulated d-limonene from emulsion-based edible films. *J. Agric. Food. Chem.*, 60, 49, 12177–12185, 2012, https://doi.org/10.1021/jf303327n.
79. Sihombing, N., Elma, M., Thala'ah, R., Simatupang, F., Pradana, E., Rahma, A., Garlic essential oil as an edible film antibacterial agent derived from nagara sweet potato starch applied for packaging of indonesian traditional food. *IOP Conf. Ser. Earth Environ. Sci.*, 999, 1, 012026, 2022, https://doi.org/10.1088/1755- 1315/999/1/012026.
80. Li, C., Pei, J., Xiong, X., Xue, F., Encapsulation of grapefruit essential oil in emulsion- based edible film prepared by plum (pruni domesticae semen) seed protein isolate and gum acacia conjugates. *Coatings*, 10, 8, 784, 2020, https://doi.org/10.3390/coatings10080784.
81. Vieira, T., Moldão-Martins, M., Alves, V., Design of chitosan and alginate emulsion-based formulations for the production of monolayer crosslinked edible films and coatings. *Foods*, 10, 7, 1654, 2021, https://doi.org/10.3390/foods10071654.
82. Marcuzzo, E., Sensidoni, A., Debeaufort, F., Voilley, A., Encapsulation of aroma compounds in biopolymeric emulsion based edible films to control flavour release. *Carbohydr. Polym.*, 80, 3, 984–988, 2010, https://doi.org/10.1016/j.carbpol.2010.01.016.
83. Flores, Z., San-Martin, D., Beldarrain-Iznaga, T., Leiva-Vega, J., Villalobos-Carvajal, R., Effect of homogenization method and carvacrol content on microstructural and physical properties of chitosan-based films. *Foods*, 10, 1, 141, 2021, https://doi.org/10.3390/foods10010141.
84. Akkaya, S., Özel, B., Yanık, D., Göğüş, F., Physical characterization of high methoxyl pectin and sunflower oil wax emulsions: a low-field 1h nmr relaxometry study. *J. Food Sci.*, 86, 1, 120–128, 2020, https://doi.org/10.1111/1750-3841.15560.
85. Benbettaïeb, N., Debeaufort, F., Karbowiak, T., Bioactive edible films for food applications: mechanisms of antimicrobial and antioxidant activity. *Crit. Rev. Food Sci. Nutr.*, 59, 21, 3431–3455, 2018, https://doi.org/10.1080/10408398.2018.1494132.
86. McClements, D., Decker, E., Weiß, J., Emulsion-based delivery systems for lipophilic bioactive components. *J. Food Sci.*, 72, 8, 109–124, 2007, https://doi.org/10.1111/j.1750-3841.2007.00507.x.
87. Sun, R., Zhang, M., Xia, Q., Improved stability of (w1/o/w2) double emulsions based on dual gelation: oleogels and hydrogels. *J. Food Process Eng*, 42, 6, 13186, 2019, https://doi.org/10.1111/jfpe.13186.
88. Piacentini, E., Encyclopedia of Membranes, in: *Encapsulation Efficiency*, Drioli, E., Giorno, L. (Ed.), pp. 706–707, Springer, Berlin, Heidelberg, 2016, https://doi.org/10.1007/978-3-662-44324-8_1945.
89. Zabot, G.L., Schaefer Rodrigues, F., Polano Ody, L., Vinícius Tres, M., Herrera, E., Palacin, H., Córdova-Ramos, J.S., Best, I., Olivera-Montenegro, L.,

Encapsulation of Bioactive Compounds for Food and Agricultural Applications. *Polym. (Basel)*, 14, 19, 4194, 2022.

90. Nsairat, H., Khater, D., Sayed, U., Odeh, F., Al Bawab, A., Alshaer, W., Liposomes: structure, composition, types, and clinical applications. *Heliyon*, 8, 5, e09394, May 13 2022.
91. Chaize, B., Colletier, J.P., Winterhalter, M., Fournier, D., Encapsulation of enzymes in liposomes: high encapsulation efficiency and control of substrate permeability. *Artif. Cells Blood Substit. Immobil. Biotechnol.*, 32, 1, 67–75, Feb. 2004, doi: 10.1081/bio-120028669.
92. Pande, S., Liposomes for drug delivery: review of vesicular composition, factors affecting drug release and drug loading in liposomes. *Artif. Cells Nanomed. Biotechnol.*, 51, 1, 428–440, 2023.
93. Poddar, M., Lakshmi, G.B.V.S., Sharma, M., Chaudhary, N., Nigam, S., Joshi, M., Solanki, P.R., Environmental friendly Polyacrylonitrile nanofiber mats encapsulated and coated with green algae mediated Titanium oxide nanoparticles for efficient oil spill adsorption. *Mar. Pollut. Bull.*, 182, 113971, 2022, https://doi.org/10.1016/j.marpolbul.2022.113971.
94. Jafari, S.M., Poor., E.A., He, Y., Bhandari, B., Encapsulation Efficiency of Food Flavours and Oils during Spray Drying. *Drying Technol.*, 26, 7, 816–835, 2008.
95. Jyothi, N.V.N., Prasanna, P.M., Sakarkar, S.N., Prabha, K.S., Ramaiah, P.S., Srawan, G.Y., Microencapsulation techniques, factors influencing encapsulation efficiency. *J. Microencapsul.*, 27, 3, 187–197, 2010, https://doi.org/10.3109/02652040903131301.
96. Kamaldeen, O.S., Ariahu, C.C., Yusufu, M.I., Application of soy protein isolate and cassava starch based film solutions as matrix for ionic encapsulation of carrot powders. *J. Food Sci. Technol.*, 57, 11, 4171–418, Nov 2020.
97. Cui, H., Yuan, L., Lin, L., Novel chitosan film embedded with liposome-encapsulated phage for biocontrol of Escherichia coli O157:H7 in beef. *Carbohydr. Polym.*, 177, 156–164, Dec 1 2017.
98. Zhang, L., Yu, D., Regenstein, J.M., Xia, W., Dong, J., A comprehensive review on natural bioactive films with controlled release characteristics and their applications in foods and pharmaceuticals. *Trends Food Sci. Technol.*, 112, 690–707, 2021.
99. Nguyen, T.T.T., Le, T.V.A., Dang, N.N., Nguyen, D.C., Nguyen, P.T.N., Tran, T.T., Thuy Nguyen Pham, T.D., Microencapsulation of essential oils by spray-drying and influencing factors. *J. Food Qual.*, 1, 5525879, 1–15, 2021, https://doi.org/10.1155/2021/5525879.
100. Zeashan, M., Afzaal, M., Saeed, F., Ahmed, A., Tufail, T., Ahmed, A., Anjum, F.M., Survival and behavior of free and encapsulated probiotic bacteria under simulated human gastrointestinal and technological conditions. *Food Sci. Nutr.*, 8, 5, 2419–2426, Apr 17 2020.

101. Peng, G., Dou, G., Hu, Y., Sun, Y., Chen, Z., Phase Change Material (PCM) Microcapsules for Thermal Energy Storage. *Adv. Polym. Technol.*, 2020, 1, 9490873, 1–20, 2020.
102. Moreno, M.A., Vallejo, A.M., Ballester, A.R., Zampini, C., Isla, M.I., López-Rubio, A., Fabra, M.J., Antifungal edible coatings containing Argentinian propolis extract and their application in raspberries. *Food Hydrocoll.*, 107, 105973, 2020, https://doi.org/10.1016/j.foodhyd.2020.105973.
103. Yao, Y., Wang, H., Wang, R., Chai, Y., Preparation and characterization of homogeneous and enhanced casein protein-based composite films via incorporating cellulose microgel. *Sci. Rep.*, 9, 1, 1221, Feb 4 2019.
104. Wang, T., Li, Y., Luo, G., Ren, D., Wu, X., Xu, D., Polylactic acid-based microcapsules for moisture-triggered release of chlorine dioxide and its application in cherry tomatoes preservation. *Int. J. Biol. Macromol.*, 258, 128662, 2020, https://doi.org/10.1016/j.ijbiomac.2023.128662.
105. Mo, C., Yuan, W., Lei, W., Shijiu, Y., Effects of Temperature and Humidity on the Barrier Properties of Biaxially-Oriented Polypropylene and Polyvinyl Alcohol Films. *J. Appl. Packag. Res.*, 6, 1, 40–46, 2014.
106. Karthik, S., Jana, A., Selvakumar, M., Venkatesh, Y., Paul, A., Shah, S.S., Singh, N.P., Coumarin polycaprolactone polymeric nanoparticles: light and tumor microenvironment activated cocktail drug delivery. *J. Mater. Chem. B*, 5, 9, 1734–1741, 2017.
107. Quirós-Sauceda, A.E., Ayala-Zavala, J.F., Olivas, G.I., González-Aguilar, G.A., Edible coatings as encapsulating matrices for bioactive compounds: a review. *J. Food Sci. Technol.*, 51, 9, 1674–85, 2014 Sep, doi: 10.1007/s13197-013-1246-x.
108. Byun, Y., Development and characterization of moisture- and heat- activated oxygen scavenging nanoparticle. *All Dissertations*, 308, 1–153, 2008.
109. Terpou, A., Papadaki, A., Lappa, I.K., Kachrimanidou, V., Bosnea, L.A., Kopsahelis, N., Probiotics in Food Systems: Significance and Emerging Strategies Towards Improved Viability and Delivery of Enhanced Beneficial Value. *Nutrients*, 11, 7, 1591, Jul 13 2019.
110. Prakoso, F.A.H., Indiarto, R., Utama, G.L., Edible Film Casting Techniques and Materials and Their Utilization for Meat-Based Product Packaging. *Polymers*, 15, 13, 2800, 2023.
111. Hamed, I., Jakobsen, A.N., Lerfall, J., Sustainable edible packaging systems based on active compounds from food processing byproducts: A review. *Compr. Rev. Food Sci. Food Saf.*, 21, 1, 198–226, 2022.
112. Chhikara, S. and Kumar, D., Edible coating and edible film as food packaging material: A review. *J. Packag. Technol. Res.*, 6, 1, 1–10, 2022.
113. Suhag, R., Kumar, N., Petkoska, A.T., Upadhyay, A., Film formation and deposition methods of edible coating on food products: A review. *Food Res. Int.*, 136, 109582, 2020.
114. Himani, Singla, M., Prabhakar, P. K., Sharma, A., Meghwal, M., Edible and Oral Thin Films: Formulation, Properties, Functions, and Application in

Food Packaging and Pharmaceutical Industry, in: *Edible Food Packaging: Applications, Innovations and Sustainability,* Springer Nature, Singapore, pp. 411–432, 2022, https://doi.org/10.1007/978-981-16-2383-7_21.
115. Nair, S. S., Trafiałek, J., Kolanowski, W., Edible packaging: a technological update for the sustainable future of the food industry. *Appl. Sci.*, 13,14, 8234, 2023, https://doi.org/10.3390/app13148234.
116. Umaraw, P. and Verma, A.K., Comprehensive review on application of edible film on meat and meat products: An eco-friendly approach. *Crit. Rev. Food Sci. Nutr.*, 57, 1270–1279, 2017.
117. Lisitsyn, A., Semenova, A., Nasonova, V., Polishchuk, E., Revutskaya, N., Kozyrev, I., Kotenkova, E., Approaches in animal proteins and natural polysaccharides application for food packaging: Edible film production and quality estimation. *Polymers*, 13, 10, 1592, 2021.
118. Fitch-Vargas, P.R., Aguilar-Palazuelos, E., de Jesús Zazueta-Morales, J., Vega-García, M.O., Valdez-Morales, J.E., Martínez-Bustos, F., Jacobo-Valenzuela, N., Physicochemical and microstructural characterization of corn starch edible films obtained by a combination of extrusion technology and casting technique. *J. Food Sci.*, 81, 9, E2224– E2232, 2016.
119. Cheng, Y., Wang, W., Zhang, R., Zhai, X., Hou, H., Effect of gelatin bloom values on the physicochemical properties of starch/gelatin–beeswax composite films fabricated by extrusion blowing. *Food Hydrocoll.*, 113, 106466, 2021.
120. Punia Bangar, S., Chaudhary, V., Thakur, N., Kajla, P., Kumar, M., Trif, M., Natural Antimicrobials as Additives for Edible Food Packaging Applications: A Review. *Foods*, 10, 10, 2282, 2021, https://doi.org/10.3390/foods10102282.
121. Ceballos, R.L., Ochoa-Yepes, O., Goyanes, S., Bernal, C., Famá, L., Effect of Yerba Mate Extract on the Performance of Starch Films Obtained by Extrusion and Compression Molding as Active and Smart Packaging. *Carbohydr. Polym.*, 244, 116495, 2020.
122. Teil, M., Regazzi, A., Harthong, B., Dumont, P.J.J., Imbault, D., Putaux, J.L., Peyroux, R., Manufacturing of starch-based materials using ultrasonic compression moulding (UCM): Toward a structural application. *Heliyon*, 7, 3, e06482, 2021.
123. Yu, F., Prashantha, K., Soulestin, J., Lacrampe, M.F., Krawczak, P., Plasticized-starch/polyethylene oxide blends prepared by extrusion. *Carbohydr. Polym.*, 91, 253–261, 2013.
124. Díaz-Montes, E. and Castro-Muñoz, R., Edible films and coatings as food-quality preservers: An overview. *Foods*, 10, 2, 249, 2021.
125. Tavassoli-Kafrani, E., Gamage, M.V., Dumée, L.F., Kong, L., Zhao, S., Edible films and coatings for shelf-life extension of mango: A review. *Crit. Rev. Food Sci. Nutr.*, 62, 9, 2432–2459, 2022.
126. Wang, H., Ding, F., Ma, L., Zhang, Y., Edible films from chitosan-gelatin: Physical properties and food packaging application. *Food Biosci.*, 40, 100871, 2021.

127. Han, J.H., A review of food packaging technologies and innovations, in: *Innovations in food packaging*, pp. 3–12, 2014.
128. Ahmad, S.U., Li, B., Sun, J., Arbab, S., Dong, Z., Cheng, F., Zhou, X., Mahfuz, S., Zhang, J., Recent advances in microencapsulation of drugs for veterinary applications. *J. Vet. Pharmacol. Ther.*, 44, 3, 298–312, 2021.
129. Anal, A.K. and Singh, H., Recent advances in microencapsulation of probiotics for industrial applications and targeted delivery. *Trends Food Sci. Technol.*, 18, 5, 240–251, 2007.
130. Dhall, R.K., Advances in edible coatings for fresh fruits and vegetables: a review. *Crit. Rev. Food Sci. Nutr.*, *53*, 5, 435–50, 2013.
131. Arnon-Rips, H. and Poverenov, E., Biopolymers-embedded nanoemulsions and other nanotechnological approaches for safety, quality, and storability enhancement of food products: active edible coatings and films, in: *Emulsions*, pp. 329–363, Academic Press, Cambridge, Massachusetts, 2016, https://doi.org/10.1016/B978-0-12-804306-6.00010-6.
132. Garcia, F. and Davidov-Pardo, G., Recent advances in the use of edible coatings for preservation of avocados: A review. *J. Food Sci.*, 86, 1, 6–15, 2021.
133. Taqi, A., Mutihac, L., Stamatin, I., Physical and Barrier Properties of Apple Pectin/Cassava Starch Composite Films Incorporating L aurus nobilis L. Oil and Oleic Acid. *J. Food Process. Preserv.*, 38, 4, 1982–1993, 2014, https://doi.org/10.1111/jfpp.12174.
134. Kumar, N., Pratibha, Trajkovska Petkoska, A., Khojah, E., Sami, R., Al-Mushhin, A.A., Chitosan edible films enhanced with pomegranate peel extract: Study on physical, biological, thermal, and barrier properties. *Materials*, 14, 12, 3305, 2021, https://doi.org/10.3390/ma14123305.
135. Chawla, R., Sivakumar, S., Kaur, H., Antimicrobial edible films in food packaging: Current scenario and recent nanotechnological advancements-a review. *Carbohydr. Polym. Technol. Appl.*, 2, 100024, 2021, https://doi.org/10.1016/j.carpta.2020.100024.
136. Iversen, L.J.L., Rovina, K., Vonnie, J.M., Matanjun, P., Erna, K.H., Aqilah, N.M.N., Funk, A.A., The emergence of edible and food-application coatings for food packaging: a review. *Molecules*, 27, 17, 5604, 2022, https://doi.org/10.3390/molecules27175604.
137. Soukoulis, C., Behboudi-Jobbehdar, S., Macnaughtan, W., Parmenter, C., Fisk, I.D., Stability of Lactobacillus rhamnosus GG incorporated in edible films: Impact of anionic biopolymers and whey protein concentrate. *Food Hydrocoll.*, 70, 345–355, 2017, https://doi.org/10.1016/j.foodhyd.2017.04.014.
138. Benbettaïeb, N., Karbowiak, T., Debeaufort, F., Bioactive edible films for food applications: Influence of the bioactive compounds on film structure and properties. *Crit. Rev. Food Sci. Nutr.*, 59, 7, 1137–1153, 2019, https://doi.org/10.1080/10408398.2017.1393384.
139. Šuput, D., Lazić, V., Popović, S., Hromiš, N., Edible films and coatings: sources, properties and application. *Food Feed Res.*, 42, 1, 11–22, 2015, https://doi.org/10.5937/ffr1501011s.

140. Costa, M., Maciel, L., Teixeira, J., Vicente, A., Cerqueira, M., Use of edible films and coatings in cheese preservation: opportunities and challenges. *Food Res. Int.*, 107, 84–92, 2018, https://doi.org/10.1016/j.foodres.2018.02.013.
141. Nogueira, G., Oliveira, R., Velasco, J., Fakhouri, F., Methods of incorporating plant-derived bioactive compounds into films made with agro-based polymers for application as food packaging: a brief review. *Polymers*, 12, 11, 2518, 2020, https://doi.org/10.3390/polym12112518.

13

Edible Material as a Sustainable Eco-Friendly Option of Food Packaging

Arushi Phillips and Mousumi Sen*

Department of Chemistry, Amity Institute of Applied Sciences, Amity University, Uttar Pradesh, India

Abstract

Environmental worries about conventional packaging materials like plastics and chemicals are prompting the industry to seek bioactive alternatives in food packaging. Due to their slow degradation and persistent environmental presence lasting for centuries, these materials contribute to pollution and adversely affect species and ecosystems. Inappropriate preservation, protection, storage, and transportation methods are cited as primary culprits behind food wastage. Hence, endeavors are being made to advocate for sustainable and enduring choices in food packaging. A crucial condition for the adoption of these materials in the food sector is that they must enjoy broad recognition as food-grade, ensuring their safety for use in direct contact with food. The advantage of active antimicrobial food packaging lies in its ability to offer protection beyond passive methods. There is a rising emphasis on edible films and coatings in the food packaging sector, driven by their ability to shield products from the environment and their biodegradable features. These solutions are crafted to actively impede or diminish microbial development on the food surface, aiming to prolong the shelf life of the goods. In conventional food packaging, edible films play a dual role: they are consumable and serve as antimicrobials. An edible coating, containing antimicrobial ingredients, is applied to the food surface as an alternative approach to directly applying antimicrobials. The incorporation of active agents into edible films can impact how consumers perceive the information provided on the product label. Edible films, being thin material-based layers, play a role in shielding food items from external influences, improving their quality, and extending shelf life. Whether in the form of thin coatings or solid sheets, these are termed as edible films when applied over or within food components. Utilizing edible materials, it is possible to create food pouches or film wraps for food products.

**Corresponding author*: mousumi1976@gmail.com

Mousumi Sen (ed.) Food Coatings and Preservation Technologies, (447–484) © 2025 Scrivener Publishing LLC

The increasing use of edible films in food protection and preservation is driven by their numerous benefits over synthetic materials, such as biodegradability and environmental friendliness. These benefits involve a strong antimicrobial effect with minimal adverse effects, enhancing consumer trust, cutting down costs, and more. The natural polymers commonly found and utilized include polysaccharides like alginate, fats like glycerol proteins like zein. An important benefit of this packaging type is its ingestible nature along with the food item. The varied properties of the composite matrix, shaped by each component, result in differing functionalities of edible film materials. In film formulation, polysaccharides or proteins offer adequate mechanical and gas barrier properties, yet they tend to exhibit less effective water vapor protective attributes. Derived from a range of sources such as plants, animals, and microbes, polysaccharides stand out as the most abundant carbohydrates in the natural world. Conversely, lipids-based films display commendable steam insulating qualities, diminished structural robustness, and elevated permeation of oxygen permeability. When integrated into film production, polysaccharides or proteins typically contribute to films with reliable mechanical and gas barrier traits but weaker water vapor barrier capabilities. In contrast, films made from lipids display significant oxygen permeability, weaker mechanical strength, and commendable water vapor barrier characteristics.

***Keywords*:** Edible packaging, edible films production, shelf life, protective coatings, water repellent materials, additives in edible films, barrier properties of edible packaging materials

13.1 Introduction

The global food business relies significantly on food packaging systems [1]. Efficient packaging minimizes food waste and ensures consistent quality throughout the product's shelf life. Furthermore, Rahman and the team stated, food packaging serves the dual purpose of enticing customers, providing product information, and safeguarding and preserving the contents [2]. The materials utilized in packaging act as a shield against diverse hazards, encompassing breakage, evaporation, microbes, light, temperature, and spoilage [3]. Robertson studied that the common types of packaging materials in the food industry include traditional options like glass, cardboard, metal, and plastics. Notably, plastics stand out as the most widely utilized materials for food packaging [4]. In food packaging, polyethylene, polypropylene, and polyethylene terephthalate are widely utilized because of their affordability, ease of shaping, and lighter weight compared to glass or metal alternatives. However, it is important to note that these plastics, being petroleum-based, are not environmentally sustainable and contribute significantly to the carbon footprint. The widespread use of non-biodegradable packaging materials is problematic due to their significant drawbacks, such as contributing to environmental hazards by emitting higher levels of toxins, disrupting the carbon dioxide balance, and creating composting difficulties. Chawla and the team remarked that the use of non-biodegradable packaging materials is unsustainable and results in environmental pollution due to package waste [5].

There is a growing interest in replacing conventional plastic materials due to their substantial pollution and non-renewable nature. Achieving a complete transition from conventional methods to biodegradable options may be challenging, but adopting biodegradable active food packaging materials offers a feasible solution, especially in sectors such as food packaging. This subject is widely discussed as these materials are pervasive in modern society, primarily owing to their affordability, ease of processing and manipulation, physiological inertness, and versatile applicability [6]. Ananda and the team observed plastics have been extensively promoted for various purposes, such as 3D printing, the food industry, and smartphones, due to their diverse qualities [7]. Despite these advantages, plastic pollution is a pressing issue affecting soil, oceans, drinking water, human and animal health, and the atmosphere. Ellen and their team stated that predictions indicate that plastic production will triple in the next 20 years, exceeding the existing capacities for waste management and recycling, raising alarming concerns [8]. The most significant environmental issue of our era is plastic pollution. The surge in demand for minimally processed food products has driven market diversification, highlighting the essential role of efficient and innovative packaging technologies in preserving and protecting food, a core component of evolving policy initiatives. Ensuring food safety is paramount to its ability to fulfill its purpose without causing harm. Various initiatives led by non-governmental organizations and civil society have played a vital role in increasing awareness, visualizing the issue, and proposing remedial actions to address this problem. In 2018, the European Union initiated efforts to create plans for reducing plastic consumption, emphasizing a commitment to environmental preservation. There are two critical issues demanding resolution: (i) The financial challenge, where only 5% of plastic materials stay in the economy after initial use, leading to yearly losses of 70 to 105 million euros; and (ii) The extended degradation time of plastic materials compared to other materials like glass, paper, or metals, as non-recycled plastic takes years to break down. Scholars delving into these issues have substantiated that early plastic residue items disposed of in the environment are recognizable. Consequently, Incombustible substances find their way into the environment. Haward and the team findings indicated the significant pollution of the oceans, estimated between 5 and 13 million tons annually, is an overly concerning consequence [9, 10]. Moreover, the extraction and incineration of plastics are believed to emit around 400 million tons of CO_2 each year. The way plastics are currently manufactured, used, and disposed of fails to capitalize on the economic advantages of a circular system, ultimately causing harm to the environment. Reevaluating and optimizing the operations of such an intricate value chain necessitates collaborative endeavors and stronger cooperation among all major participants, ranging from plastics manufacturers to recyclers, retailers, and consumers. The policy embraced by European leaders is grounded in bioeconomy theory, with overarching objectives focused on environmental preservation and the protection of the population. These objectives necessitate a decrease in the use of plastics and the recycling of all packaging in the European Union market by 2030. The effort is dedicated to reducing plastic waste, putting an end to extensive storage, and financing the creation of innovative materials [11]. For this system to evolve successfully, cooperation is essential from all parties involved plastic manufacturers, recyclers, merchants, and consumers. The aim of establishing a new plastic economy, hinges on the active participation of all significant stakeholders. In addition to emphasizing the principles of reuse and recycling, plastic design and manufacturing should strive to acquire and promote the use of more sustainable materials. Christodoulidou and the research team stated by taking these steps, the impact of plastic pollution on the environment and people can be minimized [12]. The food business has positioned itself as a significant frontrunner in research initiatives aiming to replace the use of conventional materials partially or entirely, a trend mirrored in various industries. According to the Food and Agriculture Organization (FAO), global food waste is estimated to be 1.3 billion tons per year [13].

Polluting sources encompass waste from agriculture, commerce, industry, and households. Food packaging containers, like other biological materials, commonly retain traces of the food they previously held. Unfortunately, due to economic and practical limitations, copious quantities of plastic packaging end up being wasted into the environment. The surge in these numbers is significantly influenced by extravagant utilization of non-reusable packaging. Comprising Incombustible and exhaustible resources, elements like these are unfavorable in several aspects and are identified as a significant contributor to pollution. The substantial amount of waste generated adds to the solid waste burden in urban settings, intensifying environmental distress and challenges. The escalating worry regarding the environmental consequences of packaging waste has driven consumers to insist on natural, exceptional, environmentally conscious, convenient edible contaminant that prioritizes safety. Moreover, it initiates a noteworthy obstacle to the edible goods sector for developing environmentally responsible containerization solutions while reducing ecological contamination.

Daniloski emphasizes that on meeting these demands, the food industry is dedicated to advancing biodegradable, edible, and sustainable materials. This endeavor is designed to enhance food safety and quality, prompting a surge in research and industry attention to this concern [14]. Espitia and the team highlighted the exploration of edible packaging has rapidly expanded in recent years, driven by heightened consumer interest in health, nutrition, food safety, and environmental concerns related to packaging waste [15]. Sustainable packaging incorporates edible packaging as a vital component. Zhao and the team's findings on the utilization of edible packaging significantly expands the range of packaging materials, diminishing dependency on non-renewable petroleum resources, and effectively mitigating food losses and waste [16]. Incorporating edible packaging can simplify packaging demands by transitioning from multilayer or multilevel packaging to a single-component package, promoting reduced material consumption and enhanced recyclability of the simplified packaging system while preserving protective functions, as shown in Figure 13.1. The expansion of such packaging options leads to a decrease in the carbon footprint on the environment. Moreover, edible packaging is gaining prominence as a method for optimizing food quality. A significant advantage of edible packaging systems is their seamless integration as an inherent part of the food product. Barbosa studies stated that the benefit for consumers lies in the ability to eat it directly without the necessity of unpacking and

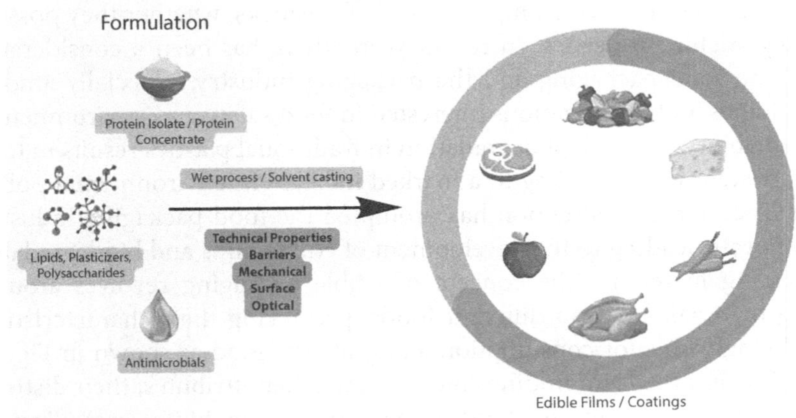

Figure 13.1 Representation of the edible film applications.

disposing of the package [17, 18]. Depending on its proximity and direct contact with the food, such packaging may be composed of triple tiers. The initial layer, promptly exposed to the food, should ensure safety throughout maintenance and shipping.

Serving as a shield against physical alterations, the second layer, which is not in direct contact with the food (such as plastic boxes or flexible bags), plays a protective role. Pallets and adhesive films, for instance, frequently include integrated films as part of the third layer. This layer enhances defense against climatic or atmospheric conditions and mechanical damage. The suggested edible packaging typically falls within the category of the first-layer film type [19]. The objective is to provide an overview of the current information on the utilization of edible packaging systems in food production.

13.2 Edible Packaging

Effective packaging is vital in the food industry, as it helps maintain the freshness and physiochemical integrity of products during transportation and until they are consumed. Non-renewable materials are the primary source of non-biodegradable plastics, which are prevalent in various industries. Additionally, these substances have a detrimental impact on the environment, as burning them results in the production of polluted gases containing sulfur oxides, nitrogen oxides, and carbon oxides. Addressing the alarming problem of pollution caused by traditional single-use materials, a practical and viable solution is the advancement of entirely decomposable, even consumable substances synthesized using bioderived polymers. Sourced from natural polymer materials, they are described as polymerized substances synthesized using sustainable feedstocks, whether they possess biodegradable properties. In recent years, there has been a considerable focus on food packaging and the packaging industry, especially among individuals and organizations interested in food safety and environmental sustainability. The lack of degradation in traditional plastics results in tons of persistent waste, leading to a marked increase in environmental pollution. The increased attention has prompted the food packaging industry to innovate, leading to the development of edible films and biodegradable packaging materials. The concept of edible packaging revolves around using materials to wrap different foods, preserving their characteristics, and being edible for consumption alongside the food as shown in Figure 13.2. Biji and the team findings indicate the other attributes, their distinctive properties, including biodegradability, renderability, cost-effective development, abundance in nature, non-toxicity, and biocompatibility,

Figure 13.2 Edible packaging.

have resulted in their intensive utilization for the creation of novel materials, as summarized in Table 13.1 [20–22]. Edible films refer to thin layers that are cohesive and composed of bio-based polymers. Foods, especially those incorporating protein-based films like wheat gluten, whey proteins, and casein, benefit nutritionally from the presence of biofilms, which provide essential amino acids. As of 2011, bio-based polymer production had already exceeded 3.2 billion tons, and post-2020 forecasts suggest it will surpass the 12-billion-ton threshold [23]. The potential of edible packaging as a packaging solution extends to many different types of food. There is evidence to support the successful implementation of edible packaging across various food types, encompassing meat, grains, nuts, cheese, bakery products, confectioneries, and fresh produce like fruits and vegetables. Insufficient packaging can compromise the quality of high-nutrient meat products, leading to spoilage caused by microbial growth and reactions with oxygen from the atmosphere. Yin stated that in the edible goods sector, biomass could be a serve as a source for bio-based polymers, polymers could be produced through the synthesis of bio derived monomers or generated through microbes. Opting for these synthetic materials evaluates the purposes the film or coating must fulfill.

Food-grade encasement can exhibit no interaction accompanying the product, or it could be classified as smart or enhanced encasement [24]. Barbosa and the team highlighted active packaging, in active packing, active components such as vitamins, antioxidants, and antimicrobials are incorporated. The controlled release of these components into the product is designed to enhance food durability and other characteristic features, such as sensory attributes and product safety [25, 26]. The sensors integrated

Table 13.1 Edible packaging over plastic packaging.

Problem	**Solution**
Globally, a portion of 78 million, equivalent to 10.92 million Tonnes of synthetic containerization recycled annually. Plastic, being lightweight and buoyant, finds its way into our oceans, contributing to an annual total of nine million tons, the majority of which comes from developing nations with insufficient infrastructure for effective management. The issue is projected to intensify as these nations experience economic growth and inevitably increase their consumption of packaged foods. At the same time, in a world increasingly driven by convenience, the continual rise in meal-kit, grocery services, and take-out foods is foreseen to contribute significantly to the generation of packaging waste. After recycling, plastics are typically shredded, melted, and repurposed into items like lumber, fleece, or carpeting. Nevertheless, these items end up in landfills. Despite attempts by manufacturers to produce thinner bottles and shrink wraps, the underlying issue persists, synthetic materials are manufactured from exhaustible commodities, primarily fossil fuels, and the majority at no time experiences an alternative existence.	A collaborative effort involving designers, engineers, biologists, investors, and recyclers is underway to develop packaging in line with the concept of a circular economy. In contrast to the linear "take, make, waste" model, a circular design employs supply chains that continually repurpose old materials into high-value products. This approach underscores the importance of durable design, remanufacturing, and encouraging reuse. The circular economy involves two separate loops for material goods: one focuses on recovering engineering elements like alloys, earth elements and synthetic compounds for second use, while several channelizes organic substances such as plant-based resources revert to the environment via composting solutions or alters via bio-gas production digestion into environmentally balanced power energy. Food-grade encasement exemplifies the utilization of sustainable and compostable resources, providing individuals with an eco-friendly substitute to traditional polymer food packaging. Food-grade encasement, derived from various eco-friendly decomposable substances, eradicates the necessity for collection, processing, recycling, or disposal; it is meant to be consumed directly!

into the films and coatings of intelligent packaging enable reactions or shifts impacting freshness, density, protection, and standard shelf-stable products. Probes, typically color-changing promoters, utilize natural pigments like anthocyanin and curcumin as pH indicators. Primarily, these sensors find application in the preservation of seafood and meat products. Rocha and the team stated the materials employed for food packaging consist of compounds that are safe for human consumption, posing no health risks. With minor modifications in material composition and structure, these materials can be adapted into several types of films and coatings by adjusting their thickness. The optimal thickness for making edible films is regarded as being less than 0.3 mm (about 0.01 in), making thickness a key physical property in the production process [27].

For an edible film to be considered typical, it should exhibit non-toxicity, desirable sensory qualities, microbial stability, reliable mechanical properties, safety for human consumption, non-pollutive attributes, and affordability. Like their conventional counterparts, edible food packaging materials need testing to confirm their adherence to quality standards for safe ingestion. An essential aspect in determining the safety of edible packaging is the assessment of its microbiological quality.

13.3 Production of Edible Film

In the realm of food-grade encasement, the primary forms are consumable films and overlays, differing within their application routine and concrete forms. Cunha and the researchers highlighted the edible films come in the form of solid laminated sheets, functioning as food wraps that can be consumed or removed. On the other hand, coatings are liquid formulations applied to foods through techniques like coverings, showering, and soaking. Upon dehumanization, the formation of the edible layer is created on the top of the food, and this layer could be directly consumed or else removed [28]. The production of edible films typically involves two processes: moistened and dried techniques [29]. The core idea of the wet process method is to use evaporation to eliminate water from solutions containing biopolymers, thereby achieving drying. Creating edible films generally follows a four-step process: dissolving biopolymers in solvents, applying heat treatment, casting the material, and finally drying it. After the solution evaporates and the film dries, it can be delaminated and put to its intended use. The hydrophilicity of polysaccharides makes wet casting the optimal choice for producing edible films, as it aligns well with the material's water-attracting properties. One major benefit of the casting method is its ease of use without requiring special equipment. Moreover, the ability to operate at lower temperatures is advantageous for food-processing materials sensitive to higher temperatures. However, the casting process has limitations such as a longer drying duration, making it less suitable for commercial manufacturing purposes. The dry formation method necessitates the presence of minimal, if any, additional substances beyond the primary constituent, as there is no liquid component involved. The underlying concept of this process is showcased, illustrating the transformation of a powdered substrate into a polymeric film through the application of heat energy. This technique is often termed a "dry process" due to its optimal operation with minimal water or solvents. During the movement through the kneading zone, the mixture experiences increased strain, temperature, and density. This is followed by heating these materials above their glass transition temperature to allow them to transition into a molten state, which is then extruded through a nozzle with the help of the rotational force from an extrusion screw. The final stage includes cooling the resulting materials to solidify them, resulting in the formation of the film, as illustrated in Figure 13.3. Essential elements utilized during manufacturing of food-grade films and coverings comprise a film-forming substance, and additional elements which may include softeners, dispersants and antibacterials agents. Detailed attributes of these elements are crucial; Coating agents must encompass proteins, polysaccharides, composites, or lipids and these agents should exhibit film-forming characteristics. The advantages offered by polysaccharides have propelled them to the forefront as the preferred material for creating edible films used in food packaging. The choice of solubilizers and softening agents is contingent upon their compatibility

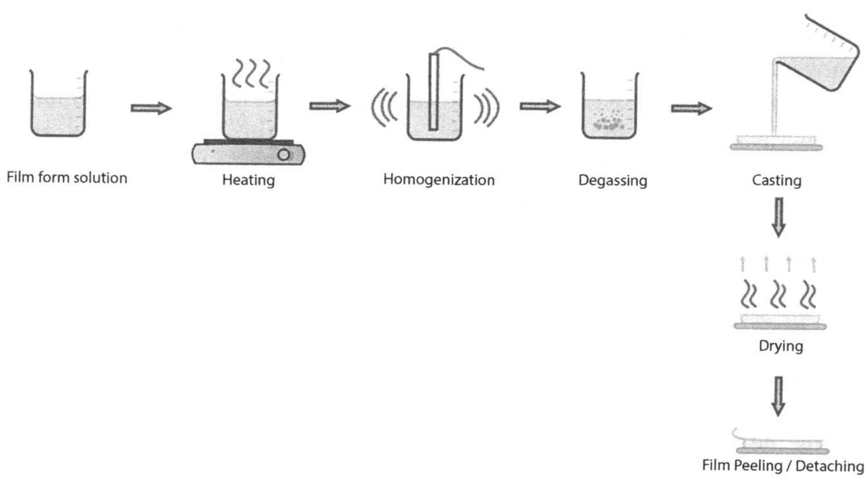

Figure 13.3 Representation of the production of edible film.

with the specified polymer. Solvents can dissolve polymers and modify their mechanical and barrier properties, with examples such as alcohol, water, citric acid, acetic acid, lactic acid, and hydrochloric acid. Plasticizers contribute to the reduction of brittleness, the enhancement of flexibility, the decrease in porosity, and the improvement of the mechanical properties of films. Garcia's studies have shown that the preference of every constituent is contingent upon a specific edible commodity employed, and it does not impact the sensory attributes of the food. Ensuring proper adhesion to the food surface relies on the chemical composition of the polymerized substance as well as the films outcome attributes.

13.4 Edible Packaging Materials

A biodegradable film is a type of primary packaging made from biodegradable polymers and food-grade additives, ensuring safety for food contact. Utilizing a thin biodegradable material, it's feasible to produce a food wrap

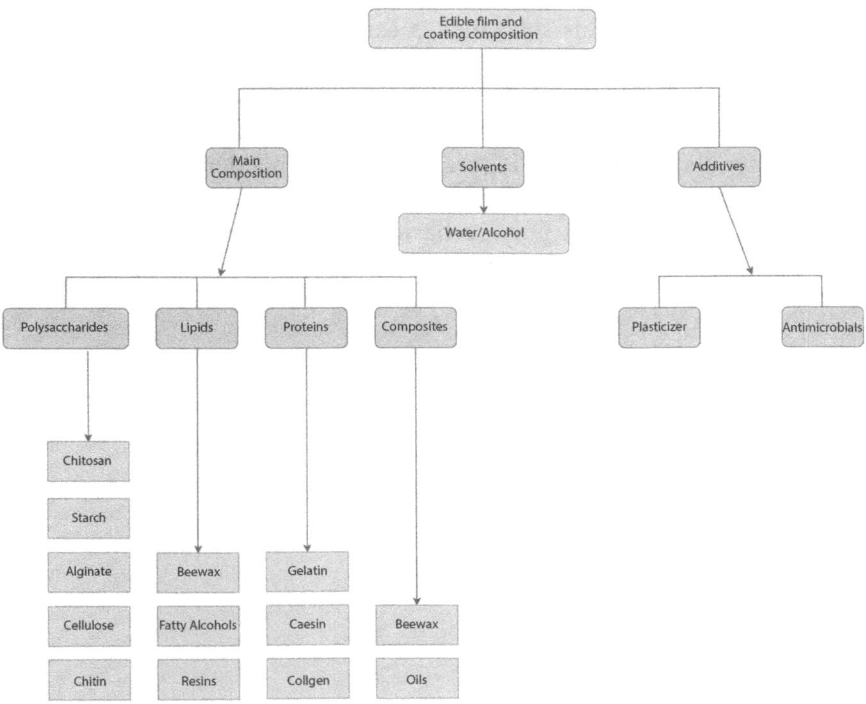

Figure 13.4 Classification of edible films and coatings.

that retains the original ingredients and processing approach [30]. Edible films and coatings, used in the creation of edible packaging, must satisfy two requirements: they should be edible, and they should have the capacity to create a continuous layer [31]. Zhao studies indicate that the food-grade films and coverings are categorized as polysaccharides, lipids, proteins and composite materials, as shown in Figure 13.4. Preferring between food-grade films or coverings necessitates evaluating required composition materials, the type of food intended for application, and the application technique. The addition of a plasticizer to the solution is a frequent practice to increase flexibility and elasticity [32]. Specific film properties and functionality can be achieved by incorporating additional additives such as antimicrobial agents, colors, and flavors into the solution, depending on the ultimate application.

13.5 Polysaccharide-Based Packaging Materials

Being the most abundant natural macromolecule in nature, polysaccharide is characterized by long chains of polymeric carbohydrates connected by glycosidic bonds. This substance is extensively employed for the creation of food-grade encasement. Food packaging's functionality is shaped by factors such as the properties and primary deterioration modes of the packaged foods, along with the compositions and properties of the packaging materials, as shown in Figure 13.5. Thus, an extensive analysis of a range of polysaccharides is necessary, as it is instrumental in evaluating the feasibility of different polysaccharides for use in food packaging, as detailed in Table 13.2. The commendable oxygen-holding capabilities of

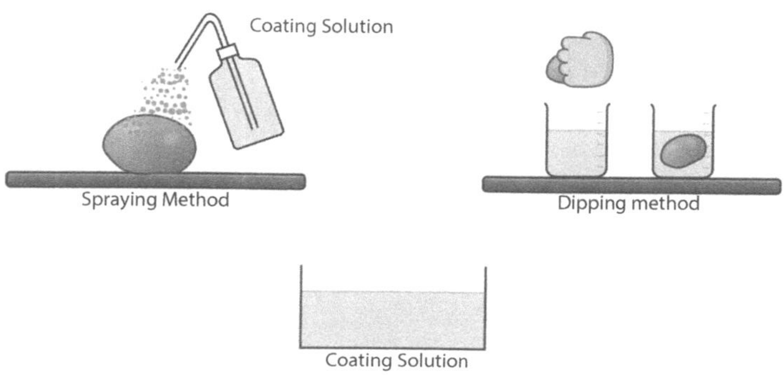

Figure 13.5 Methods of applying edible coatings.

Table 13.2 Distinct types of polysaccharides based edible material.

Polysaccharide based packaging material	
Cellulose Film	Characterized as a linear chain homopolysaccharide with anhydrous β1-4 glycosidic bonded glucose rings. cost-effective, lightweight, reliability, high tensile strength, harmlessness, well-tolerated by living systems, disposal by natural processes, recyclable, effective coating ability, resilience to chemical alterations, and simplicity in chemical derivatization. Cellulose acetate finds widespread application in wrapping fresh and baked goods.
Chitosan Film	The composition of the linear polysaccharide involves β-linked N-acetyl-glucosamine and N-glucosamine units. Noted for its biodegradable, non-toxic, and biocompatible nature, it also displays antimicrobial and antioxidant properties. Degradation occurs in chitosan-based films due to their non-thermoplastic behavior, preventing them from reaching the melting point. The hydrophilic characteristic of chitosan results in inadequate moisture barrier properties.
Alginate Film	A linear anionic polysaccharide polymer, the hetero polysaccharide is constituted by M-blocks and G-blocks in a β-(1-4)-D mannuronic and α-L-guluronic acid arrangement. Possessing biodegradability, biocompatibility, biostability, non-toxicity, and hydrophilicity. The uniform, highly water-soluble, and transparent nature of alginate edible films is evident. Utilizing divalent cations like $Ca2+$ in the cross-linking of alginate while film synthesis leads to ionic interactions that enhance water insolubility and strengthen the films.

polysaccharides are a result of their edible transformations, which introduce multiple hydrogen chains to increase density and impede gas flow. Sources of polysaccharides encompass microorganisms, plants, animals, and marine life. The renewable nature of polysaccharides emphasizes their sustainability and eco-friendliness. The natural decomposition of polysaccharides by microorganisms results in harmless compounds that are reintegrated into the ecosystem.

Polysaccharide-based coverings and films can be devoid of oil, flavorless, and transparent, boasting improved resilience to chemical changes and manufacturing flexibility in contrast with the packaging materials derived from proteins and lipids. Safety is of utmost importance in food packaging, and polysaccharides are generally considered safe for food contact applications. Polysaccharides stand out as a significant class of compounds, serving as essential elements in the fabrication of edible films. Luo and the teams contended the utilization of polysaccharide-based films and coatings have the potential to decrease dryness, minimize discoloration, and reduce oxidative spoilage, thereby possibly strengthening the shelf life of meat, fruits, vegetables and other meat products [33, 34]. Kasim stated, due to their intermolecular structure, Coatings derived from polysaccharides function as proficient barriers to oxygen, carbon dioxide under conditions of reduced to moderate humidity levels [35, 36].

Mohammed and the teams declared, by leveraging polysaccharide-based materials using their specific transmission, can further contribute to prolonging the span of quality preservation for freshly sliced vegetables through reducing metabolism rate and atmospheric interchange. Due to their hydrophilic nature, polysaccharide-based materials offer a low barrier to moisture [37].

a) Cellulose

Anhydrous β1-4 glycosidic bonded glucose rings form the basis of cellulose, which is a linear-chain homopolysaccharide. Shahidi and the researchers stated, the properties of cellulose include cost-effective, lightweight, reliability, high tensile strength, harmlessness, well-tolerated by living systems, disposal by natural processes, recyclable, effective coating ability, resilience to chemical alterations, and simplicity in chemical derivatization [38]. These elongated and unbranched chains may contain thousands of glucose units, forming a sturdy and inflexible structure. Solubility in aqueous solutions is a notable challenge for cellulose, primarily due to its unique crystalline structure that sets it apart from the typical polysaccharide composition. Shahidi and the researchers reported that the edible film manufacturing using cellulose has been restricted due to insolubility in water and numerous solubilizing agents, despite its favorable properties. The high crystallinity, tightly packed hydrogen bonds, and structural complexity of cellulose are responsible for its resistance to dissolution in water and most organic solvents as shown in Figure 13.6 [39].

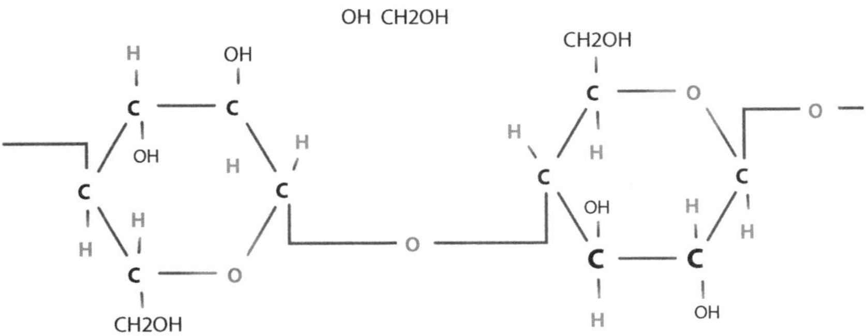

Figure 13.6 Chemical structure of cellulose.

Overcoming this limitation involves derivatization, where cellulose undergoes alkali processing with subsequent acidification stage by hydration promoters like CH3Cl, ClCH2COOH. This process generates heat-sensitive polymer. Derivatives of cellulose extensively used during development of food-grade encasement involving organic oxides like Cellulose methyl ether, Cellulose hydroxypropyl ether, Cellulose hydroxypropyl methyl ether along with others such as acetate of cellulose. While Films incorporating HPC and methylcellulose-coated films excel as barriers against O2, CO2, and lipids, they exhibit quick water vapor transmissibility. Improving water vapor repelling qualities could be possibly achieved through introducing water-repellent components like oils into the coating agent solution. Cellulose acetate is frequently used for packaging fresh and baked items.

b) Chitosan

Chitosan, the second most abundant natural polysaccharide following cellulose, is characterized by the sequential structure formed by the linkage of β-N-acetylglucosamine and N-glucosamine components as shown in Figure 13.7. The antimicrobial effectiveness of chitosan is highest against yeasts and molds, with gram-positive bacteria being more susceptible than gram-negative bacteria. The polysaccharide chitosan, obtained from marine animals like shrimp, is of animal origin. Moreover, produced through the N-deacetylation of chitin using an alkaline method. Chitin is sourced from the exoskeletons of crustaceans, insects, and fungi. Chitosan, obtained from chitin, is characterized by being environmentally friendly, harmless, and compatible with living tissue. Despite being insoluble in water and typical dissolving agents, the application of elevated alkalinity along with warmth eases conversion into chitosan. Chitosan's application in food

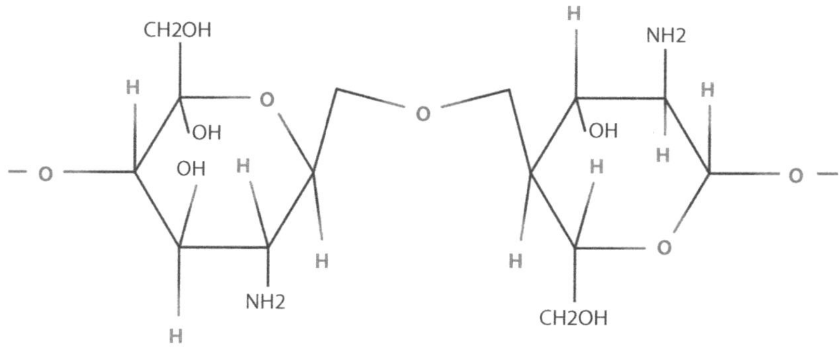

Figure 13.7 Chemical structure of Chitosan.

packaging has been extensively explored in numerous studies, emphasizing its physicochemical properties. Montero and the team studies have shown, it exhibits favorable mechanical attributes, proficient film-forming capabilities, and selective permeability to O2 and CO2. The antimicrobial and antioxidant properties of chitosan, in addition to its mechanical and selective permeability attributes [40], make chitosan-based films and coatings suitable for products like apples, pears, strawberries, cucumbers, bell peppers, peaches, and plums. When applied to meat surfaces, chitosan edible film has shown effectiveness in inhibiting the growth of *L. monocytogenes* and *E. coli*. The use of chitosan films in food products faces limitations, primarily due to the hydrophilic nature of chitosan resulting in subpar moisture barrier properties. Cazón and the team suggested several strategies such as modifying solvent type, adjusting pH, incorporating plasticizers [41], and integrating lipid components can be applied to enhance the functional attributes of chitosan. Xie and the team stated, Chitosan-based films and coatings encounter a notable challenge as they are not thermoplastic, resulting in degradation before reaching the melting point. The difficulty in extruding chitosan and the incapability of its films to be heat-sealed present limitations to their commercial-level production [42].

c) Alginate

Alginate films, sourced from marine brown seaweed (Phaeophyceae), are considered edible and natural heteropolysaccharides. Brown algae species serve as the main source for alginate, a naturally occurring polysaccharide extensively utilized in the bio industry as illustrated in Figure 13.8. Mollah and the team specified, they are consisting of alginic acid salt

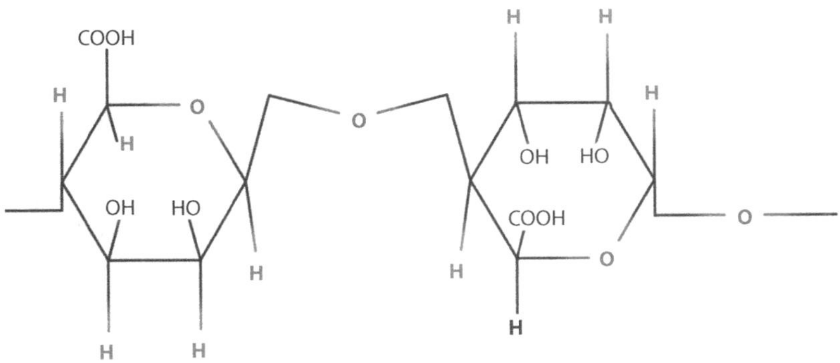

Figure 13.8 Chemical structure of alginate.

compounds, with sodium, calcium, and magnesium being prominent examples. It is characterized with the involvement of a linear and anionic polysaccharide composition composed of M-blocks with β-(1–4)-D mannuronic acid and G-blocks with α-L-guluronic acid. However, it demonstrates Gel-forming characteristics owing to the significant number of GM blocks and their interactions across chains. Alginate's special characteristics, such as non-toxicity, biocompatibility, biodegradability, bio-stability, and hydrophilic nature, make it exceptionally well-suited for cutting-edge biomaterial uses [43]. Puscaselu, Gutt and the team stated, the capacity of alginate to produce films of excellent quality includes features such as good mechanical strength, efficient moisture barriers, and cohesiveness [44, 45].

Alginate edible films stand out for their uniformity, high water solubility, and transparency. Othman contented that the incorporation of divalent cations, such as Ca2+, for cross-linking alginate in film synthesis creates ionic interactions, resulting in improved water insolubility and strengthened films [46]. According to the United States government agency responsible for food and drug regulation (FDA), Sodium alginate is categorized as Generally Recognized as Safe (GRAS) in accordance with Title 21 of the CFR [47].

13.6 Protein-Based Packaging Materials

On various food items, protein-based edible films are employed to mitigate water loss, control oxygen intake, decrease lipid migration, enhance mechanical handling characteristics, and offer physical protection. Recent years have witnessed a surge in interest in protein-based films and coatings as they are recognized as sustainable alternatives to conventional plastic packaging materials. Biodegradability, functionality, and versatility are key advantages offered by these films, making them highly desirable for a range of uses. They also serve as substitutes for synthetic packaging materials. Cazon and the team clearly stated that the Food-grade films which are protein-based operate as conveyors for antimicrobial and antioxidant substances, with common applications on food surfaces aimed at controlling the diffusion of preservatives from the surface into the interior. The inherent qualities of different protein sources are key determinants in shaping the strength, barrier effectiveness, and film-forming potential of the films. Protein-based films are subjected to extensive characterization to determine their physical, mechanical, and barrier characteristics. The utilization of protein-based edible films extends to being part of multilayer food packaging materials in conjunction with non-edible films. In this scenario, the edible film with a protein base is positioned as an interior layer, directly in contact with the food surface. Food-grade films which are protein-based, possess both mechanical and barrier characteristics, having the potential to substitute synthetic polymer films [48].

a) Casein Film

The primary composition of caseins involves being phosphoproteins, which organize into colloidal micelles in milk and are stabilized by calcium phosphate bridges. For food packaging needs, casein films offer notable strength and superb oxygen protection, though they lack elasticity and are sensitive to moisture levels. Adapting film attributes to target specific environmental conditions broadens the potential commercial uses of casein-based films. The propensity of casein films to absorb and release water molecules makes them highly moisture sensitive. This water absorption acts as a plasticizer, significantly altering the mechanical properties of the films and constraining their range of use. Chen and the researchers said, the low cysteine content in caseins results in an inability to establish extensive covalent disulfide bonds, either intermolecular or intramolecular, through thermal denaturation. Consequently, the formation of water-insoluble films is hindered [49]. The open, flexible, and random conformation of casein molecules is conducive to film formation from aqueous solutions. McHugh specified, this is made possible by their capacity to create extensive intermolecular hydrogen and electrostatic bonds, as well as hydrophobic interactions [50]. Gennadios highlighted that the casein films have been explored for their role as barrier against humidity in dissolvable bags, market-fresh goods, desiccated fruits, and cold storage foods. Notably, this outstanding oxygen barrier effectiveness of films composed of casein under modest water activity conditions have proven effective in retarding lipid oxidation in nuts. Moreover, casein films are seen as having substantial potential as carriers for flavor, nutrients, or bioactive components [51].

b) Zein film

A group of prolamins, termed alcohol-soluble proteins, is present in corn zein. Zein exhibits hydrophobic behavior due to its rich content of non-polar amino acids. Weller stated, its solubility characteristics include lack of water solubility, except under conditions of exceptionally modest or extreme pH levels, Furthermore, it persists in being insoluble within anhydrous alcoholic solutions [52]. Zein stands out for its ability to form films effectively, making it a valuable material for biodegradable film production. Formulations for Zein coating or casting are commonly formulated via the dissolution of zein in heated solutions of ethyl alcohol or isopropanol. Padua said that the formation of the film process implies the creation of a hydrophobic state, H2, and controlled disulfide connections among zein molecular chains incorporated into the film matrix as the alcohol dissipates from the film's surface [53]. As a commercially successful finishing agent, Zein is distinguished by its capacity to create a glossy finish and functioning as an oxygen barrier, lipids, and humidity in cashews, chocolates, confections, and an array of food items [54]. In the pharmaceutical industry, zein is utilized to coat tablets, providing a controlled release of active ingredients, and masking the taste of orally administered drugs [55]. The application of zein-based coatings is recommended by the Chinna and the team to diminish the oil absorption in fried foods [56]. The use of zein films with moderate gas barrier characteristics is effective in decreasing the moisture loss and respiration rate of vegetables [57].

c) Collagen

Collagen, containing abundant glycine, hydroxyproline, and proline, is a hydrophilic protein that exhibits swelling in polar liquids due to its high solvency characteristics. In the food industry, collagen is highly valued for its remarkable properties, such as its strong film-forming ability, resistance to organic solvents, biocompatibility, and stability. One example is collagen fiber films, which exhibit excellent antioxidant properties and are employed in packaging meat, fish, and other products. The collagen casing serves as a representation of one of the leading food-grade films made from proteins in commercial success. Farouk's team studies have shown that the utilization of a collagen film for wrapping chilled and thawed beef round steak demonstrated a decrease in exudation, with no significant impact on color or lipid oxidation, as cited in reference [58]. Collagen is renowned as the most commercially successful edible protein film, with a prominent presence in the meat industry, particularly in the production of edible sausage casings. Collagen is easily accessible, non-toxic, and serves as an excellent foundation for biomaterials. The use of films based on collagen is recommended for ready-to-eat sources as they effectively alleviate coagulation loss, enhance juiciness, facilitate convenient removal of mesh after smoking or cooking, and soak up moisture across various cooked meat products.

13.7 Lipid-Based Packaging Materials

Protective coatings against moisture transfer and for the purpose of adding gloss in diverse applications have utilized numerous edible lipid materials. Lipid films and coatings have played a crucial role in food preservation for centuries, preventing moisture loss. Lipids are becoming more prominent in the food industry because of their distinct features, such as hydrophobicity and their role as a barrier against water-vapor transfer. It is essential to recognize that, unlike other macromolecules, lipid and resin compounds do not qualify as biopolymers. A notable feature is the scarcity of repeating units bound by covalent bonds, preventing the creation of an extensive molecular framework in these materials. As a result, they are fragile and incapable of forming cohesive, self-supporting film structures. Greener and the team suggested a way to address these, low-polarity components such as fatty substances and resins have been incorporated into food-grade film-forming materials to add the barrier against moisture [59]. Despite this, use of lipids in the development of edible packaging will have drawbacks, like imparting a taste and texture reminiscent of wax, creating a sleek and oily surface, and the risk of potential spoilage.

a) Waxes

Kester and the team stated, Waxes are formed by the esterification of a long-chain fatty acid with a long-chain alcohol. Comprising long and medium-length carbon atom chains, waxes are characterized as hydrophobic substances. Their remarkably low polar group content makes them more resistant to water diffusion than most edible films, whether lipid or non-lipid in nature [60]. Natural waxes such as rice byproduct wax, palm tree wax, Inorganic and petroleum-based waxes find application in the formulation of protective coatings. These coatings can be applied independently or in combination with other ingredients, as detailed in reference.

b) Resin

Baldwin and team studies have shown that edible resins like shellac, terpene resin, and wood rosin serve the purpose of enhancing gloss in food products, as indicated in reference [61]. The extraction and processing techniques for shellac have developed to a mature stage, providing a clear comprehension of its structure. In addition, the issues stemming from shellac's self-polymerization and their consequences on the stability of shellac-based food systems are seldom reported, emphasizing the need to address these challenges. Overall, the utilization of shellac in the food industry demonstrates significant potential. A combination of aliphatic and alicyclic polymers forms the intricate composition of shellac and aleuritic and shellolic acids, categorized as hydroxyl acids, are part of the source material. With the capability to dissolve in both non-polar compounds and non-acidic substances, the prevalent use of shellac serves as an eatable layer for sweet delicacies and farm-fresh picks. Additionally, it serves as an enteric coating for pharmaceuticals, as highlighted in references [62].

c) Fatty acids

Acetylated monoglycerides are fats that undergo modification by replacing one of the fatty acids connected to a glycerol molecule with acetic acid. Acetic acid esters exhibit distinct properties based on the degree of acetylation and the specific type of monoglyceride utilized. An increase in the degree of acetylation was found to improve the water vapor permeability (WVP) of acetic acylglycerol. Bourlieu studies found that the proposed theory posits that variations in crystal packing are responsible for this phenomenon or the elimination of unbound OH group that could interact with moving water. A sour or bitter taste and the potential for cracking and peeling during storage are challenges associated with acetylated glycerides [63]. Alternatively, lipid coatings can be accomplished using fatty acids and sucrose esters of fatty acids.

- **Composite based packaging materials**

The inclusion of water-repellent materials to a film composition depends on polysaccharides, forming a multicomponent food-grade film, acknowledged as an effective means of enhancing the film's properties, particularly its resistance to water vapor. Complex in nature, composite films comprise multiple components that combine hydrophobic and hydrophilic compounds to enhance their functional properties significantly. Thakur and the team said enhancing the functionality of the edible film involved the incorporation of materials to create a composite structure. In the context of carrageenan-based films, the introduction of hydrophobic components such as lipids and oils are anticipated to bolster its moisture barrier properties [64]. Concurrently, fatty acids like oleic acid, palmitic acid, and stearic acid have been employed to enhance the properties of edible films [65, 66].

Amini specified that the previous studies have indicated that stearic acid displays the most hydrophobic performance compared to oleic and palmitic acid [67] and exhibits superior moisture barrier properties on the surface of edible films due to its long chain and saturated fatty acid nature [68, 69]. Luís described that Zein, an industrial protein, is extracted from the solid residues of corn flour production. Zein, the primary corn protein, is characterized by significant amounts of hydrophobic chains, neutral amino acids, and sulfur-containing amino acids, all of which play a unique role in promoting its film-forming capability [70]. Zein stands out as a promising material valued for its compactness, low allergenicity, and efficient gas barrier capabilities. However, there is a scarcity of research on green composite edible films incorporating kappa-carrageenan, zein, and stearic acid.

The focus of this study is to explore the influence of increased concentrations of zein and stearyl acid for the enhancement of humidity repellant and strength and durability characteristics in edible films utilizing kappa carrageenan. Through the solution casting method, the composite edible film was developed by incorporating varying quantity of stearic acid density and zein. These films have thicknesses ranging from 0.092 to 0.122 mm (about 0 in). It was observed that augmenting the concentration of stearic acid leads to a noteworthy improvement in both the vapor impermeability and stretching resistance of the food-grade film. Conversely, an increase in zein concentration had a marginal negative impact on the water vapor barrier properties. Subsequently, increasing the proportion of stearic acid significantly enhanced the lengthening of the created films. Moreover, there is no improvement of increase in zein proportion or the combination of the two substances. Despite this, Nuha said that the integration of stearic acid and zein into the film based on refined kappa carrageenan exhibited a remarkable boost in tensile strength (12-18%), elongation (23-27%), and water vapor barrier properties (43-44%) compared to the neat film. The analysis outcomes indicate that the most favorable film formulation consists of 10% stearic acid and 2.5% zein. Consequently, this study underscores the potential of employing stearic acid for reinforcement in food packaging applications [71].

- **Additives in edible films**

In the context of films, additives refer to compounds distinct from coating agents which are incorporated to improve properties related to mechanics, structure, and ease of operation, or to impart dynamic functionalities. Its role lies in facilitating interactions between packaging, product, and the environment, encompassing diverse physical, chemical, and biological processes aimed at altering the natural conditions of the packaged food. This helps bolster sustainability, safety, quality, and shelf life.

a) Plasticizers

In film-forming formulations, plasticizers are typically low molecular weight hydrophilic agents. Biopolymer-based edible films often display inferior mechanical properties, characterized by fragility and a tendency to crack during the drying process. The presence of cohesive forces in polymer films leads to these issues. Nevertheless, the incorporation of plasticizers into the film's composition offers a solution to overcome these challenges. These low molecular weight compounds are utilized to increase the flexibility of biopolymers. The critical characteristics of plasticizers are their ability to blend well, their effectiveness, and their lasting impact. Moreover, by weakening the intermolecular forces among polymer chains, plasticizers play a role in decreasing brittleness and boosting the flexibility and tear resistance of edible films. The plasticizer's efficiency is tied to its capacity to function well at lower concentrations and to diffuse readily throughout the polymer matrix. Their function is to enhance film's tensile properties by embedding within the polymer network and striving for hydrogen bond sites aligned with the polymer structure. While plasticizers often enhance mechanical characteristics, their strong hygroscopicity raises water vapor permeability (WVP) and diminishes the barriers to gases, moisture, and flavor compounds. Selecting appropriate plasticizers is vital to preserve the palatability of the edible film and mitigate any potential hazards to human health during consumption. The plasticizer was selected considering its compatibility with the polymer matrix, the final application requirements, and the processing technique. Sothornvit studies have shown that the plasticizers frequently utilized in comestible containerization include single, double, or multiple saccharides such as granulated sugar, polyols such as sugar alcohol as well as fatty-acid substances. Plasticizer preference involves careful consideration of factors like compatibility, effectiveness, stability, and cost-effectiveness, as indicated by Sothornvit and Krochta in their 2005 study [72].

b) Antimicrobials

The utilization of edible coatings extends beyond a protective role, serving as carriers for antibacterial and antifungal agents to enhance the shelf life of food products. Active packaging serves as a mechanism facilitating interactions among the environment, the product, and the packaging components within the system. Antimicrobials are key components in guaranteeing food safety across food manufacturing and packaging operations. Edible coatings are increasingly being utilized as carriers for antifungal and antimicrobial agents, playing a vital role in extending the shelf life of food products. Furthermore, they are utilized as carriers of nutrients to boost the nutritional content of the final processed food products. Concurrently, these coatings serve as carriers for nutrients, contributing to the enhancement of the nutritional composition in processed food items. Pooja stated, Various natural and synthetic antimicrobial agents have been devised and integrated into different edible packaging materials as an alternative means to control bacterial growth [73]. The incorporation of antimicrobial substances into films is a widespread practice, with standard preference options comprising bacteriocidal peptide, catalyst proteins, chitinase substrate, fuel, herbal extracts, shelf-life extenders, microbial inhibitors, nanostructured metal substrates and various botanical substances [74]. García specified about the antibacterial properties and antibacterial properties of chitosan extend to its efficacy against yeast, mold, and proteobacteria have explored the scarce literature on edible films and coatings treated with bacteriophages. Nutraceutical films and food-grade coverings enhanced with phages demonstrate commendable antibacterial characteristics, proving to be suitable for preserving a range of food items such as fruits, vegetables, cheese, and meats [75].

13.8 Barrier Properties of Edible Packaging Materials

Edible films contribute significantly to enhancing the shelf life of food products by providing a protective barrier. Yet, considering the respiration activity in fruits and vegetables, it is crucial for the packaging not to be entirely impervious to water, gas, and ethylene. The effectiveness of the barrier properties relies on factors such as the chemical makeup of the film, the film manufacturing process, and the application technique employed to cover the product's surface. The barrier properties of materials are influenced by their composition as well as environmental factors such as temperature and relative humidity. Several specific functional requirements must be met by edible packaging materials, and the extent of compliance is contingent on the material type, its formation process, and its designated application. The desired characteristics of edible films include good barrier properties, acceptable sensory attributes, a flexible and stretchable structure for ease of application onto food, and compliance with regulatory composition requirements. For a wide range of applications, edible films and coatings must feature a reliable moisture barrier. Evaluating future packaging applications hinges significantly on understanding the mass transfer dynamics between packaged food, packaging materials, and the surrounding atmosphere. Through their barrier properties, the edible films safeguard against chemical changes, ensuring the integrity of the product's quality and extending its shelf life.

13.8.1 Environmental Barrier

Edible films and coatings serve as environmental barriers, controlling mass transfer between the food and its surroundings. Kaur and the team reported the critical consideration in choosing edible materials for packaging is their permeability. The response of the film to temperature and moisture corresponds to the conditions in which it is intended to be used. Hence, prior to utilizing it as a food packaging material, the permeability must be gauged under specified circumstances [76].

13.8.2 Moisture Barrier

For a broad spectrum of uses, edible films should exhibit suitable moisture barrier properties. The regulation of moisture transfer within composite food products or between food and its surrounding environment remains a prominent challenge in modern food preservation practices. Incorporating edible moisture barriers presents a practical approach to stabilize fresh or processed food items, fulfilling consumer demand for products with extended shelf life that maintain good quality and nutritional standards. Employing edible moisture barriers proves to be a successful approach in stabilizing both fresh and processed food items, meeting consumer expectations for longer shelf life while maintaining excellent quality and nutritional standards. Mathlouthi stated that ensuring the preservation of distinct crunchy and soft textures in heterogeneous foods such as pizza, quiche, cakes, or biscuits, as well as in mixtures like apéritif combinations, breakfast assortments featuring dried fruit, and/or cereals with varying water activity levels, necessitates the effective reduction of moisture exchange between their compartments [77]. Alterations in the water activity (aw) of packaged food can trigger critical microbiological growth, unwanted textural modifications, and unfavorable chemical and enzymatic reactions. The sensory quality and shelf life of food are significantly influenced by water activity (aw). Ensuring robust water vapor barrier properties is vital for both fresh food items, including vegetables, to prevent dehydration, and dry foods like bread, where preventing moisture uptake from the environment is of utmost importance.

13.8.3 Oxygen Barrier

The shelf life of oxygen-sensitive foods can be prolonged, and their quality preserved by employing edible packaging with low oxygen permeability (OP). Oxygen barrier packaging, incorrectly labeled as barrier packaging, has been indispensable for centuries in ensuring the safety and longevity of food, healthcare products, and various other commodities. The occurrence of oxidative deterioration is a significant determinant of the shelf life of food products. The presence of oxygen can initiate food spoilage reactions and facilitate the proliferation of microorganisms. Janjarasskul suggested this approach which also aids in reducing dependence on costly and non-recyclable oxygen-barrier plastics. Edible films exhibit a diverse range of OP values [78].

13.8.4 Oil Barrier

A strategy to meet consumer preferences and address health concerns involves reducing the fat content in fried food products through the utilization of oil barrier coatings. This initiative has the potential to establish oil barrier coatings as a practical substitute for reducing oil absorption during frying. The utilization of coatings to decrease the fat content in fried foods serves as an alternative strategy that satisfies both health concerns and consumer preferences. Coatings on food can serve as an effective solution to reduce oil uptake during the frying process. The effectiveness of a coating is dependent on its mechanical properties, barrier performance, composition, microstructure, and the characteristics of the substrate. Garcia stated that the efficacy of a coating is contingent upon its mechanical and barrier characteristics, which are shaped by its composition, microstructure, and the substrate's parameters [79]. Liu studies have shown that fluorinated oil repellents, extensively utilized in food packaging, especially for fast-food applications, have been demonstrated to potentially carry risks for both human well-being and the environment. Liu *et al.* (2022) [80] introduced a feasible and sustainable approach to replace fluorine-free oil repellents with a microcrystalline wax emulsion as an effective oil repellent for paper.

13.8.5 Organoleptic Barrier

Edible films and coatings should possess organoleptic properties that are as neutral as possible, including features like clarity, transparency, odor lessness, and tastelessness, to ensure they are inconspicuous during consumption. Surface enhancements, like brilliance, and improvements in tactile properties, such as decreased stickiness, may be essential. Olivas and the team reported that the preservation of ideal characteristics such as flavor, color, spiciness, acidity, sweetness, saltiness, and other qualities can be achieved using films and coatings [81].

13.9 Challenges and Opportunities

The food and pharmaceutical industry, along with numerous research groups, has duly recognized the potential of edible packaging. It is considered an alternative or a synergistic addition to traditional packaging, with the goal of enhancing food protection and/or promoting the recyclability of packaging. The research landscape of edible films is dynamic, characterized by ongoing exploration of innovative raw materials, advancements in active packaging methodologies, and the incorporation of nanotechnology approaches. Both new applications and alternatives to existing technologies have been envisioned for edible films and coatings, highlighting their diverse and innovative potential. The field of edible packaging research is constantly progressing, exploring new raw materials, developing active packaging, and incorporating nanotechnology to create films with properties akin to traditional synthetic polymers. In recent decades, packaging research has prioritized the development of active films, which contribute significantly to extending the shelf life of perishable fruits and vegetables. This is achieved by incorporating diverse antimicrobial and antioxidant components into their base polymer matrix. The dry thermoplastic process is gaining traction as a feasible commercial technique for manufacturing edible packaging. Ongoing research and development efforts are centered on utilizing by-products or waste from the food industry to create innovative edible packaging materials. By focusing on improving the quality and design of food packaging, it is possible to effectively reduce food waste through sustainable material management practices. The presence of an oxygen barrier in edible packaging can prevent anaerobic respiration, resulting in a slowed ripening process for fruits and vegetables. Active edible films play an active role in interacting with the food they encapsulate, offering health benefits to consumers. At the forefront of edible packaging research and development, nanocomposites are emerging as a pivotal area of focus. Nanotechnology's application in edible films is on the rise, yet challenges such as isolating and achieving uniform dispersion of nanomaterials within matrix polymers, along with economic considerations, continue to be significant hurdles. In addition, the scarcity of information regarding the toxicological effects of nanomaterials and other film-forming components causes consumers to feel fearful and hesitant. Scientists leverage nanotechnology to design the nanostructure of packaging materials, achieving specific barriers and mechanical properties, enabling the incorporation of bioactive ingredients, and optimizing their functional performance. Despite these advancements, edible packaging encounters several hurdles that need to be addressed for substantial commercial utilization.

13.9.1 Challenges

Edible packaging has a substantial history of research, extending over several decades. However, there is a prevailing misconception among certain individuals who assume it is acceptable to dispose of the packaging, anticipating it will naturally break down. According to the United Nations report, the act of classifying a product as biodegradable was noted as a potential systematic remedy, shifting accountability away from individuals, potentially creating reluctance among consumers to embrace edible cutlery and food packaging. The inclusion of reinforcement agents in the biopolymer matrix results in the improvement of critical properties required in packaging materials, such as barrier, mechanical, thermal, antioxidant, and antimicrobial properties. Furthermore, doubts may arise regarding the sanitary aspects of edible packaging, particularly its efficacy in safeguarding food from contamination like traditional packaging. Shipping edible packaging requires special attention due to its vulnerability to temperature shifts. Most edible films and coatings degrade or nearly degrade at their melting point, which adversely affects their usability in thermal processing methods like extrusion, compression molding, and injection molding. The relatively higher cost of edible packaging materials, as opposed to conventional packaging materials, could potentially affect consumer acceptance significantly. It's worth mentioning that allergic reactions to edible food packaging, such as caseins, can be experienced by certain individuals. Gontard and the team specified a further downside is that the application of edible foils or coatings, often essential for hygiene purposes, typically demands the use of Inedible containerization. Moreover, diverse environmental factors impact the stability of food-grade containerization. It demands efficient handling during both shipment and maintenance. By and large, limited understanding acts as a hindrance in developing edible films that fulfill the required specifications. Gulibert and the researcher's studies have found that the edible packaging functions as both a packaging material and an integral part of the food and it encounters rigorous requirements and distinctive challenges. It is imperative that all components of edible packaging adhere to the requisite regulations governing food products [82]. Voilley stated, food-grade films and food-grade coverings fall within the category of consumables spectrum, flavor enhancers, substances in contact with food, or food packaging materials, depending on their specific application [83]. For pharmaceutical and nutraceutical applications, additional regulations may apply. The substances utilized in crafting edible containerization should be deemed safe under GRAS or approved by the FDA for the intended use aligning with the criteria set by the Code of Federal Regulations. When employing food-grade containerization substances and enhancers, adherence to good manufacturing practice (GMP) is imperative, necessitating their qualification as food-grade and meticulous handling as a food additive. Furthermore, they should conform to any restrictions established by the United States Food and Drug Administration (FDA) [84]. For up-to-date information on the classification and regulations of food ingredients, the EAFUS database, courtesy of the FDA, is esteemed as an asset for information on everything added to food in the United States. The label must provide a transparent account of all ingredients utilized in food-grade containerization adhering to the standards set forth by the Federal Food, Drug, and Cosmetic Act. In cases where food-grade encasement contains frequently encountered allergens including dairy beverage, soybean, seafood, explicit labeling is required to provide information to consumers with allergies or intolerances, as mandated by FALCPA regulations. The utilization of food-grade substances is potentially constrained by consumer acceptance concerns, encompassing aspects like sensory qualities, safety considerations, marketing strategies, and alignment with cultural and religious limitations on the adoption of new materials and applications. Edible packaging materials, treated as food ingredients, should either feature neutral sensory properties or be in harmony with the sensory attributes of the edible packaged foods to prevent any detection during the consumption process.

The increasing awareness of consumers extends to the scrutiny of food labels. Although the FDA enforces the listing of ingredients and allergens, bringing in new food-grade film-forming substrates or applications may encounter complications with buyer group having allergies, religious or cultural restrictions. Current marking regulations focus primarily on the disclosure of eight major allergens and lack stringent enforcement, especially in contexts like fresh produce, bulk merchandise, institutional domains, or diverse food applications. Addressing additional safety concerns, notably those linked to potential shifts in the microflora of food products with the introduction of edible packaging, is a critical consideration. The commercial success of edible packaging is intertwined with marketing dynamics, involving aspects like pricing, consumer apprehension towards embracing new materials, and the diligent consideration of any instructions pertaining to the opening, cooking, consumption of packaged/coated foods, or the disposal of the packaging. Han and the team studies have shown that edible film production is currently mostly confined to laboratory settings, and the associated expenses are considered elevated in comparison with polymer-based films. Investigating methods for financial efficiency and it is imperative to expand manufacturing capabilities to fortify the feasibility of commercially accessible food-grade containerization. Effectiveness of systems in commercial use is hinged upon elements such as the intricacy of manufacturing, considerations of investment scale in coating or film implements, possible clashes with existing food-grade wrapping systems, and the repulsion of manufacturers to incorporating modern resources [85]. Furthermore, food manufacturers demand an extended shelf life for products participating in interstate or international trade. The inherent tendency of edible packaging materials to biodegrade means that their protective functions remain stable for a shorter duration compared to conventional packaging. Consequently, a thorough investigation into the stability and safety of edible packaging under intended storage and use conditions is warranted.

13.9.2 Opportunities

The development of food-grade and decomposable coatings/films for efficient consumable containerization has been a notable surge of attention in modern days, as they are holding promise for reduction or replacing conventional, non-biodegradable plastics. With promising results, active food packaging is utilized to extend the shelf life of food by protecting it from microbial microorganisms and oxidation. Prospects for edible films and coatings are robust, given their ability to align with consumer preferences for eco-friendly and natural food products. Packaging materials for food must meet food-grade standards, contribute to maintaining or enhancing product shelf-life stability and safety, and employ minimal packaging resources. The adoption of biopolymers in food packaging has shown a substantial increase over the past few years. Minimization or substitution with eco-friendly alternatives results in efficient operational expenses while minimizing disposal pathways. The ongoing research and development activities focused on edible and biodegradable films/coatings constitute a vital and distinctive sector within food packaging, highlighting considerable potential in both commercial and environmental aspects.

References

1. Suvarna, V., Nair, A., Mallya, R., Khan, T., Omri, A., Antimicrobial Nanomaterials for Food Packaging. *Antibiotics*, 11, 729, 2022, https://www.mdpi.com/2079-6382/11/6/729.
2. Tokiman, N.A., Rahman, N.H.A., Hajar, N., Mohamad, N., Munawar, N., Rahim, N.H.A., The Effects of Different Packaging Materials on Physicochemical and Sensorial Properties of Rempeyek. *J. Academia*, 7, 20–29, 2019, https://doi.org/10.24191/jfast.v7i2.7182.
3. Goswami, B., *The Role of Food Packaging*, pp. ICGCISM/AP–18, KAAV PUBLICATIONS, Delhi, India, 2019, ISBN 978-93-86789-85-3.
4. Robertson, G.L., *Encyclopedia of Agriculture and Food Systems*, pp. 232–249, Amsterdam, Netherlands, 2014, ISBN 978-0- 08-093139-5.
5. Chawla, R., Sivakumar, S., Kaur, H., Antimicrobial edible films in food packaging: Current scenario and recent nanotechnological advancements- a review. *Carbohydr. Polym. Technol. Appl.*, 2, 100024, 2666–8939, 2021, https://doi.org/10.1016/j.carpta.2020.100024.
6. Ananda, A.P., Manukumar, H.M., Umesha, S., Soumya, G., Priyanka, D., Kumar, A.S.M., Krishnamurthy, N.B., Savitha, K.R.A., Relook at Food Packaging for Cost Effective by Incorporation of Novel Technologies. *J. Packag. Technol. Res.*, 1, 67–85, 2017, https://doi.org/10.1007/s41783-017-0011-4.
7. Andrady, A.L., *Plastics and Environmental Sustainability: Fact and Fiction*, John Wiley & Sons, Hoboken, NJ, USA, 2015, https://doi.org/10.3390/coatings10020166.
8. Ellen MacArthur Foundation, *The New Plastics Economy: Catalysing Action; Ellen MacArthur Foundation and World Economic Forum*, Davos, Switzerland, 2017, https://doi.org/10.3390/coatings10020166.
9. 2018, https://ec.europa.eu/commission/news/first-ever-europe-wide-strategy-plastics-2018- jan-16_ro.
10. Haward, M., Plastic pollution of the world's seas and oceans as a contemporary challenge in ocean governance. *Nat. Commun.*, 9, 667, 2018, https://doi.org/10.1038/s41467-018-03104-3.
11. 2019, Available online: https://ec.europa.eu/commission/presscorner/detail/en/IP_18_5.
12. Christodoulidou, A., Horváth, Z., Lodi, F., Dusemund, B., Scientific opinion on the re-evaluation of alginic acid and its sodium, potassium, ammonium,

and calcium salts (E 400–E 404) as food additives. *EFSA J.*, 15, 11, 5049, 57, 2017, https://doi.org/10.2903/j.efsa.2017.5049.
13. 2019, https://eur-lex.europa.eu/legal-content/EN/TXT/?qid=1516265440535&uri=COM:2018:28:FIN.
14. Trajkovska, P.A., Daniloski, D., D'Cunha, N.M., Naumovski, N., Broach, A.T., Edible Packaging: Sustainable Solutions and Novel Trends in Food Packaging. *Food Res. Int.*, 140, 109981, 2021, https://doi.org/10.1016/j.foodres.2020.109981.
15. Espitia, P.J.P., Du, W.-X., Avena-Bustillos, R.d.J., Soares, N.d.F.F., McHugh, T.H., Edible Films from Pectin: Physical-Mechanical and Antimicrobial Properties – A Review. *Food Hydrocolloids*, 35, 287–296, 2014, https://doi.org/10.1016/j.foodhyd.2013.06.005.
16. Zhao, Y., Li, B., Li, C., Xu, Y., Luo, Y., Liang, D., Huang, C., Comprehensive Review of Polysaccharide-Based Materials in Edible Packaging: A Sustainable Approach. *Foods*, 10, 1845, 2021, https://doi.org/10.3390/foods10081845.
17. Pooja, S.P.S., Nath, S., Kishor, D.S., Singh, S.M., A Review of Edible Packaging for Foods. *Int. J. Curr. Microbiol. Appl. Sci.*, 8, 2885–2895, 2019, https://doi.org/10.20546/ijcmas.2019.807.359.
18. Barbosa, C.H., Andrade, M.A., Vilarinho, F., Fernando, A.L., Silva, A.S., Active Edible Packaging. *Encyclopedia*, 1, 360– 370, 2021, https://doi.org/10.3390/encyclopedia1020030.
19. Tsang, Y.F., Kumar, V., Samadar, P., Yang, Y., Lee, J., Ok, Y.S., Song, H., Kim, K.-H., Kwon, E.E., Jeon, Y.J., Production of bioplastic through food waste valorization. *Environ. Int.*, 127, 625–644, 2019, https://doi.org/10.1016/j.envint.2019.03.076.
20. Hammam, A.R.A., Technological, applications, and characteristics of edible films and coatings: A review. *SN Appl. Sci.*, 1, 632, 2019, https://doi.org/10.1007/s42452-019-0660-8.
21. Vargas, M., Pastor, C., Chiralt, A., McClements, D.J., González-Martínez, C., Recent Advances in Edible Coatings for Fresh and Minimally Processed Fruits. *Crit. Rev. Food Sci. Nutr.*, 48, 496–511, 2008, https://doi.org/10.1080/10408390701537344.
22. Biji, K.B., Ravishankar, C.N., Mohan, C.O., Gopal, T.K.S., Smart packaging systems for food applications: A review. *J. Food Sci. Technol.*, 52, 6125–6135, 2015, https://doi.org/10.1007/s13197-015-1766-7.
23. Calva, E.S.J., Jiménez, F.M., Lugo-C, E., Protein-Based Films: Advances in the Development of Biomaterials Applicable to Food Packaging. *Food Eng. Rev.*, 11, 78–92, 2019, https://doi.org/10.1007/s12393-019-09189-w.
24. Yin, W., Qiu, C., Ji, H., Li, X., Sang, S., McClements, D.J., Jiao, A., Wang, J., Jin, Z., Recent Advances in Biomolecule-Based Films and Coatings for Active and Smart Food Packaging Applications. *Food Biosci.*, 52, 102378, 2023, http://dx.doi.org/10.1016/j.fbio.2023.102378.
25. Trajkovska, P.A., Daniloski, D., D'Cunha, N.M., Naumovski, N., Broach, A.T., Edible Packaging: Sustainable Solutions and Novel Trends in Food

Packaging. *Food Res. Int.*, 140, 109981, 2021, https://doi.org/10.1016/j.foodres.2020.109981.

26. Barbosa, C.H., Andrade, M.A., Vilarinho, F., Fernando, A.L., Silva, A.S., Active Edible Packaging. *Encyclopedia*, 1, 360–370, 2021, http://dx.doi.org/10.3390/encyclopedia1020030.
27. Ribeiro, A.M., Estevinho, B.N., Rocha, F., Preparation and Incorporation of Functional Ingredients in Edible Films and Coatings. *Food Bioprocess Technol.*, 14, 209–231, 2021, https://doi.org/10.1007/s11947-020-02528-4.
28. Trajkovska, P.A., Daniloski, D., D'Cunha, N.M., Naumovski, N., Broach, A.T., Edible Packaging: Sustainable Solutions and Novel Trends in Food Packaging. *Food Res. Int.*, 140, 109981, 2021, https://doi.org/10.1016/j.foodres.2020.109981.
29. Barbosa, C.H., Andrade, M.A., Vilarinho, F., Fernando, A.L., Silva, A.S., Active Edible Packaging. *Encyclopedia*, 1, 360–370, 2021, http://dx.doi.org/10.3390/encyclopedia1020030.
30. García, A.M.C., Sepulveda, D.R., Zamudio, F.P.B., Acosta, M.C.H., Bacteriophages as Additives in Edible Films and Coatings. *Trends Food Sci. Technol.*, 132, 150–161, 2023, https://doi.org/10.1016/j.tifs.2023.01.008.
31. Pooja, S.P.S., Nath, S., Kishor, D.S., Singh, S.M., A Review of Edible Packaging for Foods. *Int. J. Curr. Microbiol. Appl. Sci.*, 8, 2885–2895, 2019, https://doi.org/10.20546/ijcmas.2019.807.359.
32. Janjarasskul, T. and Krochta, J.M., Edible Packaging Materials. *Annu. Rev. Food Sci. Technol.*, 1, 415–448, 2010, https://doi.org/10.1146/annurev.food.080708.100836.
33. Zhao, Y., Li, B., Li, C., Xu, Y., Luo, Y., Liang, D., Huang, C., Comprehensive Review of Polysaccharide-Based Materials in Edible Packaging: A Sustainable Approach. *Foods*, 10, 1845, 2021, https://doi.org/10.3390/foods10081845.
34. Nešić, A., Cabrera-Barjas, G., Dimitrijević-Branković, S., Davidović, S., Radovanović, N., Delattre, C., Prospect of Polysaccharide Based Materials as Advanced Food Packaging. *Molecules*, 25, 135, 2019, https://doi.org/10.3390/molecules25010135.
35. Kasim, R. and Kasim, M.U., The Effect of Tapioca-Starch Edible Coating on Quality of Fresh-Cut Cauliflower during Storage. *J. Agric. Food Environ. Sci.*, 72, 21–28, 2018, https://doi.org/10.55302/JAFES18721021k.
36. Popyrina, T.N., Demina, T.S., Akopova, T.A., Polysaccharide-Based Films: From Packaging Materials to Functional Food. *J. Food Sci. Technol.*, 60, 11, 2736–2747, 2022, https://doi.org/10.1007/s13197-022-05595-x.
37. Mohamed, S.A.A., El-Sakhawy, M., El, S.M.A.-M., Polysaccharides, Protein and Lipid-Based Natural Edible Films in Food Packaging: A Review. *Carbohydr. Polym.*, 238, 116178, 2020, https://doi.org/10.1016/j.carbpol.2020.116178.
38. Shahidi, F. and Hossain, A., Preservation of Aquatic Food Using Edible Films and Coatings Containing Essential Oils: A Review. *Crit. Rev. Food Sci. Nutr.*, 62, 66– 105, 2022, https://doi.org/10.1080/10408398.2020.1812048.

39. Maurizzi, E., Bigi, F., Quartieri, A., De Leo, R., Volpelli, L.A., Pulvirenti, A., The Green Era of Food Packaging: General Considerations and New Trends. *Polymers*, 14, 4257, 2022, https://doi.org/10.3390/polym14204257.
40. Gómez, E.J., Montero, P., Giménez, B., Gómez-Guillén, M.C., Effect of Functional Edible Films and High-Pressure Processing on Microbial and Oxidative Spoilage in Cold-Smoked Sardine (Sardina pilchardus). *Food Chem.*, 105, 511–520, 2007, https://doi.org/10.1016/j.foodchem.2007.04.006.
41. Cazón, P., Velazquez, G., Ramírez, J.A., Vázquez, M., Polysaccharide-Based Films and Coatings for Food Packaging: A Review. *Food Hydrocoll.*, 68, 136–148, 2017, https://doi.org/10.1016/j.foodhyd.2016.09.009.
42. Xie, Q., Liu, G., Zhang, Y., Yu, J., Wang, Y., Ma, X., Active Edible Films with Plant Extracts: A Updated Review of Their Types, Preparations, Reinforcing Properties, and Applications in Muscle Foods Packaging and Preservation. *Crit. Rev. Food Sci. Nutr.*, 63, 32, 11425–11447, 2022, https://doi.org/10.1080/10408398.2022.2092058.
43. Mollah, M.Z.I., Zahid, H.M., Mahal, Z., Faruque, M.R.I., Khandaker, M.U., The Usages and Potential Uses of Alginate for Healthcare Applications. *Front. Mol. Biosci.*, 8, 719972, 2021, https://doi.org/10.3389/fmolb.2021.719972.
44. U.S. Food & Drug Administration Code for Federal Regulations Title 21 Part 184—Direct Food Substances Affirmed as Generally Recognized as Safe, https://www.ecfr.gov/current/title-21/chapter-I/subchapter-B/part-184.
45. Gheorghita, R., Gutt, G., Amariei, S., The Use of Edible Films Based on Sodium Alginate in Meat Product Packaging: An Eco-Friendly Alternative to Conventional Plastic Materials. *Coatings*, 10, 166, 2020, https://doi.org/10.3390/coatings10020166.
46. Puscaselu, R.G., Anchidin-Norocel, L., Petraru, A., Ursachi, F., Strategies and Challenges for Successful Implementation of Green Economy Concept: Edible Materials for Meat Products Packaging. *Foods*, 10, 3035, 2021, https://doi.org/10.3390/foods10123035.
47. Othman, F., Idris, S.N., Ahmad Nasir, N.A.H., Nawawi, M.A., Preparation and Characterization of Sodium Alginate-Based Edible Film with Antibacterial Additive Using Lemongrass Oil. *Sains Malays.*, 51, 485–494, 2022, http://doi.org/10.17576/jsm-2022-5102-13.
48. Cazon, P., Velasquez, G., Ramirez, J., Vazquez, M., Polysaccharide-based films, and coatings for food packaging: A review. *Food Hydrocoll.*, 68, 136–148, 2017, https://doi.org/10.3390/foods9050562.
49. Chen, H., Formation and properties of casein films and coatings, in: *Protein-Based Films and Coatings*, vol. 7, A. Gennadios (Ed.), pp. 181–212, CRC Press, New York, 2002, http://dx.doi.org/10.1201/9781420031980.ch7.
50. McHugh, T.H. and Krochta, J.M., Milk protein-based edible films and coatings. *Food Technol.*, 48, 1, 97–103, 1994, http://pascal-francis.inist.fr/vibad/index.php?action=getRecordDetail&idt=4098320.
51. Gennadios, A., McHugh, T.H., Weller, C.L., Krochta, J.M., Edible coatings and film based on proteins, in: *Edible Coatings and Films to Improve*

Food Quality, vol. 9, Krochta, J.M., Baldwin, E.A., Nisperos-Carriedo, M., Lancaster, P.A. (Eds.), pp. 201–77, Technomic, 1994, https://doi.org/10.1201/B11082.

52. Gennadios, A. and Weller, C.L., Edible films and coatings from wheat and corn proteins. *Food Technol.*, 44, 10, 63– 69, 1990.
53. Padua, G.W. and Wang, Q., Formation, and properties of corn zein films and coatings, in: *Protein-Based Films and Coatings*, vol. 2, A. Gennadios (Ed.), pp. 43–68, CRC Press, New York, 2002, http://dx.doi.org/10.1201/9781420031980.ch2.
54. Krochta, J.M. and De, M.J.C., Edible, and biodegradable polymer films: challenges and opportunities. *Food Technol.*, 51, 2, 61–73, 1997, https://typeset.io/papers/edible-and-biodegradable-polymer-films-challenges-and-odnxq6ios3.
55. G.A. Meyer and T.B. Mazer, U.S. Patent No. 5,599,556, 1997.
56. Mallikarjunan, P., Chinnan, M.S., Balasubramaniam, V.M., Phillips, R.D., Edible coatings for deep-fat frying of starchy products. *Lebensm. Wiss. Technol.*, 30, 709– 14, 1997, https://doi.org/10.1006/fstl.1997.0263.
57. Park, H.J., Chinnan, M.S., Shewfelt, R.L., Edible corn-zein film coatings to extend storage life of tomatoes. *J. Food Proc. Preserv.*, 18, 4, 317–31, 1994, http://dx.doi.org/10.1111/j.1365-2621.1994.tb05563.x.
58. Farouk, M.M., Price, J.F., Salih, A.M., Effect of an edible collagen film overwrap on exudation and lipid oxidation in beef round steak. *J. Food Sci.*, 55, 6, 1510–63, 1990, https://doi.org/10.1111/j.1365-2621.1990.tb03556.x.
59. Greener, I.K. and Fennema, O., Barrier properties and surface characteristics of edible, bilayer films. *J. Food Sci.*, 54, 6, 1393–99, 1989, http://dx.doi.org/10.1111/j.1365-2621.1989.tb05120.x.
60. Kester, J.J. and Fennema, O.R., Edible films and coatings: a review. *Food Technol.*, 198640, 12, 47–59, 1986, http://dx.doi.org/10.4236/fns.2015.63037.
61. Baldwin, E.A., Surface treatments and edible coatings in food preservation, in: *Handbook of Food Preservation*, vol. 21, M.S. Rahman (Ed.), pp. 478– 508, CRC Press, Boca Raton, FL, 2007, http://dx.doi.org/10.1201/9781420017373.ch21.
62. Rhim, J.W. and Shellhammer, T.H., Lipid-based edible films and coatings, in: *Innovations in Food Packaging*, vol. 21, J.H. Han (Ed.), pp. 362–383, Elsevier Acad, New York, 2005, http://dx.doi.org/10.1016/B978-012311632-1/50053-X.
63. Bourlieu, C., Guillard, V., Valle, P.B., Gontard, N., Edible moisture barriers for food product stabilization, in: *Food Materials Science*, J.M. Aguilera and P.J. Lillford (Eds.), pp. 547–575, Springer, New York, 2008, http://dx.doi.org/10.1007/978-0-387-71947-4_23.
64. Thakur, R., Saberi, B., Pristijono, P., Golding, J., Stathopoulos, C., Scarlett, C., Vuong, Q., Characterization of rice starch-ιcarrageenan biodegradable edible film. Effect of stearic acid on the film properties. *Int. J. Biol. Macromol.*, 93, 952–960, 2016, https://doi.org/10.1016/j.ijbiomac.2016.09.053.

65. Fakhouri, F.M., Martelli, S.M., Caon, T., Velasco, J.I., Buontempo, R.C., Bilck, A.P., Innocentini Mei, L.H., The effect of fatty acids on the physicochemical properties of edible films composed of gelatin and gluten proteins. *LWT*, 87, 293–300, 2018, http://dx.doi.org/10.1016/j.lwt.2017.08.056.
66. Hashemi, G.H., Mostaghimi, M., Ghiasi, F., Tavakoli, S., Naseri, M., Hosseini, S.M.H., The effects of fatty acids chain length on the techno-functional properties of basil seed gum-based edible films. *Int. J. Biol. Macromol.*, 160, 245–251, 2022, https://doi.org/10.1016/j.ijbiomac.2020.05.136.
67. Amini, A.M., Razavi, S.M.A., Zahedi, Y., The influence of different plasticisers and fatty acids on functional properties of basil seed gum edible film. *Int. J. Food Sci. Technol.*, 50, 5, 1137–1143, 2015, https://doi.org/10.1111/ijfs.12765.
68. Chen, H., Wu, C., Feng, X., He, M., Zhu, X., Li, Y., Teng, F., Effects of two fatty acids on soy protein isolate/sodium alginate edible films: Structures and properties. *LWT Food Sci. Technol.*, 159, 113221, 2022, https://doi.org/10.1016/j.lwt.2022.113221.
69. Seyedi, S., Koocheki, A., Mohebbi, M., Zahedi, Y., Improving the physical and moisture barrier properties of Lepidium perfoliatum seed gum biodegradable film with stearic and palmitic acids. *Int. J. Biol. Macromol.*, 77, 151–158, 2015, https://doi.org/10.1016/j.ijbiomac.2015.03.005.
70. Luís, Â., Domingues, F., Ramos, A., Production of Hydrophobic Zein-Based Films Bioinspired by The Lotus Leaf Surface: Characterization and Bioactive Properties. *Microorganisms*, 7, 8, 2019, https://doi.org/10.3390/microorganisms7080267.
71. Praseptiangga, D., Afrida, B., Mufida, N., Widyaastuti, D., Effects of Stearic Acid and Zein Incorporation on Refined Kappa Carrageenan-Based Composite Edible Film Properties. *J. Curr. Sci. Technol.*, 13, 762–773, 2023, http://dx.doi.org/10.59796/jcst.V13N3.2023.1324.
72. Sothornvit, R. and Krochta, J.M., Plasticizers in edible films and coatings, in: *Innovations in Food Packaging*, vol. 23, J.H. Han (Ed.), pp. 403–33, Elsevier Acad, New York, 2005, https://doi.org/10.1016/B978-012311632-1/50055-3.
73. Pooja, S.P.S., Nath, S., Kishor, D.S., Singh, S.M., A Review of Edible Packaging for Foods. *Int. J. Curr. Microbiol. Appl. Sci.*, 8, 2885–2895, 2019, https://doi.org/10.20546/ijcmas.2019.807.359.
74. Mahcene, Z., Hasni, S., Goudjil, M.B., Khelil, A., Food Edible Coating Systems: A Review. *Eur. Food Sci. Eng.*, 2, 26–33, 2021, https://www.elsevier.com/locate/lwt.
75. García, A.M.C., Sepulveda, D.R., Zamudio, F.P.B., Acosta, M.C.H., Bacteriophages as Additives in Edible Films and Coatings. *Trends Food Sci. Technol.*, 132, 150–161, 2023, https://doi.org/10.1016/j.tifs.2023.01.008.
76. Kaur, J., Gunjal, M., Rasane, P., Singh, J., Kaur, S., Poonia, A., Gupta, P., Edible Packaging: An Overview, in: *Edible Food Packaging*, A. Poonia, and T. Dhewa, (Eds.), pp. 3–25, Springer Nature Singapore, Singapore, 2022, https://doi.org/10.1007/978-981-16-2383-7_1.

77. Mathlouthi, M., *Food Packaging and Preservation*, Springer US, Boston, MA, 1994, ISBN 978-1- 4613-5909-8, http://citations.springer.com/item?doi=10.1007/978-1-4615-2173-0_4.
78. Janjarasskul, T. and Krochta, J.M., Edible Packaging Materials. *Annu. Rev. Food Sci. Technol.*, 1, 415–448, 2010, https://doi.org/10.1146/annurev.food.080708.100836.
79. García, M., Bifani, V., Campos, C., Martino, M.N., Sobral, P., Flores, S., Ferrero, C., Bertola, N., Zaritzky, N.E., Gerschenson, L. *et al.*, Edible Coating as an Oil Barrier or Active System, in: *Food Engineering Series*, pp. 225–241, 2008, https://doi.org/10.1007/978-0-387-75430-7_12.
80. Liu, D., Duan, Y., Wang, S., Gong, M., Dai, H., Improvement of Oil and Water Barrier Properties of Food Packaging Paper by Coating with Microcrystalline Wax Emulsion. *Polymers*, 14, 1786, 2022, https://doi.org/10.3390/polym14091786.
81. Olivas, G.I.I. and Barbosa, C.G., Edible Films and Coatings for Fruits and Vegetables, in: *Edible Films and Coatings for Food Applications*, K.C. Huber, and M.E. Embuscado, (Eds.), pp. 211–244, Springer New York, New York, NY, 2009, ISBN 978-0-387-92823-4, https://doi.org/10.1007/978-0-387-92824-1.
82. Guilbert, S. and Gontard, N., Edible, and biodegradable food packaging, in: *Foods and Packaging Materials- Chemical Interactions*, P. Ackermann, M. Jagerstad, T. Ohlsson (Eds.), pp. 159–68, The R. Soc. Chem, Cambridge, UK, 1995, https://doi.org/10.22271/tpi.2023.v12.i7Sr.21623.
83. Debeaufort, F., Quezada, G.J., Voilley, A., Edible films and coatings: tomorrow's packages: a review. *Crit. Rev. Food Sci.*, 38, 299–313, 1998, https://doi.org/10.1080/10408699891274219.
84. Krochta, J.M. and De, M.J.C., Edible and biodegradable polymer films: challenges and opportunities. *Food Technol.*, 51, 2, 61–73, 1997, https://typeset.io/papers/edible-and-biodegradable-polymer-films-challenges-and-odnxq6ios3.
85. Han, J.H. and Gennadios, A., Edible films and coatings: a review, in: *Innovations in Food Packaging*, vol. 15, J.H. Han (Ed.), pp. 239–62, Elsevier Acad, New York, 2005, http://dx.doi.org/10.1016/B978-0-12-394601-0.00009-6.

14

Edible Coating Deposition Methods: Dipping, Spraying, Fluidized Bed, and Panning

Leya B., Nivetha T. U., Freeda Blessie R.* and M. M. Pragalyaashree

Division of Food Processing Technology, School of Agricultural Sciences, Karunya Institute of Technology and Sciences, Coimbatore, India

Abstract

Food packaging is essential to deliver food products to the intended customers in a safe and hygienic environment. Since most food products are not consumed directly in the field and have to be shipped for a long duration, certain processing stages are required. The processing stage includes laborious handling, storage, and shipping operations, which could result in deterioration, loss of aesthetic and nourishment of the product. Recently, edible food packaging is gaining more attention as it is a modified and compostable packaging technique that may be ingested with the food product or ejected before consuming the edible packaged food. This packaging material is either applied as a coating or as a film. Edible coatings or edible films are substances used for enrobing multiple foods to upgrade the shelf life of the product. Since the products would be damaged if particular protection is not offered within a stipulated time, edible coatings or films can be adopted to protect the product by compensating or fortifying natural layers for a controlled interchange of vital gases including oxygen, carbon dioxide, and ethylene, which are concerned in the respiration process and maintains moisture. The use of edible coating or film methods which moreover achieves surface sterility and loss of other crucial components. They are also used to transport chemicals including antioxidants, antimicrobials, and minerals, which assist in the preservation of the quality of food products or minimally processed fruits and vegetables. Hydrocolloids (polysaccharides or proteins), hydrophobic substances (lipids or waxes), and even a consolidation of both can be used to make edible coatings. The use of composite coatings is an ingenious technique for improving coating qualities and reducing

**Corresponding author*: freedablessier@karunya.edu

Mousumi Sen (ed.) Food Coatings and Preservation Technologies, (485–514)

limits by leveraging the features of both types of components. These biopolymer coatings or films can be applied to many products like chicken breast, fresh beef, salami, roasted turkey, fresh-cut fruits, and vegetables. Normally, edible coatings or films depend upon the nature of the bio-polymers, concentration, and viscosity also they can be applied to foods with different coating methods. The most commonly used coating techniques are Dipping, Spraying, Fluidized-bed processing, and panning. Dipping is for applying extensively viscous coatings to fruits and vegetables, whereas the spraying method is used for non-viscous solutions. The fluidized bed processing is used for small pellets and granules while the panning technique involves spraying edible coatings on food items. The overall idea and conception of edible coating are covered in this chapter, along with the deposition methods and their applications.

Keywords: Edible coating, biomaterials, coating methods, antimicrobial, barrier

14.1 Introduction

Innovative methods and technologies are always being researched to widen the shelf life of perishable goods in the ever-evolving field of food preservation. The evolution and utilization of edible coatings is one such novel choice. Edible coatings have drawn a lot of interest over the past few decades because of the potential they have for amplifying the quality, preservation time, and protection of food items [1]. The coatings themselves may be created from a variety of materials, including organic polymers like chitosan, alginate, pectin, and cellulose, as well as lipids like beeswax. They can be prepared as liquid solutions, suspensions, or emulsions and deposited on food surfaces using various methods. The appeal of edible coatings to fresh fruits and vegetables has yielded good results in terms of preserving freshness, minimizing water loss, limiting microbial development, and delaying ripening. When applying edible coatings on coated fruits, studies have shown that it is critical to monitor parameters such as change in color, firmness reduction, fermentation of ethanol, degradation ratio, and decreased weight [2].

To prevent ripening while preserving postharvest quality, edible coatings were additionally applied to stone fruits such as peaches and plums. Furthermore, research has concentrated on the use of edible coatings on tropical fruits, emphasizing their potential for prolonging their shelf life and retaining quality. The advantages of edible films and coatings go beyond fruits and vegetables. They have been investigated for a variety of food products, such as meats, dairy, confectionary, bakery goods, and fast food. Edible coatings can operate as a barrier against moisture, oxygen, and other external influences, enhancing the food's look and health [3].

In addition, incorporating biodegradable polymers and bioactive substances onto edible films and coatings has been shown to improve food quality qualities in the manner that antibacterial and antioxidant capabilities. The creation and utilization of edible coatings are continuing fields of study and technology. The use of edible coatings in conjunction with nanoencapsulation as technology is also being researched for a variety of culinary applications. Furthermore, edible film and coating production procedures are constantly being refined in order to increase their qualities and functioning [4].

Panning, fluidized bed, dipping, and spraying are all ways for applying edible coatings. All of these approaches have benefits and drawbacks, and the effectiveness is mostly determined by the qualities of the foods to be overlaid as well as the physical parameters of the coating (viscosity, density, surface tension, and so on). Spray coating is the exploited method for implementing food coatings. A spray system uses nozzles to expand the surface area of the liquid and distribute it across the food surface area by forming droplets. The key advantages of this technology include uniform coating, thickness control, and the ability to use layered applications, such as rotating sodium alginate as well as calcium chloride treatments [5].

However, it is crucial to remember that employing edible coatings has several challenges. Edible coatings are an intriguing technique for enhancing food preservation, minimizing waste, and upgrading the general caliber of several food products. Recognizing edible coatings' attributes and functions enables us to capitalize on the possibilities they offer to satisfy the requirements of a responsible and focused on quality food company.

14.2 Definition and Classification of Edible Coating Materials

Edible coating comprises a thin covering of edible substance that may be put to the top of food products. This functions as an obstacle between the food and the environment, giving numerous benefits such as increasing quality, prolonging the duration of shelf life, and boosting health or nutritional qualities. Edible coatings are often created from organic, non-toxic substances like wax, paraffin, starch, or other kinds of polymers. They may be used on a broad variety of fresh items, especially fruits, vegetables, meat, poultry, fish, including milk and cheese [6].

14.2.1 Classification Based on Raw Materials

Biomaterials such as polysaccharides (starch, cellulose, chitin, chitosan, pectin, alginates, Arabic gum, xanthan gum, etc.), proteins (gelatin, casein, whey protein, wheat gluten, wheat, corn, and soy protein), lipids (waxes, acetylated triglycerides), and composites are utilized in the production of edible films and coatings. The classification of biomaterials is described below and shown in the Figure 14.1.

14.2.1.1 Polysaccharide-Based Coatings

Polysaccharide-based coatings are regarded as eco-friendly. These protective coatings generally recyclable and made using organic materials consisting of plant and animal polysaccharides. They serve as a reinstatement

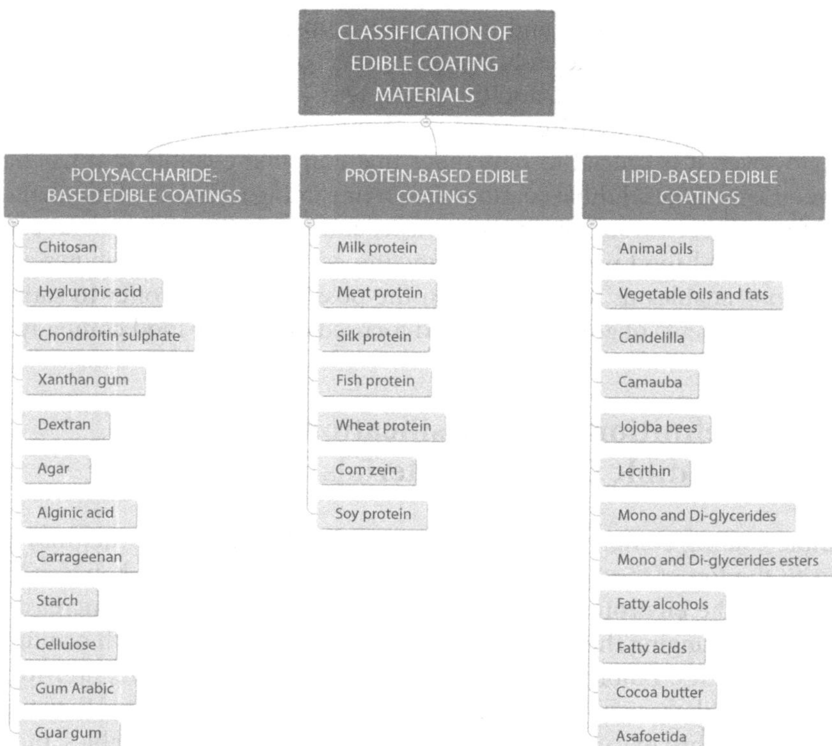

Figure 14.1 Classification edible coating materials [1].

to traditional packaging materials made of plastic, that might lead to greenhouse gas contamination [7].

Polysaccharides that consist of cellulose derivatives, chitosan, alginate, starch, and pectin are all used for producing edible films and coatings in food packaging materials. Coatings like this are able to increase the duration of storage of fruits, vegetables, and various spare perishable food products [8, 9].

14.2.1.2 Protein-Based Coatings

Proteins were simple, irregularly ordered copolymers that composed of various amino acids that form monomers. Proteins are classified differently according to their amino acid series, structure, and content. Furthermore, the proteins are categorized into three categories: coil (milk casein), fibrous (collagen), or globular (corn zein, soy protein, wheat gluten, whey protein), among others. The proteins that are fibrous are fully stretched and connected together in distinct parallel topologies via H-bonding to generate a fiber-like shape [10–12].

14.2.1.3 Lipid-Based Coatings

Lipid-based edible coatings use oil ingredients to offer adaptability, its hydrophobicity plus cohesion to the coating. These protective coatings act as good barriers versus humidity and, to a certain extent, oxygen consumption, aiding to protect the long-term shelf life of food items. Lipid-based edible coatings are capable of being utilized for covering the outermost layer of numerous food items, notably confectionery products, fresh produce, dairy products, chocolates, cereals, chicken meat, fish, frozen products, even dried goods [13]. The usage of lipid-based coatings in the food sector has been gaining popularity owing to their capability to preserve the quality of foods while enhancing the duration of storage. These coatings may function as a barrier of protection, reducing moisture loss while preserving the food's nutritional value and consistency [14].

14.2.1.4 Composite Coatings

Proteins like gelatin or whey protein are combined with polysaccharides like chitosan or alginate to form these coatings. The mixture of proteins and polysaccharides can increase the coating's mechanical and barrier qualities. These coatings combine lipids (such as beeswax or edible oils) with carbohydrates. Lipids may enhance the water vapor barrier characteristics and give the coating an attractive look. These coatings use proteins and lipids to provide a protective barrier on the food surface. For example, a composite encompassing comprised of soy protein isolate and beeswax has been used to increase the standard and expiration of fruits and vegetables [15].

These coatings incorporate proteins, polysaccharides, and lipids to form a multifunctional coating. Lipids can improve the coating's water resistance and mechanical capabilities, even with proteins and polysaccharides help with film formation and barrier qualities. Antioxidants, antimicrobials, vitamins, and probiotics are characteristics of functional additives the fact that can be used in composite coating. These additions can have additional benefits, such as protracting the shelf life of the coated food or increasing its nutritional content [16, 17].

14.2.2 Characteristics of Edible Coating

- Edible coatings contain an average thickness of not more than 0.3mm [18].
- Biopolymers as well as ingredients dispersed in aqueous media interact to produce edible coatings [19].
- Lipids, polysaccharides, and proteins are among the main elements utilized in edible coatings [18, 19].
- Edible coatings can be developed directly on the food (edible coating) or generated separately and then connected to the product [18, 20].
- Despite the distinction in techniques, both coatings and films generate sore matrices with comparable properties [22].
- Edible coatings can also alter the internal gas composition, offering an extra barrier to the product. In addition, they can include active substances into the polymer matrix, boosting safety, nutrition, and sensory properties [18, 20, 22].
- Edible coatings might be applied on raw fruits, vegetables, and various food products using various methods such as dipping and layer-by-layer treatments [21].
- Dipping requires dipping the edible item in the coating solution, enabling thorough saturation of the surface, and then drying to produce a well-intact coating [22].

- Edible coatings contribute in extensiveness the shelf life of food merchandise, enhance food safety, and protect the crispness of fresh produce. They may also assist with reducing moisture migration and function as a barrier against contaminants from the outside [18, 20, 22].

14.3 Deposition Methods of Edible Coating

There are several systems for depositing edible coatings on food products. Some of the commonly used methods include

1. Dipping
2. Spraying
3. Fluidized Bed Processing
4. Panning

14.3.1 Dipping Method

Dipping is one of the basic and widely used technique for adding edible coatings. This involves dipping the food product in the solution comprising the coating substance as depicted in the Figure 14.2. The food product is then withdrawn into the blend, enabling the remaining coating solution to drain. The leftover coating produces a thin layer over the outermost layer of the food product [23]. This procedure is frequently employed for adding edible coatings to fruits and vegetables. Fruits and vegetables have been immersed in coating solution for 5-30 seconds in this process. It is simple to use on the majority of fruits [24].

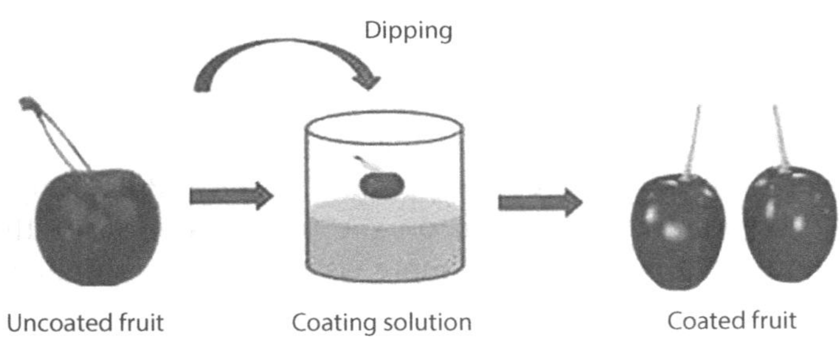

Figure 14.2 Dipping method [23].

The following steps are involved in the dipping technique used for employing edible coating on the food products: i) engrossment and residence, ii) deposition, and iii) dissipation. The material being coated is inundate during the surface coating emulsion/solution at a steady speed in the first stage, make sure sufficient solution enabling wetting the substrate and full contact between the substrate and coating matrix. Deposition is a technique utilized to create thin layers of the preceding combining onto the top layer of food products. Supplemental liquid sinks and is removed by deposition. Following the evaporation process, the chemical infusion and surplus liquid get potent away the surface of food goods via a drying and heating method. After the excess coating has been eliminated out, the object will be cured at room temperature or by using the assistance of a drier. The broadness of liquid coated films was demonstrated to be dependent on coating solution parameters such as density, viscosity, surface tension, and surface abolition pace. Dipping is a simple process that is frequently used in laboratories for film creation and coating deposition [5].

14.3.2 Spraying Method

If exterior has to be plastered and just a small amount of coating is required, the spraying approach is chosen. It is a more accurate kind of coating application than pan or fluidized bed coating since the solution is sprayed to the food surface using force (60-80 psi). Yet, following coating and curing one or both upper surfaces of the product, it must be pivot to completely supress the bottom surface. This approach is typically used for items with broad surfaces. The spray nozzle is critical in this coating process because it may affect droplet size, surface tension, spray distance and angle, flow rate, overlap speed, viscosity, temperature, and pressure of the spraying fluid [25, 29].

Spraying may be classified into numerous categories. The process in which the film establishing fluid is applied at low pressure combined with ambient air is known as air spray atomization. Pressure atomization is performed without the use of air at high pressure. Air aided airless atomization is a method of atomization that combines injecting the mixture under high pressure with smoothing with air at the end of the process for improved composition distribution over the product's surface. Processing age for improved component distribution on the product's surface. The latter two procedures can be utilized for more viscous liquids, but the required equipment is expensive [26].

The droplet technique has been utilized for coating Mozzarella cheese by using the droplet caused by static electricity with alginate, chitosan, and soy protein isolate solutions, freshly cut lotus roots by pressure atomization with xanthan gum, strawberries along regular spraying with sodium alginate, and products by electrostatic spraying with chitosan solutions of varying deacetylation degrees [27].

14.3.3 Fluidized Bed Processing Method

Fluidized bed processing is the application of a thin film of coating to dry out crumbs of very small size and/or density. The microscopic particles that should be coated are introduced to a revolving fluidized bed in this approach. The coated particles are pushed to approach the equipment's wall by centrifugal force. Furthermore, the facilities is built to allow for airflow; as a result, the integrated impacts of airflow and centrifugal force efficiently counteract the overall force exerted on the food particles [28, 29].

During a fluidized coating process, the solution for coating and suspensions are sprayed into the fluidized powder surface through a number of faucets to produce a shell-like structure. The fluidization process happens when a flow of liquid travels vertically through a bed of particles at a sufficient speed to assist those particles while amusing the particles into the liquid stream. At this stage, the particle bed absorbs the properties of the bubbling fluid, which is referred to as fluidization. There are three types of fluidized-bed process methods: top spray, bottom spray, and rotary-fluidized bed. However, in the food business, the classic top spraying approach conquering the other alternatives [5, 30].

Since the fine particles in the traditional bed do not fluidize consistently or form immoderate agglomerations at scaled down, the size of the particulate matter in the fluidized bed should be more than 100 m. The fluidized bed coating strategy's particle aggregation increases dispersion and solubility of the coating ingredient. Simultaneously, the liquid binder is utilized to refine the material's stability; the procedure is carried out under heat treatment before being sprinkled on meals. This causes particle adhesion, aggregation, and drying. It claims both batch and continuous procedures. Pneumatic or binary nozzles are diligent in fluidized-bed coating: the fluid is given at low pressure, and the air is separated into the nozzles [31].

14.3.4 Panning Method

Pan-coating is a technique for covering hard, nearly spheroidal particles in scrawny or stratified form. This coating can minimize moisture and lipid loss while also adding taste to the treated compound. The drug and pastries industry both employ the pan-coating process via apply this coating substance to their goods. The pan in this approach is a huge, spinning bowl into which the thing that needs to be coated is inserted. The coating solution is added to a spinning pan using a ladle or spray. The product spin in the pan to distribute the coating solution evenly all over the exterior part of the food product [29, 32].

Based on the features of food items, this approach is classified into three main types of application procedures. Hard panning, gentle panning, and chocolate panning are examples of these techniques. The procedure for the usance of a hard shell is hard panning, which is done by continuously spread on sugar syrup to the superficies, which dries and crystallizes. Soft panning entails coating with corn syrup and a sugar combination in the form of syrup, which is then free from moisture by adding dry sugar to make thick, soft layers with fewer layers. Chocolate agitating is the application of a fat-based coating around a core. Chocolate, cocoa- based confectionary as a choice of flavorful white chemical coating might exist employed [5].

14.3.5 Other Methods of Edible Coating

The other methods include edible coating deposition method are brushing, electrostatic deposition, vacuum impregnation and layer-by-layer method.

14.3.5.1 Knife Coating

This approach of film application was recently commercially developed. It consists of a stationary knife that meticulously applies an additional layer of film-forming substance to the flawless casting surface as well as food product underneath it. The wetting film characteristics primarily dictate the barrier qualities of the resulting film. The excess wet film substance is kept back in a puddle behind the knife. Sophisticated controls alter the length of the film and the knife's altitude over a smooth casting platform or food product [29].

14.3.5.2 Brushing

Brushing approach produces good results; edible coatings are frequently applied to beans and extremely perishable fruits and vegetables, such as strawberries and berries [24]. Brushing has been shown to be useful in a variety of food industry consideration. It has been used in the fruit and vegetable industry to extend the shelf-life of fresh food by minimizing moisture loss, delaying ripening with limiting microbial development. Brushing with a chitosan-based coating, successfully reduced the growth of spoilage microbes on strawberries, prolonging their period of storage by a maximum of 10 days [33, 34].

14.3.5.3 *Electrostatic Deposition*

Electrostatic spraying is a recent technology that is widely used. Molecules of the film cohering liquid are given an electric charge and transported in the impact of an electric field. The key distinction between this technology and traditional spraying is an evenly distributed coating layer and increased adherence as in production surface, which minimizes the dropping [26].

The experimental certification of the electrostatic spray deposition (ESD) method entails measuring droplet travel and establishing the association between spray dispersion shape and deposited thin layer thickness. Electrostatic atomization is used in this process of coating thin films. A method of estimating droplet travel utilizing photos of spray patterns captured by a vision system during the ESD process. They were able to identify the association between the quantifiable parameters for the spray's dispersion form and the thickness of the deposited thin layer by evaluating these photographs [35].

14.3.5.4 *Vacuum Impregnation*

Vacuum impregnation is a technique for applying edible coatings to food items. It entails using vacuum pressure to introduce a solution with a specified composition into the porous structure of the meal. This procedure allows the coating solution to penetrate the food, resulting in increased quality features [36, 37].

The vacuum impregnation process is a step forward from the dipping approach. The distinction is that during fruit dipping, a vacuum environment is used. Instead of dipping the food in a traditional dipping tank, the fresh food is immersed in a sealed vacuum application. While submerged in the coating solution at atmospheric pressure, the food material is exposed to atmospheric restoration [38].

14.3.5.5 *Layer-by-Layer*

The layer-by-layer (LBL) electrostatic charge deposition method entails the orderly deposition of multiple layers of various materials onto the surface of a substrate. This technology has been investigated and implemented in the field of edible coatings to boost the nutritional value and shelf life of food items [39, 40].

The different charges of the polysaccharide constituents are used for the LBL precipitation of electrostatic charges. For negatively charged xanthan gum as well as positively charged chitosan, have been employed, these materials' zeta potential values indicate their opposing charges, which facilitates electrostatic contact during the coating process. These materials have benefited such high-water solubility, biocompatibility, and the capacity to create branching polymeric chains. Furthermore, the LBL deposition approach has been used to create coatings for preserving food that are loaded with organic extracts and antibacterial compounds. The LBL layering process is being investigated and utilized in a variety of sectors, particularly food science, medication delivery systems, and nanotechnology [41].

14.4 Factors Affecting Edible Coating

Food products have been protected from physical harm by edible films and coatings against mechanical impact, pressure, vibrations, and other mechanical forces [42]. The edible coating's capability to attach to the food surface is crucial to its efficacy. It establishes how effectively the coating sticks to the meal and how well it withstands separation or cracking from mechanical stress. For edible coatings to endure mechanical deformation without breaking or cracking, they must be elastic and flexible. With the help of this characteristic, the coating can adjust to the food product's shape and size as it changes during handling or storage. The mechanical qualities of the edible coating might be affected by its thickness. Although thicker coatings can withstand physical harm better, they can also peel or shatter more easily. The ideal film thickness varies depending on the particular food product and how it needs to be handled. The mechanical properties of edible coatings may be enhanced by using nanofilms and multinanolayer films and coatings. It is possible to employ nanoparticles to improve mechanical strength, increase heat resistance, and reduce gas permeability [43].

One significant environmental aspect that can affect the characteristics and functionality of edible coatings is temperature. Increased temperatures have the capacity to quicken the coating material's deterioration and lessen its ability to preserve the food product. Some of the main factors that affects edible coatings are represented in the Figure 14.3.

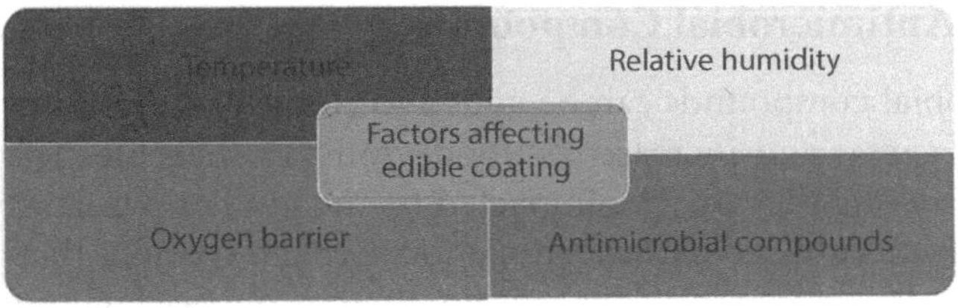

Figure 14.3 Factors affecting edible coating [42].

14.4.1 Temperature

Temperature can also have an impact on the respiration rates of the items and the permeability of the coating, which can both disrupt the food's shelf life and quality. The effectiveness of edible coatings can also be impacted by the storage area's relative humidity. Higher relative humidity levels may cause the coating material to absorb more moisture, which may weaken and damage the material's physical integrity. Relative humidity can also have an impact on the permeability of the coating, which can change how gases and moisture are exchanged among the food product and its surroundings. The efficiency of edible coatings might be impacted by the oxygen concentrations in the storage environment [44].

14.4.2 Oxygen Barrier

By functioning as oxygen barriers, edible coatings can keep food ingredients from oxidising and increase a product's shelf life. To offer adequate defence, the coating might need to have stronger oxygen barrier qualities if the oxygen levels in the storage location are essential [45]. Exposure to light, especially ultraviolet (UV) radiation, can degrade edible coatings and impact their functionality. The degradation of coating materials and the emergence of off-odors and flavours in food products can be caused by UV light. Because of this, it's crucial to take light- blocking qualities into account when creating edible coatings for food items that are sensitive to light. The number of gases like ethylene and carbon dioxide in the storage atmosphere, as well as its composition, may impact how well edible coatings work. Certain coatings that are edible may absorb gases, such as ethylene, which is produced by fruits and speeds up the ripening process. Edible coatings can contribute to the extension of fruit and vegetable shelf life by lowering ethylene levels [39].

14.4.3 Antimicrobial Compounds

Antimicrobial compounds can be added to edible coatings to prevent the growth of germs and increase the food product's shelf life. The coating's preservation qualities may be impacted by the kind and quantity of antibacterial agents utilised in the formulation. Certain coatings that are edible possess the capacity to scavenge oxygen, thereby decreasing its availability and delaying oxidative reactions that may cause food deterioration. The food's colour, flavour, and nutritional content may all be preserved in part due to the coating's ability to scavenge oxygen [45]. Effective preservation depends on the edible coating's compatibility with the food product. Strong adhesion in the middle of coating and the food surface is necessary to provide a barrier that stops moisture loss, gas exchange, and microbiological contamination. To improve the functionality and preservation qualities of edible coatings, several additives, such as flavours, colours, or preservatives, can be added. Admixtures should be carefully chosen and incorporated to ensure that they work well with the coating and effectively preserve the food product [43].

The food product's respiration rate, or how quickly it takes in oxygen and releases carbon dioxide, might affect the choice and composition of the coating. Coatings with greater oxygen barrier qualities may be necessary for foods with higher respiration rates in order to slow down respiration and prolong shelf life. Fruits and vegetables can mature and sense more quickly when exposed to ethylene, a natural plant enzyme. Coatings with ethylene-scavenging qualities may be necessary for foods that produce larger amounts of ethylene in order to prevent ripening and extend freshness. The process by which water evaporates from fruits' and vegetables' surfaces is called transpiration.

Coatings that serve as moisture barriers to stop water loss and preserve ideal moisture levels may be beneficial for foods with greater transpiration rates. The firmness, hardness, and porosity of the food product's surface can have an impact on the edible coating's adherence and efficacy. More homogeneity and smoothness in the surface might enhance coating adherence and barrier qualities. The stability and compatibility of the edible coating might be affected by the food product's pH and content. Certain coatings might work better on foods that are acidic, while others might work better on food that are neutral or alkaline. When combined with other preservation methods, edible coatings can have a synergistic impact that improves food products' overall preservation and can extend their shelf life. For instance, using vacuum packaging or modified atmosphere packaging (MAP) in conjunction with edible coatings can produce a controlled environment that aids in delaying enzymatic reactions and microbial growth, thus prolonging the product's shelf life [46–48]. It is crucial to confirm that the preservation technique of choice is appropriate for the unique qualities and attributes of the edible coating. For example, it is imperative to choose an edible coating that can tolerate high temperatures without losing its integrity or usefulness if the preservation method calls for it. When used in conjunction with other preservation methods, edible coatings can strengthen the packaging system's overall barrier qualities. During storage and transit, the coating can serve as a barrier against moisture, gases (such carbon dioxide and oxygen), and outside impurities, preserving the food product's quality and freshness [42].

It is possible to develop edible coatings with active preservation ingredients like flavour enhancers, antioxidants, or antibacterial chemicals. The release of these compounds from the coating, when paired with other preservation methods, can improve the food product's quality and preservation even further. For successful preservation, the edible covering and the particular food product must get along. When choosing and creating the edible coating, factors including the food product's pH, moisture level, and composition should be taken into account. The food's nutritional value or sensory qualities shouldn't be negatively impacted by the coating. It concerns how the edible coating is applied in conjunction with other preservation methods. It is important to apply the coating evenly to the food surface to ensure adequate adhesion and coverage. Methods like dipping, spraying, or brushing might be employed position on the type of food product and the intended result [49].

14.5 Application of Edible Coating

Perishable food items in that manner raw fruits and vegetables, meat, fish, milk, and eggs can have their shelf lives extended with the use of edible coatings [50]. By preventing the softening, browning, and flavour loss of fruits and vegetables, edible coatings help preserve their texture, colour, and flavour. By reducing nutrient deterioration during storage, they can also aid in maintaining the nutritious content of the confection. Edible coatings have the potential directed towards decreasing food waste by prolonging the shelf life of fruits and vegetables. They add another degree of defence, reducing spoiling and raising the produce's general quality. Antimicrobial substances, such as natural extracts or essential oils, can be added to edible coatings to assist prevent the growth of pathogens and bacteria that cause food spoiling on the food's surface. This may help to lower the risk of foodborne infections and enhance food safety. An eco-friendly substitute for conventional plastic packaging materials is edible coatings. They can be used directly on produce, doing away with the need for extra containers or plastic wrapping [1].

14.5.1 Fruits and Vegetables

In recent times, edible coatings have garnered substantial interest as a recently developed approach for conserving and improving the quality of fruits and vegetables. These coatings, which are derived from natural substances such proteins, polysaccharides, lipids, and waxes, have a number of advantages that help preserve quality, increase shelf life, and lower post- harvest losses. Fruits and vegetables can have their shelf life increased with the use of edible coating. By performing as a barrier of defence, the coatings create a regulated microenvironment surrounding the product. By functioning as a barrier against the diffusion of water vapour, they help to slow down the rate of transpiration and respiration, hence reducing moisture loss. also serve a vital part in preserving the fruits' and veggies' natural moisture content. These coatings aid in preventing over drying, wilting, and softening of the produce by lowering water evaporation. For fragile fruits and vegetables that are very prone to drying during storage and transit, this moisture retention is especially crucial. Edible coatings serve as a barrier of defence, preventing food from losing water and preserving its ideal texture and look. Coatings that are edible provide a barrier against contamination and microbial development. The coatings form a physical barrier that keeps bacteria from getting presumably of the fruit or vegetable. By preventing the growth of bacteria, mould, and yeasts, this barrier effect helps lower the danger of spoiling and increases the shelf life. In order to improve their capacity to prevent the growth of bacteria that cause spoiling, certain edible coatings further include antimicrobial chemicals or essential oils. Edible coatings aid in preserving the safety and quality of vegetables by lowering microbial activity [1].

Edible coatings have the ability to serve as a barrier, inhibiting the enzymatic browning process and limiting oxygen exposure. The coatings may reduce the fruit or vegetable's surface contact with the oxygen in the air by forming a protective layer that suppresses the enzyme polyphenol oxidase, which causes browning. Combining edible coatings with other preservation methods, including packaging under a modified environment or storing at a low temperature, can improve their ability to prevent enzymatic browning. When it comes to preserving the quality and prolonging the life span of fruits and vegetables, these integrated approaches can work in concert [51]. Antioxidants like ascorbic acid, phenolic compounds, or volatile oils can be added to edible coating formulations to assist regulate polyphenol oxidase activity and lessen enzymatic browning. These protective agents have the ability to scavenge free radicals and impede the oxidation processes that result in browning [39]. In order to improve edible coatings' ability to fend off enzymatic browning, nanotechnology has been investigated in this regard. To increase the stability and effectiveness of the coatings' oxygen barrier, nano emulsions, nanoparticles, and nanoliposomes can be added [52]. As a basis for edible coatings, polysaccharides like chitosan or alginate are often utilized. By lowering oxygen permeability and preventing enzymatic browning, these coatings can provide a protective layer on the surface of fruits and veggies. Fresh-cut fruit and vegetable shelf lives can be successfully increased and their quality maintained with coatings based on polysaccharides [53].

Recently cut fruits and vegetables have been coated with edible films and films composed of pectin and whey protein that has been crosslinked with transglutaminase. These coatings have demonstrated the ability to prevent browning while preserving the texture and quality of the product [54]. It has been common practice to use polysaccharides as the basis for edible coatings, such as chitosan or alginate. These coatings have the capacity to create a barrier that prevents enzymatic browning and decreases oxygen permeability on the surface of fruits and vegetables. They have proven successful in keeping fresh produce's quality and prolonging its shelf life. Coatings Based on Nano emulsion: Edible coatings have been modified to improve their ability to prevent browning by adding nano emulsions. These coatings can increase the secureness and shelf life of fruits and vegetables and offer a more effective barrier against oxygen [55]. To enhance their utility, edible coatings can be developed with a diversity of active ingredients, including vitamins, nutrients, antimicrobials, and anti-browning agents. These substances have the ability to maintain the aesthetic appeal and quality of fruits and vegetables [39].

14.5.2 Meat

By performing as barriers to carbon dioxide, oxygen, and water vapour, edible coatings can assist meat products last longer on the shelf while preventing spoiling and preserving quality. Improved food safety can result from the federation of natural antimicrobial fusion into edible coatings, which can impede the proliferation of harmful and spoilage bacteria [56]. Meat products texture can be improved by edible coatings, which act as a block to keep moisture in and prevent surface drying. These coatings, which lessen lipid oxidation and delay moisture loss, can enhance the quality of fresh, frozen, and processed meat. The meat can have a more soft and juicy texture by avoiding dryness and toughness through the retention of moisture by the edible coatings. By maintaining the meat's ideal colour, look, and sheen on the outside, the coatings can improve the meat's visual attractiveness. Because the coatings assist maintain the fresh appearance and texture, this might be especially crucial for meat items that are stored or shown for extended periods of time. Meat coatings that are edible are frequently made from a variety of natural ingredients, including proteins, chitosan, and Gelatin [57].

14.5.3 Minimally Processed Food

Edible coatings act as a boundary to prevent moisture loss, oxygen entry along with microbial growth, therefore extending the shelf life of minimally processed food products. They have the ability to lower the rate of respiration and ethylene production, two things that cause produce and fruits to deteriorate. Edible coatings are able to slow down deterioration and increased product longevity by managing these processes. Minimally processed food's quality features can be maintained with the benefit of edible coatings. They can preserve the food's nutritional qualities, transport chemicals that delay dis colouration, and seal in flavor volatile. These coatings can also improve the food's mechanical and organoleptic qualities, which will increase its overall consumer appeal. Food waste can be decreased by using edible coatings. Edible coatings can lessen the quantity of food that spoils and is wasted by distention shelf life of minimally prepared food [58].

14.5.4 Storage

Edible coatings play a crucial role in attenuating the shelf life of perishable food items, including meat, fish, milk, eggs, fruits, and vegetables [50]. Every polymer matrix of edible coatings can include a number of active substances that are eaten with the food. This makes the coated food products safer, more nutritious, and have better sensory qualities. Edible coatings have the ability to alter the internal gas composition of food, much like modified atmosphere storage. This capability aids in regulating moisture transfer, gas exchange, and oxidation processes, ultimately safeguarding the food's quality. Edible coatings can incorporate dynamic substance into the polymer matrix, which can be ingested with the food. This allows for the enhancement of safety, nutritional value, and sensory attributes of the food [1]. Edible coatings are useful for preserving fruits and vegetables for a few days or less. These coatings stop moisture loss, stop enzymatic browning, and assist preserve the desired texture. During this time, they serve as a barrier of defence, prolonging the produce's shelf life and maintaining its quality [43]. Edible coatings offer a useful alternative for medium-term preservation, which can span a few days to a few weeks. They assist in keeping fruits and vegetables fresh and of high quality while preventing enzymatic browning and microbiological growth. Produce can be coated with lipid, protein, or polysaccharide-based coatings to form a protective layer and prolong its shelf life [39].

14.5.5 Transportation and Packaging

Edible coatings can provide a protective layer that helps prevent physical damage to food products during transportation. Horticultural produce can have scars concealed, deterioration inhibited, and its beauty enhanced by them. In coated food products, edible coatings aid in controlling oil transfer and moisture content. This can stop fresh food goods from absorbing moisture, which could lead to deteriorating changes while being transported. Food goods can have their overall quality enhanced during transit by edible coatings. Functional ingredients that assist maintain the products' quality include antioxidants, antimicrobials, vitamins, minerals, anti-browning agents, enzymes, and probiotics [59]. Elasticity modulus, stretched out, and durability are among the mechanical characteristics of edible films and coatings that exists assessed. Its elasticity, stiffness, and resistance to breaking are determined by these qualities, all of which are crucial for transportation [50].

Edible coatings help to conserve the freshness and attribute of food products by acting as a barrier against gases, moisture, and solutes. They lessen the commutation of gases between the food and its encompassing, minimise oxidation, and stop moisture loss. By reducing the rate of deterioration, edible coatings aid and support the shelf life of food goods. They lessen spoilage and waste by assisting in maintaining the food's texture, flavour, and nutritional content while it is packaged. Food items can be shielded from physical harm during handling and transit by the use of edible coatings. By preventing bruising, abrasions, and mechanical injuries, they can enhance the packed food's overall quality and appearance. By serving as a barrier against microbial contamination, edible coatings can improve the safety of food. By preventing the growth of specific pathogens and spoilage microbes, they can lower the risk of foodborne infections.

Biodegradable substances like proteins, lipids, or polysaccharides are frequently used to create edible coatings. They can aid in the decrease of plastic waste and are more environmentally friendly than conventional packaging materials. It is possible to synthesise edible coatings with useful ingredients including vitamins, antimicrobials, and antioxidants. These substances can improve the packaged food's quality and safety even more [60].

14.6 Advantages of Edible Coating

Edible coatings were supposed to have mechanical and barrier properties proportionate to those of conventional packaging materials so that they can be used in the food industry with prosperity. The progression of edible films and coatings could be an environmentally friendly reverberation to packaging waste. Specifically, coatings form directly on food and are ingested along it, whereas films are stand-alone compounds a certain that stick to food and could not be eliminated before consuming. Edible films refined using the proteins and the pectin bring out from the pumpkin industry, orange peel extract, mushroom by-products, cheese whey and orange peels and, overall, the waste construct by fruit-processing operations [61].

The nutritional value of food products can be maintained with the help of edible coatings. Various beneficial compounds, such as probiotics, antioxidants, vitamins, minerals, anti- browning agents, enzymes, and antimicrobials, can be transported by them. The products' quality can be maintained by including these compounds into the coating matrix. The transmission of moisture in fresh food products can be managed via edible coatings. They can aid in preventing moisture migration into food by serving as a barrier, which can stop deteriorative changes from occurring and lower the product's quality and texture [62].

Comparing edible coatings to synthetic packaging technologies, the former are thought to be more environmentally friendly. They are frequently composed of renewable and biodegradable natural components like lipids and polysaccharides. They are therefore a more environmentally friendly substitute for conventional packaging materials. Edible coatings can be developed to fit the needs and features of certain food items. Depending on the required qualities and needs for product preservation, they can be formulated with a variety of materials and active agents. This adaptability enables the coating to be optimised and customised for various applications. Edible coatings can enhance the visual appeal of horticultural produce by imparting a shiny appearance, disguising imperfections, and inhibiting the development of physiological disorders and decay. This can improve the products' aesthetic appeal and raise consumer acceptance [1, 63].

Edible coatings have powerful prospective to aid sustainable food production by bringing down the packaging waste, extending the shelf-life of the products, and actively meshing to perpetuate the food quality. Moreover, the possibility of being united with circular economy scheme to make edible films and coatings conversant and competent of weakening the environmental impact of food systems. By reducing packaging desolation, increasing product shelf life, and diligently working together to maintain food quality, edible films and coatings have a great potential to support sustainable food production. Furthermore, edible coatings are adaptable and able to lessen the environmental impact of food systems due to their potential to be linked with circular economy techniques [64]. For mushrooms, dextran and chitosan-based bio packaging has been employed. Regarding gas permeability and moisture resistance, these films show good mechanical qualities. These films are biodegradable, composed of harmless materials, and may also have antibacterial properties. Chitosan and dextran are safe, renewable materials that are widely acknowledged as safe. These substances do not give food ingredients an unwanted taste or smell. Additionally, transparent, these films enable visual inspection of food products [65].

14.7 Limitations of Edible Coating

Edible coatings have the ability to regulate and prevent deteriorative changes in food products; however, the precise application and type of food being coated will determine how effective the coating performs. The problem of moisture migration into fresh food products, which can lead to serious quality problems, can be tackled with edible coatings. But controlling moisture can be difficult, and poor moisture control can result in microbial development, textural changes, and shortened shelf life. Certain post-harvest procedures, such heat treatment, can cause vitamin, flavor, and weight losses as well as nutritional losses. Edible coatings may not completely stop these losses, but they may reduce them. Modified atmosphere packaging (MAP) and controlled atmosphere (CA) are two examples of chemical treatments used in food preservation that can be costly and potentially hazardous. One innovative post-harvest technique that can assist get over these restrictions is edible coatings [66, 67].

In many instances, concentration determines how effective active ingredients are when added to edible coatings. The food's surface efficiency increases with the number of active molecules added. In order to prevent consumer overeating and industrial waste of money, the amount of active compound that can be added is restricted to that which is necessary for the needed action [68].

14.8 Future Perspectives

Edible coatings offer the opportunity to embrace various active constituent into the polymer matrix, which can inflate safety, nutritional value, and sensory attributes of the coated food products [69]. The development of innovative coating materials may help to raise food product quality and safety standards. For instance, using antimicrobial coatings made specifically for food packaging can help lower microbial contamination and increase the shelf- life of conventional sausages and fruits [45, 70]. Developments in coating application processes have progressed with breakthroughs in coating materials. The effectiveness and efficiency of edible coatings are being investigated through the use of techniques including vacuum impregnation, layer-by-layer approach, and spraying methods [71]. Consuming films and coatings are part of an integrated strategy that is becoming more widely acknowledged as a means of enhancing the food system's sustainability [72].

Edible coatings can benefit from the circular bio economy by using agricultural waste as a raw material source. A circular bio economy is in line with the principles of edible coatings' integration into the food system. Reusing waste materials as starting points for coating development can improve resource circularity and lessen the need for packaging made of petroleum. This strategy supports environmental sustainability, recycles waste, and lowers production costs. Packaging materials made of petroleum can have adverse impacts on the environment. However, edible coatings made from waste from the agriculture sector can assist solve these issues. It is possible to lessen the creation of non-biodegradable solid waste by substituting renewable and biodegradable materials, like coatings made from agro-waste, for synthetic ones. Perishable food items' shelf life may be increased by edible coatings made from agricultural industry waste. Fresh fruit and vegetable preservation can be enhanced by utilising agro-waste sources in conjunction with the benefits of edible coatings, such as moisture retention and barrier qualities. Waste and production expenses can be decreased by using agricultural waste for edible coatings. Waste materials can be used as raw ingredients for coating production, which can improve the edible coatings' economic feasibility [73, 74]. Edible coatings offer a sustainable substitute for packaging materials derived from petroleum, meeting the growing demand from consumers for environmentally friendly packaging solutions. Natural, biodegradable materials can be used to create edible coatings, which will lessen their negative effects on the environment and help to create a more sustainable food system. The food industry is also paying more attention to edible films, which are closely related to edible coatings. Food goods can be wrapped or enclosed with these thin, flexible films to protect them and prolong their shelf life. Similar advantages to coatings are provided by edible films, such as decreased waste, increased food safety, and improved product quality [75].

14.9 Summary and Conclusion

Edible coatings are now a common option in the food business for improving the nutritional value, physical appearance, as well as life span of numerous food products. These coatings are able to be applied using a variety of processes, comprising dipping, spraying, fluidized bed, and panning. Each approach has a unique set of benefits and limitations, which render it appropriate for various uses. Dipping is a simple and affordable process for coating food products, therefore being ideal for small-scale companies. Spraying, which is the most common approach, enables equal coating with thickness control via dispersing the coating in a collection of droplets through a spray machine. This approach enables for multilayer implementations and is widely used on commercial networks. Fluidized bed coating technique is great for coating baked goods due to the fact that the food item is suspended in a fluidized bed of coating particles, ensuring full coverage while preventing formation of clusters. Panning, on the other hand, uses the polishing motion of a pan to create an even and uniform surface on sweets, nuts, as well as certain processed fruits. The right deposition process may be chosen based on the coating materials' properties, the food product's intended effect, and economic considerations. Dipping and panning are popular for lab-scale activities because they are simple and inexpensive. Furthermore, spraying and extrusion are used on an industrial level for film creation along with coating deposition, etc. Finally, the deposition method for edible coatings is essential in defining the coating process's effectiveness and efficacy. The optimal deposition method may be chosen by taking into account a variety of aspects related to coating material qualities, food product features, and expected consequences.

References

1. Dhall, R.K., Advances in edible coatings for fresh fruits and vegetables: a review. *Crit. Rev. Food Sci. Nutr.*, 53, 5, 435–450, 2013, https://doi.org/10.1080/10408398.2010.541568.
2. Maringgal, B., Hashim, N., Tawakkal, I.S.M.A., Mohamed, M.T.M., Recent advance in edible coating and its effect on fresh/fresh-cut fruits quality. *Trends Food Sci. Technol.*, 96, 253–267, 2020, https://doi.org/10.1016/j.tifs.2019.12.024.
3. Kumar, P., Kumar, L., Tanwar, R., Singh, S., Gaikwad, K.K., Active edible coating based on guar gum with mint extract and antibrowning agents for ber (*Ziziphus mauritiana*) fruits preservation. *J. Food Meas. Charact.*, 17, 1, 129–142, 2023, https://doi.org/10.1007/s11694-022-01609-6.
4. Yadav, A., Kumar, N., Upadhyay, A., Sethi, S., Singh, A., Edible coating as postharvest management strategy for shelf-life extension of fresh tomato (Solanum lycopersicum L.): An overview. *J. Food Sci.*, 87, 6, 2256–2290, 2022, https://doi.org/10.1111/1750-3841.16145.
5. Suhag, R., Kumar, N., Petkoska, A.T., Upadhyay, A., Film formation and deposition methods of edible coating on food products: A review. *Food Res. Int.*, 136, 109582, 2022, https://doi.org/10.1016/j.foodres.2020.109582.
6. Popa, E., Popescu, P., Miteluț, A., Drăghici, M., Geicu-Cristea, M., Popa, M., Edible Coating for Functionality Improvement of Minimally Processed Organic Apples. *J. Ecoagritourism*, 18, 2, 125–131, 2022.
7. Cazón, P., Velazquez, G., Ramírez, J.A., Vázquez, M., Polysaccharide-based films and coatings for food packaging: A review. *Food Hydrocolloids*, 68, 136–148, 2017, https://doi.org/10.1016/j.foodhyd.2016.09.009.
8. Kocira, A., Kozłowicz, K., Panasiewicz, K., Staniak, M., Szpunar-Krok, E., Hortyńska, P., Polysaccharides as edible films and coatings: Characteristics and influence on fruit and vegetable quality—A review. *Agronomy*, 11, 5, 813, 2021, https://doi.org/10.3390/agronomy11050813.
9. Zhu, F., Polysaccharide based films and coatings for food packaging: Effect of added polyphenols. *Food Chem.*, 359, 129871, 2017, https://doi.org/10.1016/j.foodchem.2021.129871.
10. Nandane, A.S. and Jain, R., Study of mechanical properties of soy protein based edible film as affected by its composition and process parameters by using RSM. *J. Food Sci. Technol.*, 52, 6, 3645–50, Jun 2015, https://doi.org/10.1007%2Fs13197-014-1417-4.
11. Kandasamy, S., Yoo, J., Yun, J., Kang, H.-B., Seol, K.-H., Kim, H.-W., Ham, J.-S., Application of Whey Protein-Based Edible Films and Coatings in Food Industries: An Updated Overview. *Coatings*, 11, 9, 1056, 2021, https://doi.org/10.3390/coatings11091056.
12. Shendurse, A., Gopikrishna, G., Patel, A.C., Pandya, A.J., Milk protein based edible films and coatings–preparation, properties and food applications.

J. Nutr. Health Food Eng., 8, 2, 219–226, 2018, https://doi.org/10.15406/jnhfe.2018.08.00273.
13. Rhim, J.W. and Shellhammer, T.H., Lipid-based edible films and coatings, in: *Innovations in food packaging*, pp. 362–383, Academic Press, Cambridge, Massachusetts, 2005, https://doi.org/10.1016/B978- 012311632-1/50053-X.
14. Debeaufort, F. and Voilley, A., Lipid-Based Edible Films and Coatings, in: *Edible Films and Coatings for Food Applications*, K. Huber, and M. Embuscado (Eds.), pp. 135–168, Springer, New York, NY, 2009, https://doi.org/10.1007/978-0-387-92824-1_5.
15. Tharanathan, R.N., Biodegradable films and composite coatings: past, present and future. *Trends Food Sci. Technol.*, 14, 3, 71–78, 2003, https://doi.org/10.1016/S0924-2244(02)00280-7.
16. Ren, Y., Zhang, L., Xie, G., Li, Z., Chen, H., Gong, H., Luo, J., A review on tribology of polymer composite coatings. *Friction*, 9, 429–470, 2021, https://doi.org/10.1007/s40544-020-0446-4.
17. Guo, L., Yuan, W., Lu, Z., Li, C.M., Polymer/nanosilver composite coatings for antibacterial applications. *Colloids Surf., A*, 439, 69–83, 2013, https://doi.org/10.1016/j.colsurfa.2012.12.029.
18. Bourtoom, T., Edible films and coatings: characteristics and properties. *Int. Food Res. J.*, 15, 3, 237–24, 2008.
19. Singh, D.P. and Packirisamy, G., Biopolymer based edible coating for enhancing the shelf life of horticulture product. *Food Chem.: Mol. Sci.*, 4, 100085, 2022, https://doi.org/10.1016/j.fochms.2022.100085.
20. Nor, S.M. and Ding, Trends and advances in edible biopolymer coating for tropical fruit: A review. *Food Res. Int.*, 134, 109208, 2020, https://doi.org/10.1016/j.foodres.2020.109208.
21. Poverenov, E., Danino, S., Horev, B., Granit, R., Vinokur, Y., Rodov, V., Layer-by-layer electrostatic deposition of edible coating on fresh cut melon model: Anticipated and unexpected effects of alginate–chitosan combination. *Food Bioprocess Technol.*, 7, 1424–1432, 2014, https://doi.org/10.1007/s11947-013-1134-4.
22. Wibowo, C., Salsabila, S., Muna, A., Rusliman, D., Wasisto, H.S., Advanced biopolymer-based edible coating technologies for food preservation and packaging. *Compr. Rev. Food Sci. Food Saf.*, 23, 1, 1–31, 2024, https://doi.org/10.1111/1541-4337.13275.
23. Pham, T.T., Nguyen, L.L.P., Dam, M.S., Baranyai, L., Application of Edible Coating in Extension of Fruit Shelf Life. *Agri Eng.*, 5, 1, 520–536, 2023, https://doi.org/10.3390/agriengineering5010034.
24. Raghav, P.K., Effect of Various Packaging Materials on Sensory Quality of Herbal Edible Coated Pears. *Indian J. Nat. Sci.*, 12, 68, 34902–34910, October, 2020.

25. Nair, S.S., Trafiałek, J., Kolanowski, W., Edible Packaging: A Technological Update for the Sustainable Future of the Food Industry. *Appl. Sci.*, 13, 14, 8234, 2023, https://doi.org/10.3390/app13148234.
26. Lisitsyn, A., Semenova, A., Nasonova, V., Polishchuk, E., Revutskaya, N., Kozyrev, I., Kotenkova, E., Approaches in animal proteins and natural polysaccharides application for food packaging: Edible film production and quality estimation. *Polymers*, 13, 10, 1592, 2021, https://doi.org/10.3390/polym13101592.
27. Zhong, Y., Cavender, G., Zhao, Y., Investigation of different coating application methods on the performance of edible coatings on Mozzarella cheese. *LWT Food Sci. Technol.*, 56, 1, 1–8, 2024, https://doi.org/10.1016/j.lwt.2013.11.006.
28. Punia Bangar, S., Chaudhary, V., Thakur, N., Kajla, P., Kumar, M., Trif, M., Natural antimicrobials as additives for edible food packaging applications: A review. *Foods*, 10, 10, 2282, 2021, https://doi.org/10.3390/foods10102282.
29. Bangar, S.P, Chaudhary, V., Thakur, N., Kajla, P., Kumar, M., Trif, M., Natural antimicrobials as additives for edible food packaging applications: A Review. *Foods*, 10, 2282, 2021.
30. Coronel-Aguilera, C.P. and San Martín-González., M.F., Encapsulation of spray dried β- carotene emulsion by fluidized bed coating technolog. *LWT Food Sci. Technol.*, 62, 1, 187–193, 2015, https://doi.org/10.1016/j.lwt.2014.12.036.
31. Solís-Morales, D., Sáenz-Hernández, C.M., Ortega-Rivas, E., Attrition reduction and quality improvement of coated puffed wheat by fluidised bed technology. *J. Food Eng.*, 93, 2, 236–241, 2009, https://doi.org/10.1016/j.jfoodeng.2009.01.020.
32. Tufan, E.G., BORAZAN, A.A., Koçkar, O.M., A review on edible film and coating applications for fresh and dried fruits and vegetable. *Bilecik Şeyh Edebali Univ. Fen Bilim. Derg.*, 8, 2, 1073–1085, 2021, https://doi.org/10.35193/bseufbd.996827.
33. Ju, J., Xie, Y., Guo, Y., Cheng, Y., Qian, H., Yao, W., Application of edible coating with essential oil in food preservation. *Crit. Rev. Food Sci. Nutr.*, 59, 15, 2467–2480, 2019, https://doi.org/10.1080/10408398.2018.1456402.
34. Zhao, R., Zhang, Y., Chen, H., Song, R., Li, Y., Performance of eugenol emulsion/chitosan edible coating and application in fresh meat preservation. *J. Food Process. Preserv.*, 46, 3, e16407, 2022, https://doi.org/10.1111/jfpp.16407.
35. Lee, C.H., Woo, H.J., Kang, J.H., Song, K.B., Electrostatic Spraying of Passion Fruit (*Passiflora Edulis L.*) Peel extract for inactivation of Escherichia Coli O157: H7 and Listeria monocytogenes on fresh-cut lollo rossa and beetroot leave. *Food Bioprocess Technol.*, 14, 898–908, 2021, https://doi.org/10.1007/s11947-021-02608-z.

36. Parreidt, T.S., Müller, K., Schmid, M., Alginate-based edible films and coatings for food packaging applications. *Foods*, 7, 10, 170, 2018, https://doi.org/10.3390/foods7100170.
37. Parreidt, T.S., Schmid, M., Müller, K., Effect of dipping and vacuum impregnation coating techniques with alginate-based coating on physical quality parameters of cantaloupe melon. *J. Food Sci.*, 83, 4, 929–936, 2018, https://doi.org/10.1111/1750-3841.14091.
38. Zhang, L., Yu, X., Yagoub, A.E.A., Xia, G., Zhou, C., Effect of vacuum impregnation assisted probiotics fermentation suspension on shelf-life quality of freshly cut lotus root. *Food Chem.*, 381, 132281, 2022, https://doi.org/10.1016/j.foodchem.2022.132281.
39. Owusu-Akyaw Oduro, K., Edible Coating, in: *Postharvest Technology - Recent Advances, New Perspectives and Applications*, 2022, https://dx.doi.org/10.5772/intechopen.101283.
40. Maulida, I.D., Rismaya, R., Vifta, R.L., Enhancing Vitamin C Retention in Edible- Coated Cherry Tomatoes using Parijoto Extract: A Potential Approach for Traditional Food Preservation. *J. Tradit. Indigenous Food*, 1, 1, 1–5, 2023, https://journal.unilak.ac.id/index.php/jtif/issue/view/623.
41. Lara, G.R., Uemura, K., Khalid, N., Kobayashi, I., Takahashi, C., Nakajima, M., Neves, M.A., Layer-by-layer electrostatic deposition of edible coatings for enhancing the storage stability of fresh-cut lotus root (Nelumbo nucifera). *Food Bioprocess Technol.*, 13, 722–726, 2022, https://doi.org/10.1007/s11947-020-02410-3.
42. Trinetta, V., Edible Packaging, in: *Reference Module in Food Science*, 2016, http://dx.doi.org/10.1016/B978-0-08-100596-5.03210-8.
43. Greiner, R., Gräf, V., Burcza, A., Hetzer, B., Milsmann, J., Walz, E., Engineered Nanomaterials in the Food Sector. *Compr. Anal. Chem.*, 68, 579–616, 2015, https://doi.org/10.1016/B978-0-444-63340-8.00011-X.
44. Rossi-Márquez, G., Dávalos-Saucedo, C.A., Di Pierro, P., Edible Films and Coatings Applied in the Food Industry. *Coatings.*, 13, 4, 670, 2023, https://doi.org/10.3390/coatings13040670.
45. Nunes, C., Silva, M., Farinha, D., Sales, H., Pontes, R., Nunes, J., Edible Coatings and Future Trends in Active Food Packaging–Fruits' and Traditional Sausages' Shelf Life Increasing. *Foods.*, 12, 17, 3308, 2023, https://doi.org/10.3390/foods12173308.
46. de Lorena Ramos-Garcia, M., Bautista-Banos, S., Barrera-Necha, L.L., Bosquez-Molina, E., Alia-Tejacal, I., Estrada-Carrillo, M., Antimocrobial Compounds Added in Edible Coatings for Use in Horticultural Products. *Rev. Mex. Fitopatol.*, 28, 1, 44–57, 2010.
47. Erginkaya, Z., Kalkan, S., Ünal, E., Use of Antimicrobial Edible Films and Coatings as Packaging Materials for Food Safety, in: *Food Processing: Strategies for Quality Assessment. Food Engineering Series*, A. Malik, Z. Erginkaya, S. Ahmad, H. Erten (Eds.), pp. 261–295, Springer, New York, NY, 2014, https://doi.org/10.1007/978-1-4939-1378-7_10.

48. Cagri, A., Ustunol, Z., Ryser, E.T., Antimicrobial edible films and coatings. *J. Food Prot.*, 67, 4, 833–848, 2004, https://doi.org/10.4315/0362-028X-67.4.833.
49. Mohan, S., Unnikrishnan, T.G., Dubey, U., Ramesh, M., Panneerselvam, K., Development and Characterization of Mustard Oil Incorporated Biodegradable Chitosan Films for Active Food Packaging Applications. *J. Polym. Environ.*, 31, 5, 2190–2203, 2023, https://doi.org/10.1007/s10924-022-02719-4.
50. Díaz-Montes, E. and Castro-Muñoz, R., Edible Films and Coatings as Food-Quality Preservers: An Overview. *Foods*, 10, 2, 249, Jan 26 2021, doi: 10.3390/foods10020249.
51. Moon, K.M., Kwon, E.B., Lee, B., Kim, C.Y., Recent Trends in Controlling the Enzymatic Browning of Fruit and Vegetable Products. *Molecules*, 25, 12, 2754, Jun 15 2022, doi: 10.3390/molecules25122754.
52. Perez-Vazquez, A., Barciela, P., Carpena, M., Prieto, M.A., Edible Coatings as a Natural Packaging System to Improve Fruit and Vegetable Shelf Life and Quality. *Foods*, 12, 19, 3570, 2023, https://doi.org/10.3390/foods12193570.
53. Cruz-Monterrosa, R.G., Rayas-Amor, A.A., González-Reza, R.M., Zambrano-Zaragoza, M.L., Aguilar-Toalá, J.E., Liceaga, A.M., Application of Polysaccharide-Based Edible Coatings on Fruits and Vegetables: Improvement of Food Quality and Bioactivities. *Polysaccharides*, 4, 2, 99–115, 2023, https://doi.org/10.3390/polysaccharides4020008.
54. Liyanapathiranage, A., Dassanayake, R.S., Gamage, A., Karri, R.R., Manamperi, A., Evon, P., Jayakodi, Y., Madhujith, T., Merah, O., Recent Developments in Edible Films and Coatings for Fruits and Vegetables. *Coatings.*, 13, 7, 1177, 2023, https://doi.org/10.3390/coatings13071177.
55. de Oliveira Filho, J.G., Miranda, M., Ferreira, M.D., Plotto, A., Nanoemulsions as Edible Coatings: A Potential Strategy for Fresh Fruits and Vegetables Preservation. *Foods*, 10, 10, 2438, 2023, https://doi.org/10.3390/foods10102438.
56. Sánchez-Ortega, I., García-Almendárez, B.E., Santos-López, E.M., Amaro-Reyes, A., Barboza-Corona., J.E., Regalado, C., Antimicrobial Edible Films and Coatings for Meat and Meat Products Preservation quot. *Sci. World J.*, 2014, Article ID 248935, 18, 2014, https://doi.org/10.1155/2014/248935.
57. Gennadios., A., Hanna., M.A., Kurth, L.B., Application of Edible Coatings on Meats, Poultry and Seafoods. *LWT-Food science and Technology*, 30, 4, 337–350, 1997, https://doi.org/10.1006/fstl.1996.0202.
58. Miteluţ, A.C., Popa, E.E., Drăghici, M.C., Popescu, P.A., Popa, V., I., Bujor, O.C., Popa, M.E., Latest developments in edible coatings on minimally processed fruits and vegetables: A review. *Foods*, 10, 11, 2821, 2021, https://doi.org/10.3390/foods10112821.
59. Oduro, K.O.A., Edible coating, in: *Postharvest Technology-Recent Advances, New Perspectives and Applications*, pp. 215–229, 2021.

60. Zuhal, O.K.C.U., Yavuz, Y., Kerse, S., Edible film and coating applications in fruits and vegetables. *Alinteri J. Agric. Sci.*, 33, 2, 221–226, 2018, https://doi.org/10.28955/alinterizbd.368362.
61. Summo, C. and De Angelis, D., The Importance of Edible Films and Coatings for Sustainable Food Development. *Foods*, 11, 20, 3221, 2021, https://doi.org/10.3390/foods11203221.
62. Sharma, H.P., Chaudhary, V., Kumar, M., Importance of edible coating on fruits and vegetables: A review. *J. Pharmacogn. Phytochem.*, 8, 3, 4104–4110, 2019.
63. Park, H.J., Byun, Y.J., Kim, Y.T., Whiteside, W.S., Bae, H.J., Processes and applications for edible coating and film materials from agropolymers, in: *Innovations in food packaging*, pp. 257–275, Academic Press, Cambridge, Massachusetts, 2014, https://doi.org/10.1016/B978- 0-12-394601-0.00010-2.
64. Blancas-Benitez, F.J., Montaño-Leyva, B., Aguirre-Güitrón, L., Moreno-Hernández, C.L., Fonseca-Cantabrana, Á., del Carmen Romero-Islas, L., González-Estrada, R.R., Impact of edible coatings on quality of fruits: A review. *Food Control*, 139, 109063, 2022, https://doi.org/10.1016/j.foodcont.2022.109063.
65. Rashid, F., Ahmed, Z., Ferheen, I., Mehmood, T., Liaqat, S., Ghoneim, M.M., Rahman, A., Effect of fenugreek and flaxseed polysaccharide-based edible coatings on the quality attributes and shelf life of apple fruit during storage. *Food Sci. Nutr.*, 12, 3, 2093–2103, 2023, https://doi.org/10.1002/fsn3.3909.
66. Vargas, M., Pastor, C., Albors, A., Chiralt, A., González-Martínez, C., Development of edible coatings for fresh fruits and vegetables: possibilities and limitations. *Fresh Produce*, 2, 2, 32–40, 2008.
67. Turhan, K.N., Is edible coating an alternative to MAP for fresh and minimally processed fruits?, in: *X International Controlled and Modified Atmosphere Research Conference*, vol. 876, pp. 299–305, 2009.
68. Ghidelli, C. and Pérez-Gago, M.B., Recent advances in modified atmosphere packaging and edible coatings to maintain quality of fresh-cut fruits and vegetables. *Crit. Rev. Food Sci. Nutr.*, 58, 4, 662–679, 2018.
69. Yaashikaa, P.R., Kamalesh, R., Kumar, P.S., Saravanan, A., Vijayasri, K., Rangasamy, G., Recent advances in edible coatings and their application in food packaging. *Food Res. Int.*, 173, 113366, 2023, https://doi.org/10.1016/j.foodres.2023.113366.
70. Yousefi, M., Azizi, M., Mohammadifar, M.A., Ehsani, A., Antimicrobial coatings and films on meats: A perspective on the application of antimicrobial edible films or coatings on meats from the past to future. *Bali Med. J.*, 7, 1, 87–96, 2018.
71. Vargas, M., Pastor, C., Chiralt, A., McClements, D.J., Gonzalez-Martinez, C., Recent advances in edible coatings for fresh and minimally processed fruits. *Crit. Rev. Food Sci. Nutr.*, 48, 6, 496–511, 2008, https://doi.org/10.1080/10408390701537344.

72. Santoso, F. and Rahmat, V.A., Safety and quality assurance of tomato using aloe vera edible coating, in: *II Asia Pacific Symposium on Postharvest Research Education and Extension: APS2012 1011*, pp. 133–140, 2012, https://doi.org/10.17660/ActaHortic.2013.1011.15.
73. Dey, P., Bhattacharjee, S., Yadav, D.K., Hmar, B.Z., Gayen, K., Bhowmick, T.K., Valorization of waste biomass for synthesis of carboxy-methyl-cellulose as a sustainable edible coating on fruits: A review. *Int. J. Biol. Macromol.*, 253, 127412, 2023, https://doi.org/10.1016/j.ijbiomac.2023.127412.
74. Ruggeri, E., Kim, D., Cao, Y., Farè, S., De Nardo, L., Marelli, B., A multilayered edible coating to extend produce shelf life. *ACS Sustain. Chem. Eng.*, 8, 38, 14312–14321, 2020, https://par.nsf.gov/servlets/purl/10218047.
75. Kouhi, M., Prabhakaran, M.P., Ramakrishna, S., Edible polymers: An insight into its application in food, biomedicine and cosmetic. *Trends Food Sci. Technol.*, 103, 248–263, 2020, https://doi.org/10.1016/j.tifs.2020.05.025.

15

Biobased Antimicrobial Food Packaging Coatings

Arushi Phillips and Mousumi Sen*

Department of Chemistry, Amity Institute of Applied Sciences, Amity University, Uttar Pradesh, India

Abstract

Food spoilage driven by microbes exacerbates our existing problems, besides food insecurity and financial loss. Consequently, one of the significant obstacles to safeguarding food from external pollutants is adequate packaging and preserving its calibre. Food has traditionally been packaged in plastic, which has limited recyclability and harms the environment. The manufacturing of plastic packaging materials depends significantly on fossil fuels. The extraction and processing of these finite resources contribute to environmental degradation. This not only endangers the environment but also presents hazards to human health, as toxins from plastics can infiltrate the food chain. Integrating biodegradable elements into food packaging supports a wider effort to optimize packaging processes for greater efficiency and environmental responsibility. The practicality of biodegradable packaging extends across diverse aspects, covering environmental gains, the conservation of resources, reduction in waste, and consideration of consumer preferences.

Biodegradable elements, commonly sourced from renewable materials, represent a sustainable alternative. Unlike standard plastic packaging that endures in landfills for extended periods, biodegradable materials naturally degrade over time, lessening the enduring ecological footprint. Because of their minimal carbon footprint, biodegradability, and renewable characteristics, bio-based materials have garnered considerable attention for their potential applications in the field of food packaging. By addressing environmental challenges and making a significant contribution to sustainable development, they possess substantial potential to reduce the extensive reliance on fossil fuels and create food packaging materials with commendable qualities.

**Corresponding author*: mousumi1976@gmail.com

Mousumi Sen (ed.) Food Coatings and Preservation Technologies, (515–550) © 2025 Scrivener Publishing LLC

In the modern era, advancements in food chemistry, technology, and biotechnology enable us to develop innovative approaches that are beneficial in addressing noteworthy safety and environmental concerns related to packaging materials. Including carbohydrate (polysaccharide)-based materials, protein-based materials, lipid-based materials, antibacterial agents, and bio-based composites, which can address the problems of both environmental impact and the prevention of foodborne pathogens and spoilage microorganisms. These materials can now endure diverse temperatures, offer effective barrier properties, and uphold the integrity of the packaged product. This leads to an enhancement in the efficiency of food packaging processes, all while maintaining the quality and safety of the food, this review offers a comprehensive overview of the evolution and potential for application of new bio-based materials from various sources in antimicrobial food packaging. Furthermore, delving into the examination of antimicrobial agents incorporated into packaging materials and explores emerging trends in the field.

Keywords: Bio-active-packaging, packaging materials, antimicrobial agents, biodegradable films, biopolymers and protein, edible film applications

15.1 Introduction

In the present day, individuals are increasingly aware of the importance of a wholesome diet and are proactively educating themselves about the harmful repercussions of deliberate inclusion of synthetic additives in food products to extend shelf life. Huang and López emphasize that the spoilage of food due to microbial contamination is undoubtedly a major concern, resulting in considerable financial losses [1, 2]. It represents more than 25% of the food content before it undergoes consumption. The choice to consume natural foods, nonetheless, is of paramount importance. According to Malhotra, Keshwani, and Kharkwal, the scientific definition of food spoilage encompasses the defilement process leading to the deterioration of texture, color, and nutritional value. Concurrently, it encourages the multiplication of disease-causing microbes, diminishing the standard and turning the item into something not fit for consumption [3]. The degradation of properties in distinct food products is associated with different bacterial and fungal entities, contingent upon the commodity. The activity of the deteriorating organism relies on factors such as temperature, pH, hydration status, and the partial pressures of CO_2 and O_2. Considering the ever-increasing demand for food quality and safety, urgent efforts are being directed towards exploring novel edible packaging approaches with heightened effectiveness [4]. The ecosystem is at risk due to the detrimental impact of conventional non-degradable packaging, driven by heightened levels of hazardous pollutants alterations in the CO_2 cycle, and challenges related to composting [5].

Given its substantial contribution to solid waste in cityscapes, the environmental impact of this material is persistently increasing. Zhong describes a noteworthy waste management issue as the apparent origin of debris, predominantly resulting from the disposal of plastic packaging. The significance of incorporating decomposable resources to produce inventive and eco-conscious packaging materials is evident [6]. According to Ataei and Pareda, entirely substituting traditional petroleum-based synthetic substances with biodegradable alternatives is not practical, the utilization of these sustainable active edible containerizing materials appears to be a viable solution, especially in certain contexts such as edible packaging [7, 8]. Numerous strategies have been the focus of researchers aiming to create biodegradable packaging that is both safe and reliable.

Hassana and Vieira stated that this is spurred by the escalating concerns about ecological footprint and individual's health. Biodegradable polymer-based food packaging material has garnered considerable attention in the food industry owing to its sustainable and eco-conscious properties [9, 10]. Utilizing biopolymers in film production has demonstrated their effectiveness as a diverse structure for crafting efficient packaging substances. The incorporation of additives like nourishment, Microbial inhibitors, radical interceptors, and colors enhances their versatility.

The expansion of microbial population is curtailed by active ingredients, leading to an improvement in food quality and an extension of the product's shelf life. Rezaei and Zemljič along with the team reported an important innovation in active food packaging technology is the integration of microbial inhibitors into edible films made from biopolymers [11, 12]. According to Zhand and Mlalila, when designing and manufacturing environmentally friendly antimicrobial containerization substances, it is imperative to carefully weigh the trade-off between biodegradability and antimicrobial activity [13, 14].

Interest in antimicrobial food packaging is on the rise, this surge in interest has propelled research into the development and utilization of biodegradable materials such as proteinaceous compounds, lipidic molecules, and glycan chains, with a thorough examination of environmental implications. In this scenario, diverse antimicrobial agents, such as biocatalysts, microbial peptides, and extracted oils have been successfully integrated into films based on biopolymers, yielding significant and validated results.

By introducing these elements into the packaging matrix composed of biopolymers, it becomes possible to manage or impede microbial growth on the food surface, resulting in an extension of the product's shelf life without jeopardizing its safety or quality. Accoridng to Ibarra and the team, Given the exceptional capacity of antimicrobial agents to impede microorganisms, it is of utmost importance to establish the necessary concentrations of the suitable active agent for achieving adequate control [15]. Furthermore, the integration of antimicrobial agents into the packaging substances could trigger a radical change in antimicrobial effectiveness. Therefore, it is crucial to select a suitable antimicrobial agent specifically designed for both food and packaging, accompanied by an altered amount that acknowledges its lowest effective concentration e against organisms of interest.

15.2 Bioactive Packaging

The evolution of edible packaging has transcended its original purpose as simple containers and barriers against external factors over the past few decades. Battisti stated that in the view of the undesirable consequences stemming from traditional food preservation methods on both the nutritional profile and sensory attributes coupled with a rising consumer preference for safe and health-driven products, collaborative efforts between scientists and industry participants aim to innovate packaging materials

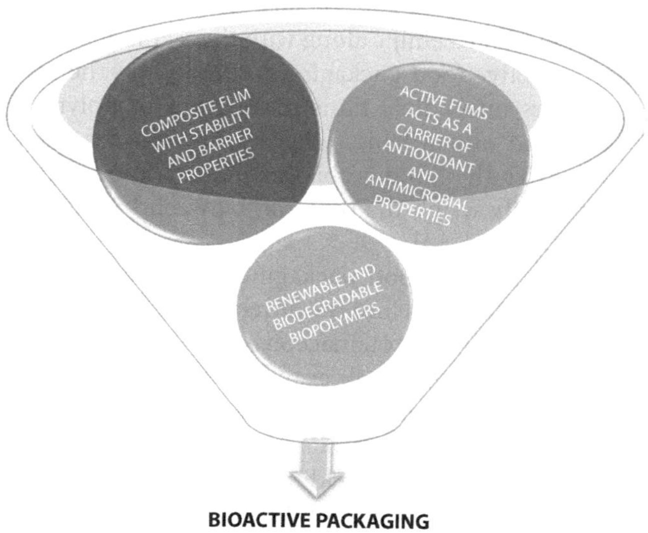

Figure 15.1 Bioactive packaging.

engagingly dedicated to guaranteeing food safety and extending the duration of its shelf life, thus sustaining its quality and flavor remain intact [16]. Valerio terms this innovative packaging approach as "active packaging" [17].

- **Active food packaging materials**

Yildirim describes 'active packaging' as signifying a transition from the defensive stance of passive packaging systems to an actively involved approach. Its function includes mediating interactions between the Item, encasement and the surrounding environment. In general, these interactions entail diverse material, molecular, and life-based actions with the goal of transforming the inherent environment of the packaged food, thereby enhancing its sustainability, safety, quality, and shelf life as illustrated in Figure 15.1 [18].

From a technical standpoint, Yam along with the team describes, active packaging involves structures to absorb or release compounds into or from the confined environment or containerized food [19]. In terms of their operational mechanism, interactive packaging technologies can be generally grouped into two classes: substances that vigorously alter the internal atmosphere of the container through either emission or absorption of gases. Scavengers, in this context, refer to substances employed for eliminating undesired elements such as odors or flavors, C2H4, CO_2, O_2, and humidity from the internal encasement environment. According to Wyrwa and team, Emitter systems are formulated to emit specific materials with desired characteristics, aiming to have a beneficial and fitting effect on the headspace of food packaging [20]. Among the numerous substances often utilized as absorbers and emitters, examples include biochemical agents, odorous elements, nutritional supplements, anti-infective and oxidation inhibitors, and extracted oils. Jideani stated that the antimicrobial-based active packaging system emerges as the most auspicious among these alternatives, enhancing both product safety and shelf life by integrating antimicrobial compounds in regulating the hazardous maturation and dissemination of spoilage microorganisms [21].

- **Antimicrobial packaging**

Moreover, various other bioactive substances are outcome-oriented and prove valuable in crafting active food packages. Owing to the constructive effects it has on food quality, safety, and the prolonged shelf life, employing antimicrobial packaging proves to be a valuable alternative technique for reducing, constraining, or inhibiting the proliferation of deterioration and illness-inducing microorganisms in food products. Within the packaging industry, there has been conspicuous scrutiny directed towards antimicrobial packaging. To produce antimicrobial packaging films, antimicrobial compounds are integrated with a polymer matrix. Sung reported that by doing so, it impedes the development of certain microorganisms that could pose a contamination risk to food [22]. The strategy is to decrease and limit pathogen multiplication, with the antimicrobial impact of these containerization solutions often centered on extending the lag phase and reducing the live cell counts of the targeted bacteria. By employing these containerization methods, one can ensure the protection of short-lived goods like animal protein, dairy products, fresh produce, and greens, often at risk of damage from microbial development and contamination. Bastarrachea and the team stated, the underlying motivation for the formation of these microbial-inhibiting packaging methods is centered around three key objectives: preserving quality, ensuring safety, and extending shelf life [23]. Sofi stated that the inhibitory potency of antimicrobial drugs against diverse microbes can fluctuate, given the unique microbial inhibition pathways and disparities in microbial physiology. Furthermore, gaining insight into the antimicrobial functions of these compounds can contribute to the assessment of their inhibitory action's efficacy and range.Within the food and packaging industries, there is a rising focus on antimicrobial packaging as a valuable alternative to conventional thermal treatments. Furthermore Zhong et al. described, it utilizes diverse conservation systems to regulate growth and prevent the dissemination of specific disease-causing and deterioration-causing microbes in food items. According to Landi and the team, to put it succinctly, antimicrobial packaging is attained by embedding an antimicrobial agent within the containerized material. This offers a possible alternative to directly introducing bioactive agents into or onto food surfaces. Such direct addition may swiftly deplete antimicrobial functionality, leading to the failure of restraining microbial growth in the medium/long term [24] According to the Asgher, this implies that antimicrobial packaging can manage the gradual release of antimicrobial substances, with migration kinetics patterns affected by variables like the molecular polymer composition and antimicrobial compounds, the material science attributes of the packaging product and the Interior and exterior states [25]. Considering this scenario, Gonçalves stated, crafting an antimicrobial packaging system is intricate, considering it entails a profound understanding of five essential factors and their interactions: the food commodity, the internal containerization ambience, the intended pathogens, the containerization material, and the bactericidal substance [26].

Chawla studies have shown a range of approaches has been explored to devise biobased antimicrobial packaging solutions. In terms of structure and production processes, antimicrobial packaging systems fall into five distinct classes [27]. In the primary class, antimicrobial sachets are placed inside the package, allowing the gradual release of the active compound during the storage period. Creating antimicrobial products in the second category involves directly blending active molecules in the polymer matrix and thus, their thermal and chemical stability should be subject to scrutiny. Antimicrobial containers in the third category are obtained by absorbing a particular matrix, acting as a conveyer for the antimicrobial additive, onto the packaging surface. This production method is designed to counter the drawbacks associated with the second class, ensuring that active substances are spared exposure to extreme thermal conditions and frictional forces during the production of the containerization material. The fixation of the antimicrobial agent onto the material matrix in the fourth category is facilitated by the formation of molecular connections among their reactive sites. For this instance, the harmony of chemical elements between macromolecules and supplements is essential, with the liberation of the active agent from the matrix predominantly influenced by the bonding type. Antimicrobial packaging in the fifth class encompasses the utilization of macromolecules endowed with inherent microbe-inhibiting features, such as chitosan. Sofi et al stated that effective inhibition in this approach necessitates direct contact between the packaging material and the food product, which might be perceived as a limiting factor for two reasons: The process of inhibition is restricted to surface contact layers; demanding that the polymer be recognized as a food additive. Explored and utilized for active packaging, Qamar stated that the compounds with antimicrobial properties fall into diverse categories of molecules, originating from either synthetic sources or extracted from plants, animals, and microbial biomasses [28].

According to Mendonca, the broad movement towards minimizing the utilization of conventional synthetic additives has prompted the widespread adoption of bio-based antimicrobial compounds. These comprise polypeptides, biogenic acids, botanical oils, antioxidants, and redox regulators [29]. Demonstrating promising results, all these classes of molecules have been effectively embedded into bio-based packaging, displaying efficacy against both pathogenic and spoilage bacteria, along with fungi [30].

15.3 Antimicrobial Agents

Essential Oils

Aromatic secondary metabolites, denoted as essential oils (EOs), are naturally occurring in different plants. Extracted from diverse plant components, like rhizomes, leafage, blossoms, outer covering these intricate oily blends consist of 20 to 60 components at different concentrations. Solvent extraction, distillation, cold pressing, and non-conventional technologies like supercritical fluids are utilized for their extraction [29]. Moeini stated, Monoterpenes and sesquiterpenes stand out as the leading quantitative components in the composition of EOs, succeeded by phenolic compounds, aldehyde derivatives, ketonic substances. EOs, harboring various active molecules with diverse functional groups, have been documented to showcase a broad spectrum of physiological effects, including microbicidal, fungistatic, and redox regulator effects [30]. According to Nazzaro, Rigorous testing of these attributes in both food and packaging research sectors highlights their promising potential as an alternative for inorganic germicidal substances. Considering the mixture of substances shaping their composition, the antimicrobial efficacy of EOs is demonstrated through various channels. Yet, there is a collective understanding that the central focus of EOs is the cytoplasmic membrane of the microbial cell, and its permeability being altered by the structure and the hydrophilic/hydrophobic properties of the compounds interacting with it [31]. Due to the hydrophobic nature of EOs, their existence triggers a transformation in the makeup and flexibility of the cell membrane. According to batiha and the team that this method of action sets in motion a series of interrelated reactions, causing internal pH disarray, change in electrical potential, and impairment of the sodium-potassium pump, ultimately leading to cell demise [32]. Guo, haghighi and the other researchers have been said that numerous investigations explored the capacity of EOs, whether in their free form or integrated into eco-friendly packaging, to hinder the proliferation of bacteria with a positive gram stain, microbes with a negative gram reaction, fungal organisms [33–36].

The findings from this investigation underscored that the antimicrobial efficiency of EOs relies on their constitution and source, in addition to the defensive mechanisms adopted by the microorganisms being examined. Mohammad, Lee and the other team's members stated that the prevailing EOs utilized as active components in food packaging include Cinnamomum, Rosmarinus officinalis, zingiber officinale, Origanum vulgare, Melaleuca alternifolia, Thymus vulgaris and Citrus varieties [37–43].

The comprehensive research outcomes revealed that, alongside the antimicrobial activity, the incorporation of EOs can significantly transform the structure of the packaging materials including films and coatings, either enhancing or degrading their functional specifications by conveying with the polymeric scaffold and the softening agents. Furthermore, their antibacterial efficacy might be jeopardized by the quick liberation of volatile active substances before the culmination of the shelf-life timeframe. According to tongnuanchan and their team, the intricate interplay of essential oils extends to a nuanced modulation of the sensory and organoleptic facets of food items, presenting a potential conundrum for their ubiquitous integration. An adept strategy for tackling these challenges is encapsulating essential oils at the micro or nano scale, followed by their subsequent integration into the polymer matrix [44].

Through this systematic modus operandi, a methodical delivery of the bioactive compounds is achieved, affording protection against the elevated temperatures inherent in the polymer-extrusion process, and precluding an excessive impact on the sensory nuances of the food entity.

15.4 Animal-Derived Polypeptides

Among animal-derived antimicrobial substances, polypeptides take precedence as the most prevalent. Pisoschi stated that their fundamental release acts as a protective measure against the microbial spread and pathogenic conditions, emphasizing lysozyme, enzymatic protein, and antimicrobial peptide, which have been extensively explored for their enhanced antibacterial effectiveness in the field of active encasing applications [45].

Lysozyme, an enzyme derived from animals, has been acknowledged as safe (Generally Recognized as Safe) for direct incorporation into food matrices, as per the FDA in 1998. It exhibits stability across extensive temperature ranges (e.g., 3.5–95°C) and various pH levels (e.g., 1.9–10). According to Davidson, thorough assessments have been carried out on lysozyme's biocidal potential against a wide range of pathogens and spoilage bacteria, confirming its principal efficacy on gram-positive bacteria such as Clostridia, clostridium tyrobutyricum, and Listeria monocytogenes [46].

Aloui stated that the disruption of bacterial cell walls' peptidoglycan layer is the mechanism through which lysozyme exerts its antibacterial effects, involving the hydrolysis of the bond linking N-Acetyl-D-GlcN and N-Acetyl-MurNAc [47]. In the case of gram-positive bacteria, lysozyme demonstrates a highly effective mechanism of action. Yet, its interaction with gram-negative bacteria is limited by the protective lipo-polysaccharidic layer, preventing it from reaching the site of action. Lèsnierowski and Yang have been showed that the potential enhancement of lysozyme activity against gram-negative bacteria has been suggested by numerous researchers. This can be achieved through various channels, such as covalent attachment of SFA compounds to lysine sites, temperature-driven denaturation, glycan attachment, disulfide bond reduction, and the chelator application [48]. Soggiu stated that modern industrial practices heavily rely on lysozyme, especially for mitigating undesired fermentation of butyric acid and late blowing caused with tyro butyric clostridia in cheeses with moderate firmness [49].

Saeed, Afzaal and the team have been shown that another animal-derived enzyme, Peroxidase in lactating fluid is produced in the epithelium cells of the mammary gland and is found in high levels present in bovine milk which is exhibiting a 20-fold greater presence in cow's milk than in human milk [50]. Enteric bacteria, including Salmonella microorganism, shigella microorganisms and Escherichia coli are highly susceptible to its potent effects, making it an exceptionally effective agent. Through catalysis, it facilitates the oxidation of thiocyanate groups with hydrogen peroxide, resulting in the generation of thiocyanogen.

Subsequently, the formed thiocyanogen undergoes hydrolysis to produce hypothiocyanite. Microbial death is provoked by the interaction of these unsteady molecules with the sulfhydryl groups found in bacterial cell membrane proteins. Buys and Sefu recommend the use of this enzyme at ambient temperature as a potent substitute to chilling for conserving raw milk [51]. Exhibiting antioxidant, anti-carcinogenic, anti-adiposity, and antimicrobial charateristics, lactoferrin is a globular glycosylated protein. Lactoferrin is distributed in the excretion of the generative system, breathing apparatus, gastrointestinal system, occurring in both humans and other mammals. According to Niaz, lactoferrin is extensively present in maternal colostrum. Its Antimicrobial efficacy stems from its capacity to chelate iron, causing the interruption of the external membrane of microbes with a negative Gram reaction [52].

Lactoferrin not only showcases biocidal activity but also exerts a bacteriostatic effect, reducing the immediacy of microorganisms to nutrients. Hence, as per the findings of Gyawali and Ibrahim, it proved exceptional in efficacy against numerous food-borne pathogenic bacteria, including Escherichia coli, Klebsiella microorganisms, and Mono cytogenic listeria [53]. With their elevated antimicrobial effectiveness and natural characteristics, Lytic enzyme, lactoperoxidase, and lactoferrin are Incorporated as enhancers in the manufacturing of biodegradable containerization designed to thwart the increase and utilization of hazardous microorganisms polluting the surfaces of conveniently prepared foods.

15.5 Antagonistic Microorganisms and Bacteriocins

Certain microorganisms, along with their metabolic products, can act to suppress the growth of others. Researchers and industries are keenly interested in the prospect of employing this ability as an "inherent protection" against the proliferation of spoilage and pathogenic microorganisms in food. The contemporary approach to food conservation involves the profound application of "microorganisms with antagonistic properties" and their derivatives, commonly referred to as " bio-antimicrobial preservation " as outlined by Quinto [54].

Lactic Acid Bacteria emerges as the leading group of antagonistic microorganisms employed extensively in food systems. Agriopoulou stated that lactic Acid Bacteria has been classified as Generally Recognized as Safe by the FDA and has received the Qualified Presumption of Safety acknowledgment from the European Food Safety Authority [55]. Investigations have been conducted into the utilization of LAB for the competition against harmful and deteriorating microorganisms in food matrices, coupled with their capacity to produce nutrients and antimicrobial metabolites. As stated by Ramos and their research counterparts, the application of these substances to fruits and vegetables, Unrefined dairy items, and processed meat yielded in successful outcomes [56, 57]. The study highlighted the diverse Lactobacillus species' ability to thrive when challenging bacteriological and fungi populations.

LAB mainly produces metabolic proteins known as bacteriocins, which function as a protective measure against other microorganism varieties. The potential utility of these methods has been evaluated across various food categories, encompassing minimally culinary vegetables and fruits, Dairy goods, meat, and seafood. When applied in conjunction with other technologies through a hurdle approach, their maximal effectiveness is achieved. Additionally, their application must conform to established sound manufacturing practices, optimal processing conditions, and impeccable storage and circulation procedures. Within the broad spectrum of bacteriocins created by LAB, nisin and pediocin seize the spotlight as the primary and most remarkable bacteriocins, warranting consideration as auspicious victuals preservatives.

Nisin, a thermostable peptide, is synthesized by specific strains of Lactococcus lactis, demonstrating resilience to high temperatures.

Possessing strong antibacterial characteristics, it effectively confronts gram-positive bacteria like Clostridium sporogenes, Staphylococci, Cereus bacilli, Botulinum bacteria, Listeria monocytogenes, Microbe clusters, and other bacterial varieties. Nevertheless, according to Tiwari, it demonstrates reduced inhibiting activity against gram-negative bacteria and fungi [58].

According to Bauer, nisin inhibits the growth of gram-positive cells by binding to specific functional groups on the cell wall, causing poration of the cell membrane and subsequent loss of intracellular constituents [59]. As a result, nisin has unveiled one of its most promising applications in managing the populations of monocytogenic listeria and Clostridium microorganisms in dairy products, both independently and in conjunction with other technologies, according to Ibarra-Sánchez [60].

Haakensen, Dobson, Hill report that pediocin is generated by different species of Microbial pedicoccus, a portion of gram-positive, homolactic fermenters within the Lactobacillus family [61]. Gabrielsen, Brede, Nes, and Diep describe that pediocin acts by forming pores in the cytoplasmic membrane of the target cells, causing a decrease in intrinsic pH, and blocking the proteins responsible for energy production [62]. In preservation studies conducted by Verma, Sood, Saini, and Saini, concentrated pediocin has been investigated for its effectiveness in preserving vegetables and dairy products. Furthermore, Castellano and the team explored its application for prolonging the freshness of prepared meat [63, 64]. Niamah notes that the effectiveness of pediocin in food is primarily governed by factors such as pH, osmotic equilibrium, enzymatic actions, and temperature [65]. The incorporation of bacteriocins into active packaging serves as an antimicrobial addition, ensuring safety, and extending the shelf life of products that directly interact with the package. For instance, Blanco Massani reported the successful application of nisin in antimicrobial films, encompassing both petroleum-derived and bio-based materials as shown in Figure 15.2 for types of bio-based materials [66]. These films are utilized in the packaging of both raw and processed meat, with testing conducted against Listeria species. Furthermore, Sun conducted a recent study to evaluate its impact on the technical properties of biodegradable films [67].

- **Biodegradable antimicrobial films**

Categorized by the source of origin, biodegradable polymers can be classified into natural, synthetic, and microbial polymers, as outlined by Del Nobile [68]. For an additional improvement in their functional attributes, one can attach antimicrobials onto the polymer matrices, which the diversity of methods can range from the uncomplicated dispersion of antimicrobial formulations onto the polymer surface to the utilization of more

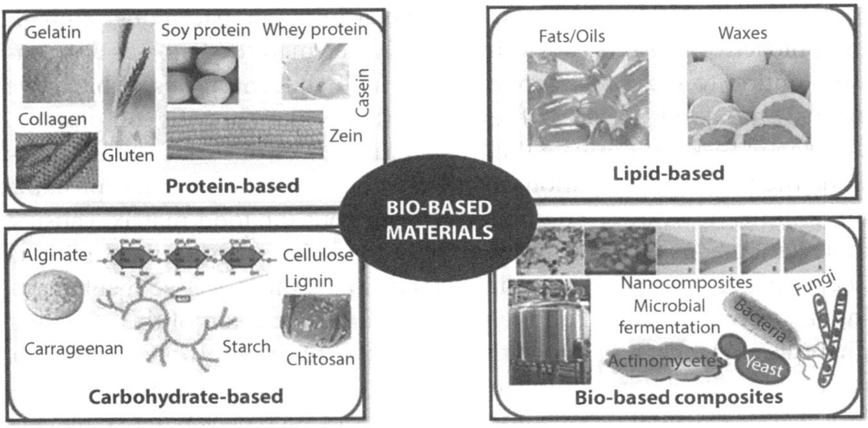

Figure 15.2 Biobased materials.

sophisticated techniques such as association of active agents with binders, which can be either cellulose-based or copolymers with acrylic components. Alternatively, active agents can be covalently linked by utilizing natural or synthetic Cross-linking reagents like Geniposide, Glutaric aldehyde, or Formic aldehyde. A significant limitation in the utilization of these chemicals arises from their toxicity, as they are known to react with tyrosine or lysine, resulting in the formation of methylene bridges. For this reason, as a precautionary measure, it is crucial to account for their use when synthesizing biodegradable materials. The assessment of the aldehyde's destiny is essential as the material approaches the end of its life. Given these considerations, Wittaya endorses the use of natural cross-linking agents such as gallic acid or tannins to improve the properties of protein films [69]. Furthermore, in some scenarios, the incorporation of plasticizers such as glycerol, polyethylene glycol, or propylene glycol is considered essential for promoting film formation. Different techniques, including solution casting, extrusion, layer-by-layer assembly, etc., can be utilized to create antimicrobial biodegradable films and coatings for food packaging applications. Notably, solution casting is recognized as the most popular method, as indicated by Dutta, Tripathi, Mehrotra [70]. Numerous biopolymers demonstrate a high level of compatibility with a broad spectrum of antimicrobial agents. These polymers have undergone thorough investigation for their ability to combat both spoilage and pathogenic microorganisms in diverse testing conditions relevant to food packaging applications.

Developing biodegradable antimicrobial films for packaging, embedding bioactive substances within the polymer matrix, stands as a promising option to the direct application of high concentrations of bioactive agents onto food surfaces through methods like spraying or dipping. This tactic seeks to mitigate the frequent instances of surface contamination. The immediate addition of antimicrobials to food products prompts rapid diffusion of these compounds throughout the food matrix, causing an instantaneous decrease in microbial populations. Yet, this expeditious deployment leads to the immediate exhaustion of the bioactive agent, proving ineffective in restricting the growth and proliferation of the surviving microflora. Hence, it is imperative to investigate the controlled release or release kinetics of these antimicrobials from the packaging to the food surface. This is essential in assessing the efficacy of such packaging systems.

According to Han, the current approach to producing biodegradable antimicrobial films involves utilizing natural polymers that incorporate either natural or synthetic antimicrobial agents [71]. Agarwal, Mahieu, Terzopoulou and the associated team members highlight that a considerable body of literature concentrates on the combination of melt-processable starches with various biodegradable polymeric esters such as caprolactone polymer, lactic acid polymer, Hydroxybutyrate-valerate copolymer, Poly (butylene succinate-adipate), Butylene adipate-terephthalate polymer, and polyether ester compound [72–74].

- **Polylactic acid (PLA)**

Recognized as highly promising, this biopolymer is characterized by its relatively low melting temperature, capacity to form compact films, and ability to increase the biocompatibility of numerous active agents. Additionally, its industrial applications position it as a viable alternative to petroleum-derived polymers. Armentano and Vodnar, along with their team, provide insight into the production of PLA, a linear aliphatic thermoplastic polyester derived entirely from renewable resources. The process involves the fermentation of biomass [75, 76]. Djidi, Ingrao, Jašo, and their team emphasize that this thermoplastic polymer is characterized by high-strength, high-modulus, good processability, and complete biodegradability. Moreover, it is compostable and biocompatible [77–79]. Rhim provides evidence that PLA is safe for food packaging applications, as it has been designated as Generally Recognized as Safe [80]. This designation ensures compliance with the requirements for direct contact with liquid-based, citric, and oils foods. As per Dutta's observations, PLA demonstrates remarkable potential for consumer products, especially in containerization, given its transparency, less-toxicity, and eco-friendly features [81]. Del Nobile underscores the scarcity of studies examining the antimicrobial effectiveness of additives like lemon extract, thymol, lysozyme, and olive leaf extract by Marcos when integrated into PLA films [82].

- **Chitosan**

Comprising chitin with 2-amino deoxy-β-D-glucose units, chitosan is a deacetylated form of chitin, ranking as the second most abundant polysaccharide in nature after cellulose. In most cases, deacetylation is not entirely achieved, making chitosan a copolymer that encompasses both deacetylated and acetylated units. Jayakumar's findings reveal that chitosan is not only non-toxic but also possesses decomposability, Bio efficacy, Compatibility with living tissue, and antimicrobial properties [83]. The antimicrobial packaging and biomaterials industry are notably intrigued by chitosan, primarily due to its cationic nature and inherent antimicrobial properties, as mentioned in Table 15.1. Chitosan holds an advantage over other biobased food packaging materials as it can integrate active and functional compounds with antimicrobial actions against Listeria monocytogenes, according to findings by Moller [84]. Chitosan's antimicrobial activity is improved over chitin, thanks to its positive charge. Despite ongoing research, the exact mechanism of chitosan and its derivatives is not yet perfectly understood, and the literature suggests various potential explanations, as highlighted by Rabea [85]. Shahidi identifies the positively charged amino group of chitosan as one of the reasons for its antimicrobial properties, attributing this to its interaction with the negatively charged cell membranes of microbes [86]. On the contrary, Cuero explains that chitosan acts as a chelating agent, choosily binding trace metals, consequently inhibiting the production of toxins and microbial growth [87]. According to Lian, the antimicrobial approach of chitosan differs between Gram-positive such as bacterium Staphylococcus aureus and Gram-negative such as Escherichia coli bacteria [88]. Lian-Ying, Z. and Zhu, J have reported the effects of chitosan on Staphylococcus aureus (Gram-positive) and Escherichia coli (Gram-negative) bacteria. In the case of Staphylococcus aureus, the antimicrobial activity showed a growth with the molecular weight of chitosan, whereas for Escherichia coli, the antimicrobial activity increased with decreasing molecular weight, as demonstrated. For the antimicrobial activity against Staphylococcus aureus, the authors suggested an explanation: chitosan forms a polymer membrane, blocking the entry of nutrients into the cell and in the case of Escherichia coli, the authors explained that chitosan entered the cell through pervasion. The interaction of oppositely charged macromolecules is involved in the preparation of chitosan nanoparticles, as detailed by de Azeredo [89]. By subjecting chitin

Table 15.1 Biopolymer with its structure, antimicrobial agent and targeted microorganisms.

Bio polymer	Structure	Target microorganism	Antimicrobials
Alginate	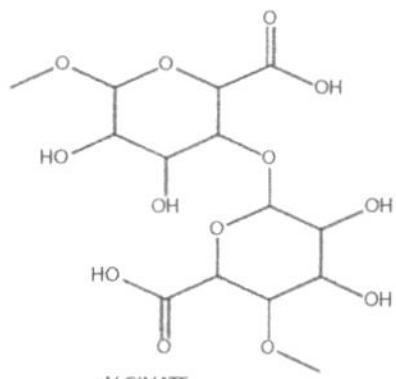 ALGINATE	**S. aureus** **B. subtilis** **E. coli O157:H7** **E. coli** **S. typhimurium** **S. aureus** **B. cereus**	**Garlic oil**
Carrageenan	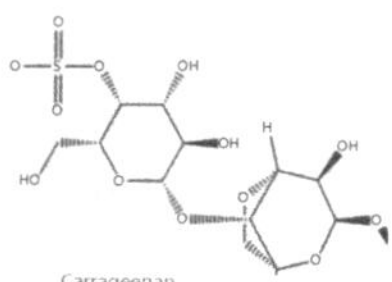 Carrageenan	**Micrococcus luteus** **L. innocua** **S. enteritides** **E. coli** **S. aureus**	**Grapefruit seed extract**
Zein	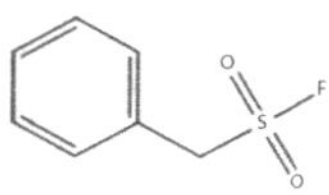 ZEIN	**L. plantarum** **B. subtilis**	**Lysozyme**
Cellulose	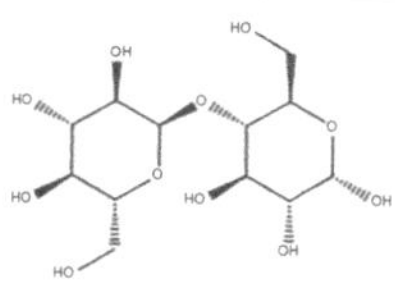 starch	**L. monocytogenes**	**pediocin**
Starch	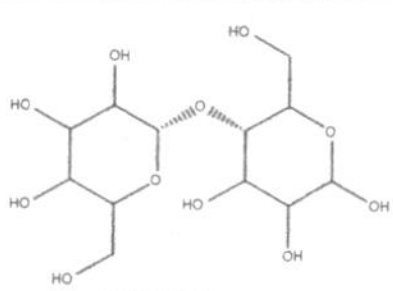 CELLULOSE	**E. coli**	**Potassium sorbate**
Chitosan	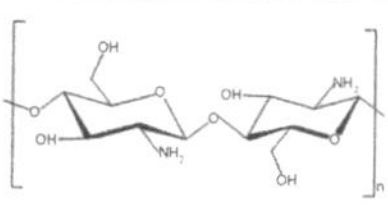 CHITOSAN	**L. monocytogenes**	**Green and black teas extract**

to acid hydrolysis, chitin whiskers in the form of crystalline nanofibrils can be prepared, and further diacetylation of these chitin whiskers leads to the formation of chitosan whiskers. Madureira emphasizes that chitosan nanoparticles and their derivatives have a broad spectrum of antimicrobial effects and reported that chitosan nanoparticles, within concentrations ranging from 0.1% to 0.7% w/v, exhibit a greater antimicrobial effect than bulk chitosan [90]. This heightened efficacy is attributed to their increased surface area and charge density, enabling enhanced interaction with the cationic surfaces of bacterial cells. T Many researchers, such as Ramani, Song, Jang, and their team, have studied the antimicrobial effect of nano-structures that integrate chitosan with other antimicrobial agents [91, 92]. The synthesis of carvacrol-loaded chitosan nanoparticles involved an oil-in-water emulsion followed by the ionic gelation of chitosan, as performed by Keawchaoon and Yoksan [93]. Carvacrol was observed to enhance the antimicrobial activity of chitosan nanoparticles against Escherichia coli, Staphylococcus aureus, and Bacillus cereus, according to the minimum inhibitory concentration results.

- **Pectin**

Playing a crucial role in tissue rigidity and integrity within the plant cell wall, pectin is recognized as one of the most intricate macromolecules in nature. This characterization is substantiated by research conducted by Bayarri, Benito-Peña, De'Nobili, and Tripathi [94–97]. Videcoq identifies apple pomace and citrus peels as the foremost sources for pectin extraction [98]. Pectin, as highlighted by Laurent and Boulenguer, is an ingredient extensively used in the food industry. It is not subject to limitations other than current good manufacturing practice [99]. The FDA recognizes it as typically safe, and its primary use in food involves serving as a coagulating, ensuring stability, or viscosity enhancer in products like fermented dairy, fruit preserves, beverages, and frozen dessert.

- **Soy protein isolate**

Characterized by a diverse range of molecular properties, it is a compound blend of proteins, presenting a size distribution extending from 200 to 600 kDa. Globulins constitute the predominant group of soybean proteins, and they can be separated into fractions of 2 S, 7 S, 11 S, and 15 S based on their sedimentation coefficients. The predominant fractions, namely 7 S and 11 S, are the main contributors, representing less than 37% and 30.9% of the total extractable protein. These fractions, as observed by Cho and Rhee, demonstrate the potential for polymerization [100]. Khan emphasizes that soy protein isolate is characterized by reliable sourcing, favorable biocompatibility, biodegradability, processability, and film-forming capability. It holds substantial promise for applications in the food industry, agriculture, bioscience, and biotechnology [101, 102].

- **Alginates**

Structural polysaccharides are derived from brown algae and are known as alginates. Molecularly, alginates comprise linear water-soluble polysaccharides with (1–4) linked units of R-Dmannuronate and β-L-guluronate, with diverse proportions and distributions within the chain as specified by Hambleton [103]. The gel-forming property of alginate, driven by cation binding, serves as the basis for its various applications. The transition from water-soluble sodium alginate to water-insoluble calcium alginate has been scrutinized by Galus and Kadzińska [104]. Edible films derived from sodium alginate offer a solution to dehydration issues in animal meat, sea food, and fruits. Additionally, alginate can be augmented with active substances, functioning as an encapsulation agent, as proposed by Pop [105].

- **Whey proteins**

Banerjee and Chen underscore the remarkable nutritional value and functional characteristics of whey protein. In addition, Banerjee and Chen observe a consistent increase in the annual production of liquid whey, which is generated in significant quantities [106]. Whey protein-derived edible films and coatings, as proposed by Ozdemir and Floros, have the competence to increase the utilization of whey, enhance the nutritional value of foods, and extend shelf life [107]. Representing the more concentrated form of whey proteins, whey protein isolates showcase promising functional aspects. In addition to their other beneficial properties, whey protein isolates, as observed by McHugh, exhibit moderate moisture permeability and oxygen barrier properties that rival those of premier synthetic polymer-based films [108].

15.6 Applications of Edible Antimicrobial Films in Food

a. Fruits and vegetables

The impressive nutritional profile of fruits and vegetables, encompassing generous amounts of Fibrous elements, vitamins, and minerals, is offset by their proclivity for rapid degradation, mainly due to elevated moisture content and subsequent higher moisture actions. Furthermore, their composition of living tissues renders them highly susceptible to issues such as microbial contamination, the generation of off flavors, enzymatic browning, and the degradation of texture. order to address the issues arising from the undesirable properties of fruits and vegetables and to meet the continuously rising expectations of consumers, and to prolong the shelf life of fresh fruits and vegetables while safeguarding them against microbiological contamination and spoilage, the food packaging sector has delved into diverse alternatives. Valdés, Ramos, Beltrán, Jiménez, and Garrigós provide insights into this strategy [109].

Investigating cold storage preservation methods, Sánchez-González developed an edible coating by combining hydroxypropylmethylcellulose (HPMC) and chitosan, enhanced with bergamot essential oil (EO) activation, specifically for grapes [110]. In their recommendations, the combination of chitosan and HPMC with bergamot essential oil (EO) stood out as the most effective formulation, outperforming the use of chitosan alone with bergamot EO. This formulation displayed the most potent inhibitory action and demonstrated optimal control over respiration rates, leading to minimal weight losses during postharvest cold storage. In separate studies focused on apples and melons, Salvia-Trujillo, Rojas-Graü, Soliva-Fortuny, and Martín-Belloso and Raybaudi-Massilia, Mosqueda-Melgar, and Martín-Belloso respectively obtained analogous outcomes [111, 112]. To enhance the microbiological quality and shelf life of avocados, Saucedo-Pompa formulated an edible film utilizing candelilla wax and incorporating ellagic acid as the active phenolic compound [113]. The 42-day storage investigation of the active films exhibited positive results in terms of product quality. This indicates a promising application for these edible films in preserving quality and prolonging the shelf life of avocados.

Jiang explored the enhancement of barrier properties, such as water vapor and gas permeability, in active films developed using biodegradable PLA polymer and Poly(3HB–4HB) copolymer [114]. The objective of the study was to enhance film characteristics for the preservation of peaches. Incorporating angelica essential oil into the film demonstrated an increased rate of free radical scavenging activity, alongside elevated water vapor permeability and gas transmission rates. This proved effective in preserving peaches, delaying oxidation and extending their shelf life beyond two weeks.Capsicum treated with coatings made from chitosan and alginate, infused with pomegranate peel extract, exhibited remarkable antifungal activity against various strains of Colletotrichum gloeosporioides, resulting in a prolonged shelf life of near to 4 weeks, Coatings, as studied by Nair, Saxena, and Kaur, demonstrated the ability to extend the shelf life of guava by up to nearly 3 weeks[115, 116]. In oranges, Kharchoufi reported that coatings comprising chitosan and locust bean gum, along with pomegranate peel extract, displayed notable effectiveness in preventing the growth and spread of Penicillium digitatum [117]. Falco's research indicates that applying alginate films containing green tea extracts and oleic acid to fruits like raspberries and strawberries significantly reduces the risk of food-borne illnesses. Importantly, this protective coating demonstrates strong antiviral efficacy against human enteric viruses such as hepatitis A virus and human noroviruses [118].

In a study by Alvarez, Ponce, and Moreira, the effectiveness of chitosan films incorporating bioactive extracts-tea tree, rosemary, pomegranate, resveratrol, and propolis against Escherichia coli and L.monocytogenes was examined [119]. The research employed both in vivo and in vitro assays, focusing on minimally processed broccoli. Research findings indicated that activated chitosan coverings not only decelerated the growth of both psychrophilic and thermophilic microorganisms but also upgraded the hue, texture, and sensory properties of broccoli. In view of this, the evidence can be utilized to construct chitosan films bolstered with bioactive components as potential choices for packaging material in the regulation of both indigenous and deteriorative microflora in broccoli. An investigation conducted by Boyacı on melon fruit indicated strong antimicrobial efficacy of zein-centric films fortified with essential oils against Listeria innocua and Escherichia coli [120]. In addition, the employment of carboxymethyl cellulose (CMC)-mixed quaternized chitosan films for an enhanced shelf life of whole bananas were associated with a deferred fruit degradation when compared to bananas encased in films with a higher proportion of quaternized chitosan, as reported by Hu, Wang & Wang [121]. Utilizing the ionic gelation technique, chitosan-quinoa protein-based antimicrobial edible films were developed by introducing chitosan-thymol nanoparticles. The purpose of these films was to extend the post-harvest life of tomato cherries and blueberries. Enhanced antimicrobial efficacy of chitosan-thymol nanoparticles against foodborne pathogens like Listeria innocua, Salmonella typhimurium, and Staphylococcus aureus was documented by the researchers. Not only did they observe a stronger inhibitory effect compared to free thymol, but they also proposed enhanced water barrier characteristics in the film with the inclusion of these nanoparticles, according to findings by Medina [122].

b. Meat and poultry

Human consumption of animal protein is predominantly satisfied by meat, encompassing both fresh and cured meat products. Nevertheless, it undergoes swift degradation due to microbial contamination, particularly by pathogens like Listeria monocytogenes. This presents a significant hurdle for the packaging sector as it grapples with the issue of microbial recontamination in diverse ready-to-eat meat products. The nutritional perspective on meat underscores the profound importance of protein quality, and effective packaging plays a crucial role in attaining this objective. Amid numerous choices in the queue, Chen predominantly selects ecological biopolymer packaging materials to guarantee the upkeep of quality and safety in meat products [123]. In refrigerated beef samples, Zinoviadou, Koutsoumanis, and Biliaderis established the antimicrobial activity of whey protein edible films incorporating oregano essential oil. The application of soy-based edible films embedded with oregano and thyme essential oils resulted in reduced microbial counts in refrigerated fresh ground beef patties, as reported by Emiroğlu, Yemiş, Coşkun, and Candoğan [124, 125].

A strong inhibitory effect against Escherichia coli and total aerobic bacterial counts in fresh chicken breasts stored at 5 °C was observed by Seol, Lim, Jang, Jo, and Lee when utilizing κ-carrageenan edible films containing a combination of ethylenediaminetetraacetic acid and ovo transferrin [126]. Neetoo, and Chen undertook a study wherein distinct active packaging coatings were created, incorporating various biopolymers including starch, alginate, xanthan gum, pectin, and κ-carrageenan [127]. Activation was achieved using diverse bioactive substances like nisin, sodium lactate, Novagard CB1, potassium sorbate, sodium diacetate, etc. Subsequently, these activated coatings were applied to turkey products to mitigate the risk of contamination by Listeria monocytogenes. Alginate coatings, in the spectrum of different biopolymers investigated, demonstrated superior effectiveness against L. monocytogenes contamination. Meanwhile, the most effective formulations for the preservation of various turkey products were identified to involve combinations of nisin, sodium acetate, and sodium diacetate.

In the instance of poached turkey, the most successful combination comprised sodium lactate paired with either nisin or sodium diacetate. In contrast, for processed deli turkey, the formulation of nisin either with sodium lactate or sodium diacetate was recommended as the most convenient. Barbiroli documented that cellulose-based active films, incorporating either lysozyme or a combination of lysozyme and lactoferrin, were the most appropriate choice as packaging material for the preservation of thin raw meat slices [128]. As per a separate report from Amjadi, the implementation of gelatin-based bio-nanocomposite films integrated with chitosan nanofibers and ZnO nanoparticles for packaging chicken fillets resulted in reduced overall bacterial counts. Additionally, the growth of pathogens such as Staphylococcus aureus and Escherichia coli was restrained compared to control samples over a 12-day storage period [129]. The research by Noori, Zeynali, and Almasi revealed similar findings, highlighting the efficacy of ginger essential oil (EO) nanoemulsions in extending the shelf life of chicken breast [130]. Significant reductions in both chemical and microbial spoilage of stored pork chops were observed when the chops were coated with active chitosan films containing either free or nanoencapsulated Paulownia Tomentosa essential oil (EO).

Lin, Zhu, and Cui devised a preventive measure for surface contamination of poultry by Campylobacter jejuni. They created a gelatin nanofibrous matrix embedded with thyme essential oil/β-cyclodextrin ε-polylysine nanoparticles using electrospinning [131]. Effective inhibition of C. jejuni was observed with the antimicrobial nanofiber coating, and the quality and sensory characteristics of the meat product were preserved. Siripatrawan and Noipha developed active chitosan films with green tea extract for pork sausage packaging, leading to the successful inhibition of microbial development in sausages stored at 4 °C [132]. Additionally, the film exhibited enhanced antimicrobial and antioxidant characteristics, contributing to improved food safety and quality.

15.7 Conclusion and Future Roadmap

The market for minimally processed food products is witnessing continuous growth, prompting the need for efficient and innovative packaging technologies. These technologies have become an integral aspect of emerging policies dedicated to the preservation and protection of food. The research focus in the packaging sector has shifted towards biopolymers as an alternative to conventional sources, driven by their ecological and non-toxic properties. Despite these advantages, the widespread industrial use of biopolymers faces challenges related to limitations in barrier, thermal, and mechanical attributes, along with high manufacturing and processing costs.

The transition to industrial-scale utilization of biopolymer-based packaging materials calls for ongoing investigations and research endeavors. The emphasis should be on improving physical and microbiological quality, with the goal of extending shelf life and enhancing nutritional value. Successful improvements in the physical characteristics and performance of these packaging materials have been achieved through a range of techniques. Notable approaches include chemical and physical alterations, polymeric blending, and the utilization of nanocomposites. While there has been a considerable number of studies focusing on improving the durability and performance of bio-based packaging materials, additional research is required for industrial application. This involves devising more effective methods and refining existing ones to overcome existing limitations.

The issue at hand could potentially be resolved by leveraging nanotechnological concepts in packaging. Many researchers have thoroughly examined the advantages of integrating nanotechnology into the food packaging industry, with a particular focus on edible films and coatings. In this application, nanotechnological features have the potential to not only mask undesirable flavors and inhibit the chemical decomposition of functional compounds but also improve their solubility and enable controlled release.

The efficacy of nanomaterials in the preparation of nanoencapsulated bioactive agents for incorporation into edible packaging materials has been firmly established by various research studies. This could have a potentially positive impact on the performance and durability of biopackaging. Deploying antimicrobial peptides or nanoparticles in packaging serves the dual purpose of preventing undesirable changes in food, such as off flavors or color alterations, and ensuring a controlled release over an extended period. This strategy enhances the quality of the monitored food, presenting customers with improved value for money. Despite the promising outlook of bio-nanocomposites in future packaging, their broader application is hampered by the combination of high process expenses and low production rates. Considering the ongoing trends in the food and packaging industry, future investigations can aim to identify and comprehend the synergistic interactions among different bioactive agents. This includes devising optimal formulations and compositions for individual polymers, employing diverse processing techniques for various bio-nanocomposites, examining interactions between food components and antimicrobial nanocarriers, and determining suitable co-nanoencapsulation systems for incorporating multiple bioactive agents within a single packaging system. Nonetheless, before integrating nanomaterials into food products, a meticulous exploration of nano-based antimicrobial packaging systems is necessary. This should be accompanied by thorough toxicology studies and the implementation of rigorous safety measures.

References

1. Huang, W., Xu, H., Xue, Y., Huang, R., Deng, H., Pan, S., Layer-by-layer immobilization of lysozyme–chitosan–organic rectorite composites on

electrospun nanofibrous mats for pork preservation. *Food Res. Int.*, *48*, 2, 784–791, 2012, https://doi.org/10.1016/j.foodres.2012.06.026.
2. López-Carballo, G., Gomez-Estaca, J., Catala, R., Hernández-Muñoz, P., Gavara, R., Active antimicrobial food and beverage packaging, 2012, Yam, K.L. and Lee, D.S., Emerging food packaging technologies, in: *Woodhead Publishing Limited eBooks*, 2012, https://doi.org/10.1533/9780857095664.
3. Malhotra, B., Keshwani, A., Kharkwal, H., Antimicrobial food packaging: potential and pitfalls. *Front. Microbiol.*, *6*, 611, 2015, https://doi.org/10.3389/fmicb.2015.00611.
4. Sofi, S.A., Singh, J., Rafiq, S., Ashraf, U., Dar, B., Nayik, G.A., A Comprehensive Review on Antimicrobial Packaging and its Use in Food Packaging. *Curr. Nutr. Food Sci.*, *14*, 4, 305–312, 2018, https://doi.org/10.2174/1573401313666170609095732.
5. Ezeoha, S.L., Production of Biodegradable Plastic Packaging Film from Cassava Starch. *IOSR J. Eng.*, *3*, 10, 14–20, 2013, https://doi.org/10.9790/3021-031051420.
6. Zhong, Y., Godwin, P.M., Jin, Y., Xiao, H., Biodegradable polymers and green-based antimicrobial packaging materials: A mini-review. *Adv. Ind. Eng. Polym. Res.*, *3*, 1, 27–35, 2020, https://doi.org/10.1016/j.aiepr.2019.11.002.
7. Ataei, S., Azari, P., Hassan, A., Murphy, B., Yahya, R., Muhamad, F., Essential Oils-Loaded Electrospun Biopolymers: A Future Perspective for active Food Packaging. *Adv. Polym. Technol.*, *2020*, 1–21, 2020, https://doi.org/10.1155/2020/9040535.
8. Pereda, M., Marcovich, N.E., Ansorena, M.R., Nanotechnology in food packaging applications: barrier materials, antimicrobial agents, sensors, and safety assessment, in: *Springer eBooks*, pp. 2035–2056, 2019, https://doi.org/10.1007/978-3-319-68255-6_30.
9. Hassan, B., Chatha, S., Hussain, A.I., Zia, K.M., Akhtar, N., Recent advances on polysaccharides, lipids and protein based edible films and coatings: A review. *Int. J. Biol. Macromol.*, *109*, 1095–1107, 2018, https://doi.org/10.1016/j.ijbiomac.2017.11.097.
10. Vieira, M.G.A., Da Silva, M.A., Santos, L.O., Beppu, M.M., Natural-based plasticizers and biopolymer films: A review. *Eur. Polym. J.*, *47*, 3, 254–263, 2011, https://doi.org/10.1016/j.eurpolymj.2010.12.011.
11. Rezaei, A., Nasirpour, A., Fathi, M., Application of cellulosic nanofibers in food science using electrospinning and its potential risk. *Compr. Rev. Food Sci. Food Saf.*, *14*, 3, 269–284, 2015, https://doi.org/10.1111/1541-4337.12128.
12. Zemljič, L.F., Tkavc, T., Vesel, A., Šauperl, O., Chitosan coatings onto polyethylene terephthalate for the development of potential active packaging material. *Appl. Surf. Sci.*, *265*, 697–703, 2013, https://doi.org/10.1016/j.apsusc.2012.11.086.
13. Mlalila, N., Hilonga, A., Swai, H., Devlieghere, F., Ragaert, P., Antimicrobial packaging based on starch, poly(3-hydroxybutyrate) and

poly(lactic-co-glycolide) materials and application challenges. *Trends Food Sci. Technol.*, *74*, 1–11, 2018, https://doi.org/10.1016/j.tifs.2018.01.015.
14. Zhang, C., Li, Y., Wang, P., Zhang, H., Electrospinning of nanofibers: Potentials and perspectives for active food packaging. *Compr. Rev. Food Sci. Food Saf.*, *19*, 2, 479–502, 2020, https://doi.org/10.1111/1541-4337.12536.
15. Ibarra, V.G., Sendón, R., De Quirós, A.R., Antimicrobial food packaging based on biodegradable materials, in: *Elsevier eBooks*, pp. 363–384, 2016, https://doi.org/10.1016/b978-0-12-800723-5.00029-2.
16. Battisti, R., Fronza, N., Júnior, A.V., Da Silveira, S.M., Damas, M.S.P., Quadri, M.G.N., Gelatin-coated paper with antimicrobial and antioxidant effect for beef packaging. *Food Packag. Shelf Life*, *11*, 115–124, 2017, https://doi.org/10.1016/j.fpsl.2017.01.009.
17. Valerio, F., Masi, M., Cimmino, A., Moeini, A., Lavermicocca, P., Antimould microbial and plant metabolites with potential use in intelligent food packaging. *Nat. Prod. Res.*, *32*, 13, 1605–1610, 2017, https://doi.org/10.1080/14786419.2017.1385018.
18. Yildirim, S., Röcker, B., Pettersen, M.K., Nilsen-Nygaard, J., Ayhan, Z., Rutkaitė, R., Radusin, T., Sumińska, P., Marcos, B., Coma, V., Active packaging applications for food. *Compr. Rev. Food Sci. Food Saf.*, *17*, 1, 165–199, 2017b, https://doi.org/10.1111/1541-4337.12322.
19. Yam, K.L. and Lee, D., Emerging food packaging technologies: an overview, in: *Elsevier eBooks*, pp. 1–9, 2012a, https://doi.org/10.1533/9780857095664.1.
20. Wyrwa, J. and Barska, A., Innovations in the food packaging market: active packaging. *Eur. Food Res. Technol.*, *243*, 10, 1681–1692, 2017b, https://doi.org/10.1007/s00217-017-2878-2.
21. Jideani, V.A. and Vogt, K.F., Antimicrobial Packaging for Extending the shelf Life of Bread—A review. *Crit. Rev. Food Sci. Nutr.*, *56*, 8, 1313–1324, 2015, https://doi.org/10.1080/10408398.2013.768198.
22. Sung, S.Y., Sin, L.T., Tee, T.T., Bee, S.T., Rahmat, A.R., Rahman, W.A.W.A., Tan, A.C., Vikhraman, M., Antimicrobial agents for food packaging applications. *Trends Food Sci. Technol.*, *33*, 2, 110–123, 2013, https://doi.org/10.1016/j.tifs.2013.08.001.
23. Bastarrachea, L.J., Dhawan, S., Sablani, S.S., Engineering Properties of Polymeric-Based Antimicrobial Films for Food Packaging: a review. *Food Eng. Rev.*, *3*, 2, 79–93, 2011, https://doi.org/10.1007/s12393-011-9034-8.
24. Landi, C., Paciello, L., De Alteriis, E., Brambilla, L., Parascandola, P., High cell density culture with S. cerevisiae CEN.PK113-5D for IL-1β production: optimization, modeling, and physiological aspects. *Bioprocess. Biosyst. Eng.*, *38*, 2, 251–261, 2014, https://doi.org/10.1007/s00449-014-1264-8.
25. Asgher, M., Qamar, S.A., Bilal, M., Iqbal, H.M., Bio-based active food packaging materials: Sustainable alternative to conventional petrochemical-based packaging materials. *Food Res. Int.*, *137*, 109625, 2020, https://doi.org/10.1016/j.foodres.2020.109625.

26. Gonçalves, A.A., Safety and quality of antimicrobial packaging applied to seafood. *MOJ Food Process. Technol.*, *4*, 1, 10–16, 2017, https://doi.org/10.15406/mojfpt.2017.04.00079.
27. Chawla, R., Sivakumar, S., Kaur, H., Antimicrobial edible films in food packaging: Current scenario and recent nanotechnological advancements-a review. *Carbohydr. Polym. Technol. Appl.*, *2*, 100024, 2021b, https://doi.org/10.1016/j.carpta.2020.100024.
28. Qamar, S.A., Asgher, M., Bilal, M., Immobilization of Alkaline Protease From Bacillus brevis Using Ca-Alginate Entrapment Strategy for Improved Catalytic Stability, Silver Recovery, and Dehairing Potentialities. *Catal. Lett.*, *150*, 12, 3572–3583, 2020, https://doi.org/10.1007/s10562-020-03268-y.
29. Mendonça, A.F., Jackson-Davis, A., Moutiq, R., Thomas-Popo, E., Use of natural antimicrobials of plant origin to improve the microbiological safety of foods, in: *Elsevier eBooks*, pp. 249–272, 2018, https://doi.org/10.1016/b978-0-12-811835-1.00014-2.
30. Moeini, A., Cimmino, A., Poggetto, G.D., Di Biase, M., Evidente, A., Lavermicocca, P., Valerio, F., Leone, A., Santagata, G., Malinconico, M., Effect of pH and TPP concentration on chemico-physical properties, release kinetics and antifungal activity of Chitosan-TPP-Ungeremine microbeads. *Carbohydr. Polym.*, *195*, 631–641, 2018, https://doi.org/10.1016/j.carbpol.2018.05.005.
31. Nazzaro, F., Fratianni, F., De Martino, L., Coppola, R., De Feo, V., Effect of essential oils on pathogenic bacteria. *Pharmaceuticals*, *6*, 12, 1451–1474, 2013, https://doi.org/10.3390/ph6121451.
32. Batiha, G.E., Hussein, D.E., Algammal, A.M., George, T.T., Jeandet, P., Al-Snafi, A.E., Tiwari, A., Pagnossa, J.P., Lima, C.M.G., Thorat, N.D., Zahoor, M., El-Esawi, M.A., Dey, A., Alghamdi, S., Hetta, H.F., Cruz-Martins, N., Application of natural antimicrobials in food preservation: Recent views. *Food Control*, *126*, 108066, 2021, https://doi.org/10.1016/j.foodcont.2021.108066.
33. Guo, Q., Du, G., Jia, H., Fan, Q., Wang, Z., Gao, Z., Yue, T., Yuan, Y., Essential oils encapsulated by biopolymers as antimicrobials in fruits and vegetables: A review. *Food Biosci.*, *44*, 101367, 2021, https://doi.org/10.1016/j.fbio.2021.101367.
34. Haghighi, H., Biard, S., Bigi, F., De Leo, R., Bedin, E., Pfeifer, F., Siesler, H.W., Licciardello, F., Comprehensive characterization of active chitosan-gelatin blend films enriched with different essential oils. *Food Hydrocoll.*, *95*, 33–42, 2019, https://doi.org/10.1016/j.foodhyd.2019.04.019.
35. Kumar, S., Mukherjee, A., Dutta, J., Chitosan based nanocomposite films and coatings: Emerging antimicrobial food packaging alternatives. *Trends Food Sci. Technol.*, *97*, 196–209, 2020, https://doi.org/10.1016/j.tifs.2020.01.002.
36. Wang, Y., Yi, Z., Shi, Y., Pan, X., Lü, Y., Cao, P., Antibacterial effects of cinnamon (Cinnamomum zeylanicum) bark essential oil on Porphyromonas gingivalis. *Microb. Pathogen.*, *116*, 26–32, 2018, https://doi.org/10.1016/j.micpath.2018.01.009.

37. Mohammadi, M., Mirabzadeh, S., Shahvalizadeh, R., Hamishehkar, H., Development of novel active packaging films based on whey protein isolate incorporated with chitosan nanofiber and nano-formulated cinnamon oil. *Int. J. Biol. Macromol.*, *149*, 11–20, 2020, https://doi.org/10.1016/j.ijbiomac.2020.01.083.
38. Choulitoudi, E., Ganiari, S., Tsironi, T., Ntzimani, A., Tsimogiannis, D., Taoukis, P., Oreopoulou, V., Edible coating enriched with rosemary extracts to enhance oxidative and microbial stability of smoked eel fillets. *Food Packag. Shelf Life*, *12*, 107–113, 2017, https://doi.org/10.1016/j.fpsl.2017.04.009.
39. Bonilla, J., Poloni, T., Lourenço, R.V., Sobral, P.J.D.A., Antioxidant potential of eugenol and ginger essential oils with gelatin/chitosan films. *Food Biosci.*, *23*, 107–114, 2018, https://doi.org/10.1016/j.fbio.2018.03.007.
40. Lee, J.Y., Garcia, C.V., Shin, G.H., Kim, J.T., Antibacterial and antioxidant properties of hydroxypropyl methylcellulose-based active composite films incorporating oregano essential oil nanoemulsions. *LWT*, *106*, 164–171, 2019, https://doi.org/10.1016/j.lwt.2019.02.061.
41. Souza, V.G.L., Da Cruz Fernando, A.L.A., Pires, J.R.A., Rodrigues, P.F., Lopes, A.A.S., Fernandes, F.M.B., Physical properties of chitosan films incorporated with natural antioxidants. *Ind. Crops Prod.*, *107*, 565–572, 2017, https://doi.org/10.1016/j.indcrop.2017.04.056.
42. Sharma, P., Ahuja, A., Izrayeel, A.M.D., Samyn, P., Rastogi, V.K., Physicochemical and thermal characterization of poly (3-hydroxybutyrate-co-4-hydroxybutyrate) films incorporating thyme essential oil for active packaging of white bread. *Food Control*, *133*, 108688, 2022, https://doi.org/10.1016/j.foodcont.2021.108688.
43. Li, Y., Tang, C., He, Q., Effect of orange (Citrus sinensis L.) peel essential oil on characteristics of blend films based on chitosan and fish skin gelatin. *Food Biosci.*, *41*, 100927, 2021, https://doi.org/10.1016/j.fbio.2021.100927.
44. Tongnuanchan, P. and Benjakul, S., Essential oils: extraction, bioactivities, and their uses for food preservation. *J. Food Sci.*, *79*, 7, 1231–1249, 2014, https://doi.org/10.1111/1750-3841.12492.
45. Pisoschi, A.M., Pop, A., Georgescu, C., Turcuş, V., Oláh, N., Máthé, E., An overview of natural antimicrobials role in food. *Eur. J. Med. Chem.*, *143*, 922–935, 2018, https://doi.org/10.1016/j.ejmech.2017.11.095.
46. Davidson, P.M., Taylor, T.M., Schmidt, S.E., Chemical preservatives and natural antimicrobial compounds, in: *ASM Press eBooks*, pp. 765–801, 2014, https://doi.org/10.1128/9781555818463.ch30.
47. Aloui, H. and Khwaldia, K., Natural antimicrobial edible coatings for microbial safety and food quality enhancement. *Compr. Rev. Food Sci. Food Saf.*, *15*, 6, 1080–1103, 2016, https://doi.org/10.1111/1541-4337.12226.
48. Leśnierowski, G. and Yang, T., Lysozyme and its modified forms: A critical appraisal of selected properties and potential. *Trends Food Sci. Technol.*, *107*, 333–342, 2021, https://doi.org/10.1016/j.tifs.2020.11.004.

49. Soggiù, A., Piras, C., Mortera, S.L., Alloggio, I., Urbani, A., Bonizzi, L., Roncada, P., Unravelling the effect of clostridia spores and lysozyme on microbiota dynamics in Grana Padano cheese: A metaproteomics approach. *J. Proteomics*, *147*, 21–27, 2016, https://doi.org/10.1016/j.jprot.2016.03.035.
50. Saeed, F., Afzaal, M., Tufail, T., Ahmad, A., Use of natural Antimicrobial agents: A safe Preservation approach, in: *IntechOpen eBooks*, 2019, https://doi.org/10.5772/intechopen.80869.
51. Buys, E.M. and Seifu, E., Enzymes indigenous to milk: lactoperoxidase, in: *Elsevier eBooks*, pp. 670–676, 2022, https://doi.org/10.1016/b978-0-12-818766-1.00252-x.
52. Niaz, B., Saeed, F., Ahmed, A., Imran, M., Maan, A.A., Khan, M.K.I., Tufail, T., Anjum, F.M., Hussain, S., Suleria, H.A.R., Lactoferrin (LF): a natural antimicrobial protein. *Int. J. Food Prop.*, *22*, 1, 1626–1641, 2019, https://doi.org/10.1080/10942912.2019.1666137.
53. Gyawali, R. and Ibrahim, S.A., Natural products as antimicrobial agents. *Food Control*, *46*, 412–429, 2014, https://doi.org/10.1016/j.foodcont.2014.05.047.
54. Quinto, E.J., Caro, I., Villalobos-Delgado, L.H., Mateo, J., De Mateo Silleras, B., Del Río, M.P.R., Food Safety through Natural Antimicrobials. *Antibiotics*, *8*, 4, 208, 2019, https://doi.org/10.3390/antibiotics8040208.
55. Agriopoulou, S., Stamatelopoulou, E., Sachadyn-Król, M., Varzakas, T., Lactic acid bacteria as antibacterial agents to extend the shelf life of fresh and minimally processed fruits and vegetables: Quality and safety aspects. *Microorganisms*, *8*, 6, 952, 2020, https://doi.org/10.3390/microorganisms8060952.
56. Ramos, B., Brandão, T.R.S., Teixeira, P., Silva, C.L.M., Biopreservation approaches to reduce Listeria monocytogenes in fresh vegetables. *Food Microbiol.*, *85*, 103282, 2020, https://doi.org/10.1016/j.fm.2019.103282.
57. Cheong, E.Y., Sandhu, A., Jayabalan, J., Le, T.T.K., Nhiep, N.T.H., Ho, H.T.M., Zwielehner, J., Bansal, N., Turner, M.S., Isolation of lactic acid bacteria with antifungal activity against the common cheese spoilage mould Penicillium commune and their potential as biopreservatives in cheese. *Food Control*, *46*, 91–97, 2014, https://doi.org/10.1016/j.foodcont.2014.05.011.
58. Tiwari, B.K., Valdramidis, V., O'Donnell, C.P., Muthukumarappan, K., Bourke, P., Cullen, P.J., Application of natural antimicrobials for food preservation. *J. Agric. Food. Chem.*, *57*, 14, 5987–6000, 2009, https://doi.org/10.1021/jf900668n.
59. Bauer, R. and Dicks, L.M.T., Mode of action of lipid II-targeting lantibiotics. *Int. J. Food Microbiol.*, *101*, 2, 201–216, 2005, https://doi.org/10.1016/j.ijfoodmicro.2004.11.007.
60. Ibarra-Sánchez, L.A., El-Haddad, N., Mahmoud, D., Miller, M.J., Karam, L., Invited review: Advances in nisin use for preservation of dairy products. *J. Dairy Sci.*, *103*, 3, 2041–2052, 2020, https://doi.org/10.3168/jds.2019-17498.
61. Haakensen, M., Dobson, C., Hill, J.E., Ziola, B., Reclassification of Pediococcus dextrinicus (Coster and White 1964) Back 1978 (Approved Lists 1980) as

Lactobacillus dextrinicus comb. nov., and emended description of the genus Lactobacillus. *Int. J. Syst. Evol. Microbiol.*, *59*, 3, 615–621, 2009, https://doi.org/10.1099/ijs.0.65779-0.
62. Gabrielsen, C., Brede, D.A., Nes, I.F., Diep, D.B., Circular Bacteriocins: biosynthesis and mode of action. *Appl. Environ. Microbiol.*, *80*, 22, 6854–6862, 2014, https://doi.org/10.1128/aem.02284-14.
63. Verma, S.K., Sood, S.K., Saini, R.K., Saini, N., Pediocin PA-1 containing fermented cheese whey reduces total viable count of raw buffalo (Bubalis bubalus) milk. *LWT*, *83*, 193–200, 2017, https://doi.org/10.1016/j.lwt.2017.02.031.
64. Castellano, P., Ibarreche, M.P., Massani, M.B., Fontana, C.A., Vignolo, G., Strategies for pathogen biocontrol using lactic acid bacteria and their metabolites: A focus on meat ecosystems and industrial environments. *Microorganisms*, *5*, 3, 38, 2017, https://doi.org/10.3390/microorganisms5030038.
65. Niamah, A.K., Structure, mode of action and application of pediocin natural antimicrobial food preservative: A review. *Basrah J. Agric. Sci.*, *31*, 1, 59–69, 2018, https://doi.org/10.33762/bagrs.2018.160126.
66. Massani, M.B., Molina, V., Sanchez, M., Renaud, V., Eisenberg, P., Vignolo, G., Active polymers containing Lactobacillus curvatus CRL705 bacteriocins: Effectiveness assessment in Wieners. *Int. J. Food Microbiol.*, *178*, 7–12, 2014, https://doi.org/10.1016/j.ijfoodmicro.2014.02.013.
67. Sun, H., Shao, X., Zhang, M., Wang, Z., Dong, J., Dong, Y., Mechanical, barrier and antimicrobial properties of corn distarch phosphate/nanocrystalline cellulose films incorporated with Nisin and ε-polylysine. *Int. J. Biol. Macromol.*, *136*, 839–846, 2019, https://doi.org/10.1016/j.ijbiomac.2019.06.134.
68. Del Nobile, M.A., Conte, A., Buonocore, G.G., Incoronato, A.L., Massaro, A., Panza, O., Active packaging by extrusion processing of recyclable and biodegradable polymers. *J. Food Eng.*, *93*, 1, 1–6, 2009, https://doi.org/10.1016/j.jfoodeng.2008.12.022.
69. Wittaya, T., Protein-Based Edible Films: Characteristics and improvement of properties, in: *InTech eBooks*, 2012, https://doi.org/10.5772/48167.
70. Dutta, P.K., Tripathi, S., Mehrotra, G.K., Dutta, J., Perspectives for chitosan based antimicrobial films in food applications. *Food Chem.*, *114*, 4, 1173–1182, 2009, https://doi.org/10.1016/j.foodchem.2008.11.047.
71. Han, J.H., Innovations in food packaging, in: *Elsevier eBooks*, 2014b, https://doi.org/10.1016/c2011-0-06876-x.
72. Matyjaszewski, K. and Möller, M., Polymer Science: A Comprehensive Reference, in: *Elsevier eBooks*, 2012, https://doi.org/10.1016/c2009-1-28406-1.
73. Mahieu, A., Terrié, C., Youssef, B., Thermoplastic starch films and thermoplastic starch/polycaprolactone blends with oxygen-scavenging properties: Influence of water content. *Ind. Crops Prod.*, *72*, 192–199, 2015, https://doi.org/10.1016/j.indcrop.2014.11.037.

74. Terzopoulou, Z., Papageorgiou, G.Z., Papadopoulou, E., Athanassiadou, E., Alexopoulou, E., Bikiaris, D.N., Green composites prepared from aliphatic polyesters and bast fibers. *Ind. Crops Prod.*, *68*, 60–79, 2015, https://doi.org/10.1016/j.indcrop.2014.08.034.
75. Armentano, I., Bitinis, N., Fortunati, E., Mattioli, S., Rescignano, N., Verdejo, R., López-Manchado, M.A., Kenny, J.M., Multifunctional nanostructured PLA materials for packaging and tissue engineering. *Prog. Polym. Sci.*, *38*, 10–11, 1720–1747, 2013, https://doi.org/10.1016/j.progpolymsci.2013.05.010.
76. Vodnar, D.C., Păucean, A., Dulf, F.V., Socaciu, C., HPLC Characterization of Lactic Acid Formation and FTIR Fingerprint of Probiotic Bacteria during Fermentation Processes. *DOAJ (DOAJ: Directory of Open Access Journals)*, 38, 2, 109–113, 2010, https://doi.org/10.15835/nbha3824752.
77. Djidi, D., Mignard, N., Taha, M., Thermosensitive polylactic-acid-based networks. *Ind. Crops Prod.*, *72*, 220–230, 2015, https://doi.org/10.1016/j.indcrop.2014.09.035.
78. Ingrao, C., Tricase, C., Cholewa-Wójcik, A., Kawecka, A., Rana, R.L., Siracusa, V., Polylactic acid trays for fresh-food packaging: A Carbon Footprint assessment. *Sci. Total Environ.*, *537*, 385–398, 2015, https://doi.org/10.1016/j.scitotenv.2015.08.023.
79. Jašo, V., Glenn, G.M., Klamczynski, A., Petrovìć, Z.S., Biodegradability study of polylactic acid/ thermoplastic polyurethane blends. *Polym. Test.*, *47*, 1–3, 2015, https://doi.org/10.1016/j.polymertesting.2015.07.011.
80. Rhim, J., Preparation and characterization of vacuum sputter silver coated PLA film. *LWT*, *54*, 2, 477–484, 2013, https://doi.org/10.1016/j.lwt.2013.06.005.
81. Dutta, P.K., Tripathi, S., Mehrotra, G.K., Dutta, J., Perspectives for chitosan based antimicrobial films in food applications. *Food Chem.*, *114*, 4, 1173–1182, 2009b, https://doi.org/10.1016/j.foodchem.2008.11.047.
82. Marcos, B., Sárraga, C., Castellari, M., Kappen, F.H., Schennink, G., Arnau, J., Development of biodegradable films with antioxidant properties based on polyesters containing α-tocopherol and olive leaf extract for food packaging applications. *Food Packag. Shelf Life*, *1*, 2, 140–150, 2014, https://doi.org/10.1016/j.fpsl.2014.04.002.
83. Jayakumar, R., Nwe, N., Tokura, S., Tamura, H., Sulfated chitin and chitosan as novel biomaterials. *Int. J. Biol. Macromol.*, *40*, 3, 175–181, 2007, https://doi.org/10.1016/j.ijbiomac.2006.06.021.
84. Möller, H., Grelier, S., Pardon, P., Coma, V., Antimicrobial and physicochemical properties of Chitosan–HPMC-Based films. *J. Agric. Food. Chem.*, *52*, 21, 6585–6591, 2004, https://doi.org/10.1021/jf0306690.
85. Rabea, E.I., Badawy, M.E.I., Stevens, C.V., Smagghe, G., Steurbaut, W., Chitosan as antimicrobial agent: applications and mode of action. *Biomacromolecules*, *4*, 6, 1457–1465, 2003, https://doi.org/10.1021/bm034130m.

86. Shahidi, F., Arachchi, J.K.V., Jeon, Y., Food applications of chitin and chitosans. *Trends Food Sci. Technol.*, *10*, 2, 37–51, 1999, https://doi.org/10.1016/s0924-2244(99)00017-5.
87. Cuero, R.G., Osuji, G.O., Washington, A., N-carboxymethylchitosan inhibition of aflatoxin production: Role of zinc. *Biotechnol. Lett*, *13*, 6, 441–444, 1991, https://doi.org/10.1007/bf01030998.
88. Lian-Ying, Z. and Zhu, J., Study on antimicrobial activity of chitosan with different molecular weights. *Carbohydr. Polym.*, *54*, 4, 527–530, 2003, https://doi.org/10.1016/j.carbpol.2003.07.009.
89. De Azeredo, H.M.C., Antimicrobial nanostructures in food packaging. *Trends Food Sci. Technol.*, *30*, 1, 56–69, 2013, https://doi.org/10.1016/j.tifs.2012.11.006.
90. Madureira, A.R., Pereira, A.B., Castro, P.M., Pintado, M., Production of antimicrobial chitosan nanoparticles against food pathogens. *J. Food Eng.*, *167*, 210–216, 2015, https://doi.org/10.1016/j.jfoodeng.2015.06.010.
91. Ramani, M., Ponnusamy, S., Muthamizhchelvan, C., Marsili, E., Amino acid-mediated synthesis of zinc oxide nanostructures and evaluation of their facet-dependent antimicrobial activity. *Colloids Surf., B*, *117*, 233–239, 2014, https://doi.org/10.1016/j.colsurfb.2014.02.017.
92. Song, J. and Jang, J., Antimicrobial polymer nanostructures: Synthetic route, mechanism of action and perspective. *Adv. Colloid Interface Sci.*, *203*, 37–50, 2014, https://doi.org/10.1016/j.cis.2013.11.007.
93. Keawchaoon, L. and Yoksan, R., Preparation, characterization and in vitro release study of carvacrol-loaded chitosan nanoparticles. *Colloids Surf., B*, *84*, 1, 163–171, 2011, https://doi.org/10.1016/j.colsurfb.2010.12.031.
94. Bayarri, M., Oulahal, N., Degraeve, P., Gharsallaoui, A., Properties of lysozyme/low methoxyl (LM) pectin complexes for antimicrobial edible food packaging. *J. Food Eng.*, *131*, 18–25, 2014, https://doi.org/10.1016/j.jfoodeng.2014.01.013.
95. Benito-Peña, E., González-Vallejo, V., Rico-Yuste, A., Barbosa-Pereira, L., Cruz, J., Bilbao, A., Alvarez-Lorenzo, C., Moreno-Bondi, M.C., Molecularly imprinted hydrogels as functional active packaging materials. *Food Chem.*, *190*, 487–494, 2016, https://doi.org/10.1016/j.foodchem.2015.05.128.
96. De'nobili, M.D., Rojas, A.M., Abrami, M., Lapasin, R., Grassi, M., Structure characterization by means of rheological and NMR experiments as a first necessary approach to study the l-(+)-ascorbic acid diffusion from pectin and pectin/alginate films to agar hydrogels that mimic food materials. *J. Food Eng.*, *165*, 82–92, 2015, https://doi.org/10.1016/j.jfoodeng.2015.05.014.
97. Tripathi, S., Mehrotra, G.K., Dutta, P.K., Preparation and physicochemical evaluation of chitosan/poly(vinyl alcohol)/pectin ternary film for food-packaging applications. *Carbohydr. Polym.*, *79*, 3, 711–716, 2010, https://doi.org/10.1016/j.carbpol.2009.09.029.
98. Videcoq, P., Garnier, C., Robert, P., Bonnin, E., Influence of calcium on pectin methylesterase behaviour in the presence of medium methylated

pectins. *Carbohydr. Polym.*, *86*, 4, 1657–1664, 2011, https://doi.org/10.1016/j.carbpol.2011.06.081.

99. Laurent, M. and Boulenguer, P., Stabilization mechanism of acid dairy drinks (ADD) induced by pectin. *Food Hydrocoll.*, *17*, 4, 445–454, 2003, https://doi.org/10.1016/s0268-005x(03)00028-6.
100. Cho, S.Y. and Rhee, C., Mechanical properties and water vapor permeability of edible films made from fractionated soy proteins with ultrafiltration. *LWT*, *37*, 8, 833–839, 2004, https://doi.org/10.1016/j.lwt.2004.03.009.
101. Khan, M.K.I., Schutyser, M., Schroën, K., Boom, R., The potential of electrospraying for hydrophobic film coating on foods. *J. Food Eng.*, *108*, 3, 410–416, 2012, https://doi.org/10.1016/j.jfoodeng.2011.09.005.
102. Khan, M.K., II, Schutyser, M., Schroën, K., Boom, R., The potential of electrospraying for hydrophobic film coating on foods. *J. Food Eng.*, *108*, 3, 410–416, 2012b, https://doi.org/10.1016/j.jfoodeng.2011.09.005.
103. Hambleton, A., Voilley, A., Debeaufort, F., Transport parameters for aroma compounds through i-carrageenan and sodium alginate-based edible films. *Food Hydrocolloids*, *25*, 5, 1128–1133, 2011, https://doi.org/10.1016/j.foodhyd.2010.10.010.
104. Galus, S. and Kadzińska, J., Food applications of emulsion-based edible films and coatings. *Trends Food Sci. Technol.*, *45*, 2, 273–283, 2015, https://doi.org/10.1016/j.tifs.2015.07.011.
105. Pop, O.L., Vodnar, D.C., Suharoschi, R., Mudura, E., Socaciu, C., L. plantarum ATCC 8014 Entrapment with Prebiotics and Lucerne Green Juice and Their Behavior in Simulated Gastrointestinal Conditions. *J. Food Process Eng*, *39*, 5, 433–441, 2015, https://doi.org/10.1111/jfpe.12234.
106. Banerjee, R. and Chen, H., Functional properties of edible films using whey protein concentrate. *J. Dairy Sci.*, *78*, 8, 1673–1683, 1995, https://doi.org/10.3168/jds.s0022-0302(95)76792-3.
107. Özdemir, M. and Floros, J.D., Optimization of edible whey protein films containing preservatives for water vapor permeability, water solubility and sensory characteristics. *J. Food Eng.*, *86*, 2, 215–224, 2008, https://doi.org/10.1016/j.jfoodeng.2007.09.028.
108. McHugh, T.H., Aujard, J., Krochta, J.M., Plasticized whey Protein Edible films: Water Vapor permeability Properties. *J. Food Sci.*, *59*, 2, 416–419, 1994, https://doi.org/10.1111/j.1365-2621.1994.tb06980.x.
109. Valdés, A., Ramos, M., Beltrán, A.M., Jiménez, A., Garrigós, M.C., State of the art of antimicrobial edible coatings for food packaging applications. *Coatings*, *7*, 4, 56, 2017, https://doi.org/10.3390/coatings7040056.
110. Sánchez-González, L., Pastor, C., Vargas, M., Chiralt, A., Cháfer, M., Effect of hydroxypropylmethylcellulose and chitosan coatings with and without bergamot essential oil on quality and safety of cold-stored grapes. *Postharvest Biol. Technol.*, *60*, 1, 57–63, 2011, https://doi.org/10.1016/j.postharvbio.2010.11.004.

131. Lin, L., Zhu, Y., Cui, H., Electrospun thyme essential oil/gelatin nanofibers for active packaging against Campylobacter jejuni in chicken. *LWT*, *97*, 711–718, 2018, https://doi.org/10.1016/j.lwt.2018.08.015.
132. Siripatrawan, U. and Noipha, S., Active film from chitosan incorporating green tea extract for shelf life extension of pork sausages. *Food Hydrocoll.*, *27*, 1, 102–108, 2012, https://doi.org/10.1016/j.foodhyd.2011.08.011.

16

Prebiotics and Probiotics Food: Future Aspects

Sankarganesh, P.[1], S. A. O. Adeyeye[1], Ashok Kumar Chakka[2,3] and A. Surendra Babu[3*]

[1]Department of Food Technology, Hindustan Institute of Technology and Science, (Deemed to be University), Rajiv Gandhi Salai, Padur, Kelambakam, Chennai, Tamil Nadu, India
[2]Department of Post Graduate Studies & Research in Food Science, St Aloysius College (Autonomous), Mangaluru, Karnataka, India
[3]Department of Food Science and Technology, School of Agricultural Sciences, Malla Reddy University, Hyderabad, Telangana, India

Abstract

Currently, the demand for functional and healthful foods based on prebiotics and probiotics has increased tremendously. Consequently, research on prebiotics and probiotics is transformed to a new frontier owing to the recent and ongoing consumer demand and advances in the science of the microbiome. Prebiotics are defined as non-digestive polysaccharide and oligosaccharides that enhances the growth of beneficial *lactic acid bacteria* and *bifidobacteria* in the gastro-intestinal tract. Prebiotics often include dietary fibres and oligosaccharide which offer a wide range of health benefits has led to the emergence of multibillion food and pharmaceutical companies for the effective utilization in designing prebiotic foods. Established prebiotics includes Fructooligosaccharides (FOS), Inulin, galactooligosaccharides (GOS), polydextrose and lactulose while polyphenols, xylo-oligosachharide, lactitol are the considered as an emerging prebiotic. Benefits of prebiotics include prevention of colitis, protection against cardiovascular disease, improving gut health, prevention of obesity, maintaining blood glucose levels, decreasing constipation etc., Also, they are reported to be used as an alternative to antibiotics used as growth enhancers in fisheries, poultry, apiary and animal husbandry. Prebiotics must be chemically stable to use in various food products, hence many advanced techniques for producing prebiotic components without

**Corresponding author*: surendrababu.postdoc@gmail.com

Mousumi Sen (ed.) Food Coatings and Preservation Technologies, (551–584) © 2025 Scrivener Publishing LLC

changing their properties are being investigated. Future emerging technologies such as microwave irradiation, high hydrostatic pressure, ultrasound, drying, radio frequency, and extrusion are being widely explored to prepare possibility of producing prebiotic foods while preserving the nutritional value, sensory attributes, and safety of food. Certain recent studies were also focused on developing nanoprebiotics using whey protein isolate/inulin nano complexes with potential prebiotic effects which is a promising method and requires further exploration.

On the other hand, the term probiotics are defined as "live microorganisms which upon ingestion in certain numbers exert health benefits beyond inherent general nutrition". The usage of probiotics in food systems have tremendously increased in the recent times. Probiotics have a wide range of health advantages for people consuming them. They are used over centuries for treating diarrhea, constipation, colon cancer, inflammatory bowel syndrome, coronary heart disease, gastrointestinal diseases and many more. Probiotic applications provide numerous difficulties. Selection of the strain, processing, and inoculation of starter cultures must all be considered in addition to viability and sensory acceptance. The probiotics sector encounters difficulties while promoting the health advantages. Novel technologies like nanotechnology finds applications in probiotics as studies have confirmed that nanoparticles of probiotics can be used to prevent and treat neuroinflammation. In addition, probiotic strains coated with engineered nanoparticles was found to have potential role in cancer immunotherapy. Particularly, a nanoparticles developed from a therapeutic strain of *Salmonella typhimurium* would holistically detect and supress prostate cancer cells when loaded in nontoxic and novel bacterium. Further, Novel probiotic strains, their mode of action, and emerging technologies for effective delivery of probiotics and their application are currently under study as it has immense potential to change the scientific understanding of their medical interventions. This chapter intends to discuss the futuristic aspects of prebiotics and probiotics and the emerging technologies to produce prebiotics as well as delivery of probiotics into the human gut.

Keywords: Prebiotics, probiotics, fructooligosaccharides, short-chain fatty acids, anti-cancer potential, COVID-19, nanotechnology

16.1 Introduction

Prebiotics

The human gut health has been an area of increasing research interest in recent years. The gastrointestinal functions are associated with a balanced microflora in combination with an optimal gut-associated lymphoid tissue (GALT) which determines the state of well-being and human health and to larger extent reduction of the risk of diseases. The human colonic microflora is a complex ecosystem which functions as a combined action of the microbes which interact with the GALT, improving nutrient energy and results in metabolic end products such as the short-chain fatty acids (SCFAs) that play an important role in cell differentiation, cell proliferation, and metabolic regulatory processes. The common potentially health-promoting bacteria are principally bifidobacteria, lactobacilli, eubacteria, and bacteroides, which are, key biota and the most important genera in humans. The bacteria are classified as prebiotics.

The concept of a 'prebiotic' was put forward in 1995 by Gibson and Roberfroid [1] in their scientific publication called "Dietary modulation of the human colonic microbiota: Introducing the concept of prebiotics" (ISAPP). The International Scientific Association for Probiotics and Prebiotics (ISAPP) in 2016 defined prebiotics as "a substrate that is selectively utilized by host microorganisms conferring a health benefit". Therefore, the concept includes three essential parts: a substance, a physiologically beneficial effect, and a microbiota-mediated mechanism [2].

Improving human health through modulation of the microbiome has taken centre stage as part of a comprehensive, holistic approach to lifestyle wellness [1, 3]. The diverse microbial ecosystems present in mucosal and cutaneous surfaces are explored as a means of improving health or to treat disease [1]. Researches have shown that it is possible to shift the composition and metabolic signatures of these microbial populations through dietary or non-dietary interventions [1, 4–6].

In the last three decades, a class of compounds, termed prebiotics, has been discovered to have capability to manipulate host microbiota to the benefit of the hosts [1, 9]. Initially, fructans (fructooligosaccharides (FOS) and inulin) and galactans (galactooligosaccharides or GOS) are believed to have effects by acting through enrichment of *Lactobacillus* and/or *Bifidobacterium* spp. However, in recent time, FOS and GOS are prominent prebiotic category based on numerous studies on their prebiotic effects [1]. Recently, due to improvement in analytical techniques in microbiome research and improved knowledge of the composition of the microbiota, more substances that affect colonization have been identified. Also, beneficial microorganisms that are affected by prebiotics and their effectiveness at extra-intestinal sites have been elucidated [1, 7]

There are several endogenous factors, like mucin secretions, that affect the microbial balance; however, human diet is the chief source of energy for their growth. Non-digestible carbohydrates are found to modify the composition and function of gut microbiota [8]. Beneficial intestinal microbes are known to ferment these non-digestible dietary substances and obtain their survival energy from degrading indigestible binds of prebiotics [8–10]. Because of this, prebiotics can selectively influence gut microbiota

[8, 11, 12]. It has also been found that the gut microbiota affects intestinal functions, like metabolism and integrity of the intestine. This helps them to suppress pathogens in healthy individuals through induction of some immunomodulatory molecules with antagonistic effects against pathogens by lactic acid that is produced by *Bifidobacterium* and *Lactobacillus* genera [8, 13–16].

Fermentation of prebiotics by gut microbiota produces short-chain fatty acids (SCFAs), such as lactic acid, butyric acid, and propionic acid. These products have multiple effects on the body and could prevent diseases and improve gut health [8, 17, 18]. SCFAs have been found to decrease the pH of colon [8, 19, 20]. Fermentation of prebiotics also results in the production of peptidoglycan, and it stimulates the innate immune system against pathogenic microorganisms [8, 17, 21]. The fermentation products are influenced by the structure of prebiotics and the bacterial composition of gut [8, 22]. Prebiotics affect human health through their degradation products by microorganisms which may play key role in human physiology, for instance, butyrate influences intestinal epithelial development [8, 23]. Because SCFAs have the ability to diffuse to blood circulation through enterocytes, they can not only the gastrointestinal tract but also other organs of the body [8, 24]

Prebiotics are several benefits and research has shown that prebiotics may improve digestion and metabolism; help regulate bowel movements; improve calcium absorption and increase bone density; regulate blood sugar and insulin resistance; stimulate the production of hormones that aid in appetite suppression; lower inflammation in the body; strengthen the immune system; help balance and maintain hormone levels; reduce risk of allergy; lower the risk of cardiovascular disease; regulate moods and reduce stress hormone levels; lower cholesterol level and reduce risk of colon cancer [8, 16, 25].

Prebiotics are present naturally in some food substances such vegetables, fruits, and legumes. Foods that are good sources of prebiotics are legumes, beans, peas, oats, bananas, berries jerusalem artichokes, asparagus, dandelion greens, garlic, leeks, onions, chicory root, barley, apples, konjac root, cocoa, burdock root, flaxseeds, jicama, wheat bran and seaweed [8].

Probiotics

Probiotics are foods or supplements that contain live microorganisms intended to maintain or improve the "good" bacteria (normal microflora) in the body [26, 27]. Probiotics are considered to be generally safe to consume, however, it could result into bacteria-host interactions and unwanted side effects in rare cases [26, 28]. Probiotic was first discovered in a certain strain of bacillus in Bulgarian yoghurt, known as Lactobacillus bulgaricus. It was discovered by Bulgarian physician and microbiologist Stamen Grigorov in 1905. However, the Russian Nobel laureate Élie Metchnikoff popularized it when he postulated around 1907 that yoghurt-consuming Bulgarian peasants lived longer [29].

The World Health Organization (WHO) defines probiotics as "live microorganisms which when administered in adequate amounts confer a health benefit on the host [30, 31]. Probiotics provide health benefits to the host when ingested in adequate amounts [32]. The most commonly used strains as probiotics are lactic acid bacteria and bifidobacteria. Probiotics have proved to possess a potential therapeutic role for some of diseases, but the mechanisms of their action are not fully elucidated [32]. However, some mechanisms involve in the antagonistic effects of probiotics on various microorganisms may include: modification of the gut microbiota, competitive adherence to the mucosa and epithelium, improving the gut epithelial barrier and through modulation of the host's immune system [32].

Research have shown that several internal and external sites of the healthy human body are colonized by a diversity of symbiotic microbes [33]. The human intestine contains most dense and diverse microbial assemblages [33]. However, the gut microbiome and their interactions with the human body, has a tremendous influence on physiological aspects of mammalian function particularly immune response of the host. Poor or absence of early-life exposure to probiotics, or changes in the composition of the gut microbiome are linked to the development and progression of several intestinal and extra-intestinal diseases and inflammatory bowel disease [33]. Modulating microbial exposure through consumption of lactic acid fermented foods represents a veritable way to prevent or reduce disease through intestinal microbial community modification [33].

Probiotics are very important to human gut health. The beneficial effects of probiotics are improving intestinal epithelial cell function, protecting against physiologic stress, modulating cytokine secretion profiles, influencing T lymphocyte populations, and enhancing antibody secretion [34]. Certain fermented foods naturally contain healthy or good bacteria that support a healthy microbiome. This occurs when lactic acid bacteria convert sugar or other carbohydrates to lactic acid. Foods that contain probiotics are yogurt, kefir, fermented cheeses, like buttermilk cheese and cottage cheese, kimchi, raw sauerkraut, tempeh, miso, kombucha, natto and lacto-fermented vegetables, like pickle.

16.2 Established Health Benefits of Prebiotics

Prebiotics have demonstrated health benefits by their selective metabolism in the gastrointestinal tract (Figure16.1). International scientific association for probiotics and prebiotics have proposed three criteria for a prebiotic effect: a) prebiotics should be resistant to digestive enzymes, stomach acids and hydrolysis b) must get fermented by probiotic microorganisms and c) selectively stimulate the growth and/or activity of intestinal microbes [1].

16.2.1 Gut Health and Maintenance

Gastroenteritis is a short-term illness caused due to consumption of food or water contaminated with pathogenic microorganisms such as *Salmonella, Shigellae, Versina enterocolitica, Escherichia coli, Vibro cholera, Campylobacter jejuni,* and *Clostridus perfringes.* These pathogens multiply within the gastrointestinal tract invading the host tissue and release toxins that interrupt the functions of intestinal mucosa, causing vomiting, diarrhea, and nausea [35]. Prebiotics contribute to the growth of beneficial bacteria in the large intestine, which may assist to prevent gastroenteritis. A study conducted with 300 adults with acute gastroenteritis demonstrated the potential role of polyphenol-based prebiotics in the prevention

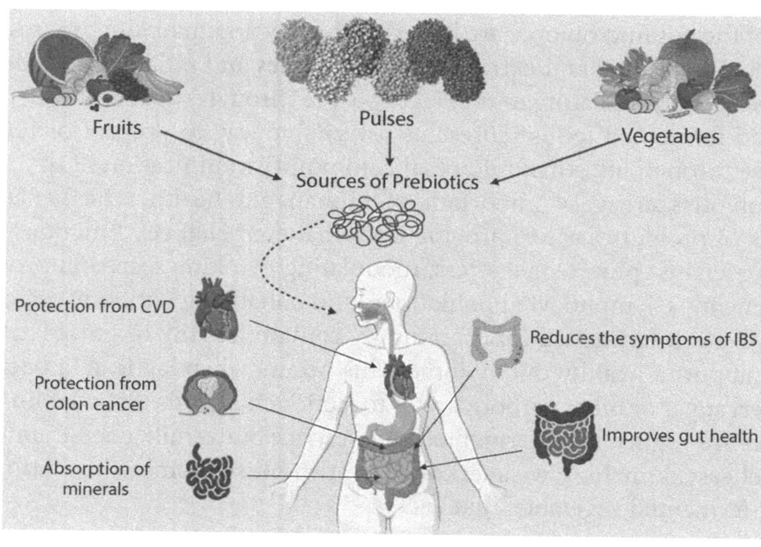

Figure 16.1 Sources of prebiotics and their potential health benefits.

of gastroenteritis [36]. Prebiotics have demonstrated a beneficial effect on gut associated lymphoid tissues (GALT). Prebiotics such as fructo-oligosaccharides, arabinogalactans, mannooligosaccharides, and inulin are reported to enhance the proliferation of probiotic microflora and thereby suspending the growth of pathogenic microorganism. Prebiotic intake can alter immunological parameters in peripheral circulation, secondary lymphoid tissues, and GALT [37, 38].

16.2.2 Inflammatory Bowel Disease and Inflammatory Bowel Syndrome

There are limited studies confirming the positive effect of prebiotics on Inflammatory bowel disease IBD and inflammatory bowel syndrome (IBS). IBD is a condition related to the disturbance of intestinal microflora and its metabolites while IBS is associated with chronic abdominal pain, discomfort, and altered bowel habits. These disorders are characterised by the decrease in the count of *Bifidobacteria, faecalibacterium prausnitzii,* and *bacteroides to firmicutes* ratio [39]. Fructo oligosaccharides and galactooligosaccharides have been reported to possess therapeutic role in treatment of IBS. A study conducted with 79 IBS patients as supplemented with short chain fructooligosaccharides showed a reduced symptoms of IBS after four weeks [40]. A recent study reported that the supplementation of 10g of 2'-fucosyllactose/lacto-N-neotetraose for four weeks have shown to increase the population of *Bifidobactrium* sps and reduced the IBS-symptom severity score, improved bowel habits, anxiety and depression [41].

16.2.3 Anticancer Potential

The by-products of prebiotics fermentation in the gut may inhibit the growth of cancer cells in the colon [42]. It is observed that prebiotics can also decrease the genotoxic enzymes. A diet rich in arabinoxylan oligosaccharides (AXOS) has demonstrated to decrease the formation of preneoplastic lesions in the colon of rats when administered with a carcinogenic drug 1, 2-dimethylhydrazine [43]. Messaoudene *et al.* [44] in a recent study noticed that castalagin, which is a polyphenol found in camu-camu (CC) berries acts as a PD-1 inhibitor in subcutaneous mouse models of sarcoma and breast cancer. Mouse that was administered with CC extract along with anti-PD-1 antibodies presented a considerable reduction in tumour volumes compared to control. The study stated that the anti-carcinogenic effect was attributed to the polyphenol castalagin. On the other hand, short chain fatty acids (SCFAs) with carbons less than six including acetate, propionate, butyrate, and valeric acid are observed to present anticancer property. SCFAs reduces the pH of the intestinal lumen which decrease the viability of enteropathogenic species while enhancing the multiplication of probiotics such as *Lactobacilli* and *Bifidobacteria*. This effect of SCFAs on the colonization of pathogenic microorganisms of the lumen could decrease the incidence of acute and chronic inflammation that could lead to DNA damage and carcinogenesis [45].

16.2.4 Cardioprotective Role

Cardiovascular disease and type 2 diabetes are directly associated with diet and obesity. Several studies have demonstrated the cardioprotective role of prebiotics in animal experimentation. Jiang *et al.* [46] studied the effect of chitosan oligosaccharides (COS) on coronary heart disease patients. COS supplementation has improved the symptoms of coronary heart disease patients by enhancing the antioxidant capacities via increased proliferation of probiotic microbes in gut such as *Lactobacillus, Lactococcus*, and *Phascolarctobacterium*. In another study, oligosaccharide from *Descurainiae Sophia* at varying concentrations (51.34 mg/kg, 102.67 mg/kg, and 205.34 mg/kg) significantly improved the cardiac function in Sprague-Dawley rats via renin-angiotensin-aldosterone system and arginine vasopressin system [47].

16.2.5 Mineral Absorption

Prebiotics can improve the absorption of minerals in gastrointestinal tract particularly calcium, and magnesium with the enhanced production of SCFA's in the colon, decreased lumen pH and increased dissolution. However, the precise mechanism of absorption was not fully understood [48]. Van den Heuvel *et al.* [49] investigated the effect of short chain fructo-oligosaccharide supplementation (10g) in adolescent girls aged between 12 to 14 with less calcium intake for 36 days. Results indicated an increased magnesium absorption by 18% after 36 days of supplementation while calcium absorption showed no significant impact. Another study has showed that the addition of galactooligosaccharides (GOS) to an iron fortified maize porridge meal in Kenyan infants did not increased the iron absorption significantly implying long-term supplementation to improve iron absorption [50]. However, GOS supplement in iron deficiency women

human cervical and colon adenocarcinoma cell lines (HeLa and HT-29, respectively). The finding of the study stated that *L. rhamnosus* supernatant and *L. crispatus* supernatant were effective in the treatment of cancer cell on HELA cell lines. Xu *et al.* [63] demonstrated that the oral administration of probiotic microorganism such as *L. rhamnosus* and *B. longum*, nanovaccine and their combinations all suppressed the tumor growth and extended the survival period of mice. Combined drug with probiotics formulation presented a synergistic effect in enhancing the survival of mice compared to nano-vaccines with naked probiotics and remaining single treatments.

16.3.5 Cholesterol-Lowering Effect

According to WHO, cardiovascular diseases (CVD) will be the leading causes of death worldwide with annual mortality of CVD expected to reach 23.6 million by 2030 [67]. People with hypercholesterolemia condition have a three-time increased risk of heart attacks than people with normal blood lipid profile. Unhealthy diets such as foods rich in fat, salt, and free sugar and refined carbohydrates, limited intake of fruits and vegetables are the major causes for the increased risk of CVD [68]. According to the findings of Bernini *et al.* (2016), consumption of probiotic milk beverage prepared using *L. plantarum* reduced low-density lipoprotein (LDL), total cholesterol, glucose, IL-6. The possible mechanism is by the production of lipid hydrolase that could deconjugate bile acids thereby reducing the absorption of cholesterol and LDL while increasing the faecal excretion. He *et al.* [70] demonstrated the anti-hypocholesterolemic effect of probiotics by supplementing yogurt containg *B. lactis*, and *L. acidophilus* along for six weeks for a group of people with mild to moderate hypercholesterolemia. The results reported a considerable reduction in blood cholesterol in the group feed with probiotic yogurt. The investigation of Dieck *et al.* [71] showed that the anti-cholestrolemic effect of probiotics are ascribed to the bile salt hydrolysis activity, inhibition of hepatic de novo synthesis of lipids and suppressing satiety hormones.

16.4 Future Aspects of Prebiotics: Prebiotics—Beyond the Fiber Revolution

The concept of prebiotics, dietary fibers selectively nourishing beneficial gut bacteria, has altered our understanding of gut health. While traditional prebiotics like inulin and fructooligosaccharides have yielded significant health benefits, the future promises a paradigm shift in diversity, functionality, and targeted applications. This section explores the exciting frontiers of prebiotics, unveiling novel sources, innovative processing technologies, the application of nanotechnology in prebiotics, and the significance of prebiotics in managing COVID-19.

16.4.1 Novel Prebiotic Sources

Recent studies reported a variety of novel prebiotics exhibiting promising health benefits. This section explores three novel prebiotic ingredients that have emerged recently.

16.4.1.1 Fucoidan

Marine algae, long admired for their nutritional powerhouse status, are emerging as a potential source of prebiotics. Fucoidan, a complex sulfated polysaccharide sourced from brown seaweeds, emerges as a promising novel prebiotic ingredient with notable potential, as evidenced by recent studies. Wang *et al.* [72] provide valuable insights into the development of fucoidan-derived functional oligosaccharides, showcasing advancements in their preparation and diverse applications in the food industry and its potential to be a prebiotic candidate. Moreover, Zhu *et al.* [73] further support this perspective by revealing that fucoidan positively modulates the growth and antibacterial capabilities of *Lactobacillus rhamnosus.* Additionally, Poveda-Castillo *et al.* [74] contribute to this narrative by highlighting the bioactivity of fucoidan as an antimicrobial agent in a functional beverage. This dual functionality, enhancing microbial proliferation and probiotic activities, underscores the multifaceted nature of fucoidan, suggesting its candidacy as a prebiotic ingredient capable of promoting gut health.

16.4.1.2 Polyphenols

Polyphenols are a group of naturally occurring compounds in various foods. These compounds are abundant in popular dietary sources such as tea, coffee, red wine, fruits, vegetables, and more [75]. The intricate relationship between polyphenols and gut microbiota revolves around producing bioactive metabolites, influencing bacterial growth, metabolism, and cell function, hindering biofilm formation, and quorum sensing in certain bacterial species [76]. Notably, polyphenols act as prebiotic substrates, fostering the growth of beneficial probiotic bacteria like *Bifidobacteriaceae* and

Lactobacillaceae while suppressing pathogenic bacteria such as *Escherichia coli, Clostridium perfringens* and *Helicobacter pylori* [75]. Research indicates that polyphenols positively impact gut microbiota composition, leading to increased levels of beneficial bacteria like *Akkermansia muciniphila* and *F. prausnitzii*, observed in studies involving red grapes and tea consumption [77]. Additionally, Vegetables like carrots and cruciferous family members contribute to the prebiotic potential of polyphenols. Beyond fruits and vegetables, cereals, pulses, nuts, and beverages like red wine and tea contribute to the prebiotic effect of polyphenols, influencing the abundance of various bacterial genera in the gut [78]. Understanding polyphenols' diverse sources and impacts on gut microbiota can pave the way for novel dietary interventions promoting a healthy gut and overall well-being.

16.4.1.3 Short-Chain Fatty Acids

Short-chain fatty acids (SCFAs), the metabolic by-products of prebiotic fermentation by gut bacteria, are increasingly recognized as potent signalling molecules influencing various aspects of human health. Research suggests that SCFAs, mainly acetate, propionate, and butyrate, can modulate gut inflammation, enhance insulin sensitivity, and improve cognitive function [79, 80]. This phenomenon opens exciting avenues for exploring SCFAs as markers of prebiotic activity and as prebiotics themselves, potentially paving the way for targeted interventions for specific health concerns.

16.4.2 Novel Food Processing Technologies for Prebiotic Enrichment: Advancing Functionality and Sustainability

Established prebiotic sources like inulin and fructooligosaccharides remain relevant, and recent advancements in food processing technologies offer exciting avenues for enhancing prebiotic functionality and broader applications. This section explores three transformative innovations that eliminate the limitations of prebiotic enrichment.

16.4.2.1 Enzymatic Hydrolysis

Plant cell walls act as intricate fortresses, often shielding valuable prebiotic fibers like arabinoxylans within their complex structures. Enzymatic hydrolysis offers a key to unlock these hidden riches. Researchers can liberate trapped prebiotics by employing specific enzymes tailored to target plant cell wall polysaccharides, significantly increasing their bioavailability [81]. Enzymatic hydrolysis could convert underutilized agricultural waste, such as wheat bran or corn stover, into valuable prebiotic ingredients, leading to sustainable resource utilization.

16.4.2.2 Probiotic Fermentation

Fermentation, an age-old microbial art form, takes centre stage in the enrichment of prebiotic ingredients. Researchers [82, 83] reported that altering certain probiotic strains was employed to enrich food products with prebiotic metabolites like short-chain fatty acids (SCFAs) and prebiotic oligosaccharides, a synergistic effect of probiotics and prebiotics, resulting in enhanced functionality and health benefits of synbiotic products. Envisage yogurt naturally enriched with SCFAs through targeted probiotic fermentation, offering potent gut health support beyond the traditional benefits of live bacteria.

16.4.2.3 3D Printing Precision

3D printing technology is poised to revolutionize prebiotic delivery with unparalleled precision. By encapsulating specific prebiotics in edible inks and utilizing 3D printers, researchers can create personalized prebiotic blends and deliver them to specific gut niches with laser-sharp accuracy, helping to perform prebiotic interventions to individual gut microbiota profiles and addressing specific health concerns with unparalleled precision [84].

16.4.2.4 Application of Nanotechnology in Prebiotics

Nanotechnology, manipulating matter at its most fundamental level, reshapes prebiotic research [85]. Its unique properties offer transformative avenues for prebiotic synthesis, delivery, and functionality, showing the way for personalized gut health interventions. Conventional prebiotic production lacked precise control, limiting its effectiveness [86]. Through techniques like reverse micelles and microfluidics, nanotechnology enables the design of targeted nanoscale reactors. These mini factories fabricate customized prebiotics with defined molecular weight, composition, and functionality, catering to specific gut microbial populations and health goals [87]. Prebiotic degradation in the harsh digestive tract hinders their efficacy. Nanotechnology offers a solution to enhance bioavailability. Encapsulation within protective nanoparticle shells, built

from polymers or lipids, significantly enhances prebiotic bioavailability and stability [88]. These "stealthy shields" safeguard prebiotics as they navigate the digestive landscape, ensuring safe and effective delivery to target bacteria [89]. Nanotechnology also allows the development of prebiotic nanoparticles functionalized with specific ligands or targeting moieties. These nanoparticles can recognize and bind to receptors on specific gut bacteria, delivering their prebiotic payload directly to the desired location [90]. Nanotechnology enables the design of prebiotic nanoparticles with controlled release mechanisms. By incorporating stimuli-responsive polymers or enzymatic triggers, researchers can program the nanoparticles to release their prebiotic cargo in response to specific pH, temperature, or enzymatic cues within the gut [91].

The application of nanotechnology in prebiotics is still in its early stages, but the possibilities are immense. Future research could explore biocompatible and biodegradable nanoparticle materials, personalized prebiotic blends based on individual gut profiles, and integration with other gut health interventions for holistic microbiome management.

While exciting, nanotechnology presents challenges. Safety concerns regarding potential toxicity and environmental impact necessitate responsible development and stringent regulatory frameworks. Cost-effective production methods are crucial for broader accessibility.

As research progresses and challenges are overcome, nanotechnology can transform the landscape of prebiotics and empower us to harness the power of the gut microbiome for optimal health and well-being.

16.4.2.5 Prebiotics and COVID-19

Probiotics and prebiotics have emerged as potential therapeutic targets in the quest for effective nutritional management of COVID-19 [92]. This section examines the specific role of prebiotics and strategies for treating COVID-19, emphasizing the role of prebiotics, which are non-digestible fibers that selectively stimulate the growth of beneficial microorganisms in the gut and are working.

Mechanism of action: Prebiotics play an essential role in mobilizing the intestinal microbiota, thus influencing the overall immune system and the gut-lung axis [93]. This effective regulation by prebiotics can prevent or prevent COVID-19. The lungs and bronchi show a two-way communication between the gut and the respiratory system, suggesting that a balanced gut microbiota can positively impact respiratory health.

16.4.2.6 *Preventing Viral Entry and Replication*

Prebiotics, such as inulin and oligosaccharides, contribute to producing short-chain fatty acids (SCFAs) by gut bacteria during fermentation. SCFAs exhibit antiviral properties by inhibiting viral entry and replication [94]. This highlights the potential application of prebiotics in preventing and treating COVID-19, suggesting that the modulation of gut microbiota composition can enhance the production of antiviral metabolites.

16.4.2.7 *Immunomodulation*

Prebiotics also play a pivotal role in modulating the host immune response. By promoting the growth of beneficial bacteria, prebiotics enhance the production of immunomodulatory molecules, such as cytokines and immunoglobulin [92]. This immune modulation is crucial for mounting an effective defence against viral infections. Integration of prebiotics into nutritional strategies holds promise for the treatment of COVID-19. By influencing the gut microbiota and the gut-lung axis, prebiotics contribute to antiviral defences and immune modulation. Future research should continue to explore the specific prebiotic formulations and dosages that optimize these effects, paving the way for targeted and effective nutritional interventions in managing COVID-19 [93, 94].

16.5 Future Aspects of Probiotics

There is emerging research on the gut-brain axis, which suggests a connection between gut health and mental well-being. Probiotics may play a role in promoting mental health and alleviating symptoms of conditions like depression and anxiety. While research is ongoing, some studies suggest that probiotics may have a beneficial impact on inflammatory bowel diseases such as Crohn's disease and ulcerative colitis, possibly by modulating inflammation and promoting gut health.

Factors such as diet, antibiotics, stress, and age can influence this balance. Probiotics can help restore or maintain a healthy microbial balance. The popularity of probiotics has led to their inclusion in a wide range of food and beverage products beyond traditional fermented foods. These include probiotic-enriched yogurts, drinks, and even non-dairy products. The efficacy of probiotics is strain-specific, meaning that different strains may have different effects [95]. It's essential to choose probiotic products based on specific strains supported by scientific evidence. Quality control and the viability of live cultures are also crucial factors in selecting probiotic supplements [96] Probiotic supplements are generally considered safe for healthy individuals. However, individuals with compromised immune systems or underlying health conditions should consult a healthcare professional before using probiotics. Regulatory standards for probiotic products vary by country [97].

16.5.1 Next Generation Probiotics with Potential Health Benefits

The host's health is greatly influenced by specific microbial strains in the gut ecosystem. These microbes are now considered biotherapeutic treatments, dietary supplements, and nutraceutical aids; they are also known as next-generation probiotics [98]. Next-generation probiotics (NGPs) are used as therapeutic agents that influence the gut microbiota and the development of diseases [99]. As per the US FDA guidelines, NGPs are living organisms, such as bacteria, that are utilized for the purpose of preventing, treating, or curing diseases or conditions in human beings. It is important to note that NGPs are distinct from vaccines [100]

Next generation probiotics (NGP) live microorganisms identified on the basis of comparative microbiota analyses that, when administered in adequate amounts, confer a health benefit on the host [101]. Until recently, the majority of microorganisms residing in the gut have remained unknown due to their predominantly anaerobic nature, making them challenging to culture [102]. But now we can detect gut microbiota microorganisms that were previously unknown, due to advances in science. Specifically, the development of next-generation sequencing (NGS), bioinformatics, and polymerase chain reaction (PCR) of the 16 S rRNA gene has allowed for the precise identification of several bacterial species. The use of PCR and NGS on different platforms, such Illumina's, has thus allowed for the detection of gut microorganisms. [103]. This allows researchers to describe microorganisms linked to obesity and obesity-related illnesses. Potential next-generation probiotics (NGPs) like *Akkermansia muciniphila, Faecalibacterium prausnitzii, Anaerobutyricum hallii, Bacteroides uniformis, Bacteroides coprocola, Parabacteroides distasonis, Parabacteroides goldsteinii, Hafnia alvei, Odoribacter laneus*, and *Christensenella minuta* have demonstrated encouraging results in preclinical studies involving obesity and related conditions [104, 105].

NGPs, specifically *Faecalibacterium prausnitzii, Akkermansia muciniphila*, and *Bacteroides fragilis*, boost the immune system in the gastrointestinal tract, preserve the integrity of the intestinal barrier, generate advantageous substances, combat harmful microorganisms, enhance the effectiveness of immunotherapy, and decrease complications related to chemotherapy and radiotherapy [101]. A new NGP, *L. plantarum* (L. plantarum)-pMG36e-GLP-1, was developed by Hu *et al.* [99] to test it in transgenic and experimentally induced mice for its ability to alleviate type 2 diabetic mellitus (T2DM). The study's findings provide hope for the treatment of type 2 diabetes by confirming the modified strain L. plantarum-pMG36e-GLP-1's anti-hyperglycemic activity [99].

SARS-CoV-2 has been hypothesized to impact gut inflammation through both direct and indirect mechanisms. It infects intestinal epithelial cells by binding to the angiotensin-converting enzyme 2 (ACE2) receptor and transmembrane protease serine 2. This infection leads to the release of pro-inflammatory chemokines and cytokines [106]. Next-generation probiotics may be designed to modulate the immune system more effectively. This could include enhancing the host's immune response to viral infections, potentially helping the body mount a more robust

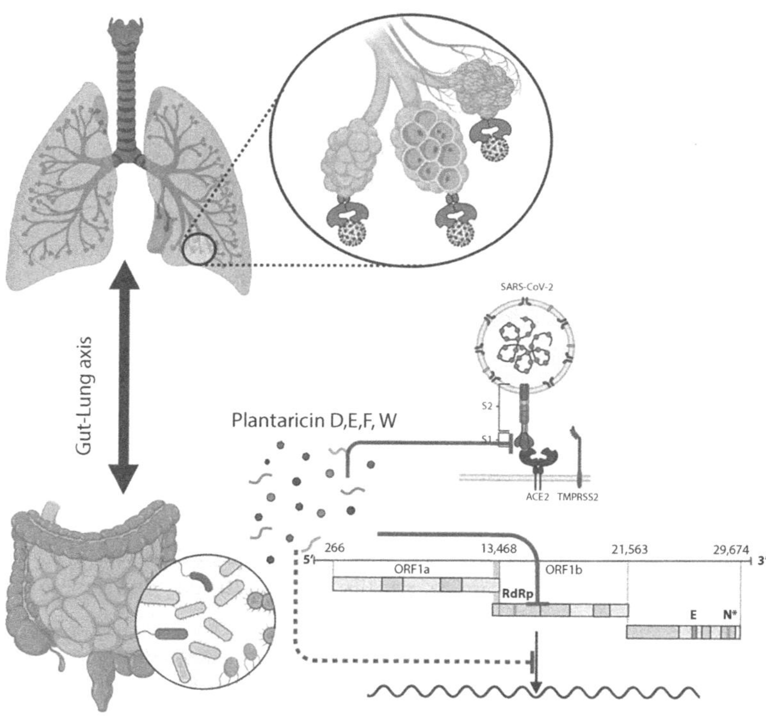

Figure 16.2 Proposed model on management of COVID19 [108] (permission obtained).

defense against viruses like SARS-CoV-2. Probiotic strains are known for their anti-inflammatory properties. However, their metabolites, which are known to have antiviral characteristics, can be utilized as an additional treatment to fight against COVID-19 [107]. Severe cases of COVID-19 are associated with an exaggerated inflammatory response, probiotics with anti-inflammatory effects could theoretically play a role in managing inflammation associated with the disease. Probiotics designed to specifically target and promote respiratory health might be explored. Brahma *et al.* [108] proposed the computational model that bioactive compounds of the probiotics srains could be considered for the management of COVID-19. Bacteriocins, such as Plantaricin, which are released by probiotic bacteria like *L. plantarum*, are hypothesized to play a role in limiting the entry and reproduction of SARS-CoV-2 (Figure 16.2).

These could potentially influence the respiratory microbiota and contribute to a healthier respiratory environment. Investigating the gut-lung axis, the bidirectional communication between the gut and the lungs, may provide insights into how probiotics could influence respiratory health and immune responses [109].

16.5.1.1 Challenges in Using NGPs

Interest in the use of novel microbes as probiotics has increased in tandem with the growing body of knowledge regarding the critical role that intestinal microbiota play in human health [110]. While the development of next-generation probiotics holds promise, there are several challenges and potential issues that researchers and developers need to address [111, 112].

Lack of Consistency and Standardization: Ensuring consistent quality and efficacy of probiotic products is challenging. There may be variations in manufacturing processes, storage conditions, and formulation, leading to inconsistent results.

Survivability and Stability: Many probiotic strains have difficulty surviving the harsh conditions of the digestive system, such as exposure to stomach acid and bile.

Limited Understanding of Mechanisms of Action: While the general benefits of probiotics are known, the specific mechanisms through which they exert their effects are not always well understood.

Individual Variability: Individual responses to probiotics can vary significantly based on factors such as genetics, diet, lifestyle, and the existing microbiome.

Regulatory Challenges: The regulation of probiotics varies globally, and there may be a lack of standardized guidelines for testing, quality control, and health claims.

Ethical and Environmental Concerns: Large-scale production of probiotics may raise concerns about the environmental impact, such as energy consumption and waste generation. Additionally, ethical concerns may arise regarding the sourcing of microbial strains.

Potential for Adverse Effects: While probiotics are generally considered safe, there is a potential for adverse effects, particularly in individuals with compromised immune systems or underlying health conditions.

Cost and Accessibility: Developing and producing advanced probiotics with specific characteristics can be costly, potentially limiting accessibility, especially in resource-limited settings.

16.5.2 Emerging Technologies for Effective Deliver of Probiotics

Several emerging technologies are being explored to enhance the effective delivery of probiotics, ensuring their survival through the digestive system and improving their efficacy. These technologies aim to protect probiotics from harsh conditions in the stomach, facilitate their targeted release in the intestines, and enhance their overall stability.

Microencapsulation involves enclosing probiotic cells within protective microcapsules, typically made of proteins, polysaccharides, or lipids. This technology shields probiotics from gastric acidity, bile salts, and other harsh conditions, promoting their survival during transit through the digestive system [113].

It is one of the most effective techniques to enhance the stability and viability of live probiotic strains from industrial processing conditions, storage [114] (Figure 16.3). Two methods that can be used for microencapsulation are emulsion and extrusion. Capsules less than 1 mm in diameter can be filled with probiotic microorganisms using the emulsion process [115, 116]. Spray drying is used to produces fine encapsulates with single

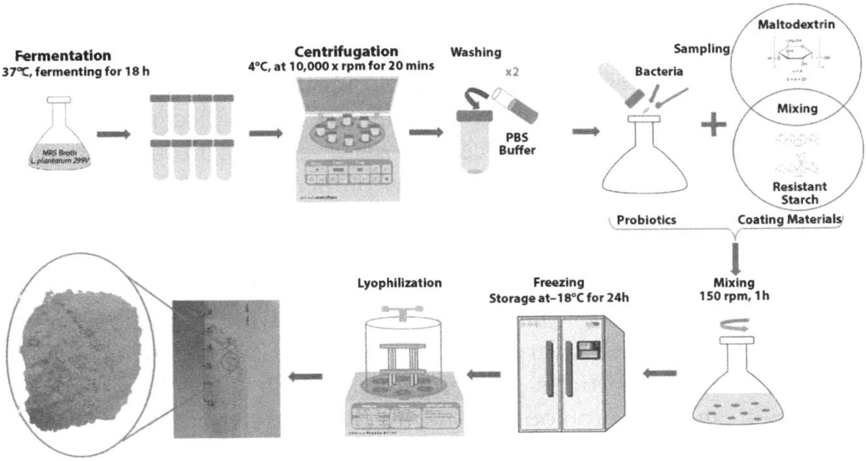

Figure 16.3 Method of microencapsulation [114] (open access).

unit operation; however, there will be reduction in viability due to higher drying temperature. Alternatively, freeze drying may retain the cell viability due to low drying temperature; however, freeze drying involved high operating cost and it produces uneven flaky structure [117]. Nanoparticles and nanocarriers are being explored to protect and deliver probiotics. Nanotechnology enables the development of nanoscale structures that can enhance the stability of probiotics, improve their bioavailability, and provide controlled release in specific regions of the gastrointestinal tract. Synthetic biology techniques are used to engineer probiotics with improved survival characteristics. Genetic modifications can enhance probiotics' resistance to gastric acid, bile salts, and other stressors, ensuring their viability until they reach the intestines [118].

Advanced coating technologies involve applying protective layers to probiotic cells. These coatings may include bioadhesive polymers that adhere to the intestinal mucosa, providing sustained release and improving the residence time of probiotics in the gut. Smart delivery systems use responsive materials that can respond to specific conditions in the digestive tract. For example, pH-responsive coatings can protect probiotics in the acidic stomach environment and release them in the neutral pH of the intestines [119]. Combining probiotics with prebiotics (synbiotics) involves coating probiotics with substances that promote their growth and activity. This dual approach supports the survival and functionality of probiotics in the gut.

3D printing allows the creation of complex structures, enabling the development of customized probiotic delivery systems. This technology can produce matrices that protect probiotics during transit and enable controlled release in specific gastrointestinal locations. Biosynthetic materials derived from natural sources can serve as carriers for probiotics. These materials provide a protective environment for probiotics and may enhance their stability and survival during digestion. Electrospinning is a technique that creates nanofibrous structures. Electrospun fibers can be used as carriers for probiotics, offering protection and controlled release. This technology allows for the creation of matrices with tailored properties for effective probiotic delivery. Magnetic guidance systems involve incorporating magnetic particles into probiotic formulations. External magnetic fields can guide probiotics to specific locations in the gastrointestinal tract, improving their targeted delivery. Improving the lyophilization process, which is commonly used for preserving probiotics, can enhance the viability and stability of probiotic cells, ensuring their effectiveness over a longer shelf life [114].

16.5.3 Application of Nanotechnology to Probiotics

Nanotechnology has the potential to revolutionize the field of probiotics by addressing challenges related to their viability, stability, and targeted delivery. Here are several applications of nanotechnology to probiotics:

Nanoencapsulated probiotics offer improved protection against environmental stressors, such as gastric acid and bile salts, during transit through the digestive system. The nanocapsules can enhance the survival and delivery of probiotics to the intestines [113]. Nanoemulsions are colloidal dispersions of nanoscale droplets in an aqueous medium. Nanoemulsion-based delivery systems can encapsulate probiotics and improve their stability. These systems can also enhance the solubility and bioavailability of lipophilic compounds associated with certain probiotic strains [120].

Nanoparticles, such as liposomes, solid lipid nanoparticles, and polymeric nanoparticles, can be used as carriers for probiotics. These nanocarriers provide a protective environment, ensuring the stability of probiotics and facilitating their controlled release in specific regions of the gastro-intestinal tract [121, 122]. Magnetic nanoparticles can be incorporated into probiotic formulations to enable targeted delivery. External magnetic fields can guide probiotics to specific locations in the gastrointestinal tract, improving their localization and interaction with the host. Bioadhesive nanoparticles can improve the adhesion of probiotics to the intestinal mucosa. This enhanced adhesion prolongs the residence time of probiotics in the gut, allowing for increased interaction with the host and promoting their beneficial effects [123, 118]. pH-responsive nanoparticles can be designed to protect probiotics in the acidic environment of the stomach and release them in the neutral pH of the intestines. This controlled release ensures that probiotics reach the target site with enhanced viability [119]. Electrospinning is a technique used to produce nanofibers. Nanofiber-based materials can serve as carriers for probiotics, offering a large surface area and promoting the controlled release of probiotics in the gut. Electrospun nanofibers can be designed to protect probiotics during transit [124]. Nanotechnology can be applied to design nanostructured probiotic foods. This includes the development of nanoscale structures within food matrices that enhance the stability and bioavailability of probiotics. Nanostructured foods may also improve the taste and texture of probiotic-containing products [125]. Nanoscale biosensors can be employed to monitor the activity and viability of probiotics in real-time. These biosensors can provide valuable insights into the behavior of probiotics in different environments, aiding in quality control and therapeutic optimization [126] While the application of nanotechnology to probiotics holds great promise, it is essential to consider safety aspects and conduct thorough research to ensure that nanomaterials used in probiotic formulations are biocompatible and do not pose risks to human health.

16.6 Concluding Remarks and Future Perspectives

Application of functional components such as prebiotics in the diet will be an alternative to improve the quality of life for the people suffering from non-communicable disorders such as cancer, obesity, hypertension, vascular disorders, and degenerative illnesses. Hence, prebiotics are predicted to eventually replace the antibiotics which are employed to treat several life-style disorders. It is necessary to validate the therapeutic potential of prebiotics, authenticate the efficacy claims, and identify the mechanism behind the claims. On the other hand, considerable research in nutritional targeting of the microbiome for therapeutic applications have extended the investigation related to prebiotics, probiotics, and related fields. Compared to conventional probiotics, which have a safe history of usage in humans, next generation probiotics represent a significant advancement in disease prevention and treatment. Scientific community faces a big challenge towards increased antibiotic resistant microorganisms as there is no method to reverse this resistance. In this scenario, next generation probiotics such as *Faecalibacterium prausnitzii, Bacteroides fragilis, Akkermansia muciniphila, Roseburia spp* and *Eubacterium hallii* have demonstrated a potential probiotic for disease intervention and treatment of metabolic, gastrointestinal, and immune system disorders. Further, nanoprebiotics and nanoprobiotics have geared great attention in pharmaceutical and food industries. The poor understanding of the mechanisms and health implications of nanoprebiotics draws attention for futuristic research. This is one of the main reasons for delaying the further application of nanoprebiotics and probiotics in the food sector. It is necessary to determine the safety and physicochemical properties of nanoprebiotics and probiotics to support tactile conclusions and policies of regulatory agencies. Further research on the antiviral potential of pre and probiotic-based medications in the prevention and treatment of COVID-19 is warranted. Preclinical and clinical experiments ought to be conducted in future to avail their benefits in the management of the COVID-19 pandemic.

References

1. Gibson, G., Hutkins, R., Sanders, M. *et al.*, Expert consensus document: The International Scientific Association for Probiotics and Prebiotics (ISAPP) consensus statement on the definition and scope of prebiotics. *Nat. Rev. Gastroenterol. Hepatol.*, 14, 491–502, 2017, https://doi.org/10.1038/nrgastro.2017.75.
2. Salminen, S., Collado, M.C., Endo, A., Hill, C., Lebeer, S., Quigley, E.M.M., Sanders, M.E., Shamir, R., Swann, J.R., Szajewska, H., Vinderola, G., The International Scientific Association of Probiotics and Prebiotics (ISAPP) consensus statement on the definition and scope of postbiotics. *Nat. Rev. Gastroenterol. Hepatol.*, 9, 649–667, 2021.
3. Prescott, S.L., Logan, A.C., Millstein, R.A., Katszman, M.A., Biodiversity, the human microbiome and mental health: moving toward a new clinical ecology for the 21st Century? *Int. J. Biodivers.*, 1, 2718275, 2016.
4. Cani, P.D. and Everard, A., Talking microbes: when gut bacteria interact with diet and host organs. *Mol. Nutr. Food Res.*, 60, 58–66, 2016.
5. David, L.A. *et al.*, Diet rapidly and reproducibly alters the human gut microbiome. *Nature*, 505, 559–563, 2014.
6. Gibson, G.R. and Roberfroid, M.B., Dietary modulation of the human colonic microbiota: introducing the concept of prebiotics. *J. Nutr.*, 125, 1401–1412, 1995.
7. Collins, S. and Reid, G., Distant site effects of ingested prebiotics. *Nutrients*, 8, E523, 2016.
8. Davani-Davari, D., Negahdaripour, M., Karimzadeh, I., Seifan, M., Mohkam, M., Masoumi, S.J., Berenjian, A., Ghasemi, Y., Prebiotics: Definition, Types,

Sources, Mechanisms, and Clinical Applications. *Foods*, 8, 3, 92, 2019, doi: 10.3390/foods8030092.
9. Gibson, G.R., Probert, H.M., Van Loo, J., Rastall, R.A., Roberfroid, M.B., Dietary modulation of the human colonic microbiota: Updating the concept of prebiotics. *Nutr. Res. Rev.*, 17, 259–275, 2004, doi: 10.1079/NRR200479.
10. Glenn, G. and Roberfroid, M., Dietary modulation of the human colonic microbiota: Introducing the concept of prebiotics. *J. Nutr.*, 125, 1401–1412, 1995.
11. Flint, H.J., Scott, K.P., Louis, P., Duncan, S.H., The role of the gut microbiota in nutrition and health. *Nat. Rev. Gastroenterol. Hepatol.*, 9, 577–589, 2012, doi: 10.1038/nrgastro.2012.156.
12. Bouhnik, Y., Raskine, L., Simoneau, G., Vicaut, E., Neut, C., Flourié, B., Brouns, F., Bornet, F.R., The capacity of nondigestible carbohydrates to stimulate fecal bifidobacteria in healthy humans: A double-blind, randomized, placebo-controlled, parallel-group, dose-response relation study. *Am. J. Clin. Nutr.*, 80, 1658–1664, 2004, doi: 10.1093/ajcn/80.6.1658.
13. Shokri, D., Khorasgani, M.R., Mohkam, M., Fatemi, S.M., Ghasemi, Y., Taheri-Kafrani, A., The inhibition effect of lactobacilli against growth and biofilm formation of pseudomonas aeruginosa. *Probiot. Antimicrob. Proteins*, 10, 34–42, 2018, doi: 10.1007/s12602-017-9267-9.
14. Morowvat, M.H., Nezafat, N., Ghasemi, Y., Zare, M.H., Mohkam, M., Probiotic potential of five lactobacillus strains isolated from traditional Persian yoghurt in Fars province, Iran: Viewing through the window of phylogenetics. *Biosci. Biotechnol. Res. Asia*, 12, 1265–1272, 2015.
15. Turroni, F., Ventura, M., Buttó, L.F., Duranti, S., O'Toole, P.W., Motherway, M.O.C., van Sinderen, D., Molecular dialogue between the human gut microbiota and the host: A lactobacillus and bifidobacterium perspective. *Cell. Mol. Life Sci.*, 71, 183–203, 2014, doi: 10.1007/s00018-013-1318-0.
16. Roberfroid, M., Health benefits of non-digestible oligosaccharides, in: *Dietary Fiber in Health and Disease*, pp. 211–219, Springer, New York, NY, USA, 1997.
17. Stinson, L.F., Payne, M.S., Keelan, J.A., Planting the seed: Origins, composition, and postnatal health significance of the fetal gastrointestinal microbiota. *Crit. Rev. Microbiol.*, 43, 352–369, 2017, doi: 10.1080/1040841X.2016.1211088.
18. Trompette, A., Gollwitzer, E.S., Yadava, K., Sichelstiel, A.K., Sprenger, N., Ngom-Bru, C., Blanchard, C., Junt, T., Nicod, L.P., Harris, N.L. *et al.*, Gut microbiota metabolism of dietary fiber influences allergic airway disease and hematopoiesis. *Nat. Med.*, 20, 159–166, 2014, doi: 10.1038/nm.3444.
19. Zhou, Z., Zhang, Y., Zheng, P., Chen, X., Yang, Y., Starch structure modulates metabolic activity and gut microbiota profile. *Anaerobe*, 24, 71–78, 2013, doi: 10.1016/j.anaerobe.2013.09.012.
20. Hernot, D.C., Boileau, T.W., Bauer, L.L., Middelbos, I.S., Murphy, M.R., Swanson, K.S., Fahey, C., *In vitro* fermentation profiles, gas production rates, and microbiota modulation as affected by certain fructans,

galactooligosaccharides, and polydextrose. *J. Agric. Food. Chem.*, 57, 1354–1361, 2009, doi: 10.1021/jf802484j.
21. Clarke, T.B., Davis, K.M., Lysenko, E.S., Zhou, A.Y., Yu, Y., Weiser, J.N., Recognition of peptidoglycan from the microbiota by nod1 enhances systemic innate immunity. *Nat. Med.*, 16, 228–231, 2010, doi: 10.1038/nm.2087.
22. Zhou, J., Du, G., Chen, J., Novel fermentation processes for manufacturing plant natural products. *Curr. Opin. Biotechnol.*, 25, 17–23, 2014.
23. Hamer, H.M., Jonkers, D., Venema, K., Vanhoutvin, S., Troost, F., Brummer, R.J., The role of butyrate on colonic function. *Aliment. Pharmacol. Ther.*, 27, 104–119, 2008, doi: 10.1111/j.1365-2036.2007.03562.x.
24. den Besten, G., van Eunen, K., Groen, A.K., Venema, K., Reijngoud, D.-J., Bakker, B.M., The role of short-chain fatty acids in the interplay between diet, gut microbiota, and host energy metabolism. *J. Lipid Res.*, 54, 2325–2340, 2013, doi: 10.1194/jlr.R036012.
25. Ashaolu, T.J., Ashaolu, J.O., Adeyeye, S.A.O., Fermentation of prebiotics by human colonic microbiota *in vitro* and short-chain fatty acids production: a critical review. *J. Appl. Microbiol.*, 130, 677–687, 2020, doi: doi.org/10.1111/jam.14843.
26. Singhi, S.C. and Kumar, S., Probiotics in critically ill children. *F1000Res.*, 5, 407, 2016, doi: 10.12688/f1000research.7630.1.
27. Doron, S. and Snydman, D.R., Risk and safety of probiotics. *Clin. Infect. Dis.*, 60, Suppl 2, S129–234, 2015, doi: 10.1093/cid/civ085.
28. Durchschein, F., Petritsch, W., Hammer, H.F., Diet therapy for inflammatory bowel diseases: The established and the new. *World J. Gastroenterol.*, 22, 7, 2179–2194, 2016, doi: 10.3748/wjg.v22.i7.2179.
29. Brown, A.C. and Valiere, A., Probiotics and Medical Nutrition Therapy. *Nutr. Clin. Care*, 7, 2, 56–68, 2004. ISSN 1096-6781.
30. Schlundt, J., Health and Nutritional Properties of Probiotics in Food including Powder Milk with Live Lactic Acid Bacteria (PDF), in: *Report of a Joint FAO/WHO Expert Consultation on Evaluation of Health and Nutritional Properties of Probiotics in Food Including Powder Milk with Live Lactic Acid Bacteria*, FAO / WHO, Cordoba, Argentina, Archived from the original (PDF) on October 22, 2012.
31. FAO, *Probiotics in food: health and nutritional properties and guidelines for evaluation*, Food and Agriculture Organization of the United Nations, World Health Organization, Rome, Food and Agriculture Organization of the United Nations, Rome, 2006, ISBN 92-5-105513-0, OCLC 70928765.
32. Bermudez-Brito, M., Plaza-Díaz, J., Muñoz-Quezada, S., Gómez-Llorente, C., Gil, A., Probiotic mechanisms of action. *Ann. Nutr. Metab.*, 61, 2, 160–74, 2012, doi: 10.1159/000342079.
33. Rauch, M. and Lynch, S.V., The potential for probiotic manipulation of the gastrointestinal microbiome. *Curr. Opin. Biotechnol.*, 23, 2, 192–201, 2012, doi: 10.1016/j.copbio.2011.11.004.

34. Thomas, C.M. and Versalovic, J., Probiotics-host communication: Modulation of signaling pathways in the intestine. *Gut Microbes*, 1, 148–63, 3, 2010, doi: 10.4161/gmic.1.3.11712.
35. Moss, P.J. and McKendrick, M.W., Bacterial gastroenteritis. *Curr. Opin. Infect. Dis.*, 10, 5, 402–407, 1997.
36. Noguera, T., Wotring, R., Melville, C.R., Hargraves, K., Kumm, J., Morton, J.M., Resolution of acute gastroenteritis symptoms in children and adults treated with a novel polyphenol-based prebiotic. *World J. Gastroenterol.*, 20, 34, 12301, 2014.
37. Patel, S. and Goyal, A., The current trends and future perspectives of prebiotics research: a review. *3 Biotech*, 2, 115–125, 2012.
38. Vandenplas, Y., Probiotics and prebiotics in infectious gastroenteritis. *Best Pract. Res. Clin. Gastroenterol.*, 30, 1, 49–53, 2016.
39. Steed, H., Macfarlane, G.T., Macfarlane, S., Prebiotics, synbiotics and inflammatory bowel disease. *Mol. Nutr. Food Res.*, 52, 8, 898–905, 2008.
40. Azpiroz, F., Dubray, C., Bernalier-Donadille, A., Cardot, J.M., Accarino, A., Serra, J., Dapoigny, M., Effects of sc FOS on the composition of fecal microbiota and anxiety in patients with irritable bowel syndrome: a randomized, double blind, placebo controlled study. *Neurogastroenterol. Motil.*, 29, 2, e12911, 2017.
41. Iribarren, C., Törnblom, H., Aziz, I., Magnusson, M.K., Sundin, J., Vigsnæs, L.K., Simrén, M., Human milk oligosaccharide supplementation in irritable bowel syndrome patients: A parallel, randomized, double-blind, placebo-controlled study. *Neurogastroenterol. Motil.*, 32, 10, e13920, 2020.
42. Munjal, U., Glei, M., Pool-Zobel, B.L., Scharlau, D., Fermentation products of inulin-type fructans reduce proliferation and induce apoptosis in human colon tumour cells of different stages of carcinogenesis. *Br. J. Nutr.*, 102, 5, 663–671, 2009.
43. Femia, A.P., Salvadori, M., Broekaert, W.F., François, I.E., Delcour, J.A., Courtin, C.M., Caderni, G., Arabinoxylan-oligosaccharides (AXOS) reduce preneoplastic lesions in the colon of rats treated with 1, 2-dimethylhydrazine (DMH). *Eur. J. Nutr.*, *49*, 127–132, 2010.
44. Messaoudene, M., Pidgeon, R., Richard, C., Ponce, M., Diop, K., Benlaifaoui, M., Routy, B., A natural polyphenol exerts antitumor activity and circumvents anti–PD-1 resistance through effects on the gut microbiota. *Cancer Discov.*, 12, 4, 1070–1087, 2022.
45. Louis, P., Flint, H.J., Michel, C., How to manipulate the microbiota: prebiotics, in: *Microbiota of the human body: Implications in health and disease*, A. Schwiertz, (Ed.), *Adv. Exp. Med. Biol*, pp. 119–142, Springer, Cham, 2016.
46. Jiang, Z., Chi, J., Li, H., Wang, Y., Liu, W., Han, B., Effect of chitosan oligosaccharide-conjugated selenium on improving immune function and blocking gastric cancer growth. *Eur. J. Pharmacol.*, 891, 173673, 2021.
47. Ke, Y., Wang, M., Li, Y., Shan, Z., Mi, W., Yuan, P., Zheng, X., Oligosaccharides composition of *Descurainiae sophia* exerts anti-heart failure by improving

heart function and water-liquid metabolism in rats with heart failure. *Biomed. Pharmacother.*, 129, 110487, 2020.
48. McCabe, L., Britton, R.A., Parameswaran, N., Prebiotic and probiotic regulation of bone health: role of the intestine and its microbiome. *Curr. Osteoporos. Rep.*, 13, 363–371, 2015.
49. van den Heuvel, E.G., Muijs, T., Brouns, F., Hendriks, H.F., Short-chain fructo-oligosaccharides improve magnesium absorption in adolescent girls with a low calcium intake. *Nutr. Res.*, 29, 4, 229–237, 2009.
50. Mikulic, N., Uyoga, M.A., Paganini, D., Mwasi, E., Stoffel, N.U., Zeder, C., Zimmermann, M.B., Consumption of a single dose of prebiotic galacto-oligosaccharides does not enhance iron absorption from micronutrient powders in Kenyan infants: a stable iron isotope study. *J. Nutr.*, 151, 5, 1205–1212, 2021.
51. Jeroense, F.M., Zeder, C., Zimmermann, M.B., Herter-Aeberli, I., Acute consumption of prebiotic galacto-oligosaccharides increases iron absorption from ferrous fumarate, but not from ferrous sulfate and ferric pyrophosphate: stable iron isotope studies in iron-depleted young women. *J. Nutr.*, 150, 9, 2391–2397, 2020.
52. Abenavoli, L., Scarpellini, E., Colica, C., Boccuto, L., Salehi, B., Sharifi-Rad, J., Capasso, R., Gut microbiota and obesity: a role for probiotics. *Nutrients*, 11, 11, 2690, 2019.
53. Daniali, M., Nikfar, S., Abdollahi, M., A brief overview on the use of probiotics to treat overweight and obese patients. *Expert Rev. Endocrinol. Metab.*, 15, 1, 1–4, 2020.
54. Zheng, D., Liwinski, T., Elinav, E., Interaction between microbiota and immunity in health and disease. *Cell Res.*, 30, 6, 492–506, 2020.
55. Yuan, Z., Xiao, S., Li, S., Suo, B., Wang, Y., Meng, L., Zhou, L., The impact of Helicobacter pylori infection, eradication therapy, and probiotics intervention on gastric microbiota in young adults. *Helicobacter*, 26, 6, e12848, 2021.
56. Agrawal, R., Probiotics: An emerging food supplement with health benefits. *Food Biotechnol.*, 19, 3, 227–246, 2005.
57. Yadav, R., Shukla, P., Probiotics for Human Health: Current Progress and Applications. In: *Recent advances in Applied Microbiology*, Shukla, P. (eds). Springer, Singapore, 133–147, 2017. https://doi.org/10.1007/978-981-10-5275-0_6.
58. Blaabjerg, S., Artzi, D.M., Aabenhus, R., Probiotics for the prevention of antibiotic-associated diarrhea in outpatients—a systematic review and meta-analysis. *Antibiotics*, 621, 4, 1–17, 2017.
59. Erdeve, O., Tiras, U., Dallar, Y., The probiotic effect of Saccharomyces boulardii in a pediatric age group. *J. Trop. Pediatr.*, 50, 4, 234–238, 2004.
60. de Vrese, M., Stegelmann, A., Richter, B., Fenselau, S., Laue, C., Schrezenmeir, J., Probiotics—compensation for lactase insufficiency. *Am. J. Clin. Nutr.*, 73, 2, 421s–429s, 2001.

61. Kumar, B.V., Vijayendra, S.V.N., Reddy, O.V.S., Trends in dairy and non-dairy probiotic products-a review. *J. Food Sci. Technol.*, 52, 6112–6124, 2015.
62. Ojetti, V., Gigante, G., Gabrielli, M., Ainora, M.E., Mannocci, A., Lauritano, E.C., Gasbarrini, A., The effect of oral supplementation with. *Eur. Rev. Med. Pharmacol. Sci.*, 14, 163–170, 2010.
63. Masoumi, S.J., Mehrabani, D., Saberifiroozi, M., Fattahi, M.R., Moradi, F., Najafi, M., The effect of yogurt fortified with Lactobacillus acidophilus and Bifidobacterium sp. probiotic in patients with lactose intolerance. *Food Sci. Nutr.*, 9, 3, 1704–1711, 2021.
64. Yadav, A., Singh, S., Sohi, H., Dang, S., Advances in delivery of chemotherapeutic agents for cancer treatment. *AAPS Pharm. Sci. Tech.*, 23, 1–14, 2022.
65. Nouri, Z., Karami, F., Neyazi, N., Modarressi, M.H., Karimi, R., Khorramizadeh, M.R., Motevaseli, E., Dual anti-metastatic and anti-proliferative activity assessment of two probiotics on HeLa and HT-29 cell lines. *Cell J.*, 18, 2, 127, 2016.
66. Xu, F., Li, Q., Wang, S., Dong, M., Xiao, G., Bai, J., Sun, X., The efficacy of prevention for colon cancer based on the microbiota therapy and the antitumor mechanisms with intervention of dietary Lactobacillus. *Microbiol. Spectr.*, 11, 5, e00189–23, 2023.
67. World Health Organization. "Cardiovascular diseases (CVDs): fact sheet no. 317." Media centre. World Health Organization, 20,3, 2009.
68. World Health Organization, *Diet, nutrition, and the prevention of chronic diseases: report of a joint WHO/FAO expert consultation*, vol. 916, World Health Organization, Geneva, Switzerland, 2003.
69. Bernini, L.J., Simão, A.N.C., Alfieri, D.F., Lozovoy, M.A.B., Mari, N.L., de Souza, C.H.B., Costa, G.N., Beneficial effects of Bifidobacterium lactis on lipid profile and cytokines in patients with metabolic syndrome: A randomized trial. Effects of probiotics on metabolic syndrome. *Nutr.*, 32, 6, 716–719, 2016.
70. He, J., Zhang, F., Han, Y., Effect of probiotics on lipid profiles and blood pressure in patients with type 2 diabetes: a meta-analysis of RCTs. *Med.*, 96, 51, e9166, 2017.
71. Dieck, H.T., Schön, C., Wagner, T., Pankoke, H.C., Fluegel, M., Speckmann, B., A synbiotic formulation comprising Bacillus subtilis DSM 32315 and L-alanyl-L-glutamine improves intestinal butyrate levels and lipid metabolism in healthy humans. *Nutr.*, 14, 1, 143, 2021.
72. Wang, M., Veeraperumal, S., Zhong, S., Cheong, K.L., Fucoidan-derived functional oligosaccharides: Recent developments, preparation, and potential applications. *Foods*, 12, 4, 878, 2023.
73. Zhu, Y., Liu, L., Sun, Z., Ji, Y., Wang, D., Mei, L., Shen, P., Li, Z., Tang, S., Zhang, H., Zhou, Q., Fucoidan as a marine-origin prebiotic modulates the growth and antibacterial ability of Lactobacillus rhamnosus. *Int. J. Biol. Macromol.*, 180, 599–607, 2021.

74. Poveda-Castillo, G.D.C., Rodrigo, D., Martínez, A., Pina-Pérez, M.C., Bioactivity of fucoidan as an antimicrobial agent in a new functional beverage. *Beverages*, 4, 3, 64, 2018.
75. Plamada, D. and Vodnar, D.C., Polyphenols - Gut microbiota interrelationship: A transition to a new generation of prebiotics. *Nutrients*, 14, 1, 137, 2021.
76. Nash, V., Ranadheera, C.S., Georgousopoulou, E.N., Mellor, D.D., Panagiotakos, D.B., McKune, A.J., Kellett, J., Naumovski, N., The effects of grape and red wine polyphenols on gut microbiota–A systematic review. *Food Res. Int.*, 113, 277–287, 2018.
77. Liu, Y.C., Li, X.Y., Shen, L., Modulation effect of tea consumption on gut microbiota. *Appl. Microbiol. Biotechnol.*, 104, 981–98, 2020.
78. Van Dorsten, F.A., Peters, S., Gross, G., Gomez-Roldan, V., Klinkenberg, M., De Vos, R.C., Vaughan, E.E., Van Duynhoven, J.P., Possemiers, S., Van de Wiele Jacobs, D.M., Gut microbial metabolism of polyphenols from black tea and red wine/grape juice is source-specific and colon-region dependent. *J. Agric. Food. Chem.*, 60, 45, 11331–11342, 2012.
79. Van der Hee, B. and Wells, J.M., Microbial regulation of host physiology by short-chain fatty acids. *Trends Microbiol.*, 29, 8, 700–712, 2021.
80. Tan, J., McKenzie, C., Potamitis, M., Thorburn, A.N., Mackay, C.R., Macia, L., The role of short-chain fatty acids in health and disease. *Adv. Immunol.*, 121, 91–119, 2014.
81. Paesani, C., Lammers, T.C., Sciarini, L.S., Moiraghi, M., Pérez, G.T., Fabi, J.P., Effect of chemical, thermal, and enzymatic processing of wheat bran on the solubilization, technological and biological properties of non-starch polysaccharides. *Carbohydr. Polym.*, 328, 121747, 2023.
82. Ashaolu, T.J., Ashaolu, J.O., Adeyeye, S.A.O., Fermentation of prebiotics by human colonic microbiota *in vitro* and short-chain fatty acids production: a critical review. *J. Appl. Microbiol.*, 130, 3, 677–687, 2021.
83. Hurtado-Romero, A., Del Toro-Barbosa, M., Garcia-Amezquita, L.E., García-Cayuela, T., Innovative technologies for the production of food ingredients with prebiotic potential: Modifications, applications, and validation methods. *Trends Food Sci. Technol.*, 104, 117–131, 2020.
84. Singh, J., Kaur, J., Rasane, P., Kaur, S., 3D printed foods-carbs from the lab for better health. *Curr. Opin. Clin. Nutr. Metab. Care*, 25, 4, 271–276, 2022.
85. Ashaolu, T.J., Emerging applications of nanotechnologies to probiotics and prebiotics. *Int. J. Food Sci. Technol.*, 56, 8, 3719–3725, 2021a.
86. Martínez-Ballesta, M., Gil-Izquierdo, A., García-Viguera, C., Domínguez-Perles, R., Nanoparticles and controlled delivery for bioactive compounds: Outlining challenges for new "smart-foods" for health. *Foods*, 7, 5, 72, 2018.
87. Nikam, P., Uprit, S., Ranvir, S., Nanomaterials for Functional Food Processing. *Res. Trends Food Technol. Nutr.*, 3, 13–30, 2018.

88. Dangi, P., Chaudhary, N., Chaudhary, V., Virdi, A.S., Kajla, P., Khanna, P., Haque, S., Nanotechnology impacting probiotics and prebiotics: A paradigm shift in nutraceuticals technology. *Int. J. Food Microbiol.*, 388, 110083, 2023.
89. Authimoolam, S.P. and Dziubla, T.D., Biopolymeric mucin and synthetic polymer analogs: Their structure, function, and role in biomedical applications. *Polymers*, 8, 3, 71, 2016.
90. Ren, Y., Nie, L., Luo, C., Zhu, S., Zhang, X., Advancement in Therapeutic Intervention of Prebiotic-Based Nanoparticles for Colonic Diseases. *Int. J. Nanomed.*, 17, 6639–6654, 2022.
91. Boostani, S. and Jafari, S.M., Controlled release of nanoencapsulated food ingredients, in: *Nanoencapsulation in the Food Industry*, vol. 5, S.M. Jafari (Ed.), pp. 27–78, Academic Press, Cambridge, 2020.
92. Batista, K.S., de Albuquerque, J.G., de Vasconcelos, M.H.A., Bezerra, M.L.R., da Silva Barbalho, M.B., Pinheiro, R.O., de Souza Aquino, J., Probiotics and prebiotics: potential prevention and therapeutic target for nutritional management of COVID-19? *Nutr. Res. Rev.*, 36, 181–198, 2023.
93. Xu, L., Yang, C.S., Liu, Y., Zhang, X., Effective regulation of gut microbiota with probiotics and prebiotics may prevent or alleviate COVID-19 through the gut-lung axis. *Front. Pharmacol.*, 13, 895193, 2022.
94. Olaimat, A.N., Aolymat, I., Al-Holy, M., Ayyash, M., Abu Ghoush, M., Al-Nabulsi, A.A., Tareq, O., Vasso, A., Liu, S.Q., Shah, N.P., The potential application of probiotics and prebiotics for the prevention and treatment of COVID-19. *npj Sci. Food*, 4, 1, 17, 2020.
95. Yan, F. and Polk, D.B., Probiotics and Probiotic-Derived Functional Factors-Mechanistic Insights Into Applications for Intestinal Homeostasis. *Front. Immunol.*, 11, 1428, 2020.
96. Bernatek, M., Żukiewicz-Sobczak, W., Lachowicz-Wiśniewska, S., Piątek, J., Factors Determining Effective Probiotic Activity: Evaluation of Survival and Antibacterial Activity of Selected Probiotic Products Using an "*In Vitro*" Study. *Nutrient.*, 14, 16, 3323, 2022.
97. Venugopalan, V., Shriner, K.A., Wong-Beringer, A., Regulatory oversight and safety of probiotic use. *Emerg. Infect. Dis.*, 16, 11, 1661–1665, 2010.
98. Barbosa, J.C., Daniela, M., Diana, A., Jose, C.A., Adriano, B., Ana, M.G., Cristina, F., Chapter 24 - Next-generation probiotics, in: *Probiotics*, pp. 483–502, 2022.
99. Hu, H., Jie, L., Ying, L., Hongyu, L., Rui, J., Shengjie, L., Jing, W., Hong, W., Chen, T., Improvement effect of a next-generation probiotic *L. plantarum*-pMG36e-GLP-1 on type 2 diabetes mellitus *via* the gut–pancreas–liver axis. *Food Funct.*, 14, 3179–3195, 2023.
100. Zhang, H., Duan, Y., Cai, F., Cao, D., Wang, L., Qiao, Z., Hong, Q., Li, N., Zheng, Y., Su, M., Liu, Z., Zhu, B., Next-Generation Probiotics: Microflora Intervention to Human Diseases. *BioMed Res. Internation*, 2022, 5633403, 2022.

101. Kaźmierczak-Siedlecka, K., Skonieczna-Żydecka, K., Hupp, T., Duchnowska, R., Marek-Trzonkowska, N., Połom, K., Next-generation probiotics - do they open new therapeutic strategies for cancer patients? *Gut Microbes*, *14*, 1, 2035659, 2022.
102. Amrane, S., Hocquart, M., Afouda, P., Kuete, E., Pham, T.P.T., Dione, N., Lagier, J.C., Metagenomic and culturomic analysis of gut microbiota dysbiosis during Clostridium difficile infection. *Sci. Rep.*, *9*, 1, 12807, 2019.
103. López-Moreno, A., Acuña, I., Torres-Sánchez, A., Ruiz-Moreno, A., Cerk, K., Rivas, A., Suárez, A., Monteoliva-Sánchez, M., Aguilera, M., Next Generation Probiotics for Neutralizing Obesogenic Effects: Taxa Culturing Searching Strategies. *Nutrient*, 13, 1617, 2021.
104. Vallianou, N.G., Kounatidis, D., Tsilingiris, D., Panagopoulos, F., Christodoulatos, G.S., Evangelopoulos, A., Karampela, I., Dalamaga, M., The Role of Next-Generation Probiotics in Obesity and Obesity-Associated Disorders: Current Knowledge and Future Perspectives. *Int. J. Mol. Sci.*, 24, 7, 6755, 2023.
105. Shetty, S.A., Zuffa, S., Bui, T.P.N., Aalvink, S., Smidt, H., De Vos, W.M., Reclassification of Eubacterium hallii as Anaerobutyricum hallii gen. nov., comb. nov., and description of Anaerobutyricum soehngenii sp. nov., a butyrate and propionate-producing bacterium from infant faeces. *Int. J. Syst. Evol. Microbiol.*, *68*, 12, 3741–3746, 2018.
106. Ceccarelli, G., Carolina, S., Francesco, P., Mastroianni, C.M., d'Ettorre, G., Probiotics and COVID-19. *Lancet*, 5, 8, P721–722, 2020.
107. Nguyen, Q.V., Chong, L.C., Hor, Y.Y., Lew, L.C., Rather, I.A., Choi, S.B., Role of Probiotics in the Management of COVID-19: A Computational Perspective. *Nutrient*, 14, 2, 274, 2022.
108. Brahma, S., Naik, A., Lordan, R., Probiotics: A gut response to the COVID-19 pandemic but what does the evidence show? *Clin. Nutr. ESPEN*, 51, 17–27, 2022.
109. Tian, Y., Ran, H., Wen, X., Fu, G., Zhou, X., Liu, R., Pan, T., Probiotics improve symptoms of patients with COVID-19 through gut-lung axis: a systematic review and meta-analysis. *Front. Nutr.*, 10, 1179432, 2023.
110. Maria, H.S., Safety aspects of next generation probiotics. *Curr. Opin. Food Sci.*, 30, 8–13, 2019.
111. Merenstein, D., Pot, B., Leyer, G., Ouwehand, A.C., Preidis, G.A., Elkins, C.A., Hill, C., Lewis, Z.T., Shane, A.L., Zmora, N., Petrova, M.I., Collado, M.C., Morelli, L., Montoya, G.A., Szajewska, H., Tancredi, D.J., Sanders, M.E., Emerging issues in probiotic safety: 2023 perspectives. *Gut Microbes*, 15, 1, 2185034, 2023.
112. Khan, M.T., Dwibedi, C., Sundh, D. *et al.*, Synergy and oxygen adaptation for development of next-generation probiotics. *Nature*, 620, 381–385, 2023.
113. Sbehat, M., Mauriello, G., Altamimi, M., Microencapsulation of Probiotics for Food Functionalization: An Update on Literature Reviews. *Microorganisms*, 10, 10, 1948, 2022.

114. Sun, W., Nguyen, Q.D., Süli, B.K., Alarawi, F., Szécsi, A., Gupta, V.K., Friedrich, L.F., Gere, A., Bujna, E., Microencapsulation and Application of Probiotic Bacteria *Lactiplantibacillus plantarum* 299v Strain. *Microorganisms*, 11, 4, 947, 2023.
115. Zanjani, K.M.A., Ghiassi, T.B., Sharifan, A., Mohammadi, N., Microencapsulation of Probiotics by Calcium Alginate-gelatinized Starch with Chitosan Coating and Evaluation of Survival in Simulated Human Gastro-intestinal Condition. *Iran. J. Pharm. Res.*, *13*, 3, 843–852, 2014.
116. Singh, S., Gupta, R., Chawla, S., Gauba, P., Singh, M., Tiwari, R.K., Upadhyay, S., Sharma, S., Chanda, S., Gaur, S., Natural sources and encapsulating materials for probiotics delivery systems: Recent applications and challenges in functional food development. *Front. Nutr.*, 9, 971784, 2022.
117. Rajam, R. and Subramanian, P., Encapsulation of probiotics: past, present and future. *Beni-Suef Univ. J. Basic Appl. Sci.*, 11, 46, 2022.
118. Gheorghita, R., Anchidin-Norocel, L., Filip, R., Dimian, M., Covasa, M., Applications of Biopolymers for Drugs and Probiotics Delivery. *Polymers*, *13*, 16, 2729, 2021.
119. Seyedehhamideh, R., Sajjad, J., Nishat, T., Deanna, L.G., Mina, H., Nanomaterial-based encapsulation for controlled gastrointestinal delivery of viable probiotic bacteria. *Nanoscale Adv.*, 3, 2699–2709, 2021.
120. Senbagam, D., Senthikumar, B., Amutha, R., Chidambaram, P., Anubarasu, K., Chapter 21 - A food-grade nanoemulsion for delivering probiotics and prebiotics, in: *Bio-based nanoemusions for agri-food applications*, pp. 347–361, 2022.
121. Noman, M., Muhammad, A., Farhan, S., Aftab, A., Ali, I., Noor, A., Aasma, A., Yasir, A.S., Huda, A., Mohammad, R.K., José, M.L., Degnet, T.A., Effect of starch-based nanoparticles on the viability and stability of probiotics under adverse conditions. *Int. J. Food Prop.*, 26, 1, 1841–1854, 2023.
122. Sachin, K. and Karn, S.K., Microbial Fabricated Nanosystems: Applications in Drug Delivery and Targeting. *Front. Chem.*, 9, 617353, 2021.
123. Nakipoglu, M., Tezcaner, A., C.H.C., Nasim, A., Nureddin, A., Bioadhesives with Antimicrobial Properties. *Adv. Mater.*, 35, 49, 2300840, 2023.
124. Meng-Xin, H., Ji-Nian, L., Qian, G., Ya-Qian, Z., Hong-Mei, N., Probiotics Biofilm-Integrated Electrospun Nanofiber Membranes: A New Starter Culture for Fermented Milk Production. *J. Agric. Food Chem.*, 3198–3208, 67 (11), 2019.
125. Sharma, S., Jaiswal, S., Duffy, B., Jaiswal, A.K., Nanostructured Materials for Food Applications: Spectroscopy, Microscopy and Physical Properties. *Bioengineer (Basel, Switzerland)*, 6, 1, 26, 2019.
126. Rottinghaus, A.G., Amrofell, M.B., Moon, T.S., Biosensing in Smart Engineered Probiotics. *Biotechnol. J.*, *15*, 10, e1900319, 2020.

17

Functional Food and Food Innovation Toward Quality and Safety Regulations

Susmita Ghosh[1], Tanmay Sarkar[2] and Runu Chakraborty[1*]

[1]*Department of Food Technology and Biochemical Engineering, Jadavpur University, Kolkata, India*

[2]*Department of Food Processing Technology, Malda Polytechnic, West Bengal State Council of Technical Education, Govt. of West Bengal, Malda, India*

Abstract

The food industry, which is one of the most significant sectors of the economy overall, is crucial to agricultural raw material processing and food production. Although traditionally the food industry is thought of as having low research intensity, innovations are now acknowledged as a critical tool for enterprises operating in the sector to differentiate themselves from rivals and meet consumer expectations. Considerably, consumer expectations in the sphere of food production have evolved, especially in the last decade. People increasingly believe that food has a direct impact on their health. Thus, foods are no longer intended solely to satisfy hunger and offer essential nutrients but also, and to enhance both physical and mental health and prevent diseases related to poor nutrition. Furthermore, the food industry has been confronted with economic and technical alterations in manufacturing and food processing, which have had an important impact on the whole food supply chain, up to the distribution of food to end consumers, and forced companies to place a premium on food products that fulfill consumers' demand for a healthy lifestyle. As a result, innovation has been extensively researched within this conventional industry as well. In recent years, food industry innovations have mostly focused on innovative scientific and technological methods of food processing, as well as the introduction of novel cuisines. In this regard, functional food exhibits an important role, as evidenced by the rising expense of healthcare, the steady growth in life expectancy, and the desire of older people for a better quality of life in their later years. As a result, researchers agree that functional food is one of the most exciting areas of innovation and research

**Corresponding author*: crunu@hotmail.com

Mousumi Sen (ed.) Food Coatings and Preservation Technologies, (585–610) © 2025 Scrivener Publishing LLC

in the food industry. The growing awareness of the strong relationship between diet and human health has significantly altered food preferences in modern countries, leading consumers to prefer one food product over another to achieve some desired health end-state. Functional products are ideal food options in this respect because they strive to increase life quality by preventing nutrition-related disorders. Functional foods play a significant role in this regard, as evidenced by the rising demand for them brought on by the rising expense of healthcare, the steadily rising life expectancy, and the desire for a better quality of life afterward. The organoleptic qualities of the finished products and, consequently, consumer approval have frequently been altered by new advancements. Combining bioactive substances could make up for this shortcoming. Functional foods have been manufactured in all food categories, although they are not evenly spread across the food industry. Consequently, consumer preferences may differ in markets. Functional foods have primarily been introduced in the bakery, dairy, soft drinks, confectionery, and baby food industries. An increasing number of consumers are becoming concerned about functional foods in the hopes of reaping extra health benefits such as lowering the risk of certain diseases or promoting optimal wellness. The concept of food innovation is that it is the result of the combined efforts of many different sectors working within the framework of a network of institutions whose functions and interactions lead to, import, and disseminate new inventions. The increasing consumer knowledge of the health advantages of foods and their nutritional functions for preventing illnesses and enhancing health is driving the global functional food and nutraceutical market. Growing demand for functional foods is also assisting producers in diversifying their agricultural and marine-based crops, as well as stimulating research and innovation. Because of increased awareness of the health benefits of functional foods and a rise in disposable income, there is a growing demand for functional foods, particularly in developed economies. Meanwhile, the functional food market continues to encounter hurdles such as country-specific legislation and health claim validation. This chapter provides an overview of functional food Innovation in terms of quality, safety regulations, problems, and future prospects.

Keywords: Functional food, food safety, health benefits, food industry

17.1 Introduction

As science and technology advance, so does our understanding of what food is—from providing for our basic needs as a species to aiding in illness prevention and maintenance of health. Concern over the connections between food and health is growing among consumers, public health professionals, and those in the food business. Nutraceuticals and functional foods have emerged as two of the hottest topics in recent years.

To guarantee the safety of the products and prevent the companies from deceiving customers, numerous governments have implemented intricate regulatory frameworks [1]. Growing knowledge of the close connection between human health and diet has significantly altered consumer preferences in industrialized countries, encouraging them to select one specific food item over another to achieve a desired level of health. As a result, functional products are great dietary choices since they work to prevent diseases linked to poor nutrition and enhance quality of life [2].

Foods classified as functional may offer health advantages over just basic nourishment. This straightforward explanation also aligns with functional foods' main selling point, which is that these foods allow customers to live better lives without altering their eating habits. Functional foods have a bright future since most consumers want to purchase healthful food but are hesitant to alter their eating habits [3]. In the worldwide food and beverage market, health is the biggest trend and innovation driver. Health is now defined as the state of being physically, mentally, and psychologically healthy in the absence of sickness. A food is referred to as a "functional food" (FF) if it offers health advantages beyond its typical nutritional content [4].

Generally speaking, the term "functional food" describes products that offer additional health benefits to those that come from basic nutrition. As with food supplements and food for medical purposes, certain terminology in the realm of food offering health advantages has been unambiguously defined by EU legislation; however, there is currently no comprehensive definition that is agreed upon for functional food [5].

Food production is becoming more balanced and standardized, which is paving the way for more straightforward diets—a phenomenon known as "nutrition transition." It is typified by the substitution of a restricted number of high-energy plant and animal sources, especially refined carbohydrates and fats, for a varied diet. Even in developing nations, there is mounting evidence linking the shift in diet to an increase in non-communicable diseases like diabetes, heart disease, obesity, and cancer [6]. The food industry is currently undergoing a systemic change in the food manufacturing chain. Consumer expectations for foods with greater nutritional content, more components that promote health, and better functional and physical qualities are just starting to emerge [7]. Regarding that, natural raw materials and food processing by-products that have the potential to be used to meet such needs are chosen over non-food sources. However, as the population grows, natural resources become more scarce and expensive. To produce food from food by-products, more focus is being placed on reprocessing agro-food industrial leftovers and by-products. Beyond the current research being conducted to include food processing leftovers for nutritional, biological, or technical purposes as food ingredients, there is potential for their repurposing in the food chain [8].

17.2 Functional Food and Nutraceutical

A new class of food products called "functional foods" is put forth with the promise of enhancing well-being, quality of life, or health. The notion of health impacts linked to functional foods is predicated on a specific product and its functional components, as opposed to a traditionally healthy

Table 17.1 Classification of types of functional food.

Class of functional food	Characteristics	Examples	Reference
Enriched products	Addition of fresh nutrients that aren't often present in the standard food product	Fruit juices with calcium	(Domínguez Díaz, L., Fernández-Ruiz, V., & Cámara, 2020; Granato *et al.*, 2020; Guiné *et al.*, 2020)
Enhanced products	Improved products, and nutritional composition altered as a result of changes in the raw materials	Fruits with enhanced content of vitamins	
Fortified products	Higher nutrient content (compared to the conventional product)	Fruit juices with added vitamin C	
Altered products	Nutritional composition altered as a result of changes in the raw materials	Application of fibers as fat releasers in foods that have high-fat content	

diet as advised by nutrition experts [9]. Functional foods are defined as "food products fortified with special constituents that possess advantageous physiological effects," a word that was first used in Japan in 1994. The increasing dietary interest of consumers has consequently led to a global expansion in the consumption of functional foods (Table 17.1). According to several studies, people buy functional foods because they understand that they have health benefits that regular foods do not. However, consumer choice is influenced by some factors, including price and flavor, in addition to the health benefits of functional foods (Figure 17.1) [2].

Foods that exhibit potentially healthy products, including any modified foods or food ingredients that provide a health benefit beyond the nutrients. In addition to adequate nutritional roles, they induce crucial roles in the target functions of the organism, significantly enhancing the health condition or reducing the risk of disease. Nutrients consumed as part of a normal diet but deliver active components within the food matrix [10]. A reliable brand with strong market awareness is necessary for functional food products to inspire consumer confidence in their purchase. For novel functional food products to be commercialized, a thorough business plan must include creating a brand to attract consumers, obtaining proprietary ingredients to guarantee sales exclusivity, and obtaining effective clinical research to support scientifically validated health claims. In the majority of

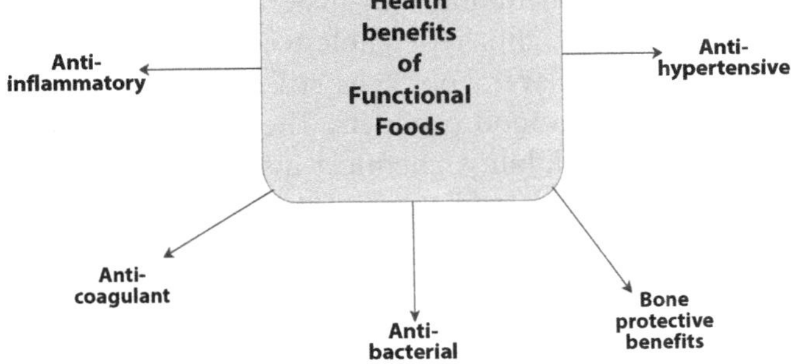

Figure 17.1 Health benefits of functional foods.

jurisdictions, authorization from regulatory bodies will also be necessary [11].

The idea of nutraceuticals is regarded as existing on the boundary between pharmaceuticals and nutrition. The phrase "nutraceutical" was coined in 1989 by Stephen L. DeFelice, the founder of the Foundation of Innovation Medicine, and is a portmanteau of the words "nutrient" and "pharmaceutical." He defined it as "a food or parts of food that provide medical or health benefits, including the prevention and/or treatment of disease." People used to associate this concept's intricacy with the purported Hippocratic ideal, "Let food be thy medicine and medicine be thy food," – which led to the widespread misconception that food was synonymous with medicine. Nutraceuticals are marketed as "food-pharma products" with vague information regarding their ability to cure or prevent illnesses, misleading consumers in the process. For nutraceuticals to be utilized "beyond the diet but before the drugs" in the management of particular diseases, a precise and thorough description, as well as an international regulation, are needed [12]. A dietary supplement known as a nutraceutical can provide a concentrated form of an assumed bioactive component from food in a non-food matrix to improve health at dosages higher than those that could be obtained from typical foods. Functional foods are those that resemble conventional foods in appearance (beverages, food matrices), are eaten as a regular part of the diet, and have physiologically beneficial biologically active ingredients that have the potential to reduce the risk of chronic disease and serve basic nutritional purposes [13]. Consumer diets are starting to include nutraceuticals regularly. The population's growing awareness of preventative health measures is mostly due to the rising occurrence of lifestyle-related disorders. In the natural products market, functional foods hold the highest proportion, followed by functional drinks and food supplements. The many substances that may be used in nutraceutical formulations have been linked to several distinct mechanisms of action, including anti-hypertensive, anti-inflammatory, anti-bacterial, anti-cancer, anti-hypercholesterolemic, anti-coagulant, and bone-protective benefits. Several obstacles still need to be overcome before they can be included into food products. These include issues with GIT absorption, low water solubility, chemical instability, sensitivity to temperature, light, and oxygen, and more. There could be several negative interactions between the drugs present and nutraceuticals [14].

Inadequate nutrition is a primary risk factor for long-term illnesses; a lot of customers want to actively participate in preventing sickness. Nutraceuticals are food-based substances that are used for sickness prevention and treatment, as opposed to medications. Nutraceuticals are used to prevent diseases and improve general health. Globally, as more and more people incorporate nutraceuticals into their regular diets, their popularity is growing. Current developments in the worldwide nutraceutical sector include an increase in non-communicable illnesses, high healthcare costs, high employee income, lifestyle changes, people's thought patterns and applications of nutraceuticals as medications, mistrust toward conventional medicines, and a demand for convenience. Several factors must be integrated to develop functionally enhanced foods, and processing has an impact on the finished food items. The food industry and research institutions are now utilizing the newest methods, which hold the potential to shorten processing times, regulate Maillard reactions, enhance product quality, and boost functionality [15].

Fundamental characteristics of functional foods include the following:

- it has a significant effect on target functions beyond basic nutrition, and it enhance well-being and provide health benefits that improve the quality of life, including psychological, physical, and behavioral performances, or reduces the risk of disease.
- It is a conventional food, and it is consumed as part of a daily diet.
- It consists of naturally occurring components in unnatural concentrations [16].

17.3 The Functional Food Innovation System

"Functional food" is important for both human health and modern nutrition. In Japan, the phrase "functional food" first appeared in the 1980s. "Foods whose consumption is associated with health benefits in terms of reducing disease risk and/or supporting the maintenance of proper health" is how the American Dietetic Association (ADA) recently defined functional foods. Functional foods are utilized to achieve significant therapeutic effects linked to the prevention of diseases related to nutrition, in addition to serving as a source of critical nutrients. These foods include products (such as cereals and oilseeds) that are rich in bioactive substances, like proteins, polyphenols, and essential fatty acids (EFAs) (Table 17.2) [17].

Nevertheless, businesses in the food sector are aware that innovations—defined as new products, procedures, or services—are crucial tools for differentiating themselves from rivals and meeting customer demands. Particularly in the past ten years, customer demands in the food production

Table 17.2 Functional foods in innovative food systems.

Food item	**Functional product**	**Health benefits**	**Reference**
Pomegranate seed powder	Bread	▪ Increased anti-oxidant function ▪ Enhanced bioavailability	(Alongi & Anese, 2021; Birch & Bonwick, 2018; Giannetti *et al.*, 2009; Pérez-Marroquín, X. A., Estrada-Fernández, A. G., García-Ceja, A., Aguirre-Álvarez, G., & León-López, 2023)
Chia oil	Bread	▪ Enhanced bioavailability ▪ Improved cardiovascular activity	
Fish oils	Bread	▪ Enhanced bioavailability ▪ Improved fatty acid profile	
Sorghum flour	Bread	▪ Reduced oxidative stress ▪ Decreased Inflammation	
Pomegranate peel	Tea	▪ Increased anti-oxidant function ▪ Enhanced bioavailability of polyphenols	
Pumpkin	Juice	▪ Improved anti-oxidant function ▪ Enhanced detoxification function	
Red potato	Red potato	▪ Improved anti-oxidant function	
Eggplant peel	Juice	▪ Increased anti-oxidant function ▪ Enhanced bioavailability of polyphenols	
Amaranth	Cookies	▪ Increase Antithrombotic property ▪ Enhanced Antihypertensive function	
Broccoli leaf powder	Sponge cake	▪ Improved anti-oxidant function	
Apple pomace	Biscuits	▪ Enhanced anti-diabetic property ▪ Improved glycaemic index	

sector have evolved significantly. A growing number of consumers now assert that food directly affects their health. Foods are now meant to avoid disorders connected to nutrition and to enhance both physical and mental health, rather than just to satiate hunger and supply essential nutrients [18].

From a technological perspective, there exist numerous methods for creating novel functional foods, and numerous varieties of functional foods can be created. There are now three primary categories into which functional foods can be categorized: Foods classified as functional are in the first category; their manufacturers assert that consuming these foods improves human health. There are several methods for creating innovative food products with health benefits: There are four ways to improve a product: 1) by adding extra nutrients to fortified foods; 2) by adding nutrients that are typically absent from the product (enriched foods); 3) by substituting more beneficial components for potentially harmful or undesirable ones (altered products); and 4) by creating enhanced commodities, which include developed products with higher concentrations of certain health-promoting ingredients [19, 20].

The process of innovation is not linear or one-sided. Within the framework of an innovation system—a network of institutions whose interactions and activities generate, import, and disseminate innovations—innovation is understood as the result of the combined efforts of many actors. Thus, every actor has an impact on the innovation process's ultimate result. The actors involved must act with limited logic and do not have access to all the facts. A network of interconnected actors with a range of interests and perspectives, including consumers, food production companies, intermediary research institutes, authorities, universities, and the supportive infrastructure, which includes investors and consulting firms, carry out the science-based innovation trajectory that is new product development, such as functional foods [21]. Food products that offer additional health benefits beyond their conventional nutritional value are referred to as functional food products. The primary driving force behind consumer purchases of functional foods is the increasing interest in using foods to either help prevent chronic illnesses like osteoporosis, Alzheimer's disease, and cardiovascular disease or to optimize health by, for example, boosting immunity, generating well-being, or increasing energy [9].

17.4 Nutritional Function

The main goal of nutrition is to satisfy the body's need for certain nutrients while also offering the customer a sense of fulfilment and well-being.

However, recent research supports the theory that diet may influence several physiological processes and may have positive or negative effects on some diseases in addition to providing for nutritional needs. The field of nutrition sciences is on the verge of a new frontier. Concepts are evolving, at least in the West, away from the previous emphasis on survival, satisfying hunger, and avoiding negative effects and toward the use of foods to extend life expectancy and older people's desire for better [22].

The phrase "nutritional functions" describes how nutrients aid in the growth, development, and upkeep of the human body's organs and systems. The basic functions of food are to provide minerals, vitamins, and energy derived from the proteins, carbohydrates, and lipids required for the well-being of the human body. Optimizing the daily diet in terms of nutrients and non-nutrients, as well as other dietary attributes that contribute to "maintenance health," is the stated goal of nutritional research [18].

17.5 Designing of Functional Food Development Cycle

Although consumers have been looking for functional foods for the past 20 years to help maintain good health and lower the risk of non-communicable diseases, particularly type II diabetes, cardiovascular

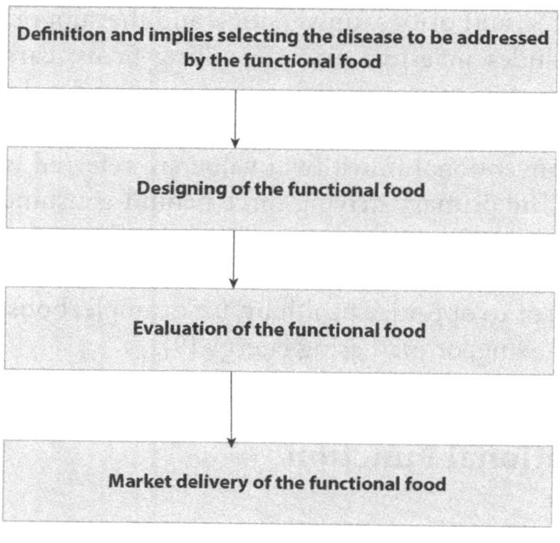

Figure 17.2 Development of functional food.

disease, and hyperlipidemia, there is no standardization on the procedures followed or the credentials of the professionals involved in the development of functional foods (Figure 17.2) [23].

Establishing a goal is the first stage in producing a functional food. This might be chosen by two diametrically opposed approaches: the first is a top-down strategy that stems from an innovative notion to address food-related health issues. This strategy, while potentially novel, may not meet consumer expectations as it fails to incorporate all relevant actors to the fullest extent. Consequently, it would be preferable to use a bottom-up strategy to ensure consumer approval while providing practical answers to food-related health problems [24].

17.5.1 Dealing with Food Functionality

When it comes to food functioning, co-ingestion of other foods during a meal with medications used to treat chronic illnesses is the first important factor to take into account. Instead of taking into account the potential positive effects, food-drug interactions are often investigated from the perspective of side effects. Certain research did, however, draw attention to the potential benefits of these interactions. Food-drug interactions have been observed to have an impact on a wide range of physiological processes, most of which are associated with the emergence of chronic non-communicable diseases. Nonetheless, the majority of research published in the literature focuses on extracts of bioactive compounds from model systems, with very few instances addressing the entire food matrix. This is because the complexity of the food matrix makes it difficult to investigate how food and medicine interact [25].

17.5.2 Formulation

From a technology perspective, customizing food formulations is the first step towards delivering health benefits through food. Specifically, the formulation can be focused on adding bioactive chemicals or changing or removing certain elements to get:

- fortified food, in which a bioactive compound has been added (e.g., prebiotics and probiotics, vitamins A, D, and calcium in vegetable beverages) or its presence has been increased (e.g., dietary fiber in bakery products, vitamins in fruit juices), respectively. These can be attained by supplementing with naturally occurring bioactive substances to strengthen their delivery.

- light food, in which the concentration of an undesired component is decreased; it results from an increase in the fraction of non-caloric foods or the elimination of unwanted ingredients [26].

17.5.3 Processing

Processing is a crucial factor that influences food functionality, along with formulation, diet, and other factors. It is generally accepted that this reduces the overall nutritional value and goes against one of the main goals of functional food design, which is to improve micronutrient intake and absorption by increasing the digestibility of macronutrients and depleting micronutrients. Just a small number of studies have examined the impact of conventional and, more significantly, unconventional processing on food functionality, although it is widely known to significantly alter matrix composition and interactions [5].

17.6 Consumer Trends and Attitudes to Functional Foods

The food industry has grown to include food ingredients, cosmetics, nutritional supplements, nutraceuticals, and of course functional foods. In the realm of economics, functional foods are crucial for the repurposing of food industry by-products that would otherwise go to waste. It is anticipated that the functional foods sector will keep expanding for many years to come. As a result, appropriate laws and consumer protection are also necessary [27].

The term "optimal nutrition" has replaced "adequate nutrition" in people's understanding of the role of nutrition plays in their health, giving rise to functional foods (FFs). In addition to being devoid of chemicals and biologically modified materials, the consumer wants his diet to provide his body with additional health benefits beyond just basic nourishment. The food industry frequently uses the health and nutritional claims made on a wide range of items. Nonetheless, a climate of misunderstanding and mistrust has developed between stakeholders and consumers, medical professionals, and businesses in recent years as a result of the quick dissemination of these health claims [28]. Functional foods have the potential to be utilized in the prevention and treatment of diseases. They may enhance general bodily conditions (like probiotics), lessen the risk of certain diseases (like cholesterol-lowering products), and more. These foods began as foods fortified with vitamins and minerals (folic acid, vitamin C, vitamin E, zinc, iron, and calcium), and over time they developed into foods that support a healthy diet and guard against several illnesses, including micronutrients like soluble fibers, phytosterols, and omega-3 fatty acids [29]. Customers agree that diet and food have an effect on both short-term and long-term health, and they have a positive perception of healthy eating. Foods that fall within the "naturalness" and "well-being" categories are actively sought by consumers. Customers are showing a growing preference for "natural" over "synthetic" pharmaceuticals, choosing food products that provide useful benefits over pills. Another essential component of a functional food is its taste. Although consumers are drawn to natural, wholesome, and nutrient-dense food products, it can be difficult to translate these preferences into actual products [2]. The growing interest in functional foods is driven by progressive research efforts to identify potential applications of nutraceutical substances, as well as by consumer demand and public interest. Nutraceuticals and functional foods offer a means of reducing the growing burden on the health care system through continuous preventive mechanisms. The present demographic and health trends are the main drivers of the functional food market's growth. The purpose of this review is to give a concise overview of the health advantages of various nutraceuticals and functional foods, with a focus on cancer, obesity, and coronary heart disease [13].

17.7 Functional Foods From Plant and Animal Sources

Since 1984, certain food products that offer alleged health benefits beyond their nutritional value have been referred to as "functional foods." The term, which is frequently used synonymously with "nutraceutical," has no accepted definition as of yet and has been given many definitions by both academic and non-academic sources. In this context, "functional foods" are defined as natural or processed foods that have one or more bioactive food elements, such as chemicals or other food ingredients that have a favorable effect on human health, such as probiotics, vitamins, carotenoids, and flavonoids. Numerous conditions have been demonstrated to be attenuated by these, including diabetes, autoimmune illnesses, obesity, cardiovascular diseases, and cancer [30].

It is essential to vary the foods one consumes since a balanced diet should give the body the nutrients it needs in sufficient proportions. Thus, there is a growing need for foods that are healthier and more nutrient-dense, which opens up new avenues for research into these foods across a range of industries [31]. A plant-based diet can lower the incidence of chronic disease, especially cancer, according to a large body of research from epidemiology (*in vivo*, *in vitro*, and clinical trial) data. Individuals who had a diet rich in fruits and vegetables had a cancer risk that was only half that of individuals who consumed fewer of these foods. It is now evident that a plant-based diet can lower cancer risk in ways other than just consuming traditional nutrients. The most significant phytochemicals are found in fruits and vegetables as phenolics and carotenoids and in cereal-based products as lignans, β-glucan, and inulin. Some of the major beneficial advantages of plant-based diets include lowering blood pressure, heart disease risk, cholesterol, tumor incidence, slowing gastric emptying, and preventing cancer and CVDs [13]. Bioactive substances included in functional food products are extra nutritional components that naturally exist in plants and have biological effects.

Foods and substances that show health advantages above and beyond their nutritional value are referred to as functional foods. Functional foods are the subject of growing research because they may be able to manage and prevent several non-communicable diseases. It is well known that whey proteins are a high-quality shice of nutrients and that they include certain bioactive ingredients. They are abundant in bioactive peptides, branched-chain amino acids (including valine, leucine, and isoleucine), and important amino acids like cysteine. Given their anti-obesity, anti-oxidant, anti-inflammatory, blood pressure-lowering, and appetite-suppressing properties that have been explored in the literature, whey proteins appear to be a suitable candidate for use as functional foods. Additionally, whey proteins have useful qualities that are known to enhance food processing as an emulsifier, fat substitute, gelling agent, and encapsulating component [32]. Consuming naturally occurring bioactive chemicals in food is linked to a lower prevalence of many chronic illnesses. Studies in epidemiology, medicine, and biochemistry have demonstrated that these bioactive substances operate in the human body in a variety of ways, including as antioxidants, antidiabetic agents, hypertensive agents, anti-Alzheimermic agents, antiproliferative agents, and antimicrobial agents. These special qualities have been connected to certain bioactive dietary ingredients or molecules that can shield against illness and advance wellness [33].

While a great deal of naturally occurring compounds that are known to improve health come from plants, animal products also contain various physiologically active ingredients that are worth considering for their potential to contribute to good health. Dairy products Dairy products are undeniably useful foods; milk is a full food in and of itself. They are among the best providers of calcium, a necessary mineral that may help prevent colon cancer and osteoporosis. However, more recent studies have concentrated on additional elements found in dairy products other than only calcium, including probiotics, which are fermented dairy products. Probiotics are live microbial feed additives that improve the balance of intestinal microbes of the host animal, which has positive effects on it [13]. For customers who are not aware of the notion of functional foods, probiotics, and prebiotics are two product categories that have been widely available in the market in recent years. The most popular category is easy-to-drink yogurts. These items have a growing body of scientific evidence supporting their beneficial effects on gut health through dietary modification of the microbiota in the human gut. The Food and Agriculture Organization (FAO)/World Health Organization (WHO) described probiotics as "live microorganisms that, when administered in adequate amounts, confer a health benefit on the host" and noted that many of them were initially isolated from the gastrointestinal system [34]. Microgreens are partially developed greens that have strong flavor, scent, texture, and nutritional qualities. They are also accepted by the senses. These superfoods' powerful flavors and desirable textures have won praise for their ability to be eaten as salads, garnishes, etc. Microgreens' functional characteristics connected to health and nutrition are enhanced by the presence of vitamins, minerals, and their precursors, such as ascorbic acid, and various other bioactive substances, such as tocopherols, carotenoids, phenols, glucosinolates, and phytosterols [35].

17.8 Quality, Safety Regulations, and Functionality of Functional Foods

Several health issues and even fatalities have been linked to the epidemic of non-infectious diseases as a result of changes in lifestyle, such as poor diet and insufficient exercise. Functional food (FF) affects particular bodily processes, can offer extra health advantages, or treat certain illnesses by increasing the quantity or inclusion of a healthy element, or can eliminate or replace a hazardous or inefficient ingredient. Food items that are incorporated into a daily diet and have the appearance of traditional foods are referred as FF. These products improve the body physiologically and/or lower the chance of developing non-communicable diseases [23].

Practical advantages providing consumers with more than just nutrition offers a chance for above-average profits, which has led to several advances in the food industry. A new inter-industry segment between foods and medications has emerged as a result of the development of functional foods and supplements. Not all inventions, nevertheless, are successful, as deciding how to position a new product between food and medication is a difficult undertaking with numerous obstacles [36].

By educating consumers about healthy eating habits that may help lower their risk of heart disease, cancer, osteoporosis, high blood pressure, dental cavities, or specific birth abnormalities, health claims aim to benefit consumers. Nonetheless, food producers may advertise and promote their proucts using health claims. Among the many kinds of claims that are permitted on food labels are health claims. They show a connection between an illness or other health issue and a nutrient or other ingredient in the diet. They can be used as dietary supplements or traditional foods. The majority of functional food researchers concur that before any product with a specific health claim is approved, it should undergo scientific examination and be labeled as "functional" and associated with particular health advantages. Many foods are linked to particular health claims in nations with strong cultural traditions of believing that certain foods and dietary ingredients can improve one's health, even though these claims may not be supported by scientific evidence [37].

Numerous social issues, including food security, business profits, environmental protection, climate change, and the expanding need for food, provide challenges for the food industry. The environment, health, economy, and food manufacturing have all changed as a result of these issues. Classic engineering, which produced food products based on raw materials and technology and put the needs of the consumer at the end of the food chain, has given way to modern designs based on reverse engineering [38]. The bioactive ingredients and nutritional value of the restructured functional foods differ from those of the foods in their original state. Finding the bioactive components and nutritional qualities of the restructured foods is important from a scientific standpoint. Destructive testing, the primary method used in traditional component detection techniques, is laborious, time-consuming, and inefficient [39]. The simplicity and speed of non-destructive testing (NDT) have made it an emergent technology. The bioactive ingredients and nutritional value of the restructured foods differ from those of the foods in their natural state. Thus, research on the nutritional qualities and bioactive elements of reconstructed functional foods has practical significance and scientific value. There are now two primary categories of restructured foods. Compound foods are the most basic kind; they are merely altered in the recipe by combining edible and medicinal homologs to obtain a different texture. The second kind is the newly developed technology in food production, which is represented by innovative external forms and internal characteristics [40].

17.9 Integrating Modern Processing Technology to Develop Novel Functional Foods

Multi-stakeholders involved in food systems have been encouraged by the growing consumer consensus to deviate from the traditional method of food production by augmenting or adding health advantages that enhance food items' functionality. Functional foods, sometimes referred to as "future foods," go above and beyond basic nutrition in terms of their health advantages. Foods must meet several requirements to be marketed as functional, such as being free to access, meeting international food safety standards, conforming to food safety regulations in the country in which they are exported, and providing evidence of the health benefits of consuming them as part of a balanced diet. Probiotics, antioxidants, prebiotics, vitamins, dietary supplements, dietary fiber, and other phytochemicals are examples of functional foods that are widely available in the market.

The production of functional foods, which involves the design, development, and optimization of various formulations as well as the manufacturing methods used on food products before their release onto the market, is crucial for food firms. For example, the use of heat processing has a significant impact on the bioavailability of nutrients and bioactive ingredients found in food. Innovative food processing technologies (such as pulsed electric fields, high hydrostatic pressure, ultrasounds, and microwaves) have become viable options for food processing over the last 20 years. These environmentally friendly technologies can be used to harvest by-products while preserving the natural nutrients found in fruits and vegetables, preventing microbial growth, and using less energy [41]. To demonstrates a novel functional food's safety and demonstrate its benefit to people, a clinical evaluation is required. *In vivo*, testing ought to be the foundation for both the functionality evaluation and the health-claim formulations of potentially functional foods and substances. These actions are required to choose appropriate biomarkers for assessing the potential's existence and extent. New ingredient efficacy and safety testing, as well as prospective functional food testing, must be included [41].

17.10 The Future of Functional Foods

Foods that offer health benefits beyond simple nourishment because they include physiologically active ingredients are referred to as functional foods. Numerous physiologically active functional dietary ingredients, known as phytochemicals from plants and zoochemicals from animals, are currently being studied for their possible involvement in disease prevention and health promotion. 'Nutraceuticals,' are defined as 'those diet supplements that provide a food's bioactive component in a concentrated form, given in a non-food matrix, and utilized to improve health at dosages higher than what could be acquired from regular food. Nutraceuticals and functional foods are dietary items that offer health advantages beyond simple nourishment. The goal of improving health, or preventing sickness (the range of functional foods), or even helping to prevent or treat specific diseases (the scope of nutraceuticals) is one of the motivations that drive food innovation. Functional foods are acknowledged to offer advantages beyond the basic roles of supplying nutrients and satisfying hunger or desire to eat, even though they lack widely recognized or legally defined criteria. Functional food development has taken place in all food categories (foods and drinks of all kinds), albeit more intensively in some [42]. A diet must provide something of value to the body beyond simple nourishment, such as a significant physiological consequence, to be classified as functional. Nutraceuticals, such as vitamin pills or fish oil capsules, are not functional foods; rather, they are nutraceuticals that must be utilized in the same way as conventional foods. This is another crucial requirement for using functional foods. Foods can be considered "functional" in several ways. Fruits, which may be strong in antioxidants and fiber, and oily fish, which have a high omega-3 fatty acid content, are two examples of these.

Over the past ten years, there has been consistent growth in the food market for both organic and functional foods. Functional food describes the nutritional value and/or health-related features of a product, whereas organic food encompasses the complete production process. Both foods have labels, which convey messages about quality and/or health that suggest consuming foods of a higher caliber or that has additional health benefits. Customers interpret these messages differently, even though they are present in both kinds of food. Moreover, the conceptual foundation of organic and functional food differs in terms of quality conceptions, reflecting the divergent underlying paradigms of these two distinct food categories [43].

The top 10 functional food trends have been determined, although there are differences between the trends and calendar years. Simplicity, sustainability, traditional nutritious foods, functional foods, immunity-boosting products, energy foods, "free-from" foods, extreme flavors, and authenticity were the top new food development trends for 2010. With digestive health, elderly nutrition, weight management, child and adolescent nutrition, obesity, and beauty-enhancing products still at the top of the list, health, indulgence, and convenience remained the main goals for new product development and innovation worldwide [22, 44].

The food industry has been quite active, reformulating and repositioning many "healthier" products in response to consumer demand, competition, and the new nutrition label's emphasis on simple comparability. It has been more challenging to pay attention to the development of functional foods, as the majority of discoveries cannot be marketed to consumers without a valid health claim. In a broader sense, consumers have been exposed to a lot of nutrition news, stronger public health messaging, and marketing campaigns, which have piqued their curiosity about the potential for such health-improving food products [45].

Nutraceuticals, food supplements, and functional foods may have a significant impact on human health by lowering the risk factors for cardiovascular disease (CD) and enhancing general health. Functional products can be added to a healthy diet or used as a means to enhance an unhealthy one for people with modest risk factors for CD. Beyond this, and with a physician's supervision, functional items may be utilized as an adjunct in patients receiving well-established medical therapy, particularly when prescribed medications are insufficient to achieve desired results [12].

Redefining the limits of food production has become necessary in recent years. Currently, we are addressing significant and opposing issues. Food security is in jeopardy due to the exponential expansion of the global population. Conversely, non-communicable diseases are on the rise and are largely attributable to imbalanced or overindulgent diets. To address these trends, a more comprehensive examination of the food production system is required. To put it another way, one must consider the food chain [10]. Consumers who care about their health are looking for functional foods more and more to take charge of their health and well-being. Still, the field of functional foods is very new.

Functional food claims concerning its health benefits have to be supported by reliable scientific evidence. Establishing a strong scientific foundation is challenging due to several factors, including the complexity of the food component, influences on the food, compensatory metabolic changes that may occur with dietary modifications, and the lack of surrogate markers of illness onset. Furthermore, it is challenging to pinpoint the right bioactive compounds and determine how they might help prevent or treat different illnesses. For such foods, further research is required to support the possible health advantages, as the diet-health linkages are not sufficiently scientifically established [13].

17.11 Conclusion

Consumer attitudes about functional food products, health-related incentives, and labeling may all have an impact on their decision to purchase. Because they enable customers to make more educated food choices, stable claims provide a high degree of consumer protection. The identification and validation of pertinent indicators and parameters for the process (just organic) and product aspects (both organic and functional) is a major issue for scientific study in both fields. Moreover, food appears to be the basic building block of nutrition rather than just individual components. Approaches that take into account the dynamic reactivity of an extremely complex system are necessary for health research examining the impact of both organic and functional foods. Nutraceuticals, food supplements, and functional foods are what sit between pharmaceuticals and nutrition. Functional items may help to maintain and improve human health, well-being, and quality of life while also lowering the risk of CD. Together with authorized medication, they may offer an additional feasible option for managing CD. Nutrition and pharmaceutical combinations must be properly communicated to prevent food-drug interactions and to substitute pharmaceutical treatments with the use of these useful products. To determine which of these combinations offers the best benefit-risk and cost-effectiveness ratio for each chronic disease, further immediate and long-term research is necessary. In the twenty-first century, consumers are more likely to experience health issues, stress, environmental degradation, and societal obstacles. Functional items may contribute to better mental and physical health and, ultimately, a greater standard of living. The analysis of the literature revealed that the two most positive factors of FF adoption are health advantages and use motivation. When it comes to health, women and the elderly are more willing to make compromises and show interest in fashion. Unaffected by socio-demographic variables, low nutritional awareness may restrict FF uptake. It is important to note that family members' health issues raise consumers' interest in useful products.

As long as the idea of functional foods is widely accepted by the general public, functional foods, drinks, and ingredients will be vital to human nutrition. Governments everywhere should provide financial incentives to enterprises dedicated to the creation of foods that have the potential to improve human health beyond basic nourishment, as a result of financial responsibilities. Food scientists and technologists need to be aware that to validate any health claims made about functional foods, *in vitro*, *in vivo* (animals), and clinical trials are necessary. If these conditions aren't met, the produced food is only nourishing and not useful. To fully comprehend and develop functional foods, science, technology, and health should work together in harmony. The production and chemical stability of these chemical compounds should be increased by using extraction technologies and optimizing experimental settings. It is important to thoroughly examine any new research that focuses on the use of natural extracts in various food matrices and how they affect rheological, functional, physicochemical, and sensory aspects. The functional food market still needs to be investigated from a technological and commercial perspective, given the recent scientific developments in human nutrition and food science.

To study the effects of foods on human metabolism and obtain a multidisciplinary perspective on food, food scientists should collaborate with experts in other domains. Robust scientific evidence has consistently demonstrated the connection between human health and diet, and modern cultures have come to expect food products that offer more than just sating hunger and providing nourishment. At the intersection of nutrition and pharmaceuticals, functional foods, food supplements, and nutraceuticals provide new avenues for the search for therapeutic solutions in the fight against nutrition-related illnesses. Nutraceuticals, food supplements, and functional foods all have health-promoting qualities and may be useful in managing chronic illnesses when used in conjunction with prescription drugs. To determine which pharmacological and nutritional combinations are best for each chronic illness, more research is necessary. The willingness of consumers to pay a premium as well as their propensity to make purchases are crucial factors in the development of new items and aid food manufacturers in determining how much profit they may anticipate from their sales. This is significantly more given that food manufacturers today allocate sizeable funds to the research and development of novel food products, such as functional foods. The degree of consumer involvement with the product, the consumer's lifestyle, the product's sensory qualities and other characteristics, price, brand, country of origin, potential health claims made on the packaging, and the product's benefits are reportedly some of the factors that affect consumers' willingness to purchase functional foods.

References

1. Patel, D., Dufour, Y., Domigan, N., Functional food and nutraceutical registration process in Japan and China: Similarities and differences. *J. Pharm. Pharm. Sci.*, *11*, 4, 1–10, 2008, https://doi.org/10.18433/j32s3n.
2. Pappalardo, G. and Lusk, J.L., The role of beliefs in purchasing process of functional foods. *Food Qual. Preference*, *53*, 151–158, 2016, https://doi.org/10.1016/j.foodqual.2016.06.009.
3. Lau, T., Chan, M., Tan, H., Kwek, C., Functional Food : A Growing Trend among the Health Conscious. *Asian Soc. Sci.*, *9*, 1, 198, 2013, https://doi.org/10.5539/ass.v9n1p198.
4. Oliveira, G.R., Vieira, L.M., de Barcellos, M.D., Hope, A., Institutional Barriers for Food Innovation: A Study of the Brazilian Functional Food Industry. *J. Oper. Supply Chain Manage.*, *7*, 1, 1–14, 2014, https://doi.org/10.12660/joscmv7n1p1-14.
5. Alongi, M. and M.A., Re-thinking functional food development through a holistic approach. *J. Funct. Foods*, *81*, 104466, 2021, https://doi.org/10.1016/j.jff.2021.104466.
6. Hermann, M., The impact of the European Novel Food Regulation on trade and food innovation based on traditional plant foods from developing countries. *Food Policy*, *34*, 6, 499–507, 2009, https://doi.org/10.1016/j.foodpol.2009.08.005.
7. Ajeeshkumar, K.K., Aneesh, P.A., Raju, N., Suseela, M., Ravishankar, C.N., Benjakul, S., Advancements in liposome technology: Preparation techniques and applications in food, functional foods, and bioactive delivery: A review. *Compr. Rev. Food Sci. Food Saf.*, *20*, 2, 1280–1306, 2021, https://doi.org/10.1111/1541-4337.12725.
8. Gemechu, F.G., Embracing nutritional qualities, biological activities and technological properties of coffee byproducts in functional food formulation. *Trends Food Sci. Technol.*, *104*, 235–261, 2020, https://doi.org/10.1016/j.tifs.2020.08.005.
9. Giannetti, V., Testani, E., Recchia, L., Food consumption and innovation: Functional foods. *J. Commod. Sci. Technol. Qual.*, *48*, 3, 213–225, 2009.
10. Alongi, M. and Anese, M., Re-thinking functional food development through a holistic approach. *J. Funct. Foods*, Elsevier Ltd, 2021, https://doi.org/10.1016/j.jff.2021.104466.
11. Khan, R.S., Grigor, J., Winger, R., Win, A., Functional food product development - Opportunities and challenges for food manufacturers. *Trends Food Sci. Technol.*, *30*, 1, 27–37, 2013, https://doi.org/10.1016/j.tifs.2012.11.004.

12. Domínguez Díaz, L., Fernández-Ruiz, V., Cámara, M., The frontier between nutrition and pharma : The international regulatory framework of functional foods , food supplements and nutraceuticals. *Crit. Rev. Food Sci. Nutr.*, *60*, 10, 1738–1746, 2020, https://doi.org/10.1080/10408398.2019.1592107.
13. Gul, K., Singh, A.K., Jabeen, R., Critical Reviews in Food Science and Nutrition Nutraceuticals and Functional Foods : The Foods for Future World. *Crit. Rev. Food Sci. Nutr.*, *56*, 16, 2617–2627, 2015, https://doi.org/10.1080/10408398.2014.903384.
14. Reque, P.M. and Brandelli, A., Encapsulation of probiotics and nutraceuticals: Applications in functional food industry. *Trends Food Sci. Technol.*, *114*, 1–10, 2021, https://doi.org/10.1016/j.tifs.2021.05.022.
15. Jayaweera, J.A.A.S., Current trends and technologies in nutraceutical industry, in: *Herbs, Spices and Their Roles in Nutraceuticals and Functional Foods*, pp. 347–360, Academic Press, Cambridge, USA, 2023.
16. Annunziata, A. and Vecchio, R., Agri-food innovation and the functional food market in Europe: Concerns and challenges. *EuroChoices*, *12*, 2, 12–19, 2013, https://doi.org/10.1111/1746-692X.12027.
17. Motyka, S., Skała, E., Ekiert, H., Szopa, A., Health-promoting approaches of the use of chia seeds. *J. Funct. Foods*, *103*, 105480, 2023, https://doi.org/10.1016/j.jff.2023.105480.
18. Bigliardi, B. and Galati, F., Innovation trends in the food industry: The case of functional foods. *Trends Food Sci. Technol.*, *31*, 2, 118–129, 2013, https://doi.org/10.1016/j.tifs.2013.03.006.
19. Moors, E.H., Donders, R., Verduijn, S., Understanding consumer needs and preferences in new product development : the case of functional food innovations, *Innovation Studies Utrecht (ISU) Working Paper Series*, 09–03, 2009.
20. Pascall, M.A., Deangelo, K., Richards, J., Arensberg, M.B., Role and Importance of Functional Food Packaging in Specialized Products for Vulnerable Populations : Implications for Innovation and Policy Development for Sustainability. *Foods*, *11*, 3043, 2022, https://doi.org/https://doi.o.
21. Moors, E.H.M., Functional foods: Regulation and innovations in the EU. *Innov.: Eur. J. Soc. Sci. Res.*, *25*, 4, 424–440, 2012, https://doi.org/10.1080/13511610.2012.726407.
22. Granato, D., Branco, G.F., Nazzaro, F., Faria, A.F., Cruz, A.G., Functional Foods and Nondairy Probiotic Food Development: Trends, Concepts, and Products. *Compr. Rev. Food Sci. Food Saf.*, *9*, 3, 292–302, 2010.
23. Granato, D., Nunes, D.S., Barba, F.J., An integrated strategy between food chemistry, biology, nutrition, pharmacology, and statistics in the development of functional foods: A proposa. *Trends Food Sci. Technol.*, *62*, 13–22, 2017, https://doi.org/10.1016/j.tifs.2016.12.010.
24. Gur, J., Mawuntu, M., Martirosyan, D., The Changing Face of Functional Foods. *Funct. Foods Health Dis.*, *8*, 7, 385–397, 2018.
25. Koziolek, M., Alcaro, S., Augustijns, P., Basit, A.W., Grimm, M., Hens, B., Trevaskis, N.L., The mechanisms of pharmacokinetic food-drug interactions

– A perspective from the UNGAP group. *Eur. J. Pharm. Sci.*, *134*, 31–59, 2019, https://doi.org/10.1016/j.ejps.2019.04.003.
26. Mcclements, D.J., Zou, L., Zhang, R., Salvia-trujillo, L., Kumosani, T., Xiao, H., Enhancing Nutraceutical Performance Using Excipient Foods : Designing Food Structures and Compositions to Increase Bioavailability. *Compr. Rev. Food Sci. Food Saf.*, *14*, 6, 824–847, 2015, https://doi.org/10.1111/1541-4337.12170.
27. Tsoupras, A., Zabetakis, I., Lordan, R., Functional foods: growth, evolution, legislation, and future perspectives, 2023, https://doi.org/https://doi.org/10.1016/B978-0-12-823811-0.00003-1.
28. Sun-waterhouse, D., The development of fruit-based functional foods targeting the health and wellness market : a review. *Int. J. Food Sci. Technol.*, *46*, 5, 899–920, 2011, https://doi.org/10.1111/j.1365-2621.2010.02499.x.
29. Karelakis, C., Zevgitis, P., Galanopoulos, K., Consumer Trends and Attitudes to Functional Foods. *J. Int. Food Agribus. Mark.*, *32*, 3, 266–294, 2019, https://doi.org/10.1080/08974438.2019.1599760.
30. Cheon, J., Haji, F., Baek, J., Wang, Q., Tam, K.C., Pickering emulsions for functional food systems. *J. Agric. Food Res.*, *11*, 100510, 2023, https://doi.org/10.1016/j.jafr.2023.100510.
31. Schmidt, D., Verruma-bernardi, M.R., Forti, V.A., Schmidt, D., Verruma-bernardi, M.R., Forti, V.A., Quinoa and Amaranth as Functional Foods : A Review. *Food Rev. Int.*, *39*, 4, 2277–2296, 2021, https://doi.org/10.1080/87559129.2021.1950175.
32. Menahem, C., Foist, M., Mansour, Y., Shtaif, B., Bar-Maisels, M., M.P., G.G.-Y., Whey proteins as a functional food: Health effects, functional properties, and applications in food. *J. Am. Nutr. Assoc.*, *42*, 8, 758–768, 2023.
33. Adefegha, S.A., Functional Foods and Nutraceuticals as Dietary Intervention in Chronic Diseases; Novel Perspectives for Health Promotion and Disease Prevention. *J. Dietary Suppl.*, *15*, 6, 977–1009, 2018, https://doi.org/10.1080/19390211.2017.1401573.
34. Birch, C.S. and Bonwick, G.A., Ensuring the future of functional foods. *Int. J. Food Sci. Technol.*, *54*, 5, 1467–1485, 2018, https://doi.org/10.1111/ijfs.14060.
35. Gupta, A., Sharma, T., Singh, S.P., Prospects of microgreens as budding living functional food : Breeding and bioforti fi cation through OMICS and other approaches for nutritional security. *Genomics Plants Phytoecosystem*, *14*, 1053810, 2023, https://doi.org/10.3389/fgene.2023.1053810.
36. Bro, S., Innovation strategies for functional foods and supplements – challenges of the positioning between foods and drugs. *Food Sci. Technol. (IFST), Sheffield*, *7*, 8, 111–123, 2016, https://doi.org/10.1616/1476-2137.15996.
37. Arvanitoyannis, I.S. and Houwelingen-Koukaliaroglou, M.V., Cons, and Current Legislation Functional Foods : A Survey of Health Claims , Pros and Cons , and Current. *Crit. Rev. Food Sci. Nutr.*, *45*, 5, 385–404, 2005, https://doi.org/10.1080/10408390590967667.

38. Pérez-Marroquín, X.A., Estrada-Fernández, A.G., García-Ceja, A., Aguirre-Álvarez, G., León-López, A., Agro-Food Waste as an Ingredient in Functional Beverage Processing: Sources, Functionality, Market and Regulation. *Foods*, *12*, 8, 15883, 2023, https://doi.org/https://doi.org/10.3390/ foods12081583.
39. Zanella, L. and Vianello, F., Potential of Microalgae as Functional Foods Applied to Mitochondria Protection and Healthy Aging Promotion. *Foods*, *3*, 119–152, 2023.
40. Shi, J., Liang, J., Pu, J., Li, Z., Zou, X., Nondestructive detection of the bioactive components and nutritional value in restructured functional foods. *Curr. Opin. Food Sci.*, *50*, 100986, 2023.
41. Granato, D., Barba, F.J., Bursać Kovačević, D., Lorenzo, J.M., Cruz, A.G., Putnik, P., Functional Foods: Product Development, Technological Trends, Efficacy Testing, and Safety. *Annu. Rev. Food Sci. Technol.*, *11*, 93–118, 2020, https://doi.org/10.1146/annurev-food-032519-051708.
42. Guiné, R.P.F., Florença, S.G., Barroca, M.J., Anjos, O., The link between the consumer and the innovations in food product development. *Foods*, *9*, 9, 3–5, 2020, https://doi.org/10.3390/foods9091317.
43. Kahl, J., Załęcka, A., Ploeger, A., S.B., M.H., Functional Food and Organic Food are Competing Rather than Supporting Concepts in Europe. *Agriculture*, *2*, 4, 316–324, 2012, https://doi.org/10.3390/agriculture2040316.
44. Amitrano, C.C., Coppola, M., Tregua, M., Bifulco, F., Knowledge Sharing in Innovation Ecosystems . *Int. J. Innov. Technol. Manage.*, *14*, 5, 1750030, 2017, https://doi.org/10.1142/S0219877017500304.
45. Childs, N.M., Functional Foods and the Food Industry. *J. Nutraceuticals Funct. Med. Foods*, *1*, 2, 25–43, 1997, https://doi.org/10.1300/J133v01n02.

18

Bacterial Cellulose Synthesis, Characterization, and Its Application in Food Packaging/Functional Foods

Poulami Mukherjee[1], Senthilkumar Sivaprakasam[1*] and Sreedeep Sekharan[2†]

[1]BioPAT Laboratory, Department of Biosciences & Bioengineering, Indian Institute of Technology Guwahati, Guwahati, Assam, India
[2]Department of Civil Engineering, Indian Institute of Technology Guwahati, Guwahati, Assam, India

Abstract

At its dawn, food packaging was mainly geared towards convenience and consumer attraction during point-of-purchase marketing. Severe disruption of economic activities and acute food crises in the post-World War era invoked industrial standardization, international regulation, and worldwide demand for food packaging materials. Traditional tin plates and paperboard-based food packaging materials could not meet the food industry's criteria and demand. They were soon replaced by polythene and polyvinyl chloride-based thermoplastics. Accessible and economically viable production schemes, massive customizability, and high barrier properties against moisture, gas, and microbes of thermoplastics or conventional plastics have revolutionized this industry. Food packaging today plays a critical role in balancing the worldwide food industry by providing means to food preservation, delivery, and storage to decrease food wastage. However, the fact that one-third of the non-biodegradable plastic waste comes from the food packaging industry is sufficient to justify the recent push towards the search for new eco-friendly biopolymers for food packaging. An effective disposal scheme, or rather a lack of it, is just one of many problems of the thermoplastic-based food packaging industry. Thermoplastics often release heavy metals, non-edible dyes, and other hazardous toxins, decreasing the packaged food's shelf life. Moreover, the modern packaging industry only partially relies on the gas and liquid resistances of

**Corresponding author*: senthilkumar@iitg.ac.in
†Corresponding author: srees@iitg.ac.in

Mousumi Sen (ed.) Food Coatings and Preservation Technologies, (611–642) © 2025 Scrivener Publishing LLC

the packaging materials to increase the shelf life of preprocessed food. Modified atmosphere packaging, edible films, coatings, and food matrices are developed using different edible, anti-microbial, functional bio-materials. These active food packaging not only protect foods for extended periods but introduce synthetic savory flavors and colors in the food. While, to date, thermoplastics have the most economical and stable production schemes for food packaging materials, they fall short in fulfilling the multi-dimensional demands of active food packaging materials in the food processing industry. Over the years, researchers have developed numerous edible/ non-biotoxic biomaterials for food packaging, but here, the focus will be on a relatively less explored polymer: bacterial cellulose (BC). BC, a non-biotoxic, chemically inert, biocompatible, biodegradable, and pure exopolysaccharide produced by bacteria, is one of the frontrunners in the search to replace plastics. BC has high tensile strength and mechanical robustness and can be combined with other biodegradable polymers and bioactive agents to form industrial-grade food packaging materials. BC is approved as "“generally recognized as safe”" (GRAS) food by the US Food and Drug Administration. It has potential use in developing rich functional foods due to its high water-holding capacity and rheological properties. Despite being endowed with much potential to be a key player in the food industry, BC has always been sidelined due to a single balking factor: the high cost of production. This situation is, however, changing rapidly in favor of BC. Researchers in the biosynthesis sector are applying metabolic and genetic engineering, culture media optimization, and other cost-reducing techniques for producing industrial-grade BC. This review focuses on the recent developments and current frontiers in BC production and highlights the industrial application of BC in the food packaging and functional food production sector.

Keywords: Bacterial cellulose, biopolymer, metabolic engineering, food packaging, functional foods, edible films, additives

18.1 Introduction

Cellulose ($(C_6H_{10}O_5)_n$), a linear polysaccharide, was first discovered by a French scientist Anselme Payen, and is composed of repeating D-glucose units connected through β-1,4-glycosidic bonds. Cellulose is the most abundant biopolymer available on the earth in terms of plant and agro biomass, and it has played an essential role in the discovery of one of the prominent materials in the evolution of modern society: Plastics. Alexander Parkes, a Birmingham-based inventor, used nitrocellulose to produce thermoplastic, the first of its kind, reported in the scientific literature [1]. Parkes' discovery, however, was not economically successful and soon replaced by thermoplastics, derived from fossil fuels. Thermoplastics have since impregnated and revolutionized every industrial sector. In the food packaging industry, plastics are essential in decreasing physical damage during long transportation, minimizing pathogen attacks, increasing shelf-life, and repackaging leftover foods in economically viable ways. Despite having so many upfront advantages, the non-biodegradability of plastics has become a headache for waste management departments across the world. According to the Plastic Waste Management Global Market Report [2], "The global plastic waste management market grew from $35.37 billion in 2022 to $36.72 billion in 2023 at a compound annual growth rate (CAGR) of 3.8%". UN Environment Program has reported that 36% of plastic waste comes from the food packaging industry [3].

This enormous economic burden of environmental costs and consumer awareness [4] has pushed industrial sectors to intensify the search for environment-friendly, biodegradable, economically viable, non-toxic food packaging material to replace plastics. Moreover, food packaging materials, by modern standards, have an extended role than just protecting packaged food by maintaining oxygen, and moisture barriers [5]. An active food packaging material should show anti-microbial activity, maintain the freshness of the packaged food for an extended period, and add sensory and savory appeal. A smart food packaging material should indicate the expiry and damage of the packaged product. Traditional plastic-based food packaging not only falls short of maintaining these modern requirements, but it is also found to leach 388 hazardous materials that lead to cancer, infertility, and other genetic diseases [6]. Biopolymer researchers have developed several alternatives of plastics for food packaging, but the 'silver bullet' remains elusive. Figure 18.1 summarizes some of the significant biodegradable candidates to replace traditional petroleum-based plastics. Some other popular biopolymers, such as pectin, alginate, carrageenan, etc., are also experimented with for active and smart food packaging in conjunction with proteins, like gelatin, casein, keratin, etc. Standard features of those materials suggest the current frontiers of mainstream research in the eco-friendly food packaging industry:

a) Reduce the usage of petroleum-based resources in favor of renewable resources.
b) Increase the packaging materials' lifetime, barrier properties, mechanical strength, and malleability while maintaining their biodegradability.
c) Increase usage of non-toxic, edible materials with active antimicrobial properties to maximize packaged food's shelf-life, and increase sensory and savory appeal.
d) Reduce the production cost.

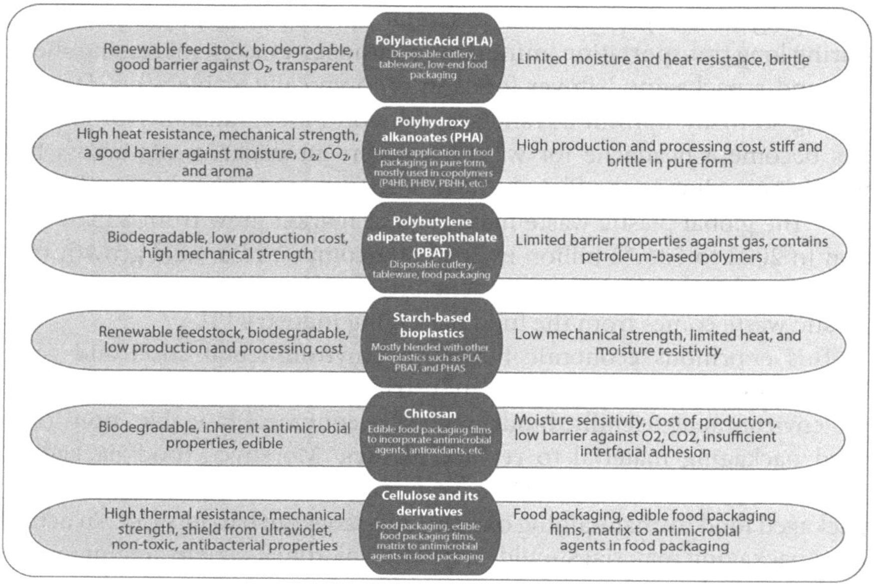

Figure 18.1 Major biodegradable polymers, their benefits, limitations, and applications towards replacing traditional petroleum-based plastics used in the food packaging industry.

The major problem behind fully adopting one of these alternatives to plastics can easily be read from Figure 18.1 itself. Biopolymers with relatively low production and processing costs lack mechanical strength and heat resistivity (such as PLA, starch-based bioplastics, etc.) or have nonrenewable, petroleum-based polymeric components (such as PBAT). Other biopolymers (such as PHA) have high mechanical strength but less malleability, and they amount to high production costs, making them less suitable for the food packaging sector. Instead of developing a single, all-purpose solution, researchers often combine different biopolymers to achieve desired properties. In this field of research, cellulose and its derivatives are among the few essential materials. The biodegradable matrix of cellulose can be easily scaffolded with other biopolymers for enhanced oxygen and moisture barriers, tensile strength, and mouldability [7]. It can further be impregnated with bioactive functional groups and antimicrobial agents to induce antimicrobial properties and can be used for active food packaging to increase shelf-life [8]. The incorporation of natural dye in cellulose can be used in smart food packaging as a freshness detector [8]. Cellulose acetate (CA), a non-toxic, water-insoluble material, is often used in biodegradable food packaging to provide mechanical strength and antibacterial properties against *E. coli, B. subtilis,* and other bacterial strains [9]. Cellulose Sulfate (CS) also offers antibacterial properties but is soluble in water at high concentrations [9]. Plant-based cellulose, the largest source of natural cellulose, however, has two significant limitations:

a) Cellulose comes with impurities, such as hemicellulose and lignin. These impurities have diverse chemical properties and often require harsh chemical treatment for purification. Some chemicals used for cleansing are toxic and are not entirely removed in the downstream processes. The presence of these chemicals in the food packaging material can be hazardous.
b) Impurities' presence decreases cellulose polymerization and crystallinity and reduces its malleability and tensile strength.

In the biopolymer industry, the acquisition of pure, crystalline, non-toxic cellulose is done in two processes: Extraction of nanocellulosic content from plant-based cellulose (PC), and production of pure cellulose using microbial culture. Extraction of cellulosic content from PC is a cumbersome, and expensive process that requires extreme chemical and physical treatment. Some of those chemicals remain in the cellulose matrix and removal of such chemicals accumulates extra cost during the downstream processing. The alternative process of using bacterial culture to produce pure cellulose uses enzymatic pathways in mild, environment-friendly conditions. Bacterial cellulose (BC) is not only free of corrosive chemicals it also shows higher mechanical strength, compactness, and thermal stability than PC-based nano cellulose [10]. Due to the higher grade of purity, BC does not need any harsh chemical treatment, and it is considered "generally recognized as safe"(GRAS) food by the US Food and Drug Administration (FDA) [11]. The non-toxicity, biodegradability, and chemical malleability due to a high percentage of hydroxyl groups, make BC a tailor-made edible food matrix, with potential use as a fat replacer in the meat industry, a rheology modifier in the dairy industry [12]. Despite the huge potential of BC in the food packaging and functional food industry, BC has a minimal footprint due to a single balking factor: the high cost of production. A rough estimation suggests that BC production should reach 15g/L from a 50-hour fermentation of 20g/L glucose [13]. At this production rate, BC production would be able to economically compete with the plant-based nano cellulose market. A techno-economic estimation for profitable BC production estimates just a 7% return on investment (ROI) [14]. In their analysis, the estimated selling cost of BC is set at US$ 25/kg, which is five times higher than PLA, another promising candidate in food packaging. Due to the high cost of synthesis, the use of BC is mostly confined to highly specialized medical and surgical appliances. This despondent picture of the BC-based industry, is, however, changing fast. The use of metabolic engineering to design high-yielding, more viable BC-producing strains, and the use of industrial and agricultural biowastes to reduce feedstock costs have dramatically reduced the cost and complexity of BC production. A recent, techno-economic feasibility study forecasted a 16.4% ROI for BC production using kombucha waste [15]. In their analysis, the selling cost for BC is set at US$ 1.63/kg, which is 1/3rd of the PLA selling cost. In light of this changed scenario, a meaningful discussion on the synthesis, *in situ*, and *ex situ* modifications, and corresponding characterization of BC in the food packaging and functional food industry, is quintessential.

Section 18.2 in this chapter first discusses some of the high-yielding BC-producing bacteria and a general metabolic pathway for BC synthesis while highlighting the major challenges and roadblocks in BC production. The use of different fermentation modules and downstream processing for quality control are discussed in Section 18.2.3 to 18.2.4. Rest of the Section 18.2 is dedicated to a brief discussion of metabolic engineering and feedstock customization to increase the techno-economic feasibility of the BC-based food packaging industry. Section 18.3 starts with a discussion of industry-grade characterization methods to asses the industry readiness of native and modified BC. Rest of the Section 18.3 focuses on the enhancement of BC characteristics and the induction of useful features for food packaging and functional food. The future potential of BC in this industry is sketched in Section 18.4.

18.2 Bacterial Cellulose Synthesis

BC has a long history rooted in the cuisines of Eastern civilization countries. Nata de Pina, a traditional eighteenth-century Philippines dessert has a chewy and translucent BC film on the surface. BC layers are found in kombucha tea, a symbiotic mixture of yeasts and acetic acid-producing bacterial culture, and a traditional drink in China. In 1949, Teódula Kalaw África used *Komagataebacter xylinus* culture to produce Nata de Coco, a dessert similar to Nata de Pina but made from coconut water. Nata de Coco has become famous in the Philippines, Indonesia, Thailand, Vietnam, Malaysia, Japan, and many other countries. Researchers have found a torrent of BC applications in food, beverage, medical, green energy, and other industrial sectors. As the era of green industry is kicking off, an industrial-scale production scheme for BC lies in the top priority list of bioengineers. Isolation of BC-producing strains, understanding the metabolic pathways for BC production, and feedstock optimization are some of the most active research fields in bioengineering sectors.

18.2.1 Bacterial Cellulose-Producing Organisms

BC is an exopolysaccharide, primarily formed by bacteria (microbes, in general) as a self-defense mechanism to maintain an aerobic environment in the air-water interface, to protect against ultraviolet exposure, draught, etc. Different aerobic bacteria, such as *Komagataeibacter* (formerly known as *Gluconacetobacter, Acetobacter*), *Achromobacter*, *Alcaligenes*, *Agrobacterium*, *Azotobacter*, *Pseudomonas*, etc., are known to be BC-producing bacteria. Some facultative anaerobic bacteria (*Escherichia coli, Salmonella sp.*), anaerobic (*Sarcina sp., Rhodobacter sp.)* bacteria, even some archaea (*Thermoplasma, Ferroplasma*), and eukaryotes (Green algae, tunicates) are also reported to produce BC [16]. The BC's chemical and mechanical properties produced by different microbes vary significantly. *Komagataeibacter, Acetobacter, and Abchromobacter* produce BC ribbons, *Aerobacter, Alcaligenes,* and *Agrobacterium* make BC fibrils, while BC produced by *Sarcina sp.* is amorphous [17].

Table 18.1 summarizes some of the beneficial bacterial strains for BC synthesis at the commercial level. While some strains have been isolated from BC-producing bacterial colonies, others are genetically engineered for efficient BC production. Almost all of the bacterial cultures produce BC in mild conditions (pH 4–7, temperature 28–31°C), compared to the extreme physical and chemical circumstances for nano-cellulose extraction. In the static fermentation modes, some of the bacterial strains give very high BC yields that in some cases cross the 15g/L threshold [13], but the usual fermentation time for BC production is very long (7–21 days). The production rate can be increased by agitating the fermentation module, but the yield, purity, and compactness of the BC degrade significantly (see Section 18.2.3). Current researchers focus on optimizing the BC production rates while maintaining the desired characteristics of BC. Another trend in the current research is to replace conventional carbon sources with agro-industrial wastes. The use of grape pomace, citrus peel, cheese whey, sugar beet molasses, tomato juice, potato juice, etc, can significantly decrease the feedstock cost without sacrificing the BC yield and productivity (see Table 18.1). Next, this section focuses on the current understanding of metabolic

Table 18.1 Bacterial cellulose-producing strains, and corresponding yields for different culture media.

Strain	Culture media module	Fermentation mode	Fermentation time	pH	Temperature	Yield (g/L)	References
G. xylinus ATCC 53524	20 g/L of carbon source with 1:3 glucose to glycerol	Static culture	7 days	6.0	30°C	8.42 g/L	[18]
K. melomenusus AV436T	Grape pomace hydrolysate	Mixed	Shake culture for 1 day and then static for 3 days	4.5	30°C	1.24 g/L	[19]
K. hansenii GA2016	Citrus peel	Static culture	21 days	4.5	28°C–32°C	2.06-3.92 %	[20]
A. xylinum ATCC 23769	Processed rice bark	Static condition followed by aeration	10 days	5.25	28°C	2.42 g/L	[21]
K. intermedius MO	Protein-based waste (cheese whey)	Static culture	7 days	7	30°C	17.8 g/L	[22]
A. xylinum 0416	Fresh oil palm frond juice	Static culture	10 days	4-6	28°C-30°C	2.88 g/L	[23]
G. xylinus ATCC 53582	Cashew apple juice with soybean molasses	Static culture	7 days	5.55	30°C	4.5 g/L	[24]
G. xylinus PTCC 1734	Sugar beet molasses and cheese whey	Static culture	14 days	5.5	28°C	14.20-25.34 g/L	[25]

(Continued)

Table 18.1 Bacterial cellulose-producing strains, and corresponding yields for different culture media. (*Continued*)

Strain	Culture media module	Fermentation mode	Fermentation time	pH	Temperature	Yield (g/L)	References
A. pasteurianus RSV-4	Tomato juice	Static culture	7 days	4.5	29°C-31°C	7.8 g/L	[26]
G. xylinum ATCC 10,245	Potato peel waste	Static culture	6 days	9	35°C	4.7 g/L	[27]
G. hansenii CGMCC 3917	Waste beer yeast	Static culture	14 days	5	30°C	7.02 g/L	[28]
K. sucrofermentas ATCC 11142	Pre-treated sugarcane straw	Static culture	15 days	6.5	29°C-31°C	2.96-4.51 g/L	[29]
K. saccharivorans BC1	Crude distillery effluent	Static culture	7 days	5	30°C	12.4 g/L	[30]
A. xylinum ATCC 23767	Tobacco waste	Shaking at 150 rpm	7 days	6.5	30°C	5.2 g/L	[31]
A. xylinum ATCC 23767	Kitchen waste	Static fermentation	15 days	5	30°C	2.07 g/L	[32]
A. xylinus ATCC 23770	Textile cotton waste	Static culture	14 days	5	30°C	10.8 g/L	[33]

pathways in BC production to highlight the motivations and logic behind the efficient design of BC-producing bacterial strains.

Next, in this section, the current understanding of metabolic pathways in BC production highlights the motivations and logic behind the efficient design of BC-producing bacterial strains.

18.2.2 Metabolic Pathway

The BC production pathway is a complicated multi-step process and can be divided into two sub-processes:

a) Fermentation, or synthesis of UDP-glucose from monosaccharide sugar molecules.
b) Fibrillation or polymerization of β-1,4-glucan chains using UDP glucose and self-assembly of β-1,4-glucan chains into BC microfibrils and fibrils.

18.2.2.1 Fermentation

Synthesis of UDP-glucose from glucose (see Figure 18.2) has three stages: Phosphorylation of glucose, isomerization of glucose-6-phosphate into glucose-1-phosphate, and synthesis of UDP-glucose. For commercial production of BC, the Hestrin-Schramm (HS) medium is the most widely used culture medium, where optimized amounts of organic and inorganic nitrogen and phosphate sources, such as yeast extracts, ammonium-sulfate, amino acids, magnesium sulfate, potassium dihydrogen phosphate, disodium hydrogen phosphate, and other mineral salts for bacterial growth. glucose in the HS medium is the most common cultivation medium for BC synthesis [34]. The use of glucose as the primary carbon source has some limitations. Many *Komagataeibacter* sp. strains have glucose-6-phosphate dehydrogenase (G6PDH) enzymes. G6PDH activity in the bacteria converts consumed glucose to phosphogluconic acid. The glucose-to-phosphogluconic acid pathway has two-way adversary effects in BC synthesis. It not only consumes accessible carbon sources and produces byproducts in the bacterial growth phase but also accumulates phosphogluconic acid in the culture medium. The accumulation of phosphogluconic acid causes a drop in the pH and affects the enzymic activity of cellulose synthase [35]. To alleviate phosphogluconic acid inhibition, glucose-based carbon sources are often used to produce glucose

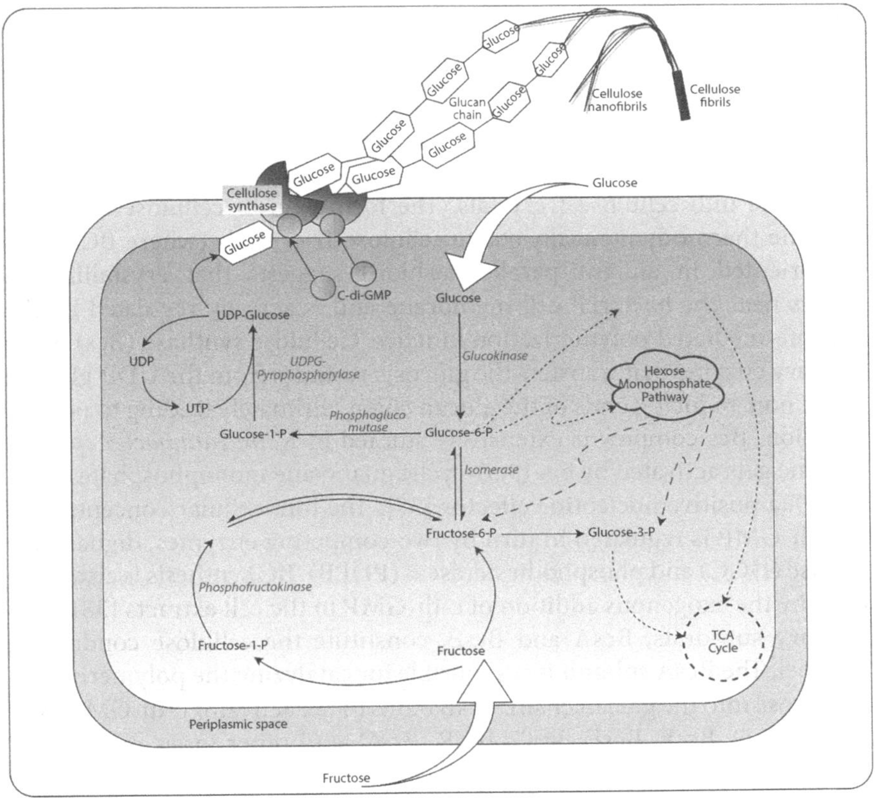

Figure 18.2 The metabolic pathway for bacterial cellulose production.

via faculative fermentations. Alternate carbon sources, such as fructose, mannitol, Sorbitol, glycerol, xylose, and sucrose do not face such phosphogluconic acid inhibition and are often considered as better feedstock components than glucose [36]. These non-glucose sources convert to fructose-6- phosphate, UDP-glucose synthesis node. Carbon flux in the UDP-glucose synthesis pathway from fructose-6- phosphate has competition with the carbon flux into the TCA cycle. BC yield from non-glucose sources is often lower than glucose as the sole carbon source [37]. Some of the carbon sources, such as sorbose, mannose, cellobiose, erythritol, ethanol, and acetate, are known to be not convertible into BC [36]. Ethanol and citric acid, however, are often added to the cultivation media to suppress phosphogluconic acid production [37]. Ethanol also balances the carbon flux into the TCA cycle to minimize the unwanted consumption of carbohydrate resources in CO_2 production [37].

18.2.2.2 Fibrillation

The polymerization and crystallization of UDP-glucose into BC fibrils could be better understood. Glucose from the UDP-glucose gets polymerized into glucan chains and secreted into the periplasmic space. Parallel glucan chains, secreted from the bacteria, assemble into nanofibers and crystallize into cellulose I crystals. The formation of cellulose I crystals over the thermodynamically stable cellulose II crystals (where BC fibrils are oriented in an anti-parallel fashion) suggests that crystallization occurs near the bacterial cell membrane and is actively regulated by the enzyme-mediated polymerization kinetics. Cellulose synthase (Bcs) is the primary enzyme that transfers the glucosyl residues from the UDP-glucose to the non-reducing ends of the glucan chain, ultimately leading to polymerization. Bcs complex is extensively studied in *Komagataibacter sp.* This enzyme gets activated by bis-(3,5)-cyclic guanosine monophosphate (c-di-GMP), a positive nucleotide effector [37]. The intracellular concentration of c-di-GMP is regulated in turn by two competing enzymes, diguanylate cyclase (DGC) and phosphodiesterase B (PDEB). BC synthesis is also escalated by the exogenous addition of c-di-GMP to the cell extracts [38]. Two primary sub-units, BcsA and BcsB, constitute the cellulose conducting channel. The BcsA subunit is responsible for catalyzing the polymerization of glucose into the glucan chain. BcsB binds to the activator c-di-GMP [39]. BcsC, BcsD, BcsE, BcsF, BcsG, BcsR, BcsQ, and other inner membrane-bound protein units are also reported. To date, the exact roles of these sub-units in BC production are an active field of research [40]. In *K. xylinus*, there are two types of constitutive operons, bcsI, and bcsII, encoding the proteins mainly responsible for cellulose biosynthesis. bcsI has four primary genes, bcsAI, bcsBI, bcsCI, and bcsDI, flanked by accessory genes (cmcAx, ccpAx, and bglAx) [37]. The other operon, bcsII, has an acetyltransferase gene responsible for acetylated cellulose [37]. Other membrane proteins, such as cellulose complementing factor (Ccp), are reported to be associated with BC production, but more information is needed about their specific role [41].

18.2.3 Fermentation Modules

All of the known BC-producing strains that are used in industry are aerobic (see Table 18.1). In a shake-flask method, the microbial culture consumes the feedstock and rapidly produces BC in the culture medium, air interface. Due to the insolubility of freshly produced BC, it starts forming a yellowish BC layer in the culture medium, and air interface. It cuts the oxygen supply to the bacterial culture. Due to a lack of oxygen, BC production slows down and finally stops prematurely, without consuming all the carbon sources. Simply put, the BC production scales up with the available surface area, not the available volume of carbon feedstock. This shortcoming poses a critical challenge in the bioreactor design for BC- production and industrial scale-up. The oxygen supply in the fermentation module can be increased artificially by pumping oxygen into the culture medium (air-lift bioreactor). Alternatively, freshly produced BC can be continuously removed from the culture medium surface (rotating disk bio-reactor). Constant stirring of the culture medium can also increase the oxygen supply in the fermentation module (stirred tank bioreactor). The culture medium gets agitated in all three cases, and the bacterial growth and feedstock consumption decrease. While the production time increases, the total yield, compactness, crystallinity, fiber thickness, crosslinking of the BC, etc, decreases significantly in these agitated culture medium. Moreover, additional electric consumption costs are incurred to constantly agitate the fermentation modules [14, 42]. The operation cost can be decreased by using lighter materials, such as plastic-based composites to prepare the agitator shaft [42]. The lifetime and reusability of the culture medium can be increased by immobilization techniques used in biofilm bioreactors [42]. For *in situ* molding of the BC film, trickle-bed bioreactors are used, where the culture medium gets a higher oxygen supply than conventional static modules, but faces less agitation-related stress compared to agitated bioreactors. Due to the complex design, and operation complexity, trickle-bed fermentation modules are mostly used in medical and surgical sectors [42, 43].

18.2.4 Downstream Processing

Due to innate insolubility, BC forms pure cellulosic films or pellicles at the surface of the culture medium, and it requires minimum labor for the isolation of BC from the culture media. Extracted BC undergoes NaOH/ KOH treatment at 100°C for 10-15 min, followed by repeated washing with hot water, and ultra-sound treatment to remove residual bacterial mass [12]. Under alkali and heat treatment, the yellowish color or the BC changes to white, and it goes under crystalline polymorphism to produce a more thermodynamic stable variant cellulose II, where cellulose fibers are arranged in an anti-parallel manner instead of the parallel organization of cellulose fibers in the native polymorph cellulose I [10]. This polymorphism can be prevented by using 2.5% (w/w) NaOCl treatment [44]. Acid hydrolysis is used to remove the amorphous BC to produce microcrystalline BC.

Acid hydrolysis can be replaced by cellulase treatment for nanocrystalline BC production. Mechanical processes, such as homogenization, grinding, and refining are used for extracting BC nanofibers from native BC [45]. The use of alkali treatment significantly increases the tensile strength of BC fibers [12].

18.2.5 Application of Metabolic Engineering in Bacterial Cellulose Synthesis

BC production, as discussed earlier in this section, goes through a complicated, multi-step, enzyme-regulated fermentation process, followed by kinematically controlled polymerization, fibrillation, and crystallization processes, regulated by a periplasmic Bcs enzyme. High feedstock cost [14], high oxygen demand during fermentation, low inoculation rate of bacterial culture, low yield, and long fermentation time (see Table 18.1) pose significant challenges to economic and industrial-scale BC production. The use of feedstock optimization, genetic overexpression, gene knockout, homologous recombination, and other genetic engineering tools can show significant improvement in metabolic pathway optimization [46], by-product reduction [37], quality control [47], stress resistance [48], etc. Most BC-producing bacterial strains are gram-negative and prone to resist any genetic makeup or modifications. Use of electroporation and other genetic engineering tools are used to make them susceptible to any genetic alteration to increase yield, substrate utilization, polymerization, fibrillation, crystallization, and fibrillation of BC. In wild *K. xylinus* strains, the phosphofructokinase enzyme is not properly expressed. It slows down the EMP pathway that, in turn, diverts the metabolic flux from the direct synthesis of glucose-6-phosphate from glucose and fructose. Heterologous expression of the *pfkA* gene from *E. coli* in the *Komagateibacter* strain activates the EMP pathway by adding up extra intracellular ATPs, and increases the production yield [46]. Excessive accumulation of phosphogluconic acid in the culture medium drops the pH level inhibiting bacterial inoculation. The phophogluconic acid production can be regulated by eliminating the *gdh* gene that expresses the membrane-bound glucose dehydrogenase enzyme. Knock-out of the *gdh* gene in *K. xylinus* BPR2001 strain doubles up the BC production in comparison to the wild one [35]. The motility of the BC-producing strains can be further enhanced by overexpression of two primary genes of flagellar motor protein (*motA* and *motB*). Such overexpression in *K. hansenii* ATCC 23769 exhibits thicker ribbon-like BC fibers with lesser compaction [47]. Disruption of the same in the *K.*

hansenii ATCC 53582 results in BC with thinner fiber and improved network density [45]. In BC synthesis a significant challenge arises from the emergence of irreversible *cel*⁻ mutants (non-BC-producing) in agitated conditions [37]. This genotypic conversion happens due to the interruption of IS (Insertion sequence) elements into the prime BC-producing gene, *bcsA*, within the Bcs operon. Hur et. al (2020) identified the putative chromosomal DNA sequence which is prone to get disrupted by the IS elements and engineered the same, affirming no change in the physical properties but a higher BC yield [50].

18.2.6 Economically Sustainable and Industrial Scale Production of Bacterial Cellulose

Wide adaptation of BC in food packaging, and synthetic food production stops short due to a single concern - production cost. Native bacterial cellulose costs around US$ 25/kg [14], which is 5 times costlier than PLA, another extensively used biopolymer. About 90% of the production costs, however, are associated with feedstock accumulation and labor wages. Optimized mass-balance analysis for the bioreactor can be used to automatization the BC-producing industry to alleviate labor costs. The use of biowastes to replace glucose as the primary carbon source can further lower the production cost by 30%. A modern estimation shows that a parallel production of BC in a kombucha drink farm can generate a staggering 23% ROI while maintaining the BC selling cost at US$ 1.63 /kg. This selling cost is just 1/3rd of the PLA. Several groups have reported high BC yield from industrial and agro wastes [51] (see Table 18.1). Industrial adaptation of biowaste-based feedstock for BC production not only enhances the economic sustainability of the BC-centric food packaging industry but also addresses waste management tasks in the parent industries that generate biowastes.

18.3 Characterization and Application of Bacterial Cellulose in the Food Industry

BC characteristics can dramatically vary depending on the culture media, fermentation module, and downstream procedures [52]. An unaltered BC in general has high crystallinity, high water retention ability thermal stability, and mechanical strength, but it lacks oxygen, and moisture barrier properties, transparency, and antimicrobial properties. *In situ* and *ex situ* chemical modification of BC can enhance or induce additional properties for its multifaceted application in food packaging and functional food industry [8, 12, 53]. To produce such industry-ready BC, a proper quantifiable characterization of BC is quintessential. In this section, some industry-grade characterization methods that are used for BC characterization have been discussed (see Table 18.2).

18.3.1 Characterization Methods

A BC usually possesses high purity, a high degree of crystallinity, a high degree of polymerization, mechanical strength, a compact network, and thermal stability. Some of these properties, however, are not compatible with each other. The crystallinity of cellulose usually decreases with its polymerization index. Thermal stability increases with the crystallinity, but the compactness of the fibrillar network increases with the degree of polymerization. While BC is modified to enhance one of its properties, changes in the competing properties must be taken care of. There are several optical (Fourier-transformed infrared spectroscopy, X-ray diffraction), mechanical (tensile tester, rheometer), and computer-assisted (Quanfima, IamgeJ) technologies are available in the industry to determine the mechanical, thermal, optical, and chemical properties of BC. Some of the available technologies and their application in the BC characterization are shown in Table 18.2.

18.3.2 Enhancement of Bacterial Cellulose Characteristics for Food Packaging

The high purity, thermal stability, tensile strength, rigidity, visco-elasticity, and non-toxicity of BC set the gold standards in food packaging and functional food industry. The use of a native BC in food packaging often falls short in two aspects - low-barrier properties against oxygen and moisture [10]. Some low-cost biopolymers, such as PLA, starch, chitosan, gelatin, etc, on the other hand, show better oxygen and moisture barrier properties but they lack mechanical strength, thermal stability, and elasticity. Chemical composites of BC with these materials and close packing increase the barrier properties, while further enhancement of the stability against mechanical and thermal stresses. The use of some hydrophobic proteins, alginate, etc in BC composites also enhances barrier properties against moisture [64]. The insolubility of BC in water or many other organic solvents poses engineering challenges for the *ex situ* production of BC composites [65].

Table 18.2 Characterization techniques used for bacterial cellulose.

Characterization	Methodology	Reference
Purity	Fourier-transformed infrared (FTIR) spectroscopy (400 cm^{-1}- 4000 cm^{-1})	[52, 54]
Crystallinity	X-ray diffraction with Kα1 and Kα2 lines from the Cu X-ray	[55]
Degree of Polymerization	Size exclusion chromatography with tetrahydrofuran (THF) eluent	[56]
Fiber Thickness and Fibrillar Network Structure	Different image analysis tools, such as Fiji (ImageJ), FIberScout, DiameterJ, FibrilTool, OrientationJ, Quanfima, etc.	[57]
Mechanical strength	Elongation and breaking point study with Universal Tensile Testing Machine or Instron Tensile Tester	[58]
Thermal stability	Thermogravimetric Analysis (TGA), Dynamic Mechanical Analysis (DMA), Differential Scanning Calorimetry (DSC), and Thermomechanical Analysis (TMA)	[59]
Water holding capacity and Water release rate	Sieve shake method	[60]
Viscosity	A rotational rheometer, with a constant shear rate	[61]
Porosity and Surface morphology	Nitrogen physisorption (for pore size diameter 0–50 nm) and mercury porosimetry (for pore size diameter 50 nm–0.5 mm)	[62]
Opacity	Spectroscopic ellipsometer	[63]

Simple immersion of BC into chitosan or gelatin solution results in the adsorption of the solvent into BC, but the adsorbed chemical leaches out of the BC films in the long run [52]. The addition of chitosan, starch, lactic acid, etc in the fermentation module forms *in situ* BC composites [66]. *In situ* modification, however, has lower control and is not suitable for certain cases. *In situ* impregnation of BC with antimicrobial agents, for example, is not suitable, as the presence of anti-microbial agents can affect the culture inoculation. *Ex situ* modification of BC is also necessary for forming multi-layered composites [52, 65, 66]. For *ex situ* modification,

Table 18.3 Enhancement of Bacterial Cellulose Characteristics in bacterial cellulose composites.

Composites	Synergistic impacts	References
BC nanofiber (BCNF) with Chitosan	Higher tensile strength in comparison to the lactic acid films	[67]
BC nanowhisker (BCNW) with PLA	Increased moisture sensitivity, hydrophobic surface, and higher oxygen permeability than native PLA	[68]
BCNF with Pectin and fruit purees	Stronger, stiffer, and water-resistant with enhanced barrier properties	[69]
BCNW with Starch and glycerol	Increased tensile strength, and improved barrier property	[70]
Regenerated BC (cellulose II) with Alginate	Tensile strength and elasticity slightly decreased with improved water adsorption capacity and water vapor transmission rate	[71]
BC suspension with Polyvinyl alcohol and bulk chitosan/ chitosan-calcium-alginate nanoparticles (CNSP)	Improved tensile strength and transparency, while the addition of CNSP has reduced water diffusion and improved UV barrier properties	[72]
BC film with Guaran, polyvinyl pyrrolidone, carboxymethyl cellulose	Enhanced elasticity and load-bearing capacity with greater hydrophobic and barrier property	[73]
BCNC suspension with Gelatin	Increased tensile strength, denser matrix with lower water vapor transmission, and increased thermal stability	[74]
Residual BC membrane Arabinogalactan and xyloglucan	Increased hydrophobicity, stability, tensile strength	[75]
BCNC with silver nanoparticles (AgNPs) and chitosan film	Compact network structure with decreased moisture content and water vapor permeability, increased hydrophobicity, UV barrier property, and opacity	[76]
BC suspension with eggshells	Increased porosity, the ability to absorb water and vegetable oil also gets enhanced	[77]

BC is dissolved in LiH/urea/thiourea solution, and electrospun with the desired compound to form BC composites [65]. The enhancement of barrier properties and thermal and mechanical stability of BC-composites are summarized in Table 18.3. While BC can achieve exceptional tensile strength, thermal stability, and barrier properties as reported in Table 18.3, the economic viability of such BC composites in replacing plastic-based food packaging materials has not been extensively studied. A low-cost production scheme of BC would be beneficial to achieve such goals [15].

18.3.3 Induced Characteristics of Bacterial Cellulose for Food Packaging

High porosity, thermal stability, and crosslinking in BC make it a perfect candidate to impregnate with various bioactive and antimicrobial agents. The introduction of such material in cellulosic matrix induces antimicrobial and antioxidant activities. Introduction of laccase, poly-sulfobetaine methacrylate in the BC composite actively scavenges oxygen and moisture from the packaged food and prevents the growth of harmful microbial colonies [78, 79]. The use of organic dye in BC can be used for pH detectors in packaged food. An extensive list of induced characteristics in modified BC are shown in Table 18.4. The use of active biomaterials in tandem with BC produces functional food packaging materials that are biodegradable and have superior functionality compared to plastic-based food packaging. The high cost of BC production, however, limits the use of such composites in the packaging of costly foods, such as mangoes [80], or highly degradable foods such as meat [81].

18.3.4 Application in Food Packaging

The primary objective of food packaging is to protect food from physical damage, during storage and transport. A good food packaging material is expected to protect packaged food from air, moisture, and pathogens. Modern-day food packaging materials, however, have a more extensive job list (see Figure 18.3). Multipurpose food packaging can be divided into three categories -

Conventional food packaging provides mechanical and thermal stability in the food packaging and protects food from physical damage. Conventional food packaging also shows a barrier effect against gas and moisture to limit the parasitic growths on the packaged food.

Table 18.4 Induced characteristics of bacterial cellulose composites.

Composites	Synergistic impacts	Reference
BCNF with Chitosan	Anti-microbial properties, non-toxic, can be used in edible packaging	[67]
BC film with Lauric acid	Inhibitory effect towards *Bacillus subtilis* and increased shelf-life	[82]
BCNC with AgNPs	Higher oxygen barrier capacity with antimicrobial properties against *S. aureus* and *E. coli*	[83]
BCNF with purple sweet potato anthocyanin and oregano essential oil	Colorimetric changes with pH, growth-inhibitory effect on *E. coli* and *L. monocytogenes*	[84]
BC with green tea and berry extract	Antioxidant properties with an increased shelf-life	[85]
BC with pomegranate peel extract, green tea extract, and rosemary extract	Antioxidant properties with enhanced oxidative stability and increased shelf-life of button mushrooms with no color degradation	[86]
BCNF with flavonoid silymarin and zein	Enhanced drug release ability for effective antibacterial and antioxidant activities, extend the shelf-life of packaged fish	[87]
Freeze-dried BC sponge with fungal laccase	Oxygen scavenging activity, prevention of the degradation of the packaged food	[78]
BC pellicle with nisin	Effective microbial activity against *L. monocytogenes* and total aerobic bacteria for packaged meat products	[88]
BC with methyl red	Colorimetric changes with pH and amine gas	[81]
BC with poly-sulfobetaine methacrylate	Conductometric changes with moisture, antimicrobial activity, moisture absorption	[79]
BC film with curcumin	Thermally stable with good antioxidant activity against many gram-positive and gram-negative bacteria	[89]

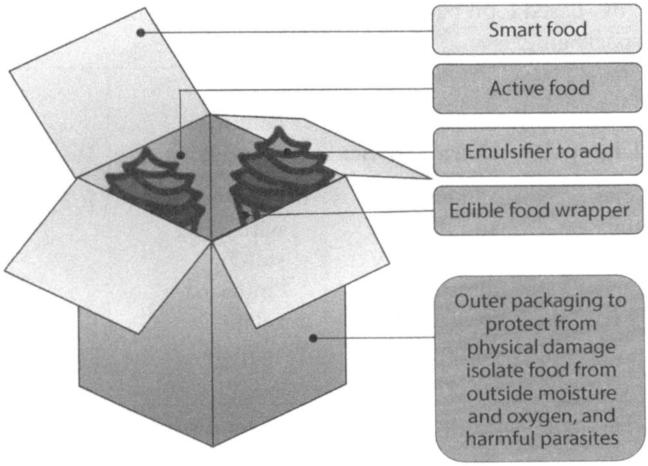

Figure 18.3 Potential use of bacterial cellulose in multilayer food packaging materials for conventional packaging (barrier against moisture, oxygen, pathogens, and physical damages), active packaging (controlled release of antimicrobial and antioxidant chemicals to increase shelf life), and smart packaging (scavenging of residual oxygen and moisture to prevent microbial infection, pH and moisture sensing). Savory and sensory-appealing chemicals can be impregnated in edible BC food wrappers. The use of BC as a rheology modifier in ice cream adds additional texture and stability [12].

Active food packaging secretes antimicrobial agents or antioxidants into the food to increase shelf life. Active food packaging can add additional flavor or taste to the food.

Smart food packaging detects the pH, moisture, and other chemical changes in the packaged food to determine the freshness of the food. Smart food packaging can also absorb harmful materials from the food to increase the shelf life of the packaged food.

BC and BC-based composites show promising potential to meet all the demands for a modern food packaging material (see Table 18.5). While there are several reports are tabulated in Table 18.5, regarding the application of BC-based composites in each of the food packaging sectors, to date, most of the applications in the conventional and active food packaging sectors have not been tested for economic viability. A cheaper production scheme for BC is quintessential for the industry-wide adaptation of BC composites.

Table 18.5 BC-based food packaging materials for conventional, active, and smart food packaging.

Food packaging	Essential characteristics	Bacterial cellulose composites	References
Conventional packaging	Mechanical and thermal stability, UV moisture, oxygen, and pathogen barrier	BC composites with PLA, starch, chitosan, gelatin, alginate, pectin, zein	[67–71]
Active packaging	Antimicrobial activity	BC composites with chitosan, AgNP, ZnO, TiO_2,	[67, 83, 84]
Active packaging	Antioxidant activity	BC composite with curcumin	[89]
Smart packaging	Oxygen scavenging	BC composite with laccase	[78]
Smart packaging	Moisture scavenging	BC with poly-sulfobetaine methacrylate	[79]
Smart packaging	pH sensing	BC composite with Methyl red, bromophenol blue (for fruit packaging)	[80, 81]
Smart packaging	Moisture sensing	poly-sulfobetaine methacrylate (for meat packaging)	[79]

18.3.5 Application in Functional Food Production

In the east-asian countries, BC-based beverages are considered among traditional cuisines. Consumption of probiotics with BC culture medium increases the survival rate of probiotics against gastric juices [90].

Digestion of cholesterol fiber lowers triglycerides and total cholesterol [91]. Due to the high viscoelasticity, BC can replace fat-based thickeners, emulsifiers, and stabilizers [12, 91]. Up to 50% replacement of fat in meatballs by BC, does not alter the taste or textures [12]. Some BC-composites with enhanced characteristics for functional food production are detailed in Table 18.6.

BCNFs being commonly used as stabilizers for Pickering emulsions due to their easier and greener production methods compared to BCNCs [92]. Combining the spray drying method with these emulsions oil microencapsulation can be improved. The microcapsules are designed to insulate hydrophobic bioactive components such as fat-soluble vitamins and essential or functional oils, allowing for efficient delivery in the bio-system [97]. The use of BC-surimi (see Table 18.), nowadays is considered a low-fat meat in eastern Asian countries [96]. BC-based kombucha tea shows some probiotic activity. However, the potential of BC in probiotic foods is yet to be commercialized at full potential [12].

Table18.6 BC-based functional foods.

Functional food	Essential characteristics	Bacterial cellulose composites	References
Pickering emulsion to encapsulate bioactive materials, probiotics	High viscosity	BC nanocrystals, BC nanofibers	[92, 93]
Probiotic immobilizer	Compact fiber network with a high number of crosslinking	BC composite with pectin	[94]
Rheology modifier in low-fat ice cream	High viscoelasticity	BC composite with soy protein	[95]
Fat replacer in meatball	High water absorption capacity	BC composite with *Mahi mahi* dolphin fish surimi	[96]
Synthetic meat for vegetarians	Fibrous network, high water retention capacity	BC composite with monascus	[91]

18.4 Conclusion and Future Prospective

High purity, tensile strength, viscosity, and modifiability- BC ticks all the right boxes to replace fuel-based plastics in the food packaging industry. Nontoxicity and presence of dietary fiber in BC hint at its untapped potential as a functional food. Over the last decade, researchers have tirelessly reported many potential applications of BC-based food packaging materials, and functional foods. To date, all the big players in the BC industry, such as Celluforce, Paperlogic, Borregaard, Inventia, Oji Paper, American Process, and Nippon Paper [98], however, focus on the surgical and high-end cosmetics market. Recent progress in biowaste utilization as alternative feedstocks, isolation, and discovery of genetically engineered stress-resistant, high-yielding BC-producing strains would like to dramatically decrease the production cost, and possibly make it cheaper than PLA [15]. This situation is possibly going to widen the BC-based market in this decade, and food packaging and functional food markets will see a momentous revolution, led by BC-based products.

References

1. Kaufman, M., *The first century of plastics: Celluloid and its sequel*, Plastics Institute, London, 1963.
2. Athanasopoulou, E., Michailidi, A., Ladakis, D., Kalliampakou, K. I., Flemetakis, E., Koutinas, A., Tsironi, T., Extraction of Fish Protein Concentrates from Discards and Combined Application with Gelatin for the Development of Biodegradable Food Packaging. *Sustainability*, 15, 15, 12062, Aug, 2023, doi:10.3390/su151512062.
3. Arijeniwa, V. F., Akinsemolu, A. A., Chukwugozie, D. C., Onawo, U. G., Ochulor, C. E., Nwauzoma, U. M., Kawino, D. A., Onyeaka, H., Closing the loop: A framework for tackling single-use plastic waste in the food and beverage industry through circular economy- a review. *J. Environ. Manage.*, 359, 120816, 2024, doi: 10.1016/j.jenvman.2024.120816.
4. Ketelsen, M., Janssen, M., Hamm, U., Consumers' response to environmentally-friendly food packaging - A systematic review. *J. Clean. Prod.*, 254, 120123, May 2020, doi: 10.1016/j.jclepro.2020.120123.
5. Ahari, H. and Soufiani, S.P., Smart and Active Food Packaging: Insights in Novel Food Packaging. *Front. Microbiol.*, 12, 657233, Jul. 2021, doi: 10.3389/fmicb.2021.657233.
6. Zimmermann, L., Scheringer, M., Geueke, B., Boucher, J. M., Parkinson, L. V., Groh, K. J., Muncke, J., Implementing the EU Chemicals Strategy for

Sustainability: The case of food contact chemicals of concern. *J. Hazard. Mater.*, 437, 129167, 2022, doi: 10.1016/j.jhazmat.2022.129167.
7. Cazón, P. and Vázquez, M., Bacterial cellulose as a biodegradable food packaging material: A review. *Food Hydrocoll.*, 113, 106530, Apr. 2021, doi: 10.1016/j.foodhyd.2020.106530.
8. Ludwicka, K., Kaczmarek, M., Białkowska, A., Bacterial Nanocellulose—A Biobased Polymer for Active and Intelligent Food Packaging Applications: Recent Advances and Developments. *Polymers*, 12, 10, 2209, Sep. 2020, doi: 10.3390/polym12102209.
9. Liu, Y. *et al.*, A review of cellulose and its derivatives in biopolymer-based for food packaging application. *Trends Food Sci. Technol.*, 112, 532–546, Jun. 2021, doi: 10.1016/j.tifs.2021.04.016.
10. Naomi, R., Bt Hj Idrus, R., Fauzi, M.B., Plant- vs. Bacterial-Derived Cellulose for Wound Healing: A Review. *Int. J. Environ. Res. Public Health*, 17, 18, 6803, Sep. 2020, doi: 10.3390/ijerph17186803.
11. Lin, D., Liu, Z., Shen, R., Chen, S., Yang, X., Bacterial cellulose in food industry: Current research and future prospects. *Int. J. Biol. Macromol.*, 158, 1007–1019, Sep. 2020, doi: 10.1016/j.ijbiomac.2020.04.230.
12. Azeredo, H.M.C., Barud, H., Farinas, C.S., Vasconcellos, V.M., Claro, A.M., Bacterial Cellulose as a Raw Material for Food and Food Packaging Applications. *Front. Sustain. Food Syst.*, 3, 7, 2019, [Online], Available: https://www.frontiersin.org/articles/10.3389/fsufs.2019.00007.
13. Donini, I.A.N. *et al.*, Biossíntese e recentes avanços na produção de celulose bacteriana. *Ecletica Quim.*, 35, 4, 165–178, Dec. 2010, doi: 10.26850/1678-4618eqj.v35.4.2010.p165-178.
14. Dourado, F., Fontão, A., Leal, M., Rodrigues, A.C., Gama, M., Process Modeling and Techno-Economic Evaluation of an Industrial Bacterial NanoCellulose Fermentation Process, in: *Bacterial Nanocellulose*, pp. 199–214, Elsevier, Amsterdam, 2016 doi: 10.1016/B978-0-444-63458-0.00012-3.
15. Behera, B., Laavanya, D., Balasubramanian, P., Techno-economic feasibility assessment of bacterial cellulose biofilm production during the Kombucha fermentation process. *Bioresour. Technol.*, 346, 126659, Feb. 2022, doi: 10.1016/j.biortech.2021.126659.
16. Raghavendran, V., Asare, E., Roy, I., Bacterial cellulose: Biosynthesis, production, and applications, in: *Advances in Microbial Physiology*, vol. 77, pp. 89–138, Elsevier, London, UK, 2020, [Online], Available: https://linkinghub.elsevier.com/retrieve/pii/S0065291120300151.
17. Jonas, R. and Farah, L.F., Production and application of microbial cellulose. *Polym. Degrad. Stab.*, 59, 1–3, 101–106, Jan. 1998, doi: 10.1016/S0141-3910(97)00197-3.
18. Wang, X. and Zhong, J.-J., Improvement of bacterial cellulose fermentation by metabolic perturbation with mixed carbon sources. *Process Biochem.*, 122, 95–102, Nov. 2022, doi: 10.1016/j.procbio.2022.08.022.

19. Gorgieva, S., Jančič, U., Cepec, E., Trček, J., Production efficiency and properties of bacterial cellulose membranes in a novel grape pomace hydrolysate by Komagataeibacter melomenusus AV436T and Komagataeibacter xylinus LMG 1518. *Int. J. Biol. Macromol.*, 244, 125368, Jun. 2023, doi: 10.1016/j.ijbiomac.2023.125368.
20. Güzel, M. and Akpınar, O., Production and Characterization of Bacterial Cellulose from Citrus Peels. *Waste Biomass Valorization*, 10, 8, 2165–2175, Aug. 2019, doi: 10.1007/s12649-018-0241-x.
21. Goelzer, F.D.E., Faria-Tischer, P.C.S., Vitorino, J.C., Sierakowski, M.-R., Tischer, C.A., Production and characterization of nanospheres of bacterial cellulose from Acetobacter xylinum from processed rice bark. *Mater. Sci. Eng., C*, 29, 2, 546–551, Mar. 2009, doi: 10.1016/j.msec.2008.10.013.
22. Lotfy, V.F., Basta, A.H., Abdel-Monem, M.O., Abdel-Hamed, G.Z., Utilization of bacteria in rotten Guava for production of bacterial cellulose from isolated and protein waste. *Carbohydr. Polym. Technol. Appl.*, 2, 100076, Dec. 2021, doi: 10.1016/j.carpta.2021.100076.
23. Said Azmi, S.N.N. *et al.*, Fresh oil palm frond juice as a novel and alternative fermentation medium for bacterial cellulose production. *Mater. Today Proc.*, 42, 101–106, 2021, doi: 10.1016/j.matpr.2020.10.220.
24. Souza, E.F., Furtado, M.R., Carvalho, C.W.P., Freitas-Silva, O., Gottschalk, L.M.F., Production and characterization of Gluconacetobacter xylinus bacterial cellulose using cashew apple juice and soybean molasses. *Int. J. Biol. Macromol.*, 146, 285–289, Mar. 2020, doi: 10.1016/j.ijbiomac.2019.12.180.
25. Salari, M., Sowti Khiabani, M., Rezaei Mokarram, R., Ghanbarzadeh, B., Samadi Kafil, H., Preparation and characterization of cellulose nanocrystals from bacterial cellulose produced in sugar beet molasses and cheese whey media. *Int. J. Biol. Macromol.*, 122, 280–288, Feb. 2019, doi: 10.1016/j.ijbiomac.2018.10.136.
26. Kumar, V., Sharma, D.K., Bansal, V., Mehta, D., Sangwan, R.S., Yadav, S.K., Efficient and economic process for the production of bacterial cellulose from isolated strain of Acetobacter pasteurianus of RSV-4 bacterium. *Bioresour. Technol.*, 275, 430–433, Mar. 2019, doi: 10.1016/j.biortech.2018.12.042.
27. Abdelraof, M., Hasanin, M.S., El -Saied, H., Ecofriendly green conversion of potato peel wastes to high productivity bacterial cellulose. *Carbohydr. Polym.*, 211, 75–83, May 2019, doi: 10.1016/j.carbpol.2019.01.095.
28. Lin, D., Lopez-Sanchez, P., Li, R., Li, Z., Production of bacterial cellulose by Gluconacetobacter hansenii CGMCC 3917 using only waste beer yeast as nutrient source. *Bioresour. Technol.*, 151, 113–119, Jan. 2014, doi: 10.1016/j.biortech.2013.10.052.
29. Dhar, P., Pratto, B., Gonçalves Cruz, A.J., Bankar, S., Valorization of sugarcane straw to produce highly conductive bacterial cellulose / graphene nanocomposite films through *in situ* fermentation: Kinetic analysis and property evaluation. *J. Clean. Prod.*, 238, 117859, Nov. 2019, doi: 10.1016/j.jclepro.2019.117859.

30. Gayathri, G. and Srinikethan, G., Bacterial Cellulose production by K. saccharivorans BC1 strain using crude distillery effluent as cheap and cost effective nutrient medium. *Int. J. Biol. Macromol.*, 138, 950–957, Oct. 2019, doi: 10.1016/j.ijbiomac.2019.07.159.
31. Ye, J. *et al.*, Bacterial cellulose production by Acetobacter xylinum ATCC 23767 using tobacco waste extract as culture medium. *Bioresour. Technol.*, 274, 518–524, Feb. 2019, doi: 10.1016/j.biortech.2018.12.028.
32. Wu, M. *et al.*, Valorizing kitchen waste through bacterial cellulose production towards a more sustainable biorefinery. *Sci. Total Environ.*, 695, 133898, Dec. 2019, doi: 10.1016/j.scitotenv.2019.133898.
33. Hong, F., Guo, X., Zhang, S., Han, S., Yang, G., Jönsson, L.J., Bacterial cellulose production from cotton-based waste textiles: Enzymatic saccharification enhanced by ionic liquid pretreatment. *Bioresour. Technol.*, 104, 503–508, Jan. 2012, doi: 10.1016/j.biortech.2011.11.028.
34. Lin, S.-P., Loira Calvar, I., Catchmark, J.M., Liu, J.-R., Demirci, A., Cheng, K.-C., Biosynthesis, production and applications of bacterial cellulose. *Cellulose*, 20, 5, 2191–2219, Oct. 2013, doi: 10.1007/s10570-013-9994-3.
35. Shigematsu, T. *et al.*, Cellulose production from glucose using a glucose dehydrogenase gene (gdh)-deficient mutant of Gluconacetobacter xylinus and its use for bioconversion of sweet potato pulp. *J. Biosci. Bioeng.*, 99, 4, 415–422, Apr. 2005, doi: 10.1263/jbb.99.415.
36. Hestrin, S., Aschner, M., Mager, J., Synthesis of Cellulose by Resting Cells of Acetobacter xylinum. *Nature*, 159, 4028, 64–65, Jan. 1947, doi: 10.1038/159064a0.
37. Singhania, R.R., Patel, A.K., Tsai, M.-L., Chen, C.-W., Di Dong, C., Genetic modification for enhancing bacterial cellulose production and its applications. *Bioengineered*, 12, 1, 6793–6807, Jan. 2021, doi: 10.1080/21655979.2021.1968989.
38. Amikam, D. and Benziman, M., Cyclic diguanylic acid and cellulose synthesis in Agrobacterium tumefaciens. *J. Bacteriol.*, 171, 12, 6649–6655, Dec. 1989, doi: 10.1128/jb.171.12.6649-6655.1989.
39. Mayer, R. *et al.*, Polypeptide composition of bacterial cyclic diguanylic acid-dependent cellulose synthase and the occurrence of immunologically crossreacting proteins in higher plants. *Proc. Natl. Acad. Sci.*, 88, 12, 5472–5476, Jun. 1991, doi: 10.1073/pnas.88.12.5472.
40. Krasteva, P.V. *et al.*, Insights into the structure and assembly of a bacterial cellulose secretion system. *Nat. Commun.*, 8, 1, 2065, Dec. 2017, doi: 10.1038/s41467-017-01523-2.
41. Sunagawa, N. *et al.*, Cellulose complementing factor (Ccp) is a new member of the cellulose synthase complex (terminal complex) in Acetobacter xylinum. *J. Biosci. Bioeng.*, 115, 6, 607–612, Jun. 2013, doi: 10.1016/j.jbiosc.2012.12.021.

42. Wang, J., Tavakoli, J., Tang, Y., Bacterial cellulose production, properties and applications with different culture methods – A review. *Carbohydr. Polym.*, 219, 63–76, Sep. 2019, doi: 10.1016/j.carbpol.2019.05.008.
43. Lu, H. and Jiang, X., Structure and Properties of Bacterial Cellulose Produced Using a Trickling Bed Reactor. *Appl. Biochem. Biotechnol.*, 172, 8, 3844–3861, Apr. 2014, doi: 10.1007/s12010-014-0795-4.
44. Gea, S. *et al.*, Investigation into the structural, morphological, mechanical and thermal behaviour of bacterial cellulose after a two-step purification process. *Bioresour. Technol.*, 102, 19, 9105–9110, Oct. 2011, doi: 10.1016/j.biortech.2011.04.077.
45. Azeredo, H.M.C., Rosa, M.F., Mattoso, L.H.C., Nanocellulose in bio-based food packaging applications. *Ind. Crops Prod.*, 97, 664–671, Mar. 2017, doi: 10.1016/j.indcrop.2016.03.013.
46. Gwon, H. *et al.*, A safe and sustainable bacterial cellulose nanofiber separator for lithium rechargeable batteries. *Proc. Natl. Acad. Sci.*, 116, 39, 19288–19293, Sep. 2019, doi: 10.1073/pnas.1905527116.
47. Jacek, P., Dourado, F., Gama, M., Bielecki, S., Molecular aspects of bacterial nanocellulose biosynthesis. *Microb. Biotechnol.*, 12, 4, 633–649, Jul. 2019, doi: 10.1111/1751-7915.13386.
48. Chien, L., Chen, H., Yang, P., Lee, C., Enhancement of Cellulose Pellicle Production by Constitutively Expressing *Vitreoscilla* Hemoglobin in *Acetobacter xylinum. Biotechnol. Prog.*, 22, 6, 1598–1603, Jan. 2006, doi: 10.1021/bp060157g.
49. Jacek, P., Kubiak, K., Ryngajłło, M., Rytczak, P., Paluch, P., Bielecki, S., Modification of bacterial nanocellulose properties through mutation of motility related genes in Komagataeibacter hansenii ATCC 53582. *New Biotechnol.*, 52, 60–68, Sep. 2019, doi: 10.1016/j.nbt.2019.05.004.
50. Hur, D.H. *et al.*, Enhanced production of cellulose in Komagataeibacter xylinus by preventing insertion of IS element into cellulose synthesis gene. *Biochem. Eng. J.*, 156, 107527, Apr. 2020, doi: 10.1016/j.bej.2020.107527.
51. Urbina, L., Corcuera, M.A., Gabilondo, N., Eceiza, A., Retegi, A., A review of bacterial cellulose: sustainable production from agricultural waste and applications in various fields. *Cellulose*, 28, 13, 8229–8253, Sep. 2021, doi: 10.1007/s10570-021-04020-4.
52. Lahiri, D. *et al.*, Bacterial Cellulose: Production, Characterization, and Application as Antimicrobial Agent. *Int. J. Mol. Sci.*, 22, 23, 12984, Nov. 2021, doi: 10.3390/ijms222312984.
53. Gregory, D.A. *et al.*, Bacterial cellulose: A smart biomaterial with diverse applications. *Mater. Sci. Eng. R Rep.*, 145, 100623, Jul. 2021, doi: 10.1016/j.mser.2021.100623.
54. Sundaram, M.K., Nehru, G., Tadi, S.R.R., Katsuno, N., Nishizu, T., Sivaprakasam, S., Bacterial cellulose production by Komagataeibacter hansenii utilizing agro-industrial residues and its application in coffee milk

stabilization. *Biomass Convers. Biorefin.*, 13, 9, 7971–7981, Jun. 2023, doi: 10.1007/s13399-021-01867-2.
55. Park, S., Baker, J.O., Himmel, M.E., Parilla, P.A., Johnson, D.K., Cellulose crystallinity index: measurement techniques and their impact on interpreting cellulase performance. *Biotechnol. Biofuels*, 3, 1, 10, Dec. 2010, doi: 10.1186/1754-6834-3-10.
56. Tahara, N., Tabuchi, M., Watanabe, K., Yano, H., Morinaga, Y., Yoshinaga, F., Degree of Polymerization of Cellulose from *Acetobacter xylinum* BPR2001 Decreased by Cellulase Produced by the Strain. *Biosci. Biotechnol., Biochem.*, 61, 11, 1862–1865, Jan. 1997, doi: 10.1271/bbb.61.1862.
57. Shkarin, R. *et al.*, Quanfima: An open source Python package for automated fiber analysis of biomaterials. *PLoS One*, 14, 4, e0215137, Apr. 2019, doi: 10.1371/journal.pone.0215137.
58. Suryanto, H., Muhajir, M., Sutrisno, T.A., Mudjiono, Zakia, N., Yanuhar, U., The Mechanical Strength and Morphology of Bacterial Cellulose Films: The Effect of NaOH Concentration. *IOP Conf. Ser. Mater. Sci. Eng.*, 515, 012053, Apr. 2019, doi: 10.1088/1757-899X/515/1/012053.
59. Pa'e, N., Salehudin, M.H., Hassan, N.D., Marsin, A.M., Muhamad, I.I., Thermal Behavior of Bacterial Cellulose-Based Hydrogels with Other Composites and Related Instrumental Analysis, in: *Cellulose-Based Superabsorbent Hydrogels*, M.I.H. Mondal, (Ed.), Polymers and Polymeric Composites: A Reference Series, pp. 763–787, Springer International Publishing, Cham, 2019, doi: 10.1007/978-3-319-77830-3_26.
60. Ul-Islam, M., Khan, T., Park, J.K., Water holding and release properties of bacterial cellulose obtained by *in situ* and *ex situ* modification. *Carbohydr. Polym.*, 88, 2, 596–603, Apr. 2012, doi: 10.1016/j.carbpol.2012.01.006.
61. Vinogradov, M.I., Makarov, I.S., Golova, L.K., Gromovykh, P.S., Kulichikhin, V.G., Rheological Properties of Aqueous Dispersions of Bacterial Cellulose. *Processes*, 8, 4, 423, Apr. 2020, doi: 10.3390/pr8040423.
62. Khamkeaw, A., Asavamongkolkul, T., Perngyai, T., Jongsomjit, B., Phisalaphong, M., Interconnected Micro, Meso, and Macro Porous Activated Carbon from Bacterial Nanocellulose for Superior Adsorption Properties and Effective Catalytic Performance. *Molecules*, 25, 18, 4063, Sep. 2020, doi: 10.3390/molecules25184063.
63. Ummartyotin, S., Juntaro, J., Sain, M., Manuspiya, H., Development of transparent bacterial cellulose nanocomposite film as substrate for flexible organic light emitting diode (OLED) display. *Ind. Crops Prod.*, 35, 1, 92–97, Jan. 2012, doi: 10.1016/j.indcrop.2011.06.025.
64. Wan, Z., Wang, L., Yang, X., Guo, J., Yin, S., Enhanced water resistance properties of bacterial cellulose multilayer films by incorporating interlayers of electrospun zein fibers. *Food Hydrocoll.*, 61, 269–276, Dec. 2016, doi: 10.1016/j.foodhyd.2016.05.024.

65. Raut, M., Asare, E., Mohamed, S.S., Amadi, E., Roy, I., Bacterial Cellulose-Based Blends and Composites: Versatile Biomaterials for Tissue Engineering Applications. *Int. J. Mol. Sci.*, 24, 2, 986, Jan. 2023, doi: 10.3390/ijms24020986.
66. Cazón, P. and Vázquez, M., Improving bacterial cellulose films by *ex situ* and in-situ modifications: A review. *Food Hydrocoll.*, 113, 106514, Apr. 2021, doi: 10.1016/j.foodhyd.2020.106514.
67. Velásquez-Cock, J. *et al.*, Influence of the acid type in the production of chitosan films reinforced with bacterial nanocellulose. *Int. J. Biol. Macromol.*, 69, 208–213, Aug. 2014, doi: 10.1016/j.ijbiomac.2014.05.040.
68. Martínez-Sanz, M., Lopez-Rubio, A., Lagaron, J.M., High-barrier coated bacterial cellulose nanowhiskers films with reduced moisture sensitivity. *Carbohydr. Polym.*, 98, 1, 1072–1082, Oct. 2013, doi: 10.1016/j.carbpol.2013.07.020.
69. Viana, R.M., Sá, N.M.S.M., Barros, M.O., Borges, M.D.F., Azeredo, H.M.C., Nanofibrillated bacterial cellulose and pectin edible films added with fruit purees. *Carbohydr. Polym.*, 196, 27–32, Sep. 2018, doi: 10.1016/j.carbpol.2018.05.017.
70. Fabra, M.J., López-Rubio, A., Ambrosio-Martín, J., Lagaron, J.M., Improving the barrier properties of thermoplastic corn starch-based films containing bacterial cellulose nanowhiskers by means of PHA electrospun coatings of interest in food packaging. *Food Hydrocoll.*, 61, 261–268, Dec. 2016, doi: 10.1016/j.foodhyd.2016.05.025.
71. Phisalaphong, M., Suwanmajo, T., Tammarate, P., Synthesis and characterization of bacterial cellulose/alginate blend membranes. *J. Appl. Polym. Sci.*, 107, 5, 3419–3424, Mar. 2008, doi: 10.1002/app.27411.
72. Ju, S., Zhang, F., Duan, J., Jiang, J., Characterization of bacterial cellulose composite films incorporated with bulk chitosan and chitosan nanoparticles: A comparative study. *Carbohydr. Polym.*, 237, 116167, Jun. 2020, doi: 10.1016/j.carbpol.2020.116167.
73. Bandyopadhyay, S., Saha, N., Brodnjak, U.V., Sáha, P., Bacterial cellulose and guar gum based modified PVP-CMC hydrogel films: Characterized for packaging fresh berries. *Food Packag. Shelf Life*, 22, 100402, Dec. 2019, doi: 10.1016/j.fpsl.2019.100402.
74. George, J. and S., High performance edible nanocomposite films containing bacterial cellulose nanocrystals. *Carbohydr. Polym.*, 87, 3, 2031–2037, Feb. 2012, doi: 10.1016/j.carbpol.2011.10.019.
75. Lucyszyn, N. *et al.*, Physicochemical and *in vitro* biocompatibility of films combining reconstituted bacterial cellulose with arabinogalactan and xyloglucan. *Carbohydr. Polym.*, 151, 889–898, Oct. 2016, doi: 10.1016/j.carbpol.2016.06.027.
76. Salari, M., Sowti Khiabani, M., Rezaei Mokarram, R., Ghanbarzadeh, B., Samadi Kafil, H., Development and evaluation of chitosan based active nanocomposite films containing bacterial cellulose nanocrystals and

silver nanoparticles. *Food Hydrocoll.*, 84, 414–423, Nov. 2018, doi: 10.1016/j.foodhyd.2018.05.037.

77. Ummartyotin, S., Pisitsak, P., Pechyen, C., Eggshell and Bacterial Cellulose Composite Membrane as Absorbent Material in Active Packaging. *Int. J. Polym. Sci.*, 2016, 1–8, 2016, doi: 10.1155/2016/1047606.
78. Chen, L., Zou, M., Hong, F.F., Evaluation of Fungal Laccase Immobilized on Natural Nanostructured Bacterial Cellulose. *Front. Microbiol.*, 6, Nov. 2015, doi: 10.3389/fmicb.2015.01245.
79. Vilela, C., Moreirinha, C., Domingues, E.M., Figueiredo, F.M.L., Almeida, A., Freire, C.S.R., Antimicrobial and Conductive Nanocellulose-Based Films for Active and Intelligent Food Packaging. *Nanomaterials*, 9, 7, 980, Jul. 2019, doi: 10.3390/nano9070980.
80. Dirpan, A., Latief, R., Syarifuddin, A., Rahman, A.N.F., Putra, R.P., Hidayat, S.H., The use of colour indicator as a smart packaging system for evaluating mangoes Arummanis (*Mangifera indica* L. var. Arummanisa) freshness. *IOP Conf. Ser. Earth Environ. Sci.*, 157, 012031, May 2018, doi: 10.1088/1755-1315/157/1/012031.
81. Kuswandi, B., Jayus, Oktaviana, R., Abdullah, A., Heng, L.Y., A Novel On-Package Sticker Sensor Based on Methyl Red for Real-Time Monitoring of Broiler Chicken Cut Freshness. *Packag. Technol. Sci.*, 27, 1, 69–81, Jan. 2014, doi: 10.1002/pts.2016.
82. Zahan, K.A., Azizul, N.M., Mustapha, M., Tong, W.Y., Abdul Rahman, M.S., Sahuri, I.S., Application of bacterial cellulose film as a biodegradable and antimicrobial packaging material. *Mater. Today Proc.*, 31, 83–88, 2020, doi: 10.1016/j.matpr.2020.01.201.
83. Wang, W., Yu, Z., Alsammarraie, F.K., Kong, F., Lin, M., Mustapha, A., Properties and antimicrobial activity of polyvinyl alcohol-modified bacterial nanocellulose packaging films incorporated with silver nanoparticles. *Food Hydrocoll.*, 100, 105411, Mar. 2020, doi: 10.1016/j.foodhyd.2019.105411.
84. Chen, S., Wu, M., Lu, P., Gao, L., Yan, S., Wang, S., Development of pH indicator and antimicrobial cellulose nanofibre packaging film based on purple sweet potato anthocyanin and oregano essential oil. *Int. J. Biol. Macromol.*, 149, 271–280, Apr. 2020, doi: 10.1016/j.ijbiomac.2020.01.231.
85. Kusznierewicz, B., Staroszczyk, H., Malinowska-Pańczyk, E., Parchem, K., Bartoszek, A., Novel ABTS-dot-blot method for the assessment of antioxidant properties of food packaging. *Food Packag. Shelf Life*, 24, 100478, Jun. 2020, doi: 10.1016/j.fpsl.2020.100478.
86. Moradian, S., Almasi, H., Moini, S., Development of bacterial cellulose-based active membranes containing herbal extracts for shelf life extension of button mushrooms (*Agaricus bisporus*). *J. Food Process. Preserv.*, 42, 3, e13537, Mar. 2018, doi: 10.1111/jfpp.13537.
87. Tsai, Y.-H., Yang, Y.-N., Ho, Y.-C., Tsai, M.-L., Mi, F.-L., Drug release and antioxidant/antibacterial activities of silymarin-zein nanoparticle/bacterial

cellulose nanofiber composite films. *Carbohydr. Polym.*, 180, 286–296, Jan. 2018, doi: 10.1016/j.carbpol.2017.09.100.
88. Nguyen, V.T., Gidley, M.J., Dykes, G.A., Potential of a nisin-containing bacterial cellulose film to inhibit Listeria monocytogenes on processed meats. *Food Microbiol.*, 25, 3, 471–478, May 2008, doi: 10.1016/j.fm.2008.01.004.
89. Luo, N., Varaprasad, K., Reddy, G.V.S., Rajulu, A.V., Zhang, J., Preparation and characterization of cellulose/curcumin composite films. *RSC Adv.*, 2, 22, 8483, 2012, doi: 10.1039/c2ra21465b.
90. Fijałkowski, K., Peitler, D., Rakoczy, R., Żywicka, A., Survival of probiotic lactic acid bacteria immobilized in different forms of bacterial cellulose in simulated gastric juices and bile salt solution. *LWT Food Sci. Technol.*, 68, 322–328, May 2016, doi: 10.1016/j.lwt.2015.12.038.
91. Shi, Z., Zhang, Y., Phillips, G.O., Yang, G., Utilization of bacterial cellulose in food. *Food Hydrocoll.*, 35, 539–545, Mar. 2014, doi: 10.1016/j.foodhyd.2013.07.012.
92. Kalashnikova, I., Bizot, H., Cathala, B., Capron, I., New Pickering Emulsions Stabilized by Bacterial Cellulose Nanocrystals. *Langmuir*, 27, 12, 7471–7479, Jun. 2011, doi: 10.1021/la200971f.
93. Zhang, X., Wang, D., Liu, S., Tang, J., Bacterial Cellulose Nanofibril-Based Pickering Emulsions: Recent Trends and Applications in the Food Industry. *Foods*, 11, 24, 4064, Dec. 2022, doi: 10.3390/foods11244064.
94. Khorasani, A.C. and Shojaosadati, S.A., Bacterial nanocellulose-pectin bionanocomposites as prebiotics against drying and gastrointestinal condition. *Int. J. Biol. Macromol.*, 83, 9–18, Feb. 2016, doi: 10.1016/j.ijbiomac.2015.11.041.
95. Guo, Y. *et al.*, Nano-bacterial cellulose/soy protein isolate complex gel as fat substitutes in ice cream model. *Carbohydr. Polym.*, 198, 620–630, Oct. 2018, doi: 10.1016/j.carbpol.2018.06.078.
96. Lin, S., Chen, L., Chen, H., PHYSICAL CHARACTERISTICS OF SURIMI AND BACTERIAL CELLULOSE COMPOSITE GEL. *J. Food Process Eng*, 34, 4, 1363–1379, Aug. 2011, doi: 10.1111/j.1745-4530.2009.00533.x.
97. Bao, C. *et al.*, The delivery of sensitive food bioactive ingredients: Absorption mechanisms, influencing factors, encapsulation techniques and evaluation models. *Food Res. Int.*, 120, 130–140, Jun. 2019, doi: 10.1016/j.foodres.2019.02.024.
98. Singhania, R. R., Patel, A. K., Tseng, Y.-S., Kumar, V., Chen, C.-W., Haldar, D., Saini, J. K., Dong, C.-D., Developments in bioprocess for bacterial cellulose production. *Bioresour. Technol.*, 344, 126343, 2022, doi: 10.1016/j.biortech.2021.12634.

Index

Printed in the USA/Agawam, MA
September 16, 2026

910661.165